PATH INTEGRALS
AND
QUANTUM PROCESSES

PATH INTEGRALS AND QUANTUM PROCESSES

Mark S. Swanson
*Department of Physics
University of Connecticut at Stamford*

DOVER PUBLICATIONS, INC.
Mineola, New York

Copyright

Copyright © 1992 by Mark S. Swanson
All rights reserved.

Bibliographical Note

This Dover edition, first published in 2014, is an unabridged republication of the work originally published by Academic Press, Inc., Boston, in 1992.

Library of Congress Cataloging-in-Publication Data

Swanson, Mark S., 1947– author.
  Path integrals and quantum processes / Mark S. Swanson, Department of Physics, University of Connecticut at Stamford.
       pages cm
  Originally published: New York : Academic Press, 1992.
  Includes bibliographical references and index.
  ISBN-13: 978-0-486-49306-0 — ISBN-10: 0-486-49306-7
  1. Path integrals. 2. Quantum theory. 3. Quantum field theory.
I. Title.
  QC174.17.P27S92 2014
  530.12—dc23
                                                              2013033825

Manufactured in the United States by Courier Corporation
49306703   2014
www.doverpublications.com

Dedicated to the memory of Meryl Katherine Swanson.
Valē anima dulcis.

Contents

Preface	xi
Chapter One — Mathematical Preliminaries	1
1.1 The Dirac Delta	2
1.2 Completeness	3
1.3 Functionals	9
1.4 Matrices	15
1.5 Gaussian Integrals	22
References	28
Chapter Two — Quantum Mechanical Path Integrals	31
2.1 Quantum Mechanics	31
2.2 The Path Integral Derived	35
2.3 The Sum over Histories	42
References	50
Chapter Three — Evaluating the Path Integral	51
3.1 The Free Particle	52
3.2 Motion with a Source	54
3.3 Continuum Techniques	62
3.4 Topological Measure	67
References	72
Chapter Four — Further Applications	75
4.1 Natural Units	75
4.2 Statistical Mechanics	77
4.3 Symmetry and Generating Functionals	83
4.4 Harmonic Oscillator Coherent States	92
4.5 Spontaneously Broken Symmetry	96
4.6 Constraints	105
References	115

Chapter Five — Grassmann Variables   117
5.1 Basic Definitions 118
5.2 Gaussian Grassmann Integrals 125
5.3 Classical Grassman Mechanics 128
5.4 Grassmann Quantum Mechanics 133
    5.4.1 The Free Particle 138
    5.4.2 The Harmonic Oscillator 139
5.5 Grassmann Path Integrals 141
5.6 Supersymmetric Quantum Mechanics 146
References 148

Chapter Six — Field Theory   151
6.1 A Mechanical Model 153
6.2 Relativity and Group Theory 158
6.3 Classical Free Fields 170
    6.3.1 The Scalar Field 171
    6.3.2 Spinor Fields 173
6.4 Symmetry and Noether's Theorem 182
    6.4.1 Translational Invariance 186
    6.4.2 Lorentz Invariance and Angular Momentum 187
    6.4.3 Phase and Chiral Invariance 189
    6.4.4 Charges as Symmetry Generators 190
6.5 Canonical Quantization 191
    6.5.1 Scalar Field Quantization 191
    6.5.2 Dirac Field Quantization 194
6.6 The S-Matrix 197
6.7 The Interaction Picture 204
6.8 The Path Integral for Field Theory 207
    6.8.1 The Scalar Field Case 207
    6.8.2 The Dirac Field Case 214
    6.8.3 Euclidean Measure 218
    6.8.4 Configuration Measure 222
References 229

Chapter Seven — Gauge Field Theory   231
7.1 The Maxwell Field 232
    7.1.1 The Classical Action 232
    7.1.2 Gauge Invariance and Gauge Fixing 235
    7.1.3 Quantization of the Free Gauge Field 239
7.2 QED as a Path Integral 249
7.3 Lie Algebras 263
7.4 Classical Yang–Mills Fields 270
7.5 Quantized Yang–Mills Fields 276
7.6 Topological Aspects of Gauge Fields 286
References 306

Chapter Eight — Perturbation Theory 311
- 8.1 Generating Functionals 312
- 8.2 Ward–Takahashi Identities 319
- 8.3 Deriving the Feynman Rules 325
- 8.4 Renormalization 341
- References 355

Chapter Nine — Nonperturbative Results 357
- 9.1 The Goldstone Theorem 358
- 9.2 The Effective Potential 360
- 9.3 The Higgs–Kibble Mechanism 372
- 9.4 The $SU(2)_L \times U(1)$ Electroweak Model 383
- 9.5 Chiral Anomalies 389
- 9.6 Classical Solutions 403
  - 9.6.1 The Kink Solution 403
  - 9.6.2 Vacuum Tunnelling 410
  - 9.6.3 Yang–Mills Instantons 416
  - 9.6.4 The Abelian Magnetic Monopole 420
- 9.7 Applications of the Effective Potential 423
  - 9.7.1 Finite Temperature and Symmetry Restoration 423
  - 9.7.2 The Coleman–Weinberg Mechanism 428
  - 9.7.3 The Gross–Neveu Model 431
- References 436

Index 441

Preface

Since the advent of modern covariant field theory shortly after the Second World War, the path integral formulation of transition amplitudes, or quantum processes, has grown steadily in popularity and use. Rather than relying on the abstract concepts of states in a Hilbert space and physical observables as operators, the traditional staples of quantum mechanics and quantum field theory, the path integral approach makes a more direct contact with classical mechanics and fields. A particularly satisfying aspect of the path integral is its incorporation of the action functional approach to the classical system under consideration into a method for calculating quantum mechanical processes in the corresponding quantized system. Such a connection brings with it numerous powerful intuitions as well as a means for a graphical visualization of quantum processes in terms of classical trajectories or paths.

As the concept and content of this book evolved, I decided that the underlying theme should be the presentation of the path integral as a logical outgrowth of the standard operator and state formulation of quantum processes. It is not the intention of this book to present an encyclopedic collection of path integral applications, nor to present the path integral in a mathematically rigorous manner. This book does not contain any detailed calculations of scattering cross sections, lifetimes, or renormalization effects; there are many excellent monographs available in which such computations are performed. This book does not contain any discussion of the less settled areas of theoretical physics, such as the quantization of gravity or superstring theory. However, this book does contain a set of logical connections and applications that are representative of the scope of problems that the path integral has been used to solve. It is hoped that the reader can use the material in this book to understand enough of the general structure of path integrals to read other articles and books in which they are used, and thereby to join the work at the frontiers of theoretical physics.

In order to facilitate this understanding, the path integral is derived for a set of general cases before being applied to specific instances. It is assumed that the reader is acquainted with the Hilbert space structure of nonrelativistic quantum mechanics, and has at least a cursory understanding of statistical mechanics. All relevant concepts of quantum field theory are developed, although some readers may find the presentation somewhat brief. Depending upon the reader's background and objectives, various parts of the book will be of greater importance or interest. The first four chapters contain applications restricted to quantum mechanics, and for that reason should be accessible to first year graduate students and precocious undergraduates. The Euclidean or Wick rotation is introduced in these early chapters and used throughout the remainder of the book. The relationship of the path integral to the partition function is discussed. Since Chapter 4 contains many of the ideas critical to later developments in the book, I urge that it be read by anyone interested in maintaining the logical continuity of the book. Chapter 5 is an attempt to place Grassmann variables on an equal footing with standard c-number variables, and presents many aspects of these anticommuting variables. Of course, the bulk of the book's more modern content is developed in Chapters 6 and 7, where quantum field theory and its associated path integrals are introduced. In particular, the quantization of gauge fields via the path integral is understood in terms of the application of constrained path integrals, and the role of the indefinite metric space in the underlying Fock space of the gauge field is stressed. The remainder of the book contains applications of path integrals to problems in field theory. The choice of topics and their manner of presentation has been made to complement existing texts and monographs. The exercises included in the book were chosen to satisfy a two-fold purpose: first, they allow the reader to become familiar with the standard manipulations of path integrals; second, they allowed the author to skip the less difficult steps in the manipulations presented. Occasionally, exercises were included that involve a fair amount of further development by the reader.

In writing this book, I have benefited immeasurably from the work of the many outstanding physicists who blazed the intellectual trails I walked. To list them all here would be impossible, and I will therefore not try. Nevertheless, it has been my greatest pleasure as a physicist to watch the profound mysteries of nature slowly yield to the insights and diligent efforts of many extremely talented people. I would like to thank everyone who made this book possible; however, a special debt of gratitude is owed to Jane Ellis of Academic Press.

Mark S. Swanson
Stamford, Connecticut

Chapter 1

Mathematical Preliminaries

Throughout this book there are numerous commonly used mathematical conventions and facts that are assumed to be familiar to the reader. This chapter reviews some aspects of these conventions and facts, but it is in no way intended as a comprehensive or systematic explication of these areas of mathematical physics. At first glance the topics covered may appear to be unrelated to each other; however, each will be essential to understanding the path integral. It is hoped that this chapter will clarify notation and refresh the reader's memory regarding the facts presented. The reader who wishes further details should consult the references [1, 2].

The central theme of this book is the functional formulation of quantum processes. At the heart of this method is the concept of a complete set of functions. In order to define what is meant by completeness and functional differentiation it is first necessary to introduce the Dirac delta and its properties. This is done in Sec. 1.1. In Sec. 1.2 the concept of a complete set of functions is developed within the specific context of Fourier series and transforms. The representations of the Dirac delta can then be written in terms of a basis set of complete functions for the general class of functions under consideration. In Sec. 1.3 functionals are defined and a method of functional differentiation is formulated. These methods are then used to give a brief review of the action functional approach to classical mechanics. In Sec. 1.4 the most important properties of matrices are outlined and extended to functions of multiple variables. The chapter closes in Sec. 1.5 with a discussion of Gaussian integrals and Jacobians, since these are of paramount importance to the standard evaluation of the path integral.

2 Mathematical Preliminaries

1.1 The Dirac Delta

The Dirac delta $\delta(x)$ is defined by its integral [3]. Given a function $f(x)$, the Dirac delta satisfies the integral relation

$$\int dx\, f(x)\delta(x-a) = f(a)\,, \tag{1.1}$$

if the range of integration contains an open interval around the point $x = a$; if it does not, then the integral vanishes. The function $f(x)$ is called a *test function*. The Dirac delta is a generalization of the concept of a function to a larger class of mappings, referred to as *distributions* [4]. For the sake of intuition the Dirac delta can be visualized as an improper function that is essentially zero everywhere except at the set of points where its argument vanishes, and at those points it is singular (infinite). This intuitive picture would lead immediately to the conclusion that the distribution $x\delta(x) = 0$, but this fails when the test function against which it is integrated is singular at the origin, e.g., $f(x) = 1/x$. Clearly, if the set of test functions does not include functions singular at the origin, then the distribution $x\delta(x)$ does vanish when it is integrated. It is apparent from this simple observation that the properties of distributions can depend crucially upon the properties of the test functions against which they are integrated.

Deferring to the next section a more substantial definition of $\delta(x)$ and the test functions, numerous properties of the Dirac delta follow from the defining relation (1.1), and these are stated without detailed proof. Integration by parts, dropping the endpoint contributions, gives

$$\int dx\, f(x)\frac{\partial^n}{\partial x^n}\delta(x-a) = (-1)^n \left.\frac{\partial^n f}{\partial x^n}\right|_{x=a}. \tag{1.2}$$

A simple change of variables in the integral gives

$$\int dx\, f(x)\,\delta(ax-b) = \frac{1}{|a|}f(b/a)\,. \tag{1.3}$$

If the argument of the Dirac delta is itself a function $g(x)$, and the set of values $x_1,\ldots,x_n$ are the n roots of $g(x)$, then it follows that

$$\int dx\, f(x)\,\delta\left(g(x)\right) = \sum_{i=1}^n \frac{f(x_i)}{g'(x_i)}\,, \tag{1.4}$$

where $g'(x_i)$ is the first derivative of g at x_i.

Exercise 1.1: Prove (1.4) by using (1.3) and a Taylor series representation of the function $g(x)$ near its roots, assuming that all the roots of $g(x)$ are simple, i.e., that the first derivative of $g(x)$ is nonzero at each root.

The previous relations may be extended to multidimensional integrals, and $\delta^n(x)$ denotes the Dirac delta for an n-dimensional volume. In Cartesian coordinates $\delta^n(x)$ is the product of n one-dimensional Dirac deltas, each of which corresponds to one of the Cartesian coordinates. Such a product of Dirac deltas satisfies

$$\int d^n x\, f(x)\, \delta^n(x - a) = f(a) , \qquad (1.5)$$

where a is a point in the n-dimensional volume. The superscript on the Dirac delta will be suppressed whenever its dimension is apparent from the context of its use.

Exercise 1.2: Given the form of the volume element in spherical coordinates in n dimensions, $d^n x = r^{n-1} dr\, d\phi\, \sin\theta_1\, d\theta_1 \cdots \sin^{n-2}\theta_{n-2}\, d\theta_{n-2}$, express $\delta^n(x)$ in spherical coordinates as a product of one-dimensional Dirac deltas.

1.2 Completeness

It will be seen that a central feature of the path integral formulation of quantum processes is its use of functionals [5]. The general concept of a functional will be defined in detail in the next section. However, as its name implies, the functional is constructed from functions, and many times these functions are arbitrary members of a general class of functions. It is therefore extremely useful to have a method of expressing an arbitrary member of a general class of functions compactly. It is the property known as completeness that allows such a function to be constructed from a subset of well-understood functions by writing it as an infinite series, or an integral, involving those functions.

As an example, let $f(x)$ be a member of the general group of real functions that are piecewise continuous and periodic on the interval $x = (-L, L)$. It is known from the theory of Fourier series [6] that such a

4 Mathematical Preliminaries

function may be written in the form

$$f(x) = \sum_{n=-\infty}^{\infty} \left\{ a_n \cos\left(\frac{n\pi x}{L}\right) + b_n \sin\left(\frac{n\pi x}{L}\right) \right\} = a_0$$

$$+ \sum_{n=1}^{\infty} \left\{ (a_n + a_{-n}) \cos\left(\frac{n\pi x}{L}\right) + (b_n - b_{-n}) \sin\left(\frac{n\pi x}{L}\right) \right\}, \qquad (1.6)$$

where the coefficients of the series are given by

$$a_n = \frac{1}{2L} \int_{-L}^{L} dx\, f(x) \cos\left(\frac{n\pi x}{L}\right), \qquad (1.7)$$

$$b_n = \frac{1}{2L} \int_{-L}^{L} dx\, f(x) \sin\left(\frac{n\pi x}{L}\right). \qquad (1.8)$$

The set of functions $\{\sin(n\pi x/L), \cos(n\pi x/L)\}$ is said to be *complete*. In other words, it is a collection of functions sufficient to construct any member of the general class of functions that are periodic and piecewise continuous over the interval.

These formulas are exactly analogous to the representation of a general vector in a vector space in terms of its components along a set of basis unit vectors. That a general vector may be represented this way is possible because every vector space possesses an inner or "dot" product. This inner product allows a determination of the projection or "overlap" of two vectors onto each other, so that two vectors **x** and **y** are *orthogonal* if and only if $\mathbf{x} \cdot \mathbf{y} = 0$. If $\{\mathbf{e}_i\}$ is a set of *orthonormal* vectors with respect to the inner product, then $\mathbf{e}_i \cdot \mathbf{e}_j = \delta_{ij}$. The symbol δ_{ij} is the Kronecker delta, which is 1 when $i = j$ and zero otherwise. If an arbitrary vector **x** in the space under consideration can be expressed as a sum of vectors along each basis vector in the form

$$\mathbf{x} = \sum_{i=1}^{n} x_i \mathbf{e}_i, \qquad (1.9)$$

where $x_i = \mathbf{x} \cdot \mathbf{e}_i$, then the set $\{\mathbf{e}_i\}$ forms a complete orthonormal basis for the vector space. In the Cartesian case the $\{x_i\}$ are the coordinates of the tip of the vector.

The extension of these simple ideas to the Fourier series lies in generalizing the inner product to include functions. Defining an inner product for functions will allow them to be treated in a manner similar to vectors. To this end, the inner product of two functions, $f(x)$ and $g(x)$, that are periodic over the interval $x = (-L, L)$ is defined as

$$f \cdot g \equiv \frac{1}{L} \int_{-L}^{L} dx\, f^*(x) g(x), \qquad (1.10)$$

where f^* is the complex conjugate of f. Such a definition satisfies the general properties of an inner product: $f \cdot f > 0$ (positive-definite norm), $(f+g) \cdot (f+g) \geq f \cdot f + g \cdot g$ (triangle inequality), and $|f \cdot g|^2 \leq (f \cdot f)(g \cdot g)$ (Cauchy–Schwarz inequality). Using this definition, it is easy to show that the sine and cosine functions used in the Fourier series (1.6) form an infinite dimensional orthonormal set of functions indexed by the integer n. This follows from the fact that the integrals of these functions satisfy

$$\frac{1}{L}\int_{-L}^{L} dx\, \sin\left(\frac{n\pi x}{L}\right) \cos\left(\frac{m\pi x}{L}\right) = 0, \tag{1.11}$$

$$\frac{1}{L}\int_{-L}^{L} dx\, \sin\left(\frac{n\pi x}{L}\right) \sin\left(\frac{m\pi x}{L}\right) = \delta_{mn}, \tag{1.12}$$

$$\frac{1}{L}\int_{-L}^{L} dx\, \cos\left(\frac{n\pi x}{L}\right) \cos\left(\frac{m\pi x}{L}\right) = \delta_{mn}, \tag{1.13}$$

for $m, n > 0$. Thus, the result of taking the inner product of any two of these functions is identical to that of the inner product of the basis vectors for a vector space. The quantities $a_n + a_{-n}$ and $b_n - b_{-n}$, obtained from (1.7) and (1.8), are the projections of the function $f(x)$ onto these basis functions. The Fourier series is then seen as the generalization of the component representation of a vector, familiar from finite dimensional vector spaces, to an infinite dimensional function space. For the space of periodic functions, the sine and cosine functions play the role of orthonormal basis vectors and span the entire function space, while the combinations $a_n + a_{-n}$ and $b_n - b_{-n}$ can be visualized as the coordinates of the function in the function space. With a few additional subtleties that will be ignored here, both the simple finite dimensional vector space discussed in the previous paragraph and the infinite dimensional function space of this paragraph are examples of a Hilbert space [7].

The Dirac delta of Sec. 1.1 may now be given a representation in terms of a Fourier series. If $x, y \in (-L, L)$, then (1.6), (1.7), and (1.8) give

$$\delta(x-y) = \frac{1}{2L}\sum_{n=-\infty}^{\infty} \left\{\cos\left(\frac{n\pi x}{L}\right)\cos\left(\frac{n\pi y}{L}\right) + \sin\left(\frac{n\pi x}{L}\right)\sin\left(\frac{n\pi y}{L}\right)\right\}. \tag{1.14}$$

It is clear that this representation of the Dirac delta is itself periodic, and therefore its test function space is limited to the set of functions periodic over the same interval. If it is integrated over the interval $(-L, L)$ against a function with a different periodicity, then it is not necessarily true that (1.1) will be satisfied.

6   Mathematical Preliminaries

> **Exercise 1.3**: Using the periodic form (1.14) for the Dirac delta, show that
> $$\int_{-\infty}^{\infty} dx\, f(x)\, \delta(x-y) = \sum_{n=-\infty}^{\infty} f(y + 2nL)\,. \qquad (1.15)$$

It is worth noting that different subclasses of functions may require a different set of basis functions. For example, the set of functions that are reflection symmetric, i.e., $f(-x) = f(x)$, do not require any sine terms in their expansion. As a result, if the test function space is restricted to this subclass, then the representation of the Dirac delta need only contain the cosine terms.

Very often a function has no periodicity. Another way of expressing this is to say that the interval of periodicity is infinite, and this is represented mathematically by taking the limit $L \to \infty$ in the previous formulas. In this limit the sum in (1.14) may be converted into an integral with the result that

$$\delta(x-y) = \frac{1}{2\pi} \int_{-\infty}^{\infty} dk\, e^{ik(x-y)}\,. \qquad (1.16)$$

> **Exercise 1.4**: Derive expression (1.16).

Using property (1.1) for a function $f(x)$ immediately gives

$$f(x) = \frac{1}{(2\pi)^{1/2}} \int_{-\infty}^{\infty} dk\, \tilde{f}(k)\, e^{ikx}\,, \qquad (1.17)$$

where $\tilde{f}(k)$ is called the Fourier transform of $f(x)$, and is given by

$$\tilde{f}(k) = \frac{1}{(2\pi)^{1/2}} \int_{-\infty}^{\infty} dx\, f(x)\, e^{-ikx}\,. \qquad (1.18)$$

Result (1.17) shows that e^{ikx} defines a continuous set of functions, and that this set is complete for the space of nonperiodic functions. It is more common in the literature to see (1.14) and (1.16) used as an assertion that the respective sets of functions are complete. The relation (1.1) is then used to define the Fourier series, or transform, of an arbitrary function by integrating the appropriate form of the Dirac delta against it. However, the two versions of the Dirac delta are quite different, and so it is more correct to make the argument in reverse, as was done here. In this respect it is important to note that e^{ikx} is a complex valued function, and that the test function space now includes complex functions. From this point forward the limits on the Fourier transform integrals will not be explicitly displayed, unless they differ from infinity.

> **Exercise 1.5**: Let $f(x)$ be a member of the class of functions that are nonsingular in an open interval around the origin $x = 0$. Show that, for such a test function,
>
> $$\lim_{\alpha \to \infty} \int_{-\infty}^{\infty} dx\, f(x) \left(\frac{\alpha}{\pi}\right)^{1/2} \exp(-\alpha x^2) = f(0),$$
>
> so that the limit of the exponential function behaves as $\delta(x)$.

The Fourier series and transform are generalized to many dimensions by simply adding additional degrees of freedom. The Dirac delta appropriate for n dimensions and nonperiodic functions is given by

$$\delta^n(x - y) = \frac{1}{(2\pi)^n} \int d^n k\, e^{ik \cdot (x-y)}, \tag{1.19}$$

where x, y, and k are vectors in the respective spaces and the integration is over the entire volume of k space. This form of the Dirac delta can be used to define the Fourier transform of a chosen function by simply integrating it against that function.

There may be many complete sets of functions that span the same function space, just as there are many choices for a basis set of unit vectors in a vector space. The choice of a set is often a matter of either calculational convenience or selection of a basis that best utilizes an underlying symmetry in the system being analyzed. However, one of the most important aspects of these complete sets of functions is that many, if not all, originate from solving the *eigenvalue* problem for some Hermitian differential operator. This result can be sketched by using $D(x)$ to denote a differential operator in one dimension, i.e., some combination of x and derivatives that has the property that, for any two suitable functions,

$$\int dx\, \Big(D^*(x) f^*(x)\Big) g(x) = \int dx\, f^*(x) \Big(D(x) g(x)\Big), \tag{1.20}$$

where D^* is the complex conjugate of D. If D satisfies (1.20), then D is said to be *Hermitian*. The functions f and g for which (1.20) is true define two function spaces. If these two function spaces coincide, then D is said

8   Mathematical Preliminaries

to be *self-adjoint*. For example, $i\partial/\partial x$ is Hermitian, while $\partial/\partial x$ is not. For many physically relevant Hermitian differential operators, in particular the second-order Hermitian linear differential operators, it is known that the solutions of the associated eigenvalue problem,

$$D(x)\varphi_n(x) = \lambda^{(n)}\varphi_n(x) \,, \qquad (1.21)$$

subject to appropriate boundary conditions on the eigenfunctions φ_n or eigenvalues $\lambda^{(n)}$, are orthonormal,

$$\int dx\, \varphi_n^*(x)\varphi_m(x) = \delta_{nm} \,, \qquad (1.22)$$

and form a complete set,

$$\sum_n \varphi_n^*(x)\varphi_n(y) = \delta(x-y) \,. \qquad (1.23)$$

In general, the proof of completeness reduces to showing that a Fourier-like series in the eigenfunctions converges to an arbitrary function in the general class. The proof will not be given here.

Exercise 1.6: Prove that the eigenvalues of a Hermitian operator are all real, and that the eigenfunctions associated with different eigenvalues are orthogonal.

For example, the Hermitian differential operator $D = \partial^2/\partial x^2$, along with the boundary condition that the eigenfunction φ_n is periodic, gives the familiar sine and cosine functions of the Fourier series, as well as the associated eigenvalues $\lambda_n = -(n\pi/L)^2$. As another example, the differential operator $D = -i\partial/\partial x$, along with the demand that the eigenvalues be real, gives the complete set of eigenfunctions e^{ikx} with the eigenvalue k. Thus, the pairing of Hermiticity and real eigenvalues leads immediately to a complete set of eigenfunctions.

As a final remark, it often happens that the set of eigenfunctions contains both a discrete and a continuous set of eigenfunctions. Such a result may be expected on purely physical grounds. For example, in quantum mechanics the set of discrete eigenfunctions of the differential operator representing the energy corresponds to bound or negative energy states, while the continuous spectrum of eigenfunctions describes the scattering or positive energy solutions. It is reasonable to expect to find quantum mechanical systems whose solutions give both discrete and continuous eigenfunctions, since there are many physically meaningful potentials that can both bind and scatter particles. If $\{\varphi_k(x)\}$ is the continuous set and $\{\varphi_n(x)\}$ the discrete set, then the general statement of completeness in such a case in one dimension is given by

$$\int dk\, \varphi_k^*(x)\varphi_k(y) + \sum_n \varphi_n^*(x)\varphi_n(y) = \delta(x-y). \tag{1.24}$$

1.3 Functionals

A real-valued function is a mapping from some space into the real numbers. According to the last section it is possible to view a function itself as a point in an infinite dimensional space. Using this analogy, a functional is defined as a map of a function or a polynomial of functions into a number. The function plays the role of the coordinates being mapped by the functional. The standard method of constructing a functional is by integrating a collection of functions and their products over some interval of their arguments.

Determining the nature of the functions that *extremize* the functional historically led to the calculus of variations [8]. By analogy to the problem of finding the extrema of a function, the determination of the form for the extremum functions requires the definition of functional differentials and differentiation. Once these definitions have been made, it is then possible to create a local or pointwise criterion that extremum functions must satisfy. In order to make this definition the simplest case is considered initially, so that the general functional of a single function $g(x)$ is written as $\mathcal{F}[g(x)]$. The functional derivative of $\mathcal{F}$ with respect to $g(y)$ is then formally defined as [9, 10]

$$\frac{\delta \mathcal{F}}{\delta g(y)} = \lim_{\epsilon \to 0} \frac{\mathcal{F}[g(x) + \epsilon \delta(x-y)] - \mathcal{F}[g(x)]}{\epsilon}, \tag{1.25}$$

where $\delta(x-y)$ is the Dirac delta of appropriate dimension to match the dimension of integration.

Exercise 1.7: Using (1.25), show that the functional, $\mathcal{F}[g(x)] = \int_a^b dx\, [g(x)]^n$, has the functional derivative

$$\frac{\delta \mathcal{F}}{\delta g(y)} = n[g(y)]^{n-1}, \tag{1.26}$$

if $y \in (a,b)$, and is zero if $y \notin (a,b)$.

---

10   Mathematical Preliminaries

The definition (1.25) has all the properties associated with the standard derivative operation. For example, if $\mathcal{F}$ and $\mathcal{G}$ are two functionals, then their product has the functional derivative

$$\frac{\delta(\mathcal{F}\mathcal{G})}{\delta g(x)} = \mathcal{G}\frac{\delta \mathcal{F}}{\delta g(x)} + \mathcal{F}\frac{\delta \mathcal{G}}{\delta g(x)}, \qquad (1.27)$$

which is the Leibniz property.

If $\mathcal{F}[g]$ is a functional that is well-behaved in the interval in function space around $g = 0$, then the functional $\mathcal{F}$ has a Taylor series representation given by

$$\mathcal{F}[g] = \sum_{n=0}^{\infty} \frac{1}{n!} \int dx_1 \cdots dx_n \, g(x_1) \cdots g(x_n) \left.\frac{\delta^n \mathcal{F}[g]}{\delta g(x_1) \cdots \delta g(x_n)}\right|_{g=0}, \qquad (1.28)$$

where the limits of integration will be suppressed from now on. Applying an arbitrary power of the functional derivative to both sides of (1.28) and evaluating the result at $g = 0$ shows that (1.28) is self-consistent. Functional Taylor series may be defined around other functions, so that

$$\mathcal{F}[g] = \sum_{n=0}^{\infty} \frac{1}{n!} \int dx_1 \cdots dx_n \, [g(x_1) - f(x_1)] \cdots$$
$$\times [g(x_n) - f(x_n)] \left.\frac{\delta^n \mathcal{F}[g]}{\delta g(x_1) \cdots \delta g(x_n)}\right|_{g=f}. \qquad (1.29)$$

Form (1.29) allows a functional to be expanded about a function for which the functional derivatives are well-defined.

Form (1.29) also allows the definition of a functional differential [11]. If $\delta g(x)$ is an infinitesimal deviation from the function $g(x)$, so that the function $g(x) + \delta g(x)$ is infinitesimally close to $g(x)$ everywhere, then the functional Taylor series (1.29) gives

$$\mathcal{F}[g + \delta g] \approx \mathcal{F}[g] + \int dx \left.\frac{\delta \mathcal{F}[g]}{\delta g(x)}\right|_{\delta g=0} \delta g(x), \qquad (1.30)$$

where terms of order $(\delta g)^2$ and higher have been dropped as irrelevant. The differential change of the functional $\mathcal{F}$ under an infinitesimal variation of its argument may then be defined as

$$\delta \mathcal{F}[g] \equiv \mathcal{F}[g + \delta g] - \mathcal{F}[g] = \int dx \, \frac{\delta \mathcal{F}[g]}{\delta g(x)} \delta g(x). \qquad (1.31)$$

The definition (1.31) is exactly analogous to the standard differential for a function of many variables. There the differential of a function $f(x)$ is defined as the change of the function under an arbitrary change in the coordinates dx, where, in the case of many variables, dx is understood to be an infinitesimal vector. The differential of $f(x)$ is then written

$$df(x) = f(x+dx) - f(x) = \nabla f(x) \cdot dx, \qquad (1.32)$$

where ∇ is the gradient operator, given in Cartesian coordinates by $\nabla = \sum_i \mathbf{e}_i (\partial/\partial x_i)$. The result (1.31) is seen to be identical to (1.32) when the form of the function space inner product (1.10) is recalled. Thus, $\delta g(x)$ plays the role of dx, and the functional derivative is effectively the gradient of the functional.

The functional form of the chain rule may be established by considering a functional $\mathcal{F}[g]$, where g is in turn a functional of f, so that $g = g[f]$. Two consecutive applications of the differential operation yields

$$\delta \mathcal{F} = \int dx \, \frac{\delta \mathcal{F}}{\delta g(x)} \delta g(x) = \int dx \, dy \, \frac{\delta \mathcal{F}}{\delta g(x)} \frac{\delta g(x)}{\delta f(y)} \delta f(y). \qquad (1.33)$$

By comparison with the standard form (1.31) for the differential, it follows that

$$\frac{\delta \mathcal{F}}{\delta f(y)} = \int dx \, \frac{\delta \mathcal{F}}{\delta g(x)} \frac{\delta g(x)}{\delta f(y)}. \qquad (1.34)$$

The extension of all these definitions to functionals of more than one function is very easy. In the general case a functional $\mathcal{F}$ has for an argument a set of functions $\{g_i(x)\}$, and these are indexed by a subscript for ease of notation. At this point it is convenient to introduce the *summation convention*. In this convention any repeated pair of indices in a formula is understood to stand for a summation over the entire range of the indices, unless one of the indices is enclosed in parentheses. As an example, the inner product of two n-dimensional real vectors, $\mathbf{x}$ and $\mathbf{y}$, can be written

$$\mathbf{x} \cdot \mathbf{y} = x_i y_i \equiv \sum_{i=1}^n x_i y_i, \qquad (1.35)$$

in terms of their Cartesian coordinates. However, the expression $\lambda^{(i)} x_i$ does *not* represent a summation, rather simply the product of $\lambda^{(i)}$ with x_i. Where there is any chance of confusion the summation will be explicitly displayed. Using the summation convention, it follows that the differential of the functional of many functions may be written

$$\delta \mathcal{F}[g_1, \ldots, g_n] = \int dx \, \frac{\delta \mathcal{F}}{\delta g_j(x)} \delta g_j(x), \qquad (1.36)$$

where the sum over j runs from 1 to n.

The Lagrangian formulation of Newtonian mechanics is one of the earliest applications of functional methods in physics [12], and it serves as an excellent example. In the simplest nontrivial case a point particle of mass m is constrained to move one dimensionally in the presence of a conservative force $F(x)$. Newton's law of motion states that

$$m\ddot{x} = F(x) = -\frac{\partial V}{\partial x}, \tag{1.37}$$

where $V(x)$ is the potential energy associated with the force, and $\ddot{x}$ is the second derivative of x with respect to the time t. Combining the solution of (1.37) with appropriate initial conditions gives a unique function $x = x(t)$, which represents the particle's trajectory. The solution to Newton's equation of motion for a given set of initial conditions will be referred to as a *classical trajectory*.

In the Lagrangian formulation of mechanics the object of importance is not the differential equation of (1.37), but rather the action functional, which is formed from the kinetic energy and the potential energy of the particle's motion. This is simply the integral of the Lagrangian density, which is defined to be the difference between the kinetic energy and the potential energy. In the simple example defined by (1.37) the action functional is given by

$$S[x(t), t_a, t_b] = \int_{t_a}^{t_b} dt\, \mathcal{L}(x, \dot{x}) = \int_{t_a}^{t_b} dt\, \left[\tfrac{1}{2} m\dot{x}^2 - V(x)\right], \tag{1.38}$$

where t_a and t_b define the time interval for the motion of the particle. At this point $x(t)$ is considered to be an arbitrary function of the time. Once a well-behaved form for $x(t)$ is chosen the action functional can be evaluated to give a value.

Newton's equation of motion is recaptured by asking for the criterion that the function $x(t)$ must satisfy in order to make the action an extremum. Exploiting the analogy between functions and functionals shows that this is equivalent to demanding that the differential of the action vanish around the extremizing form for $x(t)$. Using the result that

$$\frac{\delta \dot{x}(t')}{\delta x(t)} = \frac{d}{dt'}\delta(t' - t) \tag{1.39}$$

and the functional chain rule (1.34), it follows that the extremizing function must be chosen such that

$$\delta S[x(t)] = \int_{t_a}^{t_b} dt\, \left[\frac{\delta S}{\delta x(t)} - \frac{d}{dt}\frac{\delta S}{\delta \dot{x}(t)}\right]\delta x(t) = 0, \tag{1.40}$$

where $\delta x(t)$ is an arbitrary infinitesimal variation which vanishes at t_a and t_b. Using the results of Exercise 1.7, it follows that the form for $x(t)$ that extremizes the action must obey the Euler–Lagrange equation,

$$\frac{\partial \mathcal{L}}{\partial x} - \frac{d}{dt}\frac{\partial \mathcal{L}}{\partial \dot{x}} = 0, \tag{1.41}$$

which is the generalization of Newton's law of motion. The definitions of the functional derivative and the functional differential allow a very compact statement of the calculus of variations, which was originally developed to solve the same problem of extremizing functionals.

If the only use for the action functional approach to mechanics was to generate the already known equations of motion, then it would be considered nothing more than an oddity. However, the action approach to formulating the dynamics of a physical system contains far more information than simply the equation of motion. First, the action functional is a *global* statement about the system, from which a *local* differential equation can be derived by demanding an extremum. The action is a global object in the sense that it receives contributions from the entire trajectory of the particle, and hence it records an aspect of the "history" of the particle's motion. Second, although the action is extremized by the set of classical trajectories that solve (1.41), the action can be evaluated for *any* trajectory. It will be seen that this is of profound importance in quantum processes. Third, the action approach allows the definition of the momenta canonically conjugate to the coordinates of the particle, and the generalization of the energy known as the Hamiltonian can be constructed from these by means of a Legendre transformation. Fourth, the action contains all the information regarding the *classical* symmetries of the system, and these determine the conservation laws associated with the classical trajectory. As a result, the action approach is the most compact and systematic method to formulate the dynamics of a physical system.

In order to construct a general form for the Hamiltonian, the Lagrangian density will be written as a function of a set of n generalized coordinates, $q = \{q_i\}$, and their associated velocities, $\dot{q} = \{\dot{q}_i\}$. The momentum p_i, canonically conjugate to the coordinate q_i, is defined as

$$p_i(t) \equiv \frac{\delta S}{\delta \dot{q}_i(t)} = \frac{\partial \mathcal{L}}{\partial \dot{q}_i(t)}. \tag{1.42}$$

It is assumed that (1.42) can be solved for all the $\dot{q}_i$ in terms of the set of $2n$ variables $\{p_i, q_i\}$, so that $\dot{q}_i = \dot{q}_i(p, q)$. The Hamiltonian $H(p, q)$ is then defined by a Legendre transformation as

$$H(p, q) = p_i \dot{q}_i(p, q) - \mathcal{L}(q, \dot{q}(p, q)). \tag{1.43}$$

Using (1.41) and (1.42) shows that H is conserved if $\mathcal{L}$ has no explicit time dependence since it satisfies

$$\frac{dH}{dt} = -\frac{\partial \mathcal{L}}{\partial t}, \tag{1.44}$$

when it is evaluated along a classical trajectory. The Hamiltonian is then seen as the generalization of the energy for the classical system's motion.

The definition (1.43) may be reversed to give another variant of the action functional, expressed in terms of the $2n$ degrees of freedom p and q. In this version the action is written

$$S[p, q, t_b, t_a] = \int_{t_a}^{t_b} dt \left[p_i \dot{q}_i - H(p, q) \right] \equiv \int_{t_a}^{t_b} dt\, \mathcal{L}(p, q). \tag{1.45}$$

Again demanding that $\delta S = 0$ under simultaneous variations of p and q gives

$$\int_{t_a}^{t_b} dt \left[\left(\dot{q}_i - \frac{\partial H}{\partial p_i} \right) \delta p_i - \left(\dot{p}_i + \frac{\partial H}{\partial q_i} \right) \delta q_i \right] = 0. \tag{1.46}$$

Since the variations δp_i and δq_i are independent, the original n second-order differential equations generated by the Euler–Lagrange formula (1.41) have become the $2n$ first-order differential equations

$$\dot{q}_i = \frac{\partial H}{\partial p_i}, \quad \dot{p}_i = -\frac{\partial H}{\partial q_i}. \tag{1.47}$$

These are known as Hamilton's equations of motion.

Both form (1.38) and (1.45) represent the same dynamical system. However, there is a subtle difference between the two, and this difference can be be important. In constrained systems form (1.45) is more useful, particularly for momentum (velocity) dependent constraints. Because the form (1.45) has both p and q present, it can serve as the starting form of the action when a canonical transformation of variables is performed. A canonical transformation can be defined as any transformation on the p and q variables that has a Jacobian, to be discussed in Sec. 1.5, equal to unity.

An equivalent definition of a canonical transformation is any transformation of p and q that leaves the Poisson bracket invariant. In classical mechanics the Poisson bracket of any two functions $A(p, q)$ and $B(p, q)$ is the antisymmetric operation defined by

$$\{A, B\}_{p,q} \equiv \frac{\partial A}{\partial q_i} \frac{\partial B}{\partial p_i} - \frac{\partial B}{\partial q_i} \frac{\partial A}{\partial p_i}. \tag{1.48}$$

Using the definition of the Poisson bracket and Hamilton's equations of motion, it is straightforward to show that

$$\frac{dA}{dt} = \frac{\partial A}{\partial t} + \{A, H\}_{p,q}, \tag{1.49}$$

and

$$\{q_j, p_k\}_{p,q} = \delta_{jk}. \tag{1.50}$$

It is the Poisson bracket formalism that carries over into quantum mechanics to become the commutator (see Sec. 2.1). The Poisson bracket has the important property that it reproduces the quantum mechanical behavior of p and q from their classical forms. This will render it of great value in the functional approach.

1.4 Matrices

This section contains a very brief review of the properties of matrices and their eigenvalues [13]. A general matrix **M** is an $n \times m$ array of complex numbers called the *elements* of **M**. For the purposes of this book consideration will be limited to square $n \times n$ arrays, **M**, where n is the *rank* of the matrix, and single column $n \times 1$ arrays, **x**, where n is the *dimension*. The latter will be referred to as vectors because of their analogy with the component form of the vectors familiar from configuration space. The individual elements of a matrix will be denoted M_{ij}, and those of a vector x_i, so that i is the row and j the column of the element. A matrix is said to be *real* if every element is a real number. Two matrices are equal only if each of their respective elements are equal. The elements of the sum of two matrices **M** and **N** are given by the sum of the elements

$$(\mathbf{M} + \mathbf{N})_{ij} = M_{ij} + N_{ij}, \tag{1.51}$$

while the elements of the product of the two matrices are given by (summation convention)

$$(\mathbf{MN})_{ij} = M_{ik} N_{kj}. \tag{1.52}$$

The *commutator* of two matrices is defined as

$$[\mathbf{M}, \mathbf{N}] \equiv \mathbf{MN} - \mathbf{NM}. \tag{1.53}$$

Two matrices, **M** and **N**, are said to *commute* if $[\mathbf{M}, \mathbf{N}] = 0$.

The vectors of this section are an extension of finite dimensional real vector spaces to vectors with complex components. For that reason the inner product of two vectors **x** and **y** is defined to be $\mathbf{x} \cdot \mathbf{y} = x_i^* y_i$. Such

16 Mathematical Preliminaries

a definition again guarantees that any nonzero vector will have a positive norm. Two vectors satisfying $\mathbf{x} \cdot \mathbf{y} = 0$ are said to be orthonormal.

An indispensable object is the Levi–Civita symbol, $\varepsilon^{\alpha\beta\gamma\cdots}$, which has n superscripts that range over the values $\{1, \ldots, n\}$. It is completely antisymmetric so that it is defined to be $+1$ if the set $\{\alpha, \beta, \ldots\}$ is an even permutation of the first n integers and -1 if it is an odd permutation. Since it is completely antisymmetric, the Levi–Civita symbol vanishes if any two superscripts coincide in value. The determinant of an $n \times n$ matrix $\mathbf{M}$ is defined as

$$\det \mathbf{M} = \varepsilon^{\alpha\beta\gamma\cdots} M_{1\alpha} M_{2\beta} M_{3\gamma} \cdots . \tag{1.54}$$

Any matrix with a determinant of unity is said to be *unimodular*.

Exercise 1.8: Using definition (1.54) show that

$$\det \mathbf{M} = \varepsilon^{\alpha\beta\gamma\cdots} M_{\alpha 1} M_{\beta 2} M_{\gamma 3} \cdots , \tag{1.55}$$

and that $\det(\mathbf{MN}) = \det \mathbf{M} \det \mathbf{N}$

Another important number associated with a matrix is the *trace*. The trace of a matrix $\mathbf{M}$ is denoted $\operatorname{Tr} \mathbf{M}$, and is given by

$$\operatorname{Tr} \mathbf{M} = M_{ii} \tag{1.56}$$

The trace of the product of a set of matrices has the property that it is invariant under *cyclic* permutations of the product.

Exercise 1.9: Show that $\operatorname{Tr}(\mathbf{M_1 M_2} \cdots \mathbf{M_n}) = \operatorname{Tr}(\mathbf{M_n M_1} \cdots \mathbf{M_{n-1}})$.

The identity matrix, denoted $\mathbf{I}$, has the elements $I_{ij} = \delta_{ij}$. It is trivial to show that $\mathbf{IM} = \mathbf{MI} = \mathbf{M}$, so that $\mathbf{I}$ commutes with all other matrices. A matrix $\mathbf{M}$ has a *left inverse* matrix, denoted $\mathbf{M}_L^{-1}$, if their product satisfies $\mathbf{M}_L^{-1}\mathbf{M} = \mathbf{I}$. A *right inverse*, $\mathbf{M}_R^{-1}$, satisfies $\mathbf{M}\mathbf{M}_R^{-1} = \mathbf{I}$. In general, it will be assumed that $\mathbf{M}_R^{-1} = \mathbf{M}_L^{-1}$. Clearly, $\det \mathbf{I} = 1$ and $\operatorname{Tr}(\mathbf{I}) = n$. From Exercise 1.8 it then follows that $\det \mathbf{MM}^{-1} = \det \mathbf{M} \det \mathbf{M}^{-1} = 1$, so that $\mathbf{M}^{-1}$ exists only if $\det \mathbf{M} \neq 0$.

In complete analogy to differential operators acting on functions the square matrices can be viewed as operators acting on the set of vectors. If $\mathbf{M}$ is an $n \times n$ matrix, then an eigenvector $\mathbf{x}$ satisfies the equation

$$\mathbf{Mx} = \lambda \mathbf{x}, \tag{1.57}$$

where, in formal analogy with (1.21), λ is called an eigenvalue of $\mathbf{M}$. It can be shown that each eigenvalue must satisfy the equation

$$\det(\mathbf{M} - \lambda \mathbf{I}) = 0. \tag{1.58}$$

Since (1.58) generates an nth order polynomial, there can be at most n distinct eigenvalues associated with an $n \times n$ matrix. Once the eigenvalues of a matrix are known, it is a straightforward algebraic problem to find the eigenvectors by solving (1.57). The eigenvectors so determined will be unique only up to a multiplicative constant, but this constant may be fixed, up to an overall complex factor of modulus one hereafter referred to as a *phase*, by normalizing the eigenvectors so that $x_i^* x_i = 1$.

Given an arbitrary $n \times n$ matrix $\mathbf{M}$, two new matrices may be defined by interchanging rows and columns. The *transpose* of the matrix $\mathbf{M}$ is denoted $\mathbf{M}^T$, and its elements are given by $(\mathbf{M}^T)_{ij} = M_{ji}$. A vector, or column matrix, becomes a row matrix under the transpose operation. The *Hermitian adjoint* of the matrix $\mathbf{M}$ is denoted $\mathbf{M}^\dagger$, and its elements are given by $(\mathbf{M}^\dagger)_{ij} = M_{ji}^*$. Under the Hermitian conjugation operation a vector, or column matrix, becomes a row matrix with elements given by the complex conjugate of the original vector elements. It is easy to show that, if $\mathbf{M}$ and $\mathbf{N}$ are two equal rank square matrices, then $(\mathbf{MN})^T = \mathbf{N}^T \mathbf{M}^T$ and $(\mathbf{MN})^\dagger = \mathbf{N}^\dagger \mathbf{M}^\dagger$.

Using these definitions matrices may be classified in the following ways. A matrix $\mathbf{O}$ is said to be *orthogonal* if it obeys $\mathbf{O}\mathbf{O}^T = \mathbf{I}$, so that $\mathbf{O}^T = \mathbf{O}^{-1}$. A matrix $\mathbf{S}$ is said to be *symmetric* if it satisfies $\mathbf{S} = \mathbf{S}^T$. A matrix $\mathbf{A}$ is said to be *antisymmetric* if $\mathbf{A}^T = -\mathbf{A}$. A matrix $\mathbf{U}$ is said to be *unitary* if it obeys $\mathbf{U}\mathbf{U}^\dagger = \mathbf{I}$, so that $\mathbf{U}^\dagger = \mathbf{U}^{-1}$. A matrix $\mathbf{H}$ is said to be *Hermitian* if it satisfies $\mathbf{H} = \mathbf{H}^\dagger$.

Given a unitary matrix $\mathbf{U}$, another $n \times n$ matrix $\mathbf{M}$ may be used to create the matrix $\mathbf{M}'$ by a *unitary transformation*, defined by

$$\mathbf{M}' = \mathbf{U}^\dagger \mathbf{M} \mathbf{U}. \tag{1.59}$$

If the matrix $\mathbf{M}$ is unitary, orthogonal, or Hermitian, then the unitarily transformed matrix $\mathbf{M}'$ will retain these properties. Using the previous results shows that $\det \mathbf{M}' = \det \mathbf{M}$ and $\operatorname{Tr} \mathbf{M} = \operatorname{Tr} \mathbf{M}'$.

The set of n normalized eigenvectors that solve (1.57) are denoted $\{\mathbf{x}^{(j)}\}$, and each $\mathbf{x}^{(j)}$ has n components, denoted $x_i^{(j)}$.

Exercise 1.10: Show that, if $\mathbf{M}$ is a Hermitian matrix, then its eigenvalues are real and its eigenvectors belonging to different eigenvalues are orthonormal.

18   Mathematical Preliminaries

> **Exercise 1.11**: Show that the eigenvalues of a real antisymmetric matrix are pure imaginary numbers, and that the determinant of a real $n \times n$ antisymmetric matrix vanishes if n is an odd number.

The results of Exercise 1.10 show that the matrix $\mathbf{U}$, formed from the elements of the eigenvectors of the Hermitian matrix $\mathbf{M}$ and given by $U_{ij} = x_i^{(j)}$, is unitary. It follows that the elements of the unitary transformation of $\mathbf{M}$ by $\mathbf{U}$ are given by

$$(\mathbf{U}^\dagger \mathbf{M} \mathbf{U})_{ij} = \lambda^{(j)} \delta_{ij}, \tag{1.60}$$

where $\lambda^{(j)}$ is the jth eigenvalue of $\mathbf{M}$. The unitarily transformed matrix has nonzero elements only along its diagonal; all other elements vanish. From relation (1.60), and the invariance of the determinant and the trace under a unitary transformation, it follows that

$$\det \mathbf{M} = \prod_{j=1}^{n} \lambda^{(j)}, \tag{1.61}$$

$$\operatorname{Tr} \mathbf{M} = \sum_{j=1}^{n} \lambda^{(j)}. \tag{1.62}$$

Thus, the necessary condition for the existence of the inverse of a matrix is simply the absence of any zero eigenvalues.

It is possible to define functions of matrices by using the Taylor series representation of the function. For example, a matrix may be exponentiated by using the power series

$$\exp \mathbf{M} = \sum_{n=0}^{\infty} \frac{1}{n!} \mathbf{M}^n, \tag{1.63}$$

where $\mathbf{M}^n$ represents a product of n factors of $\mathbf{M}$. It should be obvious from this example that the elements of the function of the matrix, derived by inserting the original matrix into the power series, will coincide with the function of the elements of the original matrix *only if* the original matrix is diagonal. In the event that the power series is not well defined this definition of the function of a matrix allows an expansion to be made about another matrix in order to ensure that the power series is sensible. For example, the logarithm function is not defined for the argument zero.

For this reason the power series representation of ln M must be defined by expanding it around another nonzero matrix, typically the identity matrix I. The ln M function can then be written

$$\ln \mathbf{M} = \sum_{n=1}^{\infty} \frac{1}{n}(-1)^{n+1}(\mathbf{M} - \mathbf{I})^n. \tag{1.64}$$

Exercise 1.12: Show that a Hermitian matrix M satisfies the relation

$$\det \mathbf{M} = \exp(\text{Tr} \ln \mathbf{M}). \tag{1.65}$$

A useful result is the Baker–Campbell–Hausdorff theorem [14]. If **C** denotes the commutator of the two matrices **A** and **B**, so that **C** = [**A**, **B**], and if **C** commutes with both **A** and **B**, so that [**C**, **A**] = [**C**, **B**] = 0, then the theorem states

$$\exp(\mathbf{A} + \mathbf{B}) = \exp(\mathbf{A})\exp(\mathbf{B})\exp(-\tfrac{1}{2}\mathbf{C}). \tag{1.66}$$

This theorem exploits the fact that $\exp(\mathbf{A})\exp(-\mathbf{A}) = \mathbf{I}$. The first step in proving the theorem is to use the result that, for the conditions of the theorem,

$$[\mathbf{A}^n, \mathbf{B}] = n[\mathbf{A}, \mathbf{B}]\mathbf{A}^{n-1} = n\mathbf{C}\mathbf{A}^{n-1}. \tag{1.67}$$

Using the Taylor series representation of the exponential function along with result (1.67) shows that, for α a real variable,

$$\exp(-\alpha \mathbf{A})\mathbf{B}\exp(\alpha \mathbf{A}) = -\alpha[\mathbf{A}, \mathbf{B}]. \tag{1.68}$$

Then the matrix, defined by

$$\mathbf{F}(\alpha) = \exp(-\alpha \mathbf{B})\exp(-\alpha \mathbf{A})\exp(\mathbf{A} + \mathbf{B}), \tag{1.69}$$

satisfies the equation

$$\frac{d\mathbf{F}(\alpha)}{d\alpha} = -\alpha \mathbf{C}\, \mathbf{F}(\alpha). \tag{1.70}$$

This differential equation has the solution

$$\mathbf{F}(\alpha) = \exp(-\tfrac{1}{2}\alpha^2 \mathbf{C}), \tag{1.71}$$

so that, at $\alpha = 1$, the solution reduces to

$$\mathbf{F}(\alpha = 1) = \exp(-\mathbf{B})\exp(-\mathbf{A})\exp(\mathbf{A} + \mathbf{B}) = \exp(-\tfrac{1}{2}\mathbf{C}), \tag{1.72}$$

Mathematical Preliminaries

thus demonstrating the Baker–Campbell–Hausdorff theorem.

It is very useful to complete the analogy between functions, vectors, and matrices. In this sense it is possible to view a function of a single variable, defined over the interval $(-L, L)$, as a vector. In order to demonstrate this the inner product of two functions, defined by (1.10), is written as a Riemann sum by choosing a discrete set of points over the interval $(-L, L)$. This gives

$$f \cdot g = \int_{-L}^{L} dx\, f^*(x) g(x) = \lim_{N \to \infty} \sum_{j=1}^{N} \left(\frac{2L}{N}\right) f^*(x_j) g(x_j), \qquad (1.73)$$

where $x_j = -L + j(2L/N)$. Relation (1.73) formally resembles the inner product of two infinite dimensional matrix vectors, defined in terms of the vector components and weighted by the measure factor $2L/N$.

This analogy to vectors and matrices carries over into functions of two variables. If $f(x, y)$ and $g(x, y)$ are two arbitrary functions, then it is possible to define a third function $h(x, z)$ by writing

$$h(x, z) = \int_{-L}^{L} dy\, f(x, y) g(y, z). \qquad (1.74)$$

By again making the interval $(-L, L)$ discrete and writing (1.74) as a Riemann sum,

$$h(x, z) = \lim_{N \to \infty} \sum_{j=1}^{N} \left(\frac{2L}{N}\right) f(x, y_j) g(y_j, z), \qquad (1.75)$$

a form resembling the multiplication of two matrices is obtained. In this entirely nonrigorous manner the results of matrix analysis may be taken over into the realm of functions and functionals. The function $f(x, y)$ can be viewed as the x, y element of a continuous matrix f.

In such a scheme the identity matrix can be represented functionally by a Dirac delta, since

$$\int_{-L}^{L} dy\, f(x, y)\, \delta(y - z) = f(x, z). \qquad (1.76)$$

Through this analogy with matrices the trace of the power of a function of two variables is defined as

$$\text{Tr}\, f^n = \int_{-L}^{L} dx_1 \cdots dx_n\, f(x_1, x_2) \cdots f(x_n, x_1). \qquad (1.77)$$

A *caveat* is in order, since the action of integrating the function changes the units associated with the result by a factor of the units associated with the measure of the integral. In the cases presented so far the units of the measure dx appearing in the integrals have been length. The units of the Dirac delta are (length)$^{-1}$, so that any power series representation of a function and Dirac deltas indicates that the function itself should also have units of (length)$^{-1}$. In the formal manipulations that follow it will be assumed that this condition is met.

Using (1.64), (1.65), and (1.77), the formal determinant of a commonly occurring type of function may be evaluated. In this case the function is assumed to have the form

$$f(x,y) = \sum_n a_n \varphi_n^*(x) \varphi_n(y), \qquad (1.78)$$

where the functions $\{\varphi_n\}$ form a complete orthonormal set, satisfying

$$\int dx\, \varphi_n^*(x) \varphi_m(x) = \delta_{nm}, \qquad (1.79)$$

while a_n is an arbitrary dimensionless real number. It is to be noted that the limits on the sum and integral have been left unindicated to accommodate a large set of cases. Completeness of these functions means that

$$f(x,y) - \delta(x-y) = \sum_n (a_n - 1)\varphi_n^*(x)\varphi_n(y). \qquad (1.80)$$

Carrying over the analogy of functions to matrices allows f to be viewed as a Hermitian matrix, since $f(x,y) = f^*(y,x)$. The determinant of f is then given by the form

$$\det f = \exp \operatorname{Tr} \ln f = \exp\left\{ \sum_{j=1}^{\infty} \frac{1}{j} (-1)^{j+1} \operatorname{Tr}(f - \delta)^j \right\}. \qquad (1.81)$$

The trace operation is readily evaluated using the orthonormality of the φ_n functions. The jth term in the sum becomes

$$\operatorname{Tr}(f - \delta)^j = \sum_n (a_n - 1)^j. \qquad (1.82)$$

Placing result (1.82) back into (1.81) gives the form

$$\det f = \exp\left[\sum_n \ln a_n \right] = \prod_n a_n. \qquad (1.83)$$

This result can be related to the action of a differential operator $D(x)$ on a function f. This is written in matrix notation as

$$D(x)f(x) = \int dy\, D(x)\delta(x-y)f(y), \qquad (1.84)$$

so that the matrix $D = D(x)\delta(x-y)$ represents the differential operator. If $D(x)$ is Hermitian and has a complete set of orthonormal eigenfunctions $\{\varphi_n\}$ satisfying (1.79), then

$$D \equiv D(x)\delta(x-y) = \sum_n \lambda^{(n)} \varphi_n^*(x)\varphi_n(y) \qquad (1.85)$$

is a Hermitian matrix of the form (1.78) (up to a dimensionful factor). It follows from (1.83) that

$$\det(D) = \prod_n \lambda^{(n)}, \qquad (1.86)$$

which is identical in form to the finite dimensional matrix result (1.61).

1.5 Gaussian Integrals

It will be seen that the evaluation of many path integrals reduces to evaluating a product of Gaussian integrals. The Gaussian integral has already been encountered in Exercise 1.5, and is given by

$$\int_{-\infty}^{\infty} dx\, e^{-\alpha x^2} = \sqrt{\frac{\pi}{\alpha}}, \qquad (1.87)$$

where α is a real and positive number. However, the Gaussian integrals appearing in some forms of the path integral will have the corresponding α pure imaginary. The problem is then to evaluate this form of the Gaussian integral. In order to do this methods of complex analysis must be used, and this presents the opportunity to review briefly a few of the most important of these [15].

The complex variable z is written $z = x + iy$, and the complex plane is formed from the two variables x and y. A function of a complex variable may then be written

$$f(z) = u(x,y) + i\, v(x,y). \qquad (1.88)$$

The function f is said to be *analytic* at a point z if it satisfies the Cauchy-Riemann conditions,

$$\frac{\partial u}{\partial x} = \frac{\partial v}{\partial y}, \quad \frac{\partial u}{\partial y} = -\frac{\partial v}{\partial x}, \qquad (1.89)$$

at that point. If a function f is analytic at a point, then a unique value for the derivative, df/dz, exists at that point.

A function of complex variables may be integrated along a path or contour in the complex plane, where the line element along the path is given by $dz = dx + i\, dy$. Of particular interest are closed contours. Cauchy's theorem states that if the function $f(z)$ is analytic everywhere in the part of the complex plane inside the closed right-handed contour C, then

$$\oint_C dz\, \frac{f(z)}{z - z_0} = 2\pi i\, f(z_0)\,, \qquad (1.90)$$

if z_0 is a point inside the contour. If z_0 lies outside the contour, then the integral vanishes. Goursat's theorem states that if $f(z)$ is analytic on and in the region bounded by C, then

$$\oint_C dz\, f(z) = 0\,. \qquad (1.91)$$

Exercise 1.13: Using Cauchy's theorem, show that, if $f(z)$ is analytic in the part of the complex plane enclosed by the contour C, and the point z_0 lies inside the contour, then

$$\oint_C dz\, \frac{f(z)}{(z - z_0)^{n+1}} = 2\pi i\, \frac{1}{n!}\, \frac{d^n f(z)}{dz^n}\bigg|_{z=z_0} \qquad (1.92)$$

A function $f(z)$ may be expanded in the neighborhood of a point z_0 in a Laurent series of the form

$$f(z) = \sum_{n=-\infty}^{\infty} a_n (z - z_0)^n\,. \qquad (1.93)$$

The results listed so far can be used to prove the residue theorem,

$$\oint_C dz\, f(z) = 2\pi i\, a_{-1}\,. \qquad (1.94)$$

for a contour enclosing z_0 and contained entirely in the region where the Laurent series converges.

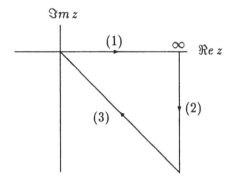

Fig. 1.1. The contour for evaluating $e^{-i\alpha z^2}$.

The Gaussian integral of (1.87) may be evaluated for imaginary α by *analytic continuation*. This is accomplished by extending the function $\exp(-i\alpha x^2)$, where α is both real and positive, to a function that is analytic everywhere in the complex plane, but one that reduces to the original function on the real line. Such a function is given by $\exp(-i\alpha z^2)$. The Cauchy–Riemann conditions demonstrate that the function $\exp(-i\alpha z^2)$ is analytic everywhere in the complex plane, and it reduces to the original function when $z = x$. If α is positive, then this function will be integrated along the contour depicted in Fig. 1.1. The path (1) lies along the real axis and extends to infinity, while path (3) lies along the line described by $y = -x$. Along path (1) $\exp(-i\alpha z^2) = \exp(-i\alpha x^2)$ and $dz = dx$. Along path (3) $\exp(-i\alpha z^2) = \exp(-2\alpha x^2)$ and $dz = (1-i)dx = (2/i)^{1/2}dx$. Path (2) contributes nothing to the contour integral since it is damped by the factor $\exp(-2\alpha x|y|)$, where $x \to \infty$. Since $f(z)$ is analytic everywhere, Goursat's theorem states that its integral around the closed contour must vanish, and this relates the integrals along paths (1) and (3). It follows that

$$\int_0^\infty dx\, e^{-i\alpha x^2} = \sqrt{\frac{2}{i}} \int_0^\infty dx\, e^{-2\alpha x^2} = \frac{1}{2}\sqrt{\frac{\pi}{i\alpha}}. \tag{1.95}$$

Extending the integration to the whole real axis gives the final result

$$\int_{-\infty}^\infty dx\, e^{-i\alpha x^2} = \sqrt{\frac{\pi}{i\alpha}}. \tag{1.96}$$

This is the same result as if α had been analytically continued to $i\alpha$ in (1.87). If α is negative, then the contour must lie in the upper quadrant; however, the result of the integration is identical to (1.96).

Sec. 1.5 Gaussian Integrals   25

The generalization of the single Gaussian integration to n real variables is given by

$$I(a) = \int_{-\infty}^{\infty} dx_1 \cdots dx_n \exp(-x_i a_{ij} x_j), \qquad (1.97)$$

where the a_{ij} are the elements of some matrix $\mathbf{A}$. If $I(a)$ is to be a real number, then the matrix $\mathbf{A}$ must be assumed to be symmetric and real, thereby making $\mathbf{A}$ Hermitian.

The evaluation of (1.97) begins by changing variables by means of a unitary transformation $\mathbf{U}$, so that $x_i = U_{ij} y_j$. The $\mathbf{U}$ is chosen so that $\mathbf{A}' = \mathbf{U}^\dagger \mathbf{A} \mathbf{U}$ is a diagonal matrix. Since $\mathbf{A}$ is Hermitian, the results of Sec. 1.4 show that such a $\mathbf{U}$ can always be constructed from the eigenvectors of $\mathbf{A}$, and that the resulting matrix $\mathbf{A}'$ has the eigenvalues of $\mathbf{A}$ along the diagonal and zeros everywhere else. In terms of the new variables the integral becomes

$$I(a) = \int_{-\infty}^{\infty} dy_1 \cdots dy_n \, J(y) \exp\left\{-\sum_{i=1}^{n} \lambda^{(i)} y_i^2\right\}, \qquad (1.98)$$

where $\lambda^{(i)}$ is the ith eigenvalue of $\mathbf{A}$, and $J(y)$ is the Jacobian of the transformation.

Before proceeding further the properties of the Jacobian will be discussed. In order to preserve the volume of integration when the variables in an integral are changed, the measure, written in terms of the new variables, must be multiplied by the Jacobian of the change of variables. To make this explicit, the old variables are denoted x_i, where it is assumed there are n of them. Under a change of variables the x_i are considered to be functions of the new variables y_i, so that

$$x_i = x_i(y_1, \ldots, y_n), \qquad (1.99)$$

for all i. Under such a change of variables the measure, or infinitesimal volume of integration, becomes

$$dx_1 \cdots dx_n = J(y_1, \ldots, y_n) dy_1 \cdots dy_n, \qquad (1.100)$$

where the Jacobian J is given by

$$J(y_1, \ldots, y_n) = \det \frac{\partial x}{\partial y}, \qquad (1.101)$$

so that $\partial x_i / \partial y_j$ is the ijth element of the matrix of which the Jacobian is the determinant.

For a unitary transformation, $x_i = U_{ij} y_j$, the Jacobian is simply $J = \det \mathbf{U}$. From the properties of determinants and unitary transformations it follows that

$$1 = \det(\mathbf{U}^\dagger \mathbf{U}) = \det \mathbf{U}^\dagger \det \mathbf{U} = |\det \mathbf{U}|^2 \ . \qquad (1.102)$$

This shows that $\det \mathbf{U} = 1$, up to an irrelevant phase factor (i.e., a complex number with modulus one). The phase factor is irrelevant here because it can always be absorbed into the definition of $\mathbf{U}$ without affecting any of the properties of $\mathbf{U}$.

Using this result, $I(a)$ becomes a product of simple Gaussian integrals with the result

$$I(a) = \pi^{n/2} \prod_{i=1}^{n} (\lambda^{(i)})^{-\frac{1}{2}} = \pi^{n/2} (\det \mathbf{A})^{-\frac{1}{2}} = (\pi^n \det \mathbf{A}^{-1})^{1/2} \ , \qquad (1.103)$$

where (1.61) has been used. Of course, buried in this form is the assumption that each of the n simple Gaussian integrals is well defined. This will be true if and only if all the eigenvalues appearing in (1.103) are nonzero and positive. A necessary, but not sufficient, condition for this to be true is $\det \mathbf{A} > 0$.

If a factor of i appears in the Gaussian (1.97), then the integral must be defined by analytic continuation in the following way. An overall factor of α is introduced into the real integral,

$$I(\alpha, a) = \int_{-\infty}^{\infty} dx_1 \cdots dx_n \exp(-\alpha x_i a_{ij} x_j) \ , \qquad (1.104)$$

so that the integral of interest is found by the analytic continuation $I(\alpha = i, a)$. In the event that the real integral is well-defined, the analytic continuation is found from (1.103),

$$\int_{-\infty}^{\infty} dx_1 \cdots dx_n \exp(-i x_i a_{ij} x_j) = \left[(-i\pi)^n \det \mathbf{A}^{-1}\right]^{1/2} \ . \qquad (1.105)$$

This analytic continuation is valid as long as the eigenvalues of $\mathbf{A}$ are nonzero and positive.

A final extension of the formulas presented so far is to include in the exponential a term that is linear in the variable of integration. Because the limits on the integration are $\pm\infty$, it follows that the integral may be performed by completing the square, so that

$$\int_{-\infty}^{\infty} dx \, e^{-\alpha x^2 \pm \beta x} = \exp\left(\frac{\beta^2}{4\alpha}\right) \int_{-\infty}^{\infty} dx \, e^{-\alpha(x \pm \frac{\beta}{2\alpha})^2}$$

$$= \sqrt{\frac{\pi}{\alpha}} \exp\left(\frac{\beta^2}{4\alpha}\right) \ . \qquad (1.106)$$

The complex version of (1.106) is found, from (1.96), to be

$$\int_{-\infty}^{\infty} dx \, e^{-i\alpha x^2 \pm i\beta x} = \sqrt{\frac{\pi}{i\alpha}} \exp\left(\frac{i\beta^2}{4\alpha}\right). \tag{1.107}$$

It is straightforward to extend result (1.103) to include a term linear in the x_i. This is most easily accomplished by translating the x_i that appears in (1.97) by a constant $\beta_i/2$, so that β defines an n-dimensional vector. Because of the limits on the integrations, the integral is invariant under this change, and therefore

$$I(a) = \int_{-\infty}^{\infty} dx_1 \cdots dx_n \, \exp(-x_i a_{ij} x_j \pm \beta_i a_{ij} x_j - \frac{1}{4}\beta_i a_{ij} \beta_j). \tag{1.108}$$

Defining $b_i = \beta_j a_{ji}$, and assuming the matrix $\mathbf{A}$ has an inverse $\mathbf{A}^{-1}$, which must also be Hermitian, it is easy to see that

$$\beta_i a_{ij} \beta_j = b_i (\mathbf{A}^{-1})_{ij} b_j. \tag{1.109}$$

Inserting this result back into (1.108) gives the n-dimensional extension of (1.106),

$$\int_{-\infty}^{\infty} dx_1 \cdots dx_n \, \exp(-\mathbf{x}^T \mathbf{A} \mathbf{x} \pm \mathbf{b}^T \mathbf{x})$$

$$= \left(\frac{\pi^n}{\det \mathbf{A}}\right)^{1/2} \exp\left(\frac{1}{4}\mathbf{b}^T \mathbf{A}^{-1} \mathbf{b}\right), \tag{1.110}$$

where matrix notation has been used. A similar form can be found for the complex version of (1.108).

Exercise 1.14: Show that the Gaussian integral over many complex variables, $z_n = x_n + iy_n$, is given by

$$\int_{-\infty}^{\infty} dz_1 \, dz_1^* \cdots dz_n \, dz_n^* \, \exp(iz_j^* a_{jk} z_k + ib_j^* z_j + ib_j z_j^*)$$

$$= \frac{(2\pi)^n}{\det \mathbf{A}} \exp(i\mathbf{b}^\dagger \mathbf{A}^{-1} \mathbf{b}), \tag{1.111}$$

where $\mathbf{A}$ is assumed to be an invertible Hermitian matrix.

It is important to note that all the results of this section regarding Gaussian integrals can be obtained in the following way. The argument

28 Mathematical Preliminaries

of the exponential that results from performing the integrations is given exactly by evaluating the argument of the exponential being integrated at its extremum. This can be summarized by writing

$$\int_{-\infty}^{\infty} \mathcal{D}x \, \exp\left[F(x)\right] \propto \exp\left[F(x_c)\right] \,, \tag{1.112}$$

where $F(x)$ is one of the Gaussian arguments considered in this section, while $\mathcal{D}x$ represents an integration over all Gaussian variables present. The form for x_c appearing in (1.112) is determined by

$$\left.\frac{\partial F(x)}{\partial x}\right|_{x=x_c} = 0 \,. \tag{1.113}$$

This technique is referred to as a *saddle-point* evaluation and will be used throughout what follows.

Exercise 1.15: Verify that result (1.112) is true for the cases considered in this section.

References

[1] G. Arfken, *Mathematical Methods for Physicists*, Academic Press, New York, 1966.

[2] R. Courant and D. Hilbert, *Methods of Mathematical Physics*, Interscience, New York, 1953.

[3] P.A.M. Dirac, *Quantum Mechanics, Fourth Edition*, Oxford University Press, London, 1958.

[4] A mathematically rigorous definition of the Dirac delta and a presentation of the general theory of distributions and test function spaces can be found in I. Halperin, *Introduction to the Theory of Distributions*, University of Toronto Press, Toronto, 1952. See also [2].

[5] The concept of functionals stems from the work of Volterra; see V. Volterra, *Opere Matematiche*, 5 vols., Accademia Nazionale dei Lincei, 1954–62. See also [8].

[6] A standard mathematical treatise on Fourier series and integrals as well as completeness is E.C. Titchmarsh, *Introduction to the Theory of Fourier Integrals*, Oxford University Press, Oxford, 1948.

[7] Hilbert space theory is discussed in any standard text on functional analysis; see for example the excellent treatise by M. Reed and B. Simon, *Methods of Modern Mathematical Physics I: Functional Analysis*, Academic Press, New York, 1980.

[8] The development of a rigorous formulation of the calculus of variations is discussed in the outstanding series by M. Kline, *Mathematical Thought from Ancient to Modern Times*, 3 vols., Oxford University Press, New York, 1972.

[9] The definition of functional differentiation used in this book has its origins in the functional methods for quantum field theory pioneered by Schwinger and Symanzik. See for example J. Schwinger, Proc. Nat. Sci. Acad. **37**, 651 (1951); K. Symanzik, Z. Natürforschung **9A**, 10 (1954).

[10] Functional differentiation techniques are discussed in H. Fried, *Functional Methods and Models in Quantum Field Theory*, MIT Press, Cambridge, Massachusetts, 1972.

[11] Functionals and functional differentials are discussed in L.S. Schulman, *Techniques and Applications of Path Integration*, Wiley, New York, 1981.

[12] The formulation of classical mechanics in terms of a variational principle began with the work of Lagrange and Hamilton. The standard presentation for physicists is found in H. Goldstein, *Classical Mechanics, Second Edition*, Wiley, New York, 1983.

[13] For a detailed introduction see H. Greub, *Linear Algebra*, Springer-Verlag, Berlin, 1967.

[14] The proof of the Baker–Campbell–Hausdorff theorem presented here may be found in J.D. Bjorken and S.D. Drell, *Relativistic Quantum Fields*, McGraw-Hill, New York, 1965.

[15] For a detailed introduction to complex analysis see for example R.V. Churchill, *Complex Variables and Applications*, McGraw-Hill, New York, 1960.

Chapter 2

Quantum Mechanical Path Integrals

In this chapter the standard version of nonrelativistic quantum mechanics will be reformulated in terms of a path integral. The path integral will be derived as a limiting expression from the fundamental structure of nonrelativistic quantum mechanics, and will be valid for a wide variety of systems. In this sense the path integral is a representation of quantum mechanical amplitudes equivalent to the usual wave mechanical or matrix formulations of these same amplitudes. This is extremely useful since there is no ambiguity in the path integral's definition. For this reason the form of the path integral derived in this case serves as a motivating form for a generalization of the path integral to all quantum processes.

In Sec. 2.1 the relevant aspects of basic quantum mechanics are reviewed. In Sec. 2.2 these are used to find the path integral form for a quantum mechanical amplitude. In Sec. 2.3 the idea of the path integral as a "sum over histories" is presented as a conceptual generalization of the results of Sec. 2.2. Some of the formidable mathematical difficulties associated with this generalization are sketched.

2.1 Quantum Mechanics

The Hamiltonian formulation of classical mechanics, discussed in Sec. 1.3, serves as the starting point for the operator form of quantum mechanics [1, 2, 3]. In the Hamiltonian formalism the generalized coordinates, q_j, and their canonically conjugate momenta, p_j, are the fundamental mechanical "observables" of a particle or system. In quantum mechanics these become operators defined on an abstract Hilbert space $\mathcal{H}$ that represents all possible quantum mechanical configurations in which the system may be observed. These operators, denoted Q_j and P_j, are assumed to obey the commutation relation

$$[Q_j, P_k] \equiv Q_j P_k - P_k Q_j = i\hbar \delta_{jk} , \qquad (2.1)$$

where $\hbar$ is Planck's constant. The elements of the Hilbert space $\mathcal{H}$ are called *states* and are written $|\psi\rangle$. The operators Q and P are assumed to have a complete set of eigenstates $|q\rangle$ and $|p\rangle$ such that

$$Q_j|q\rangle = q_j|q\rangle , \quad P_j|p\rangle = p_j|p\rangle , \qquad (2.2)$$

so that q_j and p_j are the eigenvalues of the respective operators.

The Hilbert space $\mathcal{H}$ is equipped with an inner product. If $|\psi\rangle$ and $|\phi\rangle$ are any two states in $\mathcal{H}$, then their inner product is denoted $\langle\psi|\phi\rangle$, and obeys

$$\langle\phi|\psi\rangle = \langle\psi|\phi\rangle^* , \qquad (2.3)$$

so that its form is analogous to the inner product for the functions of Sec. 1.3 or the vectors of Sec. 1.4. The inner products of the $|p\rangle$ and $|q\rangle$ states are assumed to obey

$$\langle q|q'\rangle = \delta^n(q - q'), \quad \langle p|p'\rangle = \delta^n(p - p') , \qquad (2.4)$$

so that these states are orthonormal in the continuum sense. The Dirac delta appearing in (2.4) is understood to have the test function space L^2, the space of all square integrable functions. It is also assumed that these states are complete, so that the states $|p\rangle$ and $|q\rangle$ both span the Hilbert space $\mathcal{H}$. In the continuum normalization of (2.4) this means that

$$\int d^n p\, |p\rangle\langle p| = 1, \quad \int d^n q\, |q\rangle\langle q| = 1 , \qquad (2.5)$$

where the limits on the integrals of (2.5) must run over the entirety of the phase space available to the system. Coupling the algebra of (2.1) with the inner product (2.4) gives the coordinate representation of the momentum operator

$$\langle q|P_j|q'\rangle = -i\hbar \frac{\partial}{\partial q_j} \delta^n(q - q') . \qquad (2.6)$$

This, in turn, gives the inner product

$$\langle q|p\rangle = \frac{1}{(2\pi\hbar)^{n/2}} \exp\left(\frac{i}{\hbar} p \cdot q\right) . \qquad (2.7)$$

Sec. 2.1 Quantum Mechanics   33

The factors present in (2.7) are necessary in order that

$$\langle q | q' \rangle = \int_{-\infty}^{\infty} d^n p \, \langle q | p \rangle \langle p | q' \rangle$$
$$= \int_{-\infty}^{\infty} \frac{d^n p}{(2\pi\hbar)^n} e^{ip\cdot(q-q')/\hbar} = \delta^n(q - q') \,. \tag{2.8}$$

The formal similarity of (2.1) to (1.50) indicates that the classical mechanical Poisson brackets of two observables, $A(p,q)$ and $B(p,q)$, must be replaced in quantum mechanics by commutators in the manner

$$\{A(p,q), B(p,q)\}_{p,q} \rightarrow \frac{1}{i\hbar}[A(P,Q), B(P,Q)] \,. \tag{2.9}$$

If O is a quantum mechanical observable, i.e., some function of Q, P, and possibly t, then (1.49) and the formal identity (2.9) gives

$$i\hbar \frac{dO}{dt} = i\hbar \frac{\partial O}{\partial t} + [O, H] \,, \tag{2.10}$$

where $H = H(P, Q, t)$ is the Hamiltonian of the system under analysis. Thus, the time development of a quantum mechanical observable is driven by the Hamiltonian of the system, just as it is in classical mechanics.

The physical system is represented by a state $|\psi\rangle$ in the Hilbert space $\mathcal{H}$, and the state is assumed to be of unit length, so that $\langle \psi | \psi \rangle = 1$. It follows from (2.10) that the time development of the state $|\psi\rangle$ is given by the Schrödinger equation

$$H|\psi, t\rangle_s = i\hbar \frac{\partial}{\partial t} |\psi, t\rangle_s \,. \tag{2.11}$$

The standard differential version of the Schrödinger equation used in wave mechanics may be obtained from (2.11) by forming the inner product with the position eigenstate $|q\rangle$, and identifying the wave-function $\psi(q,t) = \langle q | \psi, t \rangle_s$. This results in the differential equation

$$H(-i\hbar \frac{\partial}{\partial q}, q) \, \psi(q,t) = i\hbar \frac{\partial}{\partial t} \psi(q,t) \,. \tag{2.12}$$

When the system is in the state $|\psi\rangle$, the expectation value of a measurement of the observable $O(P,Q)$ is given by

$$\langle O \rangle = \langle \psi | O(P,Q) | \psi \rangle = \int dq \, \psi^*(q,t) \, O(-i\hbar \frac{\partial}{\partial q}, q) \, \psi(q,t) \,, \tag{2.13}$$

leading to the interpretation of $|\psi(q)|^2$ as a probability density. The normalization of the state,

$$\langle \psi | \psi \rangle = \int dq \, |\psi(q,t)|^2 = 1 ,\qquad(2.14)$$

is simply a reflection of the fact that the total probability of observing the particle must be unity.

In the Schrödinger picture of quantum mechanics the states, $|\psi\rangle_s$, are manifestly time dependent, while the observables are not, so that, if H has no explicit time dependence, the state at time t evolves from the state at $t=0$ according to

$$|\psi,t\rangle_s = \exp\left(-\frac{i}{\hbar}Ht\right) |\psi\rangle_s .\qquad(2.15)$$

If H is time dependent, then (2.15) must be replaced with the time-ordered integral of H. This is denoted by

$$|\psi,t\rangle_s = T\left\{\exp\left[-\frac{i}{\hbar}\int_0^t d\tau \, H(\tau)\right]\right\} |\psi\rangle_s .\qquad(2.16)$$

Time ordering means expanding the exponential in (2.16) so that the operators are ordered sequentially according to their time argument, with the latest at the left and earliest at the right. This is most easily expressed by introducing the step function $\theta(\tau)$, which is defined as

$$\theta(\tau) = \begin{cases} 1 & \text{if } \tau > 0, \\ 0 & \text{if } \tau < 0. \end{cases} \qquad(2.17)$$

Exercise 2.1: Show that $\theta(\tau)$ has the integral representation

$$\theta(\tau) = \lim_{\epsilon \to 0} \int_{-\infty}^{\infty} \frac{d\omega}{2\pi i} \frac{e^{i\omega\tau}}{\omega - i\epsilon} ,\qquad(2.18)$$

and that

$$\frac{\partial}{\partial t}\theta(t) = \delta(t) .\qquad(2.19)$$

Using the step function, the time-ordered product of two time-dependent operators can be written

$$T\{A(t_1)B(t_2)\} = \theta(t_1 - t_2)A(t_1)B(t_2) + \theta(t_2 - t_1)B(t_2)A(t_1) .\qquad(2.20)$$

The generalization of (2.20) to the time-ordered product of many operators is obvious.

In the Heisenberg picture of quantum mechanics the observables, O_H, are time dependent, and from (2.10), they satisfy

$$i\hbar \frac{d}{dt} O_H = i\hbar \frac{\partial}{\partial t} O_H + [O_H, H] . \quad (2.21)$$

The equality of the matrix elements in the respective pictures gives

$$_H\langle \psi | O_H(t) | \phi \rangle_H = {}_S\langle \psi, t | O_S | \phi, t \rangle_S , \quad (2.22)$$

so that differentiation of (2.22) gives

$$i\hbar \frac{\partial O_H}{\partial t} = [O_H, H] . \quad (2.23)$$

If H has no explicit time dependence, then the Heisenberg picture operators of (2.23) at time t can written in terms of the operators at $t = 0$ in the form

$$O_H(t) = e^{iHt/\hbar} O_H e^{-iHt/\hbar} . \quad (2.24)$$

If H is explicitly time dependent, then it is necessary to use time ordering, as in (2.16). The Heisenberg picture states, $|\psi\rangle_H$, are time independent. It is clear that the Schrödinger picture and Heisenberg picture can be chosen to coincide at some specific time. In what follows it will be assumed that this time is $t = 0$, and any state with no explicit time dependence displayed can be considered to be either a Heisenberg picture state or a Schrödinger picture state at $t = 0$.

In the path integral formulation of quantum mechanics the object of utmost interest is the transition amplitude. If the system is in the Schrödinger picture state $|\phi, t_a\rangle$ at the time t_a, then the transition amplitude to the state $|\psi, t_b\rangle$ at the time t_b is defined as

$$Z(\psi, \phi) = \langle \psi, t_b | \phi, t_a \rangle . \quad (2.25)$$

The transition amplitude then gives the probability of the system transiting between the respective states as $P = |Z|^2$, so that Z describes a *quantum process*. In the next section the transition amplitude is given a path integral representation.

2.2 The Path Integral Derived

In this section the transition amplitude will be given a path integral representation by applying the results of the previous section. The pivotal

36 Quantum Mechanical Path Integrals

idea, first noted by Dirac [4], is that the infinitesimal quantum mechanical transition amplitude is governed by the value of the classical action. Using fairly intuitive arguments, this idea was developed by Feynman [5, 6] into the path integral representation of the finite transition amplitude, and later presented in detail in the book by Feynman and Hibbs [7]. The path integral derived in this section will be identical in form to that first developed by Feynman. However, a derivation based on the canonical structure of quantum mechanics has the advantage of demonstrating that the path integral must yield results identical to those obtained by the standard methods of wave mechanics or matrix manipulation. For that reason it is an immensely valuable heuristic device for testing generalizations of the path integral form to systems where a rigorous derivation is impossible.

It is therefore very important to list the assumptions that go into this derivation, so that they may be kept in mind during any generalization. First, it will be assumed that the limits on the statement of completeness (2.5) are $\pm\infty$. Second, it will be assumed that the Hamiltonian H has no explicit time dependence, although this is not necessary. Third, and this is for simplicity of notation, the system will be considered to be that of a one-dimensional single particle moving in a potential.

The first step in the derivation of the path integral [3, 8, 9] is to construct the "instantaneous" eigenstates, $|q,t\rangle$ and $|p,t\rangle$, of the Heisenberg picture operators $Q(t)$ and $P(t)$. These are defined by

$$|q,t\rangle = e^{iHt/\hbar}|q\rangle, \quad |p,t\rangle = e^{iHt/\hbar}|p\rangle. \quad (2.26)$$

It is to be noted that these states are *not* Schrödinger picture states. However, these states are complete, since it follows from (2.5) that

$$\int_{-\infty}^{\infty} dq\, |q,t\rangle\langle q,t|$$
$$= \exp(iHt/\hbar)\left(\int_{-\infty}^{\infty} dq\,|q\rangle\langle q|\right)\exp(-iHt/\hbar) = 1. \quad (2.27)$$

The state $|q,t\rangle$ is an eigenstate of the Heisenberg picture operator $Q(t)$ in the sense that

$$Q(t)|q,t\rangle = q|q,t\rangle. \quad (2.28)$$

These states also have the valuable property that, for $|\psi,t\rangle$ a Schrödinger picture state,

$$\langle q,-t|\psi,t\rangle = \langle q|\psi\rangle = \psi(q). \quad (2.29)$$

A similar pair of statements holds for the instantaneous momentum eigenstates.

Sec. 2.2 The Path Integral Derived   37

These states can be used to derive the path integral form for the transition element between Schrödinger picture states. The object of interest, Z, is defined as the inner product of the instantaneous eigenstates at different times, so that
$$Z(q_a, t_a, q_b, t_b) = \langle q_b, t_b | q_a, t_a \rangle, \qquad (2.30)$$
where it is assumed that $t_b > t_a$. Knowledge of the form for Z allows the calculation of the more general form (2.25), since it follows that the transition element between Schrödinger picture states is given by

$$\langle \psi, -t_b | \phi, -t_a \rangle$$
$$= \int_{-\infty}^{\infty} dq_a \, dq_b \, \langle \psi, -t_b | q_b, t_b \rangle \langle q_b, t_b | q_a, t_a \rangle \langle q_a, t_a | \phi, -t_a \rangle$$
$$= \int_{-\infty}^{\infty} dq_a \, dq_b \, \psi^*(q_b) \phi(q_a) Z(q_a, t_a, q_b, t_b), \qquad (2.31)$$

where properties (2.27) and (2.29) have been used. Once the initial and final states of the system are specified by the normalized forms for $\psi(q)$ and $\phi(q)$, the system propagates in time from ψ to ϕ through the function Z. For this reason the transition element (2.30) is sometimes referred to as the *propagator*, since it contains all the information regarding the time development of the system.

The time interval $t_b - t_a$ is first partitioned into N infinitesimal steps of duration $\epsilon = (t_b - t_a)/N$, where the limit $N \to \infty$ is understood in everything that follows. Next, $N-1$ complete sets of intermediate instantaneous $|q\rangle$ eigenstates are inserted into the transition element sequentially at each of the respective times $t_n = t_a + n\epsilon$. This gives

$$Z(q_a, t_a, q_b, t_b) = \int_{-\infty}^{\infty} dq_1 \cdots dq_{N-1} \, \langle q_b, t_b | q_{N-1}, t_{N-1} \rangle$$
$$\times \langle q_{N-1}, t_{N-1} | \cdots | q_1, t_1 \rangle \langle q_1, t_1 | q_a, t_a \rangle, \qquad (2.32)$$

so that the whole transition element has been reduced to the product of N transition elements, which are infinitesimal in the sense that their time difference approaches zero. Each one of the infinitesimal transition elements may now be analyzed. Since the time difference is ϵ between the two states, it follows from the definition (2.26) that the jth element is given by

$$\langle q_{j+1}, t_{j+1} | q_j, t_j \rangle = \langle q_{j+1} | e^{-i\epsilon H(P,Q)/\hbar} | q_j \rangle. \qquad (2.33)$$

In order to proceed further it is necessary to select a convention that will be used consistently to reduce all the infinitesimal elements. To evaluate the matrix element of the exponentiated Hamiltonian, the exponential must be expanded in a power series. After the expansion all the P operators will be moved to the left and all the Q operators will be moved to the right. For want of a better name this will be referred to as *coordinate ordering*. The final result of coordinate ordering a product of P's and Q's will be that all the P operators lie on the left side of the expression and all the Q operators will lie on the right side of the expression. The coordinate ordering operation will be denoted by $C\{\cdots\}$, so that, for example,

$$C\{PQP^2\} = P^3Q . \qquad (2.34)$$

Obviously, the opposite convention can be selected, and it is natural to consider the possibility that the existence of different ordering schemes might induce some ambiguity in the resulting path integral [10]. That the same form results is given as Exercise 2.4. The advantage of coordinate ordering is that an arbitrary function of Q and P, denoted $f(Q,P)$, has the matrix element

$$\langle p | C\{f(Q,P)\} | q \rangle = f(q,p) \langle p | q \rangle , \qquad (2.35)$$

so that a coordinate-ordered function of the operators may be reduced to a c-number (classical number) function by taking its matrix element. Result (2.35) can be demonstrated by using a power series representation of the c-number function f and by noting that coordinate ordering suppresses the presence of any commutators between Q and P.

The next step in evaluating (2.33) is to observe that

$$e^{-i\epsilon H(P,Q)/\hbar} = C\left\{ e^{-i\epsilon H(P,Q)/\hbar} \right\} + O(\epsilon^2) . \qquad (2.36)$$

The proof of this statement for a general Hamiltonian is given as Exercise 2.2. However, a demonstration for the particularly simple but physically meaningless form $H = aP + bQ$ is instructive because of its use of the Baker–Campbell–Hausdorff theorem. The operator version of this theorem is identical to the form proved in Sec. 1.4 for matrices. Using this theorem shows that

$$e^{-i\epsilon(aP+bQ)/\hbar} = e^{-i\epsilon aP/\hbar} e^{-i\epsilon bQ/\hbar} e^{i\epsilon^2 ab/2\hbar} , \qquad (2.37)$$

so that the effects of the commutator (2.1) are $O(\epsilon^2)$ and are therefore irrelevant in the $\epsilon \to 0$ limit. Since the commutator represents quantum effects, or *quantum corrections*, this verifies the intuitive notion that, for infinitesimal time periods, the time development of the system is overwhelmingly dominated by classical dynamics. The generalization of this result, showing that the $O(\epsilon^2)$ term in (2.36) is ignorable, can be accomplished by using the Trotter product formula [11, 12]. If A and B are any two bounded operators, the Trotter product formula gives

$$\exp[t(A+B)] = \lim_{n\to\infty} \left[\exp\left(\frac{t}{n}A\right)\exp\left(\frac{t}{n}B\right)\right]^n . \quad (2.38)$$

Exercise 2.2: Prove (2.38) and adapt the theorem to demonstrate that, in the $\epsilon \to 0$ limit, the commutators generated by a Hamiltonian of the form $H = P^2/2m + V(Q)$ are irrelevant to the evaluation of (2.33).

These results show that the infinitesimal transition element (2.33) becomes, for $\epsilon \approx 0$,

$$\langle q_{j+1} | e^{-i\epsilon H(P,Q)/\hbar} | q_j \rangle$$
$$= \int_{-\infty}^{\infty} dp_j \, \langle q_{j+1} | p_j \rangle \langle p_j | e^{-i\epsilon H(P,Q)/\hbar} | q_j \rangle$$
$$\approx \int_{-\infty}^{\infty} dp_j \, e^{-i\epsilon H(p_j,q_j)/\hbar} \langle q_{j+1} | p_j \rangle \langle p_j | q_j \rangle . \quad (2.39)$$

This infinitesimal element can be simplified further by using (2.7) to give

$$\langle q_{j+1} | p_j \rangle \langle p_j | q_j \rangle = \frac{1}{2\pi\hbar} e^{ip_j(q_{j+1}-q_j)/\hbar} . \quad (2.40)$$

Using the fact that q_j is the coordinate value associated with the state at time t_j allows the *formal* identification

$$\lim_{\epsilon\to 0} \frac{1}{\epsilon}(q_{j+1} - q_j) = \frac{dq_j}{dt} \equiv \dot{q}_j . \quad (2.41)$$

Using this identification, the infinitesimal matrix element can be written

$$\langle q_{j+1}, t_{j+1} | q_j, t_j \rangle \approx \int_{-\infty}^{\infty} \frac{dp_j}{2\pi\hbar} \exp\left\{-\frac{i}{\hbar}\epsilon\left[p_j\dot{q}_j - H(p_j, q_j)\right]\right\}$$
$$= \int_{-\infty}^{\infty} \frac{dp_j}{2\pi\hbar} \exp\left[\frac{i}{\hbar}\epsilon \mathcal{L}(p_j, q_j)\right] , \quad (2.42)$$

where

$$\mathcal{L}(p_j, q_j) = p_j\dot{q}_j - H(p_j, q_j) . \quad (2.43)$$

First discussed in Sec. 1.3, $\mathcal{L}(p_j, q_j)$ is the Lagrangian density in Hamilton's formulation of the classical mechanical system.

The finite transition element Z can finally be written as the product of the N infinitesimal elements, giving

$$\langle q_b, t_b | q_a, t_a \rangle = \int_{-\infty}^{\infty} \frac{dp_0}{2\pi\hbar} \cdots \frac{dp_{N-1}}{2\pi\hbar} dq_1 \cdots dq_{N-1} \exp\left[\frac{i}{\hbar} \sum_{j=0}^{N-1} \epsilon \mathcal{L}(p_j, q_j)\right], \quad (2.44)$$

where the identifications $q_0 = q_a$ and $q_N = q_b$ are implicit in (2.44). The argument of the exponential in (2.44) has the form of a Riemann sum, enabling the identification

$$\lim_{\epsilon \to 0} \sum_{j=0}^{N-1} \epsilon \mathcal{L}(p_j, q_j) = \int_{t_a}^{t_b} dt\, \mathcal{L}(p, q) = S[p(t), q(t), t_a, t_b], \quad (2.45)$$

Thus, the classical action has appeared in the quantum mechanical transition element. The *path integral measure* appearing in (2.44) is written formally as

$$\lim_{N \to \infty} \frac{dp_0}{2\pi\hbar} \cdots \frac{dp_{N-1}}{2\pi\hbar} dq_1 \cdots dq_{N-1} \equiv \mathcal{D}p\, \mathcal{D}q. \quad (2.46)$$

Form (2.46), when combined with the exponential of the action, technically does not meet the mathematical criteria required for a probability measure [13]; nevertheless, it will be referred to as the path integral measure throughout what follows. This problem is discussed further in Sec. 2.3.

The final form of the transition element is then given by

$$\langle q_b, t_b | q_a, t_a \rangle = \int_{q_a}^{q_b} \mathcal{D}p\, \mathcal{D}q \, \exp\left\{\frac{i}{\hbar} \int_{t_a}^{t_b} dt\, \mathcal{L}(p, q)\right\}, \quad (2.47)$$

where the limits on the q integrals are present to remind the user that q_0 and q_N are identified as q_a and q_b. Expression (2.47) is the fundamental form for the path integral version of the propagator. The measure appearing in (2.47) ranges over the entire *phase space* available to the particle as it propagates from q_a to q_b.

Exercise 2.3: Consider a Hamiltonian H that is explicitly time dependent in the sense that time-dependent c-number functions (*not* velocity-dependent potentials) may appear in the Lagrangian density. Use (2.16) to show that the resulting path integral representation of the transition element is unchanged in form.

Exercise 2.3 gives the important result that Lagrangian densities describing systems with time-dependent parameters can immediately be given a path integral representation with no difficulty.

Exercise 2.4: Consider a definition of coordinate ordering where the P operator is moved to the right and the Q operator is moved to the left. Derive the form of the path integral for this definition.

Exercise 2.5: Extend the path integral formalism to a system with n degrees of freedom.

Some applications of the path integral found in the literature do not have $\mathcal{D}p$ appearing in the measure. This is not necessarily incorrect since it is possible to integrate all the p_j appearing in (2.42) for a large class of Lagrangians. An example of this occurs when the Lagrangian density of the system has the form

$$\mathcal{L}(p,q) = p\dot{q} - \frac{p^2}{2m} - V(q) , \qquad (2.48)$$

Since this gives rise to an integral Gaussian in p, it is possible to evaluate exactly the p integration in all the infinitesimal transition elements (2.42). It follows from (1.107) that

$$\int_{-\infty}^{\infty} \frac{dp_j}{2\pi\hbar} \exp \frac{i}{\hbar}\left[p_j(q_{j+1} - q_j) - \epsilon \frac{p_j^2}{2m} - \epsilon V(q_j)\right]$$

$$= \sqrt{\frac{m}{2\pi i\hbar\epsilon}} \exp \frac{i}{\hbar}\epsilon \left[\frac{1}{2}m\left(\frac{q_{j+1}-q_j}{\epsilon}\right)^2 - V(q_j)\right]$$

$$= \sqrt{\frac{m}{2\pi i\hbar\epsilon}} \exp\left[\frac{i}{\hbar}\epsilon \mathcal{L}(\dot{q},q)\right] , \qquad (2.49)$$

where the identification (2.41) has again been made. In this case the path integral has become

$$\langle q_b, t_b | q_a, t_a \rangle = \int_{q_a}^{q_b} \mathcal{D}q \, \exp\left\{\frac{i}{\hbar}\int_{t_a}^{t_b} dt \, \mathcal{L}(\dot{q},q)\right\} , \qquad (2.50)$$

where

$$\mathcal{L}(\dot{q},q) = \tfrac{1}{2}m\dot{q}^2 - V(q) . \qquad (2.51)$$

The factors resulting from the dp integrations have been absorbed into the measure, giving the definition

$$\overline{\mathcal{D}}q \equiv \lim_{N \to \infty} \left(\frac{m}{2\pi i\hbar\epsilon}\right)^{N/2} dq_1 \cdots dq_{N-1} . \tag{2.52}$$

The form (1.38) of the classical mechanical action has emerged in this case. It is form (2.50) which will be used to generalize the path integral in the next section, although (2.47) is a more general and useful form than (2.50).

The path integral form (2.50) has been derived assuming that the potential $V(q)$ appearing in the Lagrangian density is not velocity dependent. Some physical systems are characterized by a velocity-dependent potential. It is still possible to derive a path integral for the propagator; however, there may be ambiguities in the final form. The demonstration of this is left as the following exercise.

Exercise 2.6: Consider a one-dimensional system with a point mass moving in the velocity-dependent potential

$$V(\dot{q}, q) = \tfrac{1}{2}\alpha \dot{q}^2 q^2 .$$

Derive the form of the path integral equivalent to (2.50) and discuss any ambiguities that are present in the coordinate ordering problem for this potential.

2.3 The Sum over Histories

In the previous section the path integral expression for the transition amplitude was derived from the standard operator formulation of quantum mechanics. Although the measure in the second form (2.50) of the path integral is singular in the $\epsilon \to 0$ limit, the path integral, as defined in the previous section, must yield results that are identical to any other valid approach to the same problem. That this is true will be seen in detail in the next chapter. However, for the moment the explicit evaluation of the path integral will be deferred and attention will be focussed on generalizing the concepts contained in, and indicated by, the form of the path integral in hand. It will be argued that such a generalization is not rigorously defensible at this time, and that it is well to remember this when applying path integrals to new problems. In generalizing the concepts of the path integral it is also necessary to remember the assumptions that went into the derivation of the previous section.

It is the outstanding feature of the path integral that the classical action of the system has appeared in a quantum mechanical expression, and it is this feature that is considered central to any extension of the path integral formalism. It is the path integral of (2.50), obtained by integrating the momenta p_j, that will serve as the form for motivating the conceptual generalization. In the first step, the integrations over the q_j will be replaced by a Riemann sum of the form

$$\int_{-\infty}^{\infty} dq_j \rightarrow \sum_{n_j=-\infty}^{\infty} \epsilon_q , \qquad (2.53)$$

where ϵ_q is understood to be the infinitesimal measure element on all the q_j integrations. The path integral (2.50) then becomes

$$\langle q_b, t_b | q_a, t_a \rangle \rightarrow \left(\frac{m}{2\pi i \hbar \epsilon}\right)^{N/2} \sum_{n_1=-\infty}^{\infty} \epsilon_q \cdots \sum_{n_{N-1}=-\infty}^{\infty} \epsilon_q$$

$$\times \exp\left[\frac{i}{\hbar} \sum_{j=0}^{N-1} \epsilon \mathcal{L}\left([n_{j+1} - n_j]\frac{\epsilon_q}{\epsilon}, n_j \epsilon_q\right)\right] , \qquad (2.54)$$

where n_o and n_{N-1} are defined by $q_a = n_o \epsilon_q$ and $q_b = n_N \epsilon_q$.

This form of the path integral then represents the sum over all possible sets of values for the $N-1$ integer variables $\{n_j\}$. Each set of values is weighted by the exponential of the value of the action for that set of values. Specifying a set of $\{n_j\}$ is equivalent to specifying a set of $\{q_j\}$, and these values represent the intermediate values of the particle's position as it moves from q_a to q_b. In effect, specifying a set of values $\{n_j\}$ defines a *discrete path* from q_a to q_b, i.e., a set of $N-1$ intermediate positions for the particle. In the $\epsilon_q \rightarrow 0$ limit any piecewise continuous path from q_a to q_b can be represented in this manner. Thus, in the limit, the path integral can be viewed as a sum over all piecewise continuous paths from q_a to q_b in the time interval $t_b - t_a$, with each path receiving a weighting factor given by the exponential of the *classical* action along that path. In this way every piecewise continuous path represents a possible "history" of the particle's motion, and the path integral representation of the transition element is a weighted sum over all possible histories.

It is this observation that is the starting point for the generalization of the path integral, derived in the last section, to systems whose quantization by other techniques may not be well understood. In this approach to describing a quantum process, one begins by writing the classical action S for the system under consideration. This specifies the dynamical variables of the system, e.g., $q(t)$ in the case of quantum mechanics. The transition amplitude Z of the *quantized* system is then assumed to be given by

$$Z = \sum_{\text{paths}} e^{iS/\hbar}, \qquad (2.55)$$

where the paths must be chosen to go from some initial configuration of the dynamical variables to some final configuration. In (2.55) the formal derivatives of (2.41) are taken to be the exact statements, and in that sense (2.55) is a *continuum* version of the path integral derived in the previous section. The advantage of such a picture for quantum processes is the intuitive power that it brings. In the path integral formulation the classical paths of Newton have reappeared in the quantized system, allowing a graphical interpretation of quantum processes. However, the deceptive simplicity of statement (2.55) cloaks many subtle, and as yet unresolved, mathematical difficulties, and at least a cursory discussion of some of these must be made.

First, there are numerous classical systems for which no sensible quantum theory exists. Cases of this are very easy to find, even in simple one-dimensional quantum mechanical systems. A simple example is the potential $V = \alpha q^2 - \beta q^4$. If β is a positive constant the potential is unbounded from below. Even though it is possible, at least formally, to write down classical solutions to the equation of motion, the quantum theory cannot be stable since there is no normalizable ground state. Another problem associated with converting a classical mechanical system to a quantum mechanical path integral is that it is not obvious, at first glance, how to implement classical constraints on motion in a quantum mechanically consistent manner. At an even more subtle level, it will be seen that some field theories cannot be consistently quantized because of quantum mechanical anomalies in their conservation laws. The path integral approach cannot possibly improve such situations, and it may even obscure some of the problems if it is indiscriminately applied.

Second, the motivating form (2.54) indicates that the only difference in weighting each path receives is the exponential of the value of the action associated with the path, and that the prefactors, i.e., the ϵ factors in (2.54), are the same for all paths. That this is true for all quantum systems, or rather might be an artifact of the assumptions made in deriving the form (2.50), is not *a priori* obvious. For example, the assumption was made that the limits on the intermediate q_j integrations were $\pm\infty$. If, instead, the system under consideration is a particle in a one-dimensional box of width L, then clearly the integrations over the q_j should have the range 0 to L, and that in itself will give a different result for (2.47). Obtaining a form similar to (2.50) raises additional difficulties. Later in this section a similar system, a point mass on a circle, will be analyzed, and these points will become manifestly clear. There it will not be apparent that the path integral derived from the quantum mechanical solution to the problem has any reference to the classical action for the system.

Third, to say that there are "many" paths from q_a to q_b would be something of an understatement. However, there are many pathological paths that cannot contribute. For example, if t_a is a rational number and t_b is an irrational number, then a possible path would be given by the Dirichlet function,

$$q(t) = \begin{cases} q_a & \text{if } t \text{ is rational,} \\ q_b & \text{if } t \text{ is irrational.} \end{cases} \quad (2.56)$$

Such a path is not piecewise continuous and by that criterion should be excluded from the sum. In addition, the set of all piecewise continuous paths contains both discontinous and nondifferentiable paths, and these are, in some sense, "far away" from the typically smooth and completely continuous classical trajectories that extremize the action. Intuitively, paths far from the classical trajectories would be expected to contribute little to the path integral, since a gross deviation from classical behavior violates the idea that quantum corrections to most physical processes are small. Of course, quantum mechanics counters this intuition with the presence of a discrete set of bound states. This is far from the classical behavior of the system, which allows a continuum of bound states to occur in attractive potentials. In the path integral it is unclear how the bound state structure is manifested in a sum over histories, since a discrete set of bound states corresponds classically to a discrete set of bounded trajectories.

That paths associated with large values for the action do not contribute significantly to the path integral can be inferred from the Riemann–Lebesgue lemma [14]. This states that, if $f(\alpha)$ is an integrable function,

$$\lim_{t \to \infty} \int d\alpha\, f(\alpha) \sin(\alpha t) = 0 \,. \quad (2.57)$$

The Riemann–Lebesgue lemma can be proved by treating $\lim_{t \to \infty} \sin(\alpha t)$ as a distribution. It then follows that

$$\lim_{t \to \infty} \sin(\alpha t) = \int_0^\infty dt\, \frac{d}{dt} \sin(\alpha t) = \tfrac{1}{2} \alpha \int_{-\infty}^\infty dt\, e^{i\alpha t} = \pi \alpha\, \delta(\alpha) \,, \quad (2.58)$$

46 Quantum Mechanical Path Integrals

which agrees with result (2.57).

> **Exercise 2.7**: Show that $\lim_{t\to\infty} \cos(\alpha t) = 0$.

These results show that, as a distribution,

$$\lim_{t\to\infty} e^{i\alpha t} = \pi i \alpha \, \delta(\alpha) \,, \qquad (2.59)$$

which vanishes when integrated against a well-behaved function. In this sense, paths with infinite values for the action will not contribute to the path integral.

These remarks serve to draw attention to the major difficulty in generalizing the path integral to a sum over histories, and that is the question of precisely how this sum is to be performed or defined in a sensible way. The idea of summing or integrating over paths can be made mathematically rigorous only by the formal definition of a *path measure* on the space of all paths [15, 16]. This measure must be defined consistently with the weighting factor of the exponentiated action, since that is the functional being integrated. To motivate this with a familiar example, it is recalled that the dt appearing in the Riemann sum of the exponential is an example of a type of measure called the Jordan measure and has the form $t_{j+1} - t_j$. This is by no means the most general form of measure, even for the real line. To demonstrate the relation between the function being integrated and the measure used to integrate it, one need only point out that the function $q(t)$ defined by (2.56) is not an integrable function using Jordan measure. If such a pathological function is to be rendered integrable, then a more general measure, referred to as the *Lebesgue measure*, must be defined. The discussion of formal measure theory and, in particular, the problem of defining a measure over the space of paths is very difficult and requires a level of mathematics well beyond the scope of this book [17]. However, some aspects of summing over paths can be made clearer with a modicum of effort.

One of the chief difficulties in constructing a rigorous measure for the continuum version of the path integral is the presence of the i in its definition. The oscillatory nature of the exponential creates difficulties in including "unruly" paths, i.e., nondifferentiable and discontinuous paths, since the exponentiated action tends toward a distribution. A simple exclusion of the unruly paths would be incorrect. This is so because, while the differentiable and continuous paths are everywhere *dense* in the set of all paths, it can be shown that, taken alone, they are inadequate to give a correct evaluation of the path integral. In point of fact, a careful analysis of the situation shows that these paths form a set of measure zero [18]. The situation is much like that of the real number line, where the set of all rational numbers can be shown to be a set of measure zero, even though they are everywhere dense in the real line.

Unfortunately, because the exponentiated action tends toward a distribution, the path integral defined by (2.55) prevents the definition of a well-behaved measure, and therefore an alternate scheme must be found if a mathematical definition is to be made. It is the oscillatory nature of the path integral that gives rise to the distributions. If the oscillations were suppressed, then it might be possible to define a sensible measure for paths. It is with this hope that much of the rigorous work done on path integrals defines them for Euclidean time. This means that the integrand of the action undergoes the so-called Wick rotation, which amounts to replacing t with $-i\tau$. In effect, the path integral is analytically continued to imaginary time, evaluated, and analytically continued back by the inverse Wick rotation, $\tau \to it$, to yield the final result. The necessity of this analytic continuation arises from the need to evaluate the generalization of the oscillatory Gaussian integrals first encountered in Sec. 1.5. There, the integration was defined by an analytic continuation identical to the Wick rotation. Under the Wick rotation the form (2.50) for the path integral becomes

$$Z = \int_{q_a}^{q_b} \overline{\mathcal{D}}q \, e^{-S_E/\hbar} \;, \tag{2.60}$$

where, for the simple case (2.48), the *Euclidean action* S_E is given by

$$S_E = \int_{t_a}^{t_b} d\tau \left\{ \frac{1}{2} m \left(\frac{dq}{d\tau}\right)^2 + V(q) \right\} . \tag{2.61}$$

The factors in the measure of the path integral become real under the Wick rotation, so that

$$\overline{\mathcal{D}}q = \lim_{N \to \infty} \left(\frac{m}{2\pi \hbar \epsilon_\tau}\right)^{N/2} dq_1 \cdots dq_{N-1} \;, \tag{2.62}$$

where ϵ_τ is effectively $d\tau$. The Euclidean action (2.61) is identical in form to the integral of the Hamiltonian, written in terms of $\dot{q}$ instead of p. It will be shown in Chapter 4 that this is no coincidence. However, for now it is necessary only to notice that, unless the potential is unbounded from below, large deviations from the classical trajectory, as well as unruly paths, are exponentially suppressed. The Wick rotation makes a mathematically meaningful measure possible.

48   Quantum Mechanical Path Integrals

It is worth noting that, under a Wick rotation, the differential form of the Schrödinger equation, for $V = 0$, becomes

$$\frac{\hbar^2}{2m}\nabla^2\psi = \hbar\frac{d}{d\tau}\psi \,, \tag{2.63}$$

which is the diffusion equation that governs Brownian motion. Wiener [19] was the first to construct a path integral representation for Brownian motion and to show that a well-behaved measure could be defined. It has been argued, using sophisticated techniques, that the path integral has a well-defined measure in the Euclidean region, and this argument will be taken over in the remainder of this book. The interested reader is recommended to the books by Rivers [13] and by Glimm and Jaffe [20].

At best, these remarks may convey the difficulty in making the continuum path integral of (2.55) a well-defined mathematical object. In practice it is common to approximate the sum over paths by using the subset of piecewise continuous paths that are differentiable and continuous and that yield a finite action. In particular the classical trajectory is believed to represent the maximum weight path. It is hoped, but unsubstantiated in most cases, that this approximation gives the most important properties of the quantum transition amplitude. In Chapter 3 the role of the classical trajectory in the quantum transition amplitude will be discussed for several extremely simple cases.

This section will close with the derivation of the path integral representation for the transition amplitude in a simple one-dimensional quantum system with properties significantly different than those originally assumed. The system is that of a point mass constrained to move on a circle of radius R. The classical action for this system, in terms of the angular variable θ, is given by

$$\mathcal{L}(\dot\theta, \theta) = \tfrac{1}{2}mR^2\dot\theta^2 - V(\theta) \,. \tag{2.64}$$

The potential V is assumed to be periodic, so that $V(\theta + 2\pi n) = V(\theta)$, where n is an arbitrary integer. The Hamiltonian has the standard form

$$H(p_\theta, \theta) = \frac{1}{2m}p_\theta^{\,2} + V(\theta) \,. \tag{2.65}$$

The position eigenstates of the system are $|\theta\rangle$, and these are complete in the sense that

$$\int_0^{2\pi} d\theta\,|\theta\rangle\langle\theta| = 1 \,. \tag{2.66}$$

The configuration space representation of the momentum operator is

$$\langle\theta|P|\theta'\rangle = -i\frac{\hbar}{R}\frac{\partial}{\partial\theta}\delta(\theta - \theta') \,. \tag{2.67}$$

Because the system is on a circle, the eigenfunctions of all observables must be periodic. It is straightforward to show that the eigenstates of the observable p_θ are no longer continuous, but are indexed by an integer n, and are therefore denoted $|p_n\rangle$. It follows from (2.67) that the normalized momentum eigenfunctions for the system are given by

$$\langle\theta|p_n\rangle = \frac{1}{\sqrt{2\pi}}e^{in\theta}. \tag{2.68}$$

These eigenfunctions are orthonormal

$$\langle p_n|p_m\rangle = \int_0^{2\pi}\frac{d\theta}{2\pi}e^{i(m-n)\theta} = \delta_{mn}, \tag{2.69}$$

and complete in the sense that

$$\sum_{n=-\infty}^{\infty}\langle\theta|p_n\rangle\langle p_n|\theta'\rangle = \frac{1}{2\pi}\sum_{n=-\infty}^{\infty}e^{in(\theta-\theta')} = \delta(\theta-\theta'), \tag{2.70}$$

where the Dirac delta is clearly periodic. These results can be used to derive the path integral form for the transition element $\langle\theta_b,t_b|\theta_a,t_a\rangle$.

Exercise 2.8: Use the previous results to show that the path integral representation of the transition amplitude is given by

$$\langle\theta_b,t_b|\theta_a,t_a\rangle =$$

$$\lim_{N\to\infty}\int_{\theta_a}^{\theta_b}d\theta_1\cdots d\theta_{N-1}\frac{1}{(2\pi)^N}\sum_{n_0=-\infty}^{\infty}\cdots\sum_{n_{N-1}=-\infty}^{\infty}$$

$$\times\exp\left\{\frac{i}{\hbar}\sum_{j=0}^{N-1}\left[n_j\hbar(\theta_{j+1}-\theta_j)-\epsilon\frac{n_j^2\hbar^2}{2mR^2}-\epsilon V(\theta_j)\right]\right\}, \tag{2.71}$$

where the integrations over the θ_j are from 0 to 2π.

There are formal similarities between (2.71) and the general form (2.47) if the classical momentum is identified as $p = n\hbar/R$. However, this forces the classical momentum to be discrete, and such a constraint is never present in classical mechanics. Furthermore, it is not clear at this stage that a form similar to (2.50) can be obtained or, if it can be, what the limits on the θ integrations should be. Thus, passing over to a path integral of the form (2.55) for this system seems, at this point, a questionable step. This problem will be resolved in the next chapter when (2.71) is analyzed to reveal its underlying structure.

References

[1] A. Messiah, *Quantum Mechanics*, Wiley, New York, 1966.
[2] L. Schiff, *Quantum Mechanics, Third Edition*, McGraw-Hill, New York, 1968.
[3] R. Shankar, *Principles of Quantum Mechanics*, Plenum Press, New York, 1980.
[4] P.A.M. Dirac, Physikalische Z. der Sowjetunion **3**, 64 (1933).
[5] R.P. Feynman, Rev. Mod. Phys. **20**, 367 (1948).
[6] The interaction between Dirac and Feynman is described in S. Schweber, Rev. Mod. Phys. **58**, 449 (1986).
[7] R.P. Feynman and A.R. Hibbs, *Quantum Mechanics and Path Integrals*, McGraw-Hill, New York, 1965.
[8] W. Tobocman, Nuovo Cimento **3**, 1213 (1956).
[9] E.S. Abers and B.W. Lee, Phys. Rep. **9**, 1 (1973).
[10] Various aspects of ordering problems are discussed in J.S. Dowker, J. Math. Phys. **17**, 1873 (1976); B. Gaveau and L.S. Schulman, J. Math. Phys. **30**, 3019 (1989).
[11] Applications of the Trotter product formula to path integral structures is found in E. Nelson, J. Math. Phys. **5**, 332 (1964).
[12] An alternative derivation of the quantum mechanical path integral from the Trotter product formula can be found in L.S. Schulman, *Techniques and Applications of Path Integration*, Wiley, New York, 1981.
[13] See, for example, R.J. Rivers, *Path Integral Methods in Quantum Field Theory*, Cambridge University Press, New York, 1987.
[14] See, for example, E.C. Titchmarsh, *Introduction to the Theory of Fourier Integrals*, Oxford University Press, Oxford, 1948.
[15] L.L. Lee, J. Math. Phys. **17**, 1988 (1976).
[16] A. Truman, J. Math. Phys. **19**, 1742 (1978).
[17] P.R. Halmos, *Measure Theory*, Van Nostrand, New York, 1950.
[18] J.R. Klauder, Acta. Phys. Aust., Suppl. **XI**, 341 (1973).
[19] N. Wiener, J. Math. and Phys. **2**, 131 (1923); N. Wiener, Acta Math. **55**, 117 (1930); see also M. Kac, *Probability and Related Topics in the Physical Sciences*, Interscience, New York, 1959.
[20] A. Glimm and A. Jaffe, *Quantum Mechanics—A Functional Integral Point of View*, Springer-Verlag, Berlin, 1981.

Chapter 3

Evaluating the Path Integral

In this chapter various forms of the path integral will be evaluated, thereby demonstrating different techniques. Each form considered will be useful to later analyses and will illustrate the utility and validity of various methods of evaluating the path integral. The basic types of approach to evaluating the path integral form of the transition amplitude can be broken into two groups: discrete methods and continuum methods. In the discrete method the integral of the action is replaced with a Riemann sum, and the intermediate values are integrated using the measure defined by (2.46) or (2.52). In the continuum approach the action is treated as a functional, and it is expanded around a set of classical trajectories using Fourier series. The second approach is more difficult to implement exactly, and some aspects of this approach must be inferred from exact results found by other methods.

In Sec. 3.1 the free particle propagator is calculated from the corresponding form (2.50) and shown to be the exact result found by the operator approach. In Sec. 3.2 the free particle is given a time-dependent driving force, referred to as a *source*, and the corresponding form (2.47) is analyzed. The resulting form can be combined with functional derivatives to analyze, in a very compact way, time-ordered products of the Heisenberg picture position operator. This gives rise to the important concept of a generating functional [1]. It is shown that this functional can be used to analyze the motion of a particle in a more complex potential by defining a perturbation series representation of the propagator. In Sec. 3.3 the harmonic oscillator is analyzed using quasi-continuous methods, and a careful analysis of the Jacobian associated with the change of variables is made.

In Sec. 3.4 the path integral for a particle on a ring, derived in Sec. 2.4, is analyzed, using Poisson resummation techniques, to reveal its underlying topological structure.

3.1 The Free Particle

In this section the simplest possible system, that of a free point mass moving one-dimensionally, is analyzed using the path integral. The Lagrangian is given by

$$\mathcal{L}(\dot{q},q) = \tfrac{1}{2}m\dot{q}^2 . \tag{3.1}$$

This system can be analyzed very easily by operator techniques, since the associated Hamiltonian is given by

$$H(P,Q) = \frac{P^2}{2m} . \tag{3.2}$$

Using completeness of the $|p\rangle$ eigenstates and the form of the inner product (2.5), the transition element can be written

$$\begin{aligned}
Z(q_b,t_b,q_a,t_a) &= \langle q_b | e^{-iH(t_b-t_a)/\hbar} | q_a \rangle \\
&= \int_{-\infty}^{\infty} dp \exp\left[-i\frac{p^2}{2m\hbar}(t_b-t_a)\right] \langle q_b | p \rangle \langle p | q_a \rangle \\
&= \int_{-\infty}^{\infty} \frac{dp}{2\pi\hbar} \exp\left[-i\frac{p^2}{2m\hbar}(t_b-t_a) + i\frac{p}{\hbar}(q_b-q_a)\right] .
\end{aligned} \tag{3.3}$$

This is a Gaussian integral of the form (1.107), so that the final result is

$$Z(q_a,t_a,q_b,t_b) = \left[\frac{m}{2\pi i\hbar(t_b-t_a)}\right]^{\frac{1}{2}} \exp\left[\frac{i}{\hbar}\frac{m(q_b-q_a)^2}{2(t_b-t_a)}\right] . \tag{3.4}$$

Since the path integral form for the transition amplitude was derived from the operator formalism, it must also yield (3.4). In order to demonstrate that this is true the action appearing in the path integral will be treated as a discrete sum. The equivalent version of (2.44) for this action gives

$$Z(q_a,t_a,q_b,t_b) =$$
$$\int_{q_a}^{q_b} \left(\frac{m}{2\pi i\hbar\epsilon}\right)^{N/2} dq_1 \cdots dq_{N-1} \exp\left[\frac{i}{\hbar} \sum_{j=0}^{N-1} \frac{m}{2\epsilon}(q_{j+1}-q_j)^2\right] . \tag{3.5}$$

Following Feynman and Hibbs [2], the intermediate integrations are performed in sequence. The first integration over q_1 includes the q_0 and q_2 factors. Singling out only the terms that are involved gives

$$\int_{-\infty}^{\infty} dq_1 \exp\left\{\frac{i}{\hbar}\frac{m}{2\epsilon}\left[(q_2-q_1)^2+(q_1-q_a)^2\right]\right\}$$

$$=\int_{-\infty}^{\infty} dq_1 \exp\left\{\frac{i}{\hbar}\left[\frac{m}{4\epsilon}(q_2-q_a)^2+\frac{m}{\epsilon}\left(q_1-\frac{1}{2}q_2-\frac{1}{2}q_a\right)^2\right]\right\}$$

$$=\sqrt{\frac{\pi i\hbar\epsilon}{m}}\exp\left[\frac{i}{\hbar}\frac{m}{4\epsilon}(q_2-q_a)^2\right]. \quad (3.6)$$

The first integration has produced another Gaussian integral. It is left as an exercise to show that the result of the nth integration has the form

$$\int_{-\infty}^{\infty} dq_n \exp\left\{\frac{i}{\hbar}\left[\frac{m}{2n\epsilon}(q_n-q_a)^2+\frac{m}{2\epsilon}(q_{n+1}-q_n)^2\right]\right\}$$

$$=\sqrt{\frac{2\pi i\hbar n\epsilon}{(n+1)m}}\exp\left[\frac{i}{\hbar}\frac{m}{2(n+1)\epsilon}(q_{n+1}-q_a)^2\right]. \quad (3.7)$$

Exercise 3.1: Demonstrate result (3.7).

As a result, after the $N-1$ integrations have been performed and the identification $q_N = q_b$ is made, the path integral reduces to

$$Z(q_a,t_a,q_b,t_b)$$

$$=\exp\left[\frac{i}{\hbar}\frac{m(q_b-q_a)^2}{2N\epsilon}\right]\left(\frac{m}{2\pi i\hbar\epsilon}\right)^{N/2}\prod_{n=1}^{N-1}\sqrt{\frac{2\pi i\hbar n\epsilon}{(n+1)m}}$$

$$=\left(\frac{m}{2\pi i\hbar N\epsilon}\right)^{1/2}\exp\left[\frac{i}{\hbar}\frac{m(q_b-q_a)^2}{2N\epsilon}\right]. \quad (3.8)$$

Recalling that $N\epsilon = t_b - t_a$, the $N \to \infty$ limit reproduces the result (3.4), so that the path integral is identical in content to the operator formalism for this specific definition of action and measure.

For this simple case the path integral could scarcely be viewed as a calculational improvement over the operator formalism. However, some points of interest emerge from closer examination of the propagator. First, the Green's function for the time-dependent Schrödinger equation and the propagator are intimately related. The Green's function $G(x, x', t, t')$ for the one-dimensional form of Schrödinger's equation is defined as the function that satisfies

$$\left(H(-i\hbar\frac{d}{dx},x) - i\hbar\frac{d}{dt}\right) G(x,x',t,t') = i\hbar\delta(x-x')\delta(t-t') \ . \qquad (3.9)$$

The Green's function is related to the propagator by

$$G(x,x',t,t') = \theta(t'-t)Z(x',t',x,t) \ , \qquad (3.10)$$

where $\theta(t'-t)$ is the step function defined by (2.17). Relation (3.9) can be readily demonstrated by using (2.19) and the definition of the propagator (2.30). The action of the time derivative gives

$$\begin{aligned} i\hbar\frac{d}{dt}G(x,x',t,t') &= -i\hbar\delta(t'-t)Z(x',t',x,t) \\ &\quad + \theta(t'-t)\langle x | He^{iH(t'-t)/\hbar} | x' \rangle \ . \end{aligned} \qquad (3.11)$$

From the fact that $Z(x',t,x,t) = \langle x | x' \rangle = \delta(x-x')$, relation (3.9) is immediately obtained.

The second point is the role of the classical action in the overall result (3.4). The classical equation of motion, $\ddot{q} = 0$, is trivial to solve for the boundary conditions that $q(t_a) = q_a$ and $q(t_b) = q_b$. It follows that the classical trajectory, denoted $q_c(t)$, is given by

$$q_c(t) = q_a + \left(\frac{q_b - q_a}{t_b - t_a}\right)(t - t_a) \ . \qquad (3.12)$$

It is straightforward to show that the value of the action for the classical trajectory is given by

$$S_c = \int_{t_a}^{t_b} dt \ \tfrac{1}{2}m\dot{q}_c^2 = \frac{m(q_b - q_a)^2}{2(t_b - t_a)} \ . \qquad (3.13)$$

By comparing this result to (3.4) it is apparent that the argument of the exponential appearing in the propagator is given precisely by $i/\hbar$ times the value of the action along the classical trajectory. In Sec. 3.3 this will be shown to be a general feature of all quadratic actions.

3.2 Motion with a Source

In this section an evaluation will be given for the path integral describing the system of a one-dimensional point mass simultaneously moving in the

Sec. 3.2 Motion with a Source

potential $V(q)$ and undergoing an arbitrary external time-dependent force $J(t)$. The Lagrangian of the system is given by

$$\mathcal{L}(\dot{q}, q) = \tfrac{1}{2}m\dot{q}^2 - V(q) + J(t)q \ . \tag{3.14}$$

The function $J(t)$ serves as a source for the particle's acceleration, and, as a result, $J(t)$ is referred to as the *source function* [1]. The source function allows a definition of the generating functional, and this will be discussed later.

It will be expeditious to begin with the simpler case of $V(q) = 0$ and return to the case of a nonzero potential afterwards. The method of evaluating the path integral will be an application of the discrete method to the form equivalent to (2.47). The appropriate Lagrangian density is given by

$$\mathcal{L}(p, q) = p\dot{q} - \frac{p^2}{2m} + J(t)q \ . \tag{3.15}$$

The measure (2.46), derived for this form in Sec. 2.2, is appropriate since the particle's motion can range over all phase space.

The transition element is given by the path integral

$$Z(q_a, t_a, q_b, t_b) = \int_{q_a}^{q_b} dq_1 \cdots dq_{N-1} \frac{dp_0}{2\pi\hbar} \cdots \frac{dp_{N-1}}{2\pi\hbar}$$

$$\times \exp\left\{\frac{i}{\hbar} \sum_{j=0}^{N-1} \left[p_j(q_{j+1} - q_j) - \epsilon\frac{p_j^2}{2m} + \epsilon J(t_j)q_j\right]\right\} \ . \tag{3.16}$$

Rather than integrating the p_j variables to create a path integral that is Gaussian in the q_j, it is simpler to perform the q_j integrations first. The result is a product of $N - 1$ Dirac deltas, reducing the path integral to

$$Z(q_a, t_a, q_b, t_b) =$$
$$\frac{1}{2\pi\hbar} \int dp_0 \cdots dp_{N-1}\, \delta(\epsilon J(t_0) + p_0 - p_1) \cdots \delta(\epsilon J(t_{N-1}) + p_{N-2} - p_{N-1})$$

$$\times \exp\left\{\frac{i}{\hbar}\left[p_{N-1}q_b - p_0 q_a + \epsilon J(t_0)q_a\right]\right\} \exp\left[-\frac{i\epsilon}{2m\hbar} \sum_{j=0}^{N-1} p_j^2\right] \ . \tag{3.17}$$

The first $N - 1$ integrations are now trivial to perform. The resulting form for the path integral becomes equivalent to the single integral

Evaluating the Path Integral

$$Z(q_a, t_a, q_b, t_b) = \exp\left[\frac{i}{\hbar} q_a \sum_{j=0}^{N-1} \epsilon J(t_j)\right] \int \frac{dp_{N-1}}{2\pi\hbar} \exp\left\{\frac{i}{\hbar} p_{N-1}(q_b - q_a)\right.$$
$$\left. - \frac{i}{\hbar} \sum_{j=0}^{N-1} \frac{\epsilon}{2m} \left[p_{N-1} - \sum_{k=j+1}^{N-1} \epsilon J(t_k)\right]^2 \right\}. \quad (3.18)$$

Expression (3.18) is actually simpler when the continuum limit is taken. Denoting the integral over $J(t)$ as the impulse $I(t)$,

$$I(t) \equiv \int_t^{t_b} d\tau\, J(\tau) = \int_{t_a}^{t_b} d\tau\, \theta(\tau - t) J(\tau), \quad (3.19)$$

and writing $p_{N-1} = p$, in the limit $\epsilon \to 0$ the path integral becomes

$$Z(q_a, t_a, q_b, t_b) = \exp\left[\frac{i}{\hbar} q_a I(t_a)\right] \int_{-\infty}^{\infty} \frac{dp}{2\pi\hbar} \exp\left\{\frac{i}{\hbar} p(q_b - q_a)\right.$$
$$\left. - \frac{i}{\hbar} \int_{t_a}^{t_b} dt \left[\frac{p^2}{2m} - \frac{p}{m} I(t) + \frac{1}{2m} I^2(t)\right] \right\}. \quad (3.20)$$

Expression (3.20) is easily rewritten as a Gaussian integral of the general form (1.107),

$$Z(q_a, t_a, q_b, t_b) = \exp\left\{\frac{i}{\hbar}\left[q_a I(t_a) - \frac{1}{2m} \int_{t_a}^{t_b} dt\, I^2(t)\right]\right\}$$
$$\times \int_{-\infty}^{\infty} \frac{dp}{2\pi\hbar} \exp\left\{-\frac{i}{\hbar} \frac{p^2}{2m}(t_b - t_a)\right.$$
$$\left. + \frac{i}{\hbar} p\left[(q_b - q_a) + \frac{1}{m} \int_{t_a}^{t_b} dt\, I(t)\right] \right\}. \quad (3.21)$$

Performing this final integration gives [2]

$$Z(q_a, t_a, q_b, t_b) = \left[\frac{m}{2\pi i \hbar(t_b - t_a)}\right]^{\frac{1}{2}} \exp\left\{\frac{i}{\hbar}\left[\frac{m(q_b - q_a)^2}{2(t_b - t_a)}\right.\right.$$
$$+ \frac{(q_b - q_a)}{(t_b - t_a)} \int_{t_a}^{t_b} dt\, I(t) + \frac{1}{2m(t_b - t_a)} \left[\int_{t_a}^{t_b} dt\, I(t)\right]^2$$
$$\left.\left. + q_a I(t_a) - \frac{1}{2m} \int_{t_a}^{t_b} dt\, I^2(t)\right]\right\}. \quad (3.22)$$

Sec. 3.2 Motion with a Source

Although the final expression (3.22) appears quite complicated, it was relatively easy to obtain by the path integral technique.

It is a simplifying step to consider expression (3.22) for the case of a constant force, $J(t) = f$. The integrations appearing in (3.19) and (3.22) can be readily evaluated for this case, and (3.22) reduces to

$$Z(q_a, t_a, q_b, t_b) = \left[\frac{m}{2\pi i \hbar (t_b - t_a)}\right]^{\frac{1}{2}} \exp\left\{\frac{i}{\hbar}\left[\frac{m(q_b - q_a)^2}{2(t_b - t_a)} + \frac{1}{2}(q_b + q_a)(t_b - t_a)f - \frac{(t_b - t_a)^3 f^2}{24m}\right]\right\}. \quad (3.23)$$

Exercise 3.2: Show that the argument of the exponential appearing in (3.23) is $i/\hbar$ times the value of the action (3.15) evaluated along its classical trajectory for the case $J(t) = f$.

Exercise 3.3: Use (3.23) to show that the transition element between two instantaneous momentum eigenstates, $\langle p_b, t_b | p_a, t_a \rangle$, is nonzero only if the classical impulse relation $p_b = p_a + f(t_b - t_a)$ is met.

Inspection of (3.22) shows that it reduces to result (3.4) for the case that $J(t) = 0$. By definition, $Z(q_a, t_a, q_b, t_b)$, given by (3.22), is a functional of $J(t)$, so for convenience it will be denoted $Z_{ab}[J]$. This functional has the important property that its functional derivatives give the matrix elements of products of the Heisenberg picture operator $Q(t)$ between the two instantaneous eigenstates.

Exercise 3.4: Show that

$$-i\hbar \frac{\delta Z_{ab}[J]}{\delta J(t)} = \int_{q_a}^{q_b} \mathcal{D}p\,\mathcal{D}q\, q(t) \exp\left\{\frac{i}{\hbar}\int_{t_a}^{t_b} dt\, \mathcal{L}(p,q)\right\}$$
$$= \langle q_b, t_b | Q(t) | q_a, t_a \rangle_J, \quad (3.24)$$

where the subscript J serves as a reminder that the propagator is to be calculated for nonvanishing source.

From Exercise 3.4 it should be apparent that expectation values of $Q(t)$, for the free (sourceless) theory discussed in the previous section, can be obtained by simply evaluating the functional derivatives at $J(t) = 0$. Using the result that

$$\frac{\delta I(t')}{\delta J(t)} = \theta(t-t'),\qquad (3.25)$$

and the functional chain rule, the expectation value of $Q(t)$ is given by

$$\langle q_b, t_b | Q(t) | q_a, t_a \rangle_{J=0} = -i\hbar \left.\frac{\delta Z_{ab}[J]}{\delta J(t)}\right|_{J=0}$$
$$= \left(q_a + \frac{(q_b - q_a)}{(t_b - t_a)}(t - t_a)\right) Z_{ab}[0]. \quad (3.26)$$

Therefore, the nth order functional derivative yields the time-ordered product of n factors of $Q(t)$.

Exercise 3.5: Show that

$$-\hbar^2 \frac{\delta^2 Z_{ab}[J]}{\delta J(t_1)\delta J(t_2)} = \langle q_b, t_b | T\{Q(t_1)Q(t_2)\} | q_a, t_a \rangle_J, \qquad (3.27)$$

where T stands for the time-ordering operator defined by (2.20).

This result can be made explicit for the specific example of (3.22). An application of the functional chain rule gives

$$\langle q_b, t_b | T\{Q(t_1)Q(t_2)\} | q_a, t_a \rangle_{J=0} = -\hbar^2 \left.\frac{\delta^2 Z_{ab}[J]}{\delta J(t_1)\delta J(t_2)}\right|_{J=0}$$
$$= \left\{ i\hbar\theta(t_2 - t_1)\frac{(t_b - t_2)(t_1 - t_a)}{m(t_b - t_a)} + i\hbar\theta(t_1 - t_2)\frac{(t_b - t_1)(t_2 - t_a)}{m(t_b - t_a)}\right.$$
$$\left. + \left[q_a + \frac{(q_b - q_a)}{(t_b - t_a)}(t_1 - t_a)\right]\left[q_a + \frac{(q_b - q_a)}{(t_b - t_a)}(t_2 - t_a)\right]\right\} Z_{ab}[0]. \quad (3.28)$$

As a result, higher-order functional derivatives generate higher-order time-ordered products. Therefore, the form for $Z_{ab}[J]$ given by (3.22) serves as the *generating functional* for time-ordered products of the free (zero potential) theory.

Examinination of (3.28) shows that the time-ordered product of two powers of $Q(t)$ has broken into two pieces, one proportional to $\hbar$, the other with $\hbar$ absent. The piece with no reference to $\hbar$ is simply the product of the expectation values of $Q(t_1)$ and $Q(t_2)$, while the piece proportional to $\hbar$ is unique to the time-ordered product. In effect, the piece proportional to $\hbar$ reflects the quantum nature of the system, and arises because $Q(t_1)$ does not commute with $Q(t_2)$ even in the free theory. Because the piece proportional to $\hbar$ arises due to quantum effects, it will be referred to as the *connected piece* of the time-ordered product. The significance of this name will be made clearer when graphical representation of perturbation series for quantum field theory is developed in Chapter 8.

Because the piece of the time-ordered product with no reference to $\hbar$ contains no new information, it would be useful to eliminate it. This can be effected by defining a new generating functional [3, 4], given by

$$W_{ab}[J] = -i\hbar \ln Z_{ab}[J] . \tag{3.29}$$

The functional $W_{ab}[J]$ is the generator of the connected pieces of time-ordered products. At each order of differentiation of $W_{ab}[J]$ the pieces corresponding to lower order derivatives are removed and the value is normalized to unity.

This can be illustrated by examining the first several derivatives. The first derivative gives

$$\frac{\delta W_{ab}[J]}{\delta J(t)} = -i\hbar \frac{1}{Z_{ab}[J]} \frac{\delta Z_{ab}[J]}{\delta J(t)} \equiv \langle Q(t) \rangle_J . \tag{3.30}$$

By comparing this result with (3.26), it is apparent that $\langle Q(t) \rangle_J$ is the normalized expectation value of $Q(t)$, i.e., the factor corresponding to the propagator $Z_{ab}[J]$ has been divided out. Continuing on, the second derivative gives

$$-i\hbar \frac{\delta^2 W_{ab}[J]}{\delta J(t_1) \delta J(t_2)} =$$
$$-\hbar^2 \frac{1}{Z_{ab}[J]} \frac{\delta^2 Z_{ab}[J]}{\delta J(t_1) \delta J(t_2)} + \hbar^2 \left(\frac{1}{Z_{ab}[J]}\right)^2 \frac{\delta Z_{ab}[J]}{\delta J(t_1)} \frac{\delta Z_{ab}[J]}{\delta J(t_2)} . \tag{3.31}$$

However, comparison with (3.30) shows that the second term on the right-hand side of (3.31) is

$$\hbar^2 \left(\frac{1}{Z_{ab}[J]}\right)^2 \frac{\delta Z_{ab}[J]}{\delta J(t_1)} \frac{\delta Z_{ab}[J]}{\delta J(t_2)} = -\langle Q(t_1) \rangle_J \langle Q(t_2) \rangle_J , \tag{3.32}$$

so that expression (3.31) becomes

$$-i\hbar \frac{\delta^2 W_{ab}[J]}{\delta J(t_1) \delta J(t_2)} = \langle T\{Q(t_1)Q(t_2)\} \rangle_J - \langle Q(t_1) \rangle_J \langle Q(t_2) \rangle_J . \tag{3.33}$$

Comparing (3.33) with (3.28) shows that only the connected piece of the time-ordered product remains.

The process of taking higher-order derivatives could be continued to demonstrate inductively the general structure of $W_{ab}[J]$. However, inspection of (3.22) shows that, for the $V(q) = 0$ case, only the first and second derivatives of $W_{ab}[J]$ are nonzero. It will be seen that interactions, i.e., some nontrivial form for $V(q)$, must be present to create nonzero higher derivatives. In order to demonstrate this the nonzero $V(q)$ case must now be considered. While the general case of interactions, in particular nonlinear potentials, is a formidable problem to analyze, the results obtained up to this point allow a representation of the interacting case in terms of a formal perturbation series.

Returning to (3.15), the Lagrangian density is now assumed to have the form

$$\mathcal{L}(p,q) = p\dot{q} - \frac{p^2}{2m} - V(q) + J(t)q \,. \tag{3.34}$$

Of course, there are many restrictions on $V(q)$ that are necessary in order to have a sensible quantum theory. In what follows it will be assumed that these conditions are met. It should, however, be borne in mind that the manipulations presented here are of a formal nature and can disguise many of the difficulties that may be encountered in defining interacting theories.

The propagator for the Lagrangian density (3.34) is given by the path integral

$$Z_{ab}[J] = \int_{q_a}^{q_b} \mathcal{D}p\mathcal{D}q \, \exp\left\{\frac{i}{\hbar} \int_{t_a}^{t_b} dt \left[p\dot{q} - \frac{p^2}{2m} - V(q) + J(t)q\right]\right\} \,. \tag{3.35}$$

If the potential $V(q)$ has terms higher than quadratic in q, i.e., it is nonlinear, then the path integral is no longer Gaussian, and obtaining a closed form expression for it is usually impossible. In order to make headway the path integral of (3.34) is rewritten as a power series expansion. This is done by introducing the functional differential operator

$$V\left(-i\hbar\frac{\delta}{\delta J(t)}\right) \,, \tag{3.36}$$

which is simply the potential $V(q)$ with $q(t)$ replaced by $-i\hbar\delta/\delta J(t)$. Using this functional differential operator, a power series expansion of the exponential, and the rules of functional differentiation, it is straightforward to show that $Z_{ab}[J]$ can be written

$$Z_{ab}[J] = \exp\left\{-\frac{i}{\hbar}\int_{t_a}^{t_b} dt\, V\left(-i\hbar\frac{\delta}{\delta J(t)}\right)\right\}$$
$$\times \int_{q_a}^{q_b} \mathcal{D}p\,\mathcal{D}q\, \exp\left\{\frac{i}{\hbar}\int_{t_a}^{t_b} dt\, \left[p\dot{q} - \frac{p^2}{2m} + J(t)q\right]\right\}. \quad (3.37)$$

The path integral appearing in expression (3.37) is nothing more than the $V(q) = 0$ case treated earlier in this section. Its evaluation gives the propagator (3.22), which is now denoted $Z^0_{ab}[J]$ and is, in effect, the "basis" or lowest-order propagator for the interacting theory. The propagator for the interacting case can then be written

$$Z_{ab}[J] = \exp\left\{-\frac{i}{\hbar}\int_{t_a}^{t_b} dt\, V\left(-i\hbar\frac{\delta}{\delta J(t)}\right)\right\} Z^0_{ab}[J], \quad (3.38)$$

which is understood as a formal functional power series expansion of the exponential applied to the generating functional. Expression (3.38) is a perturbative representation of the propagator and encapsulates all the rules and combinatorics present in other derivations of perturbation theory [5]. Of course, implicit in the formal expression (3.38) is the assumption that each term in the perturbation series is well defined and that the series converges to a sensible result. These are assumptions, and whether or not they are valid must be determined on a case by case basis.

Exercise 3.6: Show that the lowest-order nonvanishing functional derivative of $W_{ab}[J]$ above quadratic matches the lowest-order power of q appearing in $V(q)$, or a power series representation thereof.

There are two points worth noting about the perturbation series representation. The first is the case where $J = 0$ while $V \neq 0$. Such a case is understood as the power series (3.38) evaluated at $J = 0$, which in turn means starting with the form (3.35) before the limit is taken. However, writing such a path integral as the $J \to 0$ limit of (3.35) assumes that the path integral of (3.35) is well defined in the limit $J \to 0$. It happens that there are forms of $V(q)$ for which the path integral of (3.35) is not analytic at $J = 0$. For such cases, there are possibly more than one value for the $J \to 0$ limit of the path integral, and other criteria must be used to determine which value corresponds to the physical system. This unusual aspect of path integrals will be discussed again when the quantum mechanical effective action, to be defined in the next chapter, is analyzed for spontaneous breakdown of symmetry in Sec. 4.5. The second point is that, in order for perturbation theory to have a chance to be valid, $V(q)$

should contain only terms higher than quadratic. The reason for this is simple. Terms quadratic and linear in q should be retained in the "basis" path integral $Z_{ab}^0[J]$ since they define a Gaussian integral. In the next section a very general technique for analyzing Lagrangian densities that are quadratic will be presented.

3.3 Continuum Techniques

In this section the path integral action will be considered to be a continuum functional, rather than a Riemann sum, and continuum techniques will be developed. Throughout the previous two sections it has been pointed out that the argument of the exponential appearing in the final form for the evaluated path integral is given by the value of the action along the classical trajectory. In this section it will be seen that this result is exact for all Lagrangian densities that are quadratic. The use of continuum techniques allows a rapid evaluation of the exponential and the path integral measure becomes an integration over the eigenmodes of the differential operator appearing in the action.

The starting point is the path integral form (2.50), where the measure, denoted $\overline{\mathcal{D}}q$, is given by (2.52). Because of the limits on the $N-1$ intermediate integrations, each one of the variables can be translated by a number without inducing a nontrivial Jacobian determinant in the measure. If $q_c(t)$ is the classical trajectory from $q = q_a$ at $t = t_a$ to $q = q_b$ at $t = t_b$, then the new variables of integration x_j are defined by

$$q_j = q_c(t_j) + x_j \; . \tag{3.39}$$

Under this change of variables the measure simply becomes $\overline{\mathcal{D}}x$, while the Lagrangian density can then be expanded around q_c. The result is

$$\int_{t_a}^{t_b} dt \, \mathcal{L}(\dot{q}, q) = \int_{t_a}^{t_b} dt \left[\mathcal{L}(\dot{q}_c, q_c) + x \frac{\partial}{\partial q_c} \mathcal{L}(\dot{q}_c, q_c) + \dot{x} \frac{\partial}{\partial \dot{q}_c} \mathcal{L}(\dot{q}_c, q_c) + \mathcal{L}_R(\dot{x}, x) \right] \; , \tag{3.40}$$

where $\mathcal{L}_R$ represents the remainder of terms in the expansion that are of order quadratic and higher in the new variable x. In this respect it is easy to see that $\mathcal{L}_R(\dot{x}, x)$ coincides with $\mathcal{L}(\dot{x}, x)$ if the original Lagrangian density had no terms higher than quadratic. Integrating by parts in (3.40) and using the Euler–Lagrange equation (1.41) for $q_c(t)$ immediately reduces the action to

$$\int_{t_a}^{t_b} dt\, \mathcal{L}(\dot{q}, q) = \int_{t_a}^{t_b} dt\, \mathcal{L}(\dot{q}_c, q_c) + \int_{t_a}^{t_b} dt\, \mathcal{L}_R(\dot{x}, x)\,. \tag{3.41}$$

Under the change of variables the path integral has become

$$\int_{q_a}^{q_b} \overline{\mathcal{D}}q \exp\left(\frac{i}{\hbar}\int_{t_a}^{t_b} dt\, \mathcal{L}(\dot{q}, q)\right) =$$
$$\exp\left[\frac{i}{\hbar}\int_{t_a}^{t_b} dt\, \mathcal{L}(\dot{q}_c, q_c)\right]\int_0^0 \overline{\mathcal{D}}x \exp\left[\frac{i}{\hbar}\int_{t_a}^{t_b} dt\, \mathcal{L}_R(\dot{x}, x)\right]\,. \tag{3.42}$$

The limits of 0 on the second path integral arise from combining (3.39) with the fact that the original integration variables satisfied $q_0 = q_a$ and $q_N = q_b$. By construction q_c satisfies $q_c(t_a) = q_a$ and $q_c(t_N) = q_b$, and therefore, considered as a function of t, $x(t)$ must vanish at $t = t_a$ and $t = t_b$. The advantage of employing classical solutions in the path integral is now manifest; such a technique allows an easy accommodation of the boundary conditions, i.e., the initial and final states.

If the original Lagrangian density was quadratic in the variables, then $\mathcal{L}_R = \mathcal{L}$. For such a case the remaining path integral in (3.42) has no reference to q_a and q_b and can therefore depend upon only the values t_a, t_b, and any parameters that appear in the Lagrangian density. Thus, for the case of Lagrangian densities that are quadratic and linear the exact path integral will always evaluate to the general form

$$Z(q_a, t_a, q_b, t_b) = F(t_a, t_b)e^{iS_c/\hbar}\,, \tag{3.43}$$

where S_c is the action evaluated along the classical trajectory. The prefactor $F(t_a, t_b)$ is given by the reduced path integral appearing in (3.42) and will contain no reference to q_a or q_b. This property breaks down if terms cubic or higher in x are present in $\mathcal{L}_R$.

In order to evaluate the prefactor $F(t_a, t_b)$ appearing in (3.43), the variables of integration will be changed to a more useful set. Because $x(t)$ vanishes at the times t_a and t_b, and because there are $N-1$ variables of integration at intermediate times, $x(t)$ will be written as a Fourier sine series of the form

$$x(t) = \sum_{n=1}^{N-1} a_n \sin\frac{n\pi}{T}(t - t_a)\,, \tag{3.44}$$

where $T = t_b - t_a$ and the a_n are arbitrary constants. In the $N \to \infty$ limit (3.44) allows the representation of an arbitrary piecewise continuous path that begins and ends at $x = 0$.

64   Evaluating the Path Integral

The a_n constitute new variables of integration and (3.44) should therefore be viewed as the definition of a change of variables. The relationship of x_j to the new variables is

$$x_j = \sum_{n=1}^{N-1} a_n \sin \frac{n\pi}{T}(t_j - t_a) . \qquad (3.45)$$

Result (1.100) states that such a change of variables must be accompanied by a factor in the measure corresponding to the Jacobian of the transformation. For the change of (3.45) the Jacobian is given by

$$J \equiv J(a_1, \ldots, a_{N-1}) = \det\left[\sin \frac{n\pi}{T}(t_j - t_a)\right] , \qquad (3.46)$$

where n and j are treated as the indices of the matrix whose determinant gives the Jacobian.

A direct computation of this determinant is difficult, and it is far easier to infer its value from previous results. In order to do this, the continuum technique will be applied to the free particle Lagrangian of Sec. 3.1 since the exact form for that path integral is already known. Using the orthogonality of the cosine functions the remainder term in the action becomes

$$\int_{t_a}^{t_b} dt\, \mathcal{L}_R(\dot{x}, x) = \int_{t_a}^{t_b} dt\, \tfrac{1}{2} m\dot{x}^2 = \sum_{n=1}^{N-1} m \frac{a_n^2 \pi^2 n^2}{4T} , \qquad (3.47)$$

Result (3.46) shows that the Jacobian is independent of the a_n. From results (3.43) and (3.47) it follows that the path integral, when subjected to this change of variables, must satisfy

$$\left(\frac{m}{2\pi i\hbar T}\right)^{1/2} =$$
$$J \int \left(\frac{Nm}{2\pi i\hbar T}\right)^{N/2} da_1 \cdots da_{N-1} \exp\left(\frac{i}{\hbar} \sum_{n=1}^{N-1} m \frac{a_n^2 \pi^2 n^2}{4T}\right) , \qquad (3.48)$$

where $\epsilon = T/N$ has been used. Performing the $N-1$ Gaussian integrals yields the relation

$$\left(\frac{m}{2\pi i\hbar T}\right)^{1/2} = \left(\frac{mN}{2\pi i\hbar T}\right)^{N/2} J \prod_{n=1}^{N-1} \left(\frac{2\pi i\hbar T}{m\pi^2}\right)^{1/2} \cdot \frac{1}{n} , \qquad (3.49)$$

from which it follows that

$$J(a_1, \ldots, a_{N-1}) = N^{-N/2} \pi^{N-1} (N-1)! . \qquad (3.50)$$

Sec. 3.3 Continuum Techniques   65

The Jacobian is therefore independent of T.

This Jacobian can now be used to evaluate the propagator for the harmonic oscillator by making the same change of variables. The Lagrangian density is assumed to have the form

$$\mathcal{L} = \tfrac{1}{2}m\dot{q}^2 - \tfrac{1}{2}m\omega^2 q^2 \,. \tag{3.51}$$

Since the Lagrangian density is quadratic, the previous analysis can be taken over completely. The classical solution appropriate to the boundary conditions is given by

$$q_c(t) = A\sin(\omega t + \delta)\,, \tag{3.52}$$

where the values of A and δ are determined from the boundary conditions by

$$\begin{aligned} \cot\delta &= \frac{q_b}{q_a \sin\omega T} - \cot\omega T\,, \\ A &= \frac{q_a}{\sin\delta}\,. \end{aligned} \tag{3.53}$$

Using these results shows that the value of the action for the classical trajectory (3.52) is given by

$$\begin{aligned} S_c &= \int_{t_a}^{t_b} dt\,[\tfrac{1}{2}m\dot{q}_c^2 - \tfrac{1}{2}m\omega^2 q_c^2] \\ &= \frac{m\omega}{2\sin\omega T}\left[(q_a^2 + q_b^2)\cos\omega T - 2q_a q_b\right]\,. \end{aligned} \tag{3.54}$$

All that remains to calculate is the prefactor appearing in (3.43). This is best calculated by using the change of variables (3.44). The action appearing in the path integral that determines the prefactor becomes

$$\frac{i}{\hbar}\int_{t_a}^{t_b} dt\,[\tfrac{1}{2}m\dot{x}^2 - \tfrac{1}{2}m\omega^2 x^2] = \frac{i}{\hbar}\sum_{n=1}^{N-1}\tfrac{1}{2}m\left(\frac{n^2\pi^2}{T} - \omega^2 T\right)a_n^2\,. \tag{3.55}$$

The prefactor is then given by a product of simple Gaussian integrals of the form

$$\begin{aligned} F(T) &= \left(\frac{mN}{2\pi i\hbar T}\right)^{N/2} J(a_1,\ldots,a_{N-1}) \\ &\quad \times \int da_1\cdots da_{N-1}\,\exp\left[\frac{i}{\hbar}\sum_{n=1}^{N-1}\frac{m}{2}\left(\frac{n^2\pi^2}{T} - \omega^2 T\right)a_n^2\right]\,. \end{aligned} \tag{3.56}$$

Evaluating the Path Integral

Result (3.56) shows the danger inherent in the complex form for the path integral. The coefficient of a_n^2 vanishes if $\omega T = n\pi$, rendering the integration over the nth mode undefined. The situation is remedied by analytic continuation using the Wick rotation, $T \to -iT$. The argument of the exponential becomes

$$\frac{i}{\hbar}\sum_{n=1}^{N-1}\frac{m}{2}\left(\frac{n^2\pi^2}{T}-\omega^2 T\right)a_n^2 \xrightarrow{W} -\frac{1}{\hbar}\sum_{n=1}^{N-1}\frac{m}{2}\left(\frac{n^2\pi^2}{T}+\omega^2 T\right)a_n^2, \quad (3.57)$$

which is defined for all modes and all times.

Using the form (3.50) for the Jacobian, as well as the standard result for real Gaussian integrals, gives the Wick-rotated prefactor:

$$F(T) = \left(\frac{m}{2\pi\hbar T}\right)^{\frac{1}{2}}\prod_{n=1}^{N-1}\left(1+\frac{\omega^2 T^2}{n^2\pi^2}\right)^{-\frac{1}{2}}. \quad (3.58)$$

In the limit $N \to \infty$, (3.58) is simplified by the identity

$$\prod_{n=1}^{\infty}\left(1+\frac{\omega^2 T^2}{n^2\pi^2}\right)^{-\frac{1}{2}} = \left(\frac{\omega T}{\sinh \omega T}\right)^{\frac{1}{2}}. \quad (3.59)$$

The final result for the Wick-rotated prefactor is

$$F(T) = \left(\frac{m\omega}{2\pi\hbar \sinh \omega T}\right)^{\frac{1}{2}}. \quad (3.60)$$

Analytically continuing (3.60) back to real time by using the inverse Wick rotation, $T \to iT$, as well as the identity $\sinh i\omega T = i \sin \omega T$, gives the final result for the prefactor:

$$F(T) = \left(\frac{m\omega}{2\pi i\hbar \sin \omega T}\right)^{\frac{1}{2}}. \quad (3.61)$$

Combining results (3.54) and (3.61) into equation (3.43) gives the correct propagator for the harmonic oscillator.

The generating functional $Z_{ab}[J]$ for the harmonic oscillator with a time-dependent source term of the form $J(t)q$ present is another example of a quadratic Lagrangian. The propagator for this case is given by the combination of the prefactor (3.61) and the classical action [2]

$$S_c = \frac{m\omega}{2\sin\omega T} \left\{ \frac{2q_a}{m\omega} \int_{t_a}^{t_b} dt\, J(t) \sin\omega(t - t_a) \right.$$
$$- \frac{2}{m^2\omega^2} \int_{t_a}^{t_b} dt\, dt'\, J(t)J(t')\theta(t - t')\sin\omega(t_b - t)\sin\omega(t' - t_a)$$
$$\left. + (q_a^2 + q_b^2)\cos\omega T - 2q_a q_b + \frac{2q_b}{m\omega} \int_{t_a}^{t_b} dt\, J(t)\sin\omega(t_b - t) \right\}. \quad (3.62)$$

Exercise 3.7: Verify (3.62).

It is interesting to note that the general result of this section, (3.43), shows that the dominant path for the quadratic Lagrangian density is the classical trajectory. The effect of all the other paths, unruly or not, is simply to generate the prefactor in (3.43). When this result is combined with the generator $W_{ab}[J]$, defined by (3.29), it is apparent that the only path that matters in a quadratic theory is the classical trajectory. This is true since the prefactor has no reference to the source $J(t)$, and all the J dependence is contained in the value of the action along the classical trajectory. This will remain true when generating functionals for time-ordered products are evaluated in field theory.

3.4 Topological Measure

In this section the path integral (2.71) for a point mass moving on a ring of circumference R, derived in Sec. 2.3, will be analyzed. This path integral is given by

$$\langle \theta_b, t_b | \theta_a, t_a \rangle = \lim_{N\to\infty} \int_{\theta_a}^{\theta_b} d\theta_1 \cdots d\theta_{N-1} \frac{1}{(2\pi)^N} \sum_{n_0=-\infty}^{\infty} \cdots \sum_{n_{N-1}=-\infty}^{\infty}$$
$$\times \exp\left\{ \frac{i}{\hbar} \sum_{j=0}^{N-1} \left[n_j \hbar(\theta_{j+1} - \theta_j) - \epsilon \frac{n_j^2 \hbar^2}{2mR^2} - \epsilon V(\theta_j) \right] \right\}, \quad (3.63)$$

where the integrations limits for the θ_j are 0 to 2π.

The immediate concern is to understand how the result (3.63) is equivalent to a path integral of the form (2.50) with the Lagrangian density given by

$$\mathcal{L}(\dot\theta, \theta) = \tfrac{1}{2} mR^2 \dot\theta^2 - V(\theta). \quad (3.64)$$

68 Evaluating the Path Integral

If the potential $V(\theta)$ were not present, then the angular integrations could readily be performed to reveal the structure of the path integral. However, the presence of a nontrivial form for $V(\theta)$ precludes this.

The answer is to apply the Poisson resummation to the path integral [6]. The Poisson resummation technique is possible because the path integral (3.63) is a product of Jacobi theta functions of the third kind, denoted Θ_3. The Jacobi theta function of the third kind is defined, for complex arguments, as

$$\Theta_3(A|s) = \sum_{n=-\infty}^{\infty} \exp\left(-i\pi A n^2 + 2i\pi n s\right), \qquad (3.65)$$

so that the formal similarity to the path integral of (3.63) is obvious.

The Poisson resummation technique uses the result that the function Θ_3 may be rewritten as

$$\Theta_3(A|s) = \frac{1}{\sqrt{iA}} \sum_{j=-\infty}^{\infty} \exp\left[i\frac{\pi}{A}(s+j)^2\right]. \qquad (3.66)$$

The proof of (3.66) is straightforward and stems from the result of Exercise 1.3. Setting $y = 0$ and $2L = 1$ in Exercise 1.3 gives the identity

$$\sum_{n=-\infty}^{\infty} f(n) = \sum_{j=-\infty}^{\infty} \int_{-\infty}^{\infty} dn\, f(n) e^{i2\pi n j}. \qquad (3.67)$$

After inserting the identity (3.67) into (3.65), the Jacobi theta function becomes

$$\Theta_3(A|s) = \sum_{j=-\infty}^{\infty} \int_{-\infty}^{\infty} dn\, \exp\left[-i\pi A n^2 + i2\pi n(s+j)\right]. \qquad (3.68)$$

The integral over n in (3.68) is a standard Gaussian of the form (1.107), and its evaluation immediately yields (3.66).

The Poisson resummation method can now be applied to the path integral of (3.63). Since there are N original summations, the result is

$$\langle \theta_b, t_b | \theta_a, t_a \rangle =$$

$$\int_{\theta_a}^{\theta_b} \left(\frac{mR^2}{2\pi i\hbar\epsilon}\right)^{N/2} d\theta_1 \cdots d\theta_{N-1} \sum_{n_0=-\infty}^{\infty} \cdots \sum_{n_{N-1}=-\infty}^{\infty}$$

$$\times \exp\left\{\frac{i}{\hbar} \sum_{j=0}^{N-1} \left[\frac{mR^2}{2\epsilon}(\theta_{j+1} - \theta_j + 2\pi n_j)^2 - \epsilon V(\theta_j)\right]\right\}. \qquad (3.69)$$

Sec. 3.4 Topological Measure

Now all but one of the intermediate sums can be absorbed into extending the limits of the θ_j integrations. In order to see this, the first integration over θ_1 is considered. The periodicity of the potential allows the limits on the θ_1 integration to become $-\infty$ to ∞ through an absorption of the summation over n_1,

$$\sum_{n_0=-\infty}^{\infty} \sum_{n_1=-\infty}^{\infty} \int_0^{2\pi} d\theta_1 \exp\left\{\frac{imR^2}{2\hbar\epsilon}\left[(\theta_1 - \theta_a + 2\pi n_0)^2 \right.\right.$$

$$\left.\left. + (\theta_2 - \theta_1 + 2\pi n_1)^2\right] - \frac{i\epsilon}{\hbar}V(\theta_1)\right\}$$

$$= \sum_{n_0=-\infty}^{\infty} \int_{-\infty}^{\infty} d\theta_1 \exp\left\{\frac{imR^2}{2\hbar\epsilon}\left[(\theta_1 - \theta_a + 2\pi n_0)^2 + (\theta_2 - \theta_1)^2\right] \right.$$

$$\left. - \frac{i\epsilon}{\hbar}V(\theta_1)\right\}. \quad (3.70)$$

Next, the θ_2 integration is considered, with the similar result that

$$\sum_{n_0=-\infty}^{\infty} \sum_{n_2=-\infty}^{\infty} \int_{-\infty}^{\infty} d\theta_1 \int_0^{2\pi} d\theta_2 \exp\left\{\frac{imR^2}{2\hbar\epsilon}\left[(\theta_1 - \theta_a + 2\pi n_0)^2 \right.\right.$$

$$\left.\left. + (\theta_2 - \theta_1)^2 + (\theta_3 - \theta_2 + 2\pi n_2)^2\right] - \frac{i\epsilon}{\hbar}\left[V(\theta_2) + V(\theta_1)\right]\right\}$$

$$= \sum_{n_0=-\infty}^{\infty} \int_{-\infty}^{\infty} d\theta_1 \int_{-\infty}^{\infty} d\theta_2 \exp\left\{\frac{imR^2}{2\hbar\epsilon}\left[(\theta_1 - \theta_a + 2\pi n_0)^2 \right.\right.$$

$$\left.\left. + (\theta_2 - \theta_1)^2 + (\theta_3 - \theta_2)^2\right] - \frac{i\epsilon}{\hbar}\left[V(\theta_2) + V(\theta_1)\right]\right\}, \quad (3.71)$$

where both the periodicity of $V(\theta)$ and the translational invariance of the θ_1 integration are key properties. These steps are repeated until $N-1$ of the summations are absorbed into the $N-1$ intermediate angular integrations. The end result of this process is

$$\langle \theta_b, t_b | \theta_a, t_a \rangle = \sum_{n=-\infty}^{\infty} \int_{\theta_a - 2\pi n}^{\theta_b} \left(\frac{mR^2}{2\pi i\hbar\epsilon}\right)^{N/2} d\theta_1 \cdots d\theta_{N-1}$$

$$\times \exp\left\{\frac{i}{\hbar} \sum_{j=0}^{N-1} \epsilon \left[\frac{1}{2}mR^2 \frac{(\theta_{j+1} - \theta_j)^2}{\epsilon^2} - V(\theta_j)\right]\right\}, \quad (3.72)$$

where the integrations over the θ_j have the limits $\pm\infty$. In the continuum limit the path integral has the classical action (3.64) appearing in the exponential.

The path integral of (3.72) is an infinite number of copies of the path integral associated with the motion of a particle along an infinite straight line. Each term in the sum is characterized by a different transition angle $\theta_a - 2\pi n$. The question naturally arises as to the meaning of the integer n. This integer occurs because the underlying space or *manifold*, denoted $\mathcal{M}$, for the particle's original motion is a circle, and a circle is topologically nontrivial. A manifold can be viewed as topologically nontrivial if it requires more than one open set to cover it. Using this criterion a circle is topologically nontrivial, since it requires two open sets to cover it.

Another way of characterizing a nontrivial topological space is through its *homotopy* classes [7, 8, 9]. Homotopy classes characterize the different ways that closed paths and closed surfaces can be imbedded or mapped into a manifold. In the simplest case to visualize closed loops are mapped into the manifold under consideration. All closed loops that can be continuously deformed into each other are placed into an equivalence class. Continuous deformation means that the loop is to be treated like an infinitely pliable rubber band that can be stretched or shrunk in any way, but not broken. Mathematically, this means that there must exist a *homeomorphism* between any two members of the equivalence class. The set of all equivalence classes necessary to characterize closed loops in a particular manifold is called π_1, the fundamental or first homotopy group of the manifold. In order to explain why it is called a *group* requires discussion found later in this book. For example, if the manifold is two-dimensional and possesses a hole, then a loop that wraps around the hole cannot be continuously deformed into a loop that does not wrap around the hole. If all closed paths passing through all points of the manifold can be continuously shrunk to a point, then the fundamental group of the manifold is said to be trivial.

In the case of a circle any continuous path between two points on the circle can be characterized by the number of times the path winds around the circle. This integer is appropriately called the *winding number*. Any mapping of the real line into a circle can be characterized by the winding number of the map, i.e., how many times it wraps the real line around the circle. Paths with different winding numbers cannot be smoothly deformed into each other because of the underlying nature of the circle, and so there is no homeomorphism between these different paths. Paths with different winding numbers therefore belong to different equivalence classes, and these classes are disjoint from each other. It follows that π_1 for a circle is characterized by $\mathcal{Z}$, the set of integers.

Higher-order homotopy groups refer to the ways in which higher-dimensional closed surfaces, such as spheres and hyperspheres, can be imbedded in the manifold. These higher-dimensional closed surfaces are denoted S_n, where S_1 is the circle and S_2 is the sphere. The higher-dimensional objects, in particular S_3, are discussed again in Sec. 7.6. The higher-order homotopy groups of the manifold $\mathcal{M}$ are denoted $\pi_n(\mathcal{M})$, and the results give the higher-dimensional extension of the concept of the winding number. A particularly important set of homotopy classes are those where the manifold $\mathcal{M}$ is itself the hypersphere S_n. It can be established that

$$\pi_n(S_n) = \mathcal{Z}, \quad \pi_n(S_m) = 0 \ (n > m), \quad \pi_n(S_1) = 0 \ (n > 1). \qquad (3.73)$$

As a result of the circle's nontrivial topological structure and the fact that the path integral is a weighted sum of paths, including loops, it is no surprise that the path integral exhibits the winding numbers of the paths between two points on the circle. In effect, by using the Poisson resummation technique, the original path integral on the circle has been rewritten in terms of a path integral that maps the real line onto the circle. Therefore, since the path integral contains all possible mappings, all possible winding numbers must occur. Translating the angle θ_a by $2\pi n$ corresponds to considering those paths between θ_a and θ_b with the winding number n. Upon examination of result (3.72), it is apparent that the path integral measure has broken into a sum over all the topologically disjoint homotopy classes available to the set of paths on the circle.

The path integral may be evaluated exactly for the case $V(\theta) = 0$ by using result (3.4) for the propagator of a free particle on the real line. The result is

$$Z(\theta_a, t_a, \theta_b, t_b) = \sum_{n=-\infty}^{\infty} \left[\frac{mR^2}{2\pi i\hbar(t_b - t_a)}\right]^{\frac{1}{2}} \exp\left[\frac{i}{\hbar}\frac{mR^2(\theta_b - \theta_a + 2\pi n)^2}{2(t_b - t_a)}\right]. \qquad (3.74)$$

Result (3.74) is again indexed by an integer, related to the winding number. Demonstrating this is left as an exercise.

Exercise 3.8: Show that the argument of the exponential in (3.74) is given by the value of the classical action for a classical trajectory of winding number n.

The result of Exercise 3.8 indicates that, especially if the prefactor is irrelevant, the path integral is well approximated by summing over all topologically different classical sectors available to the theory. This result can be generalized to more complicated systems in the following way. A path integral, assumed to describe a system without bound states, may initially be defined on some *compact* manifold, such as a circle, a sphere, or a torus. A compact manifold is closed and bounded. The path integral measure begins with limits on some or all of the integration variables. The generalization of the result obtained in this section allows the measure limits to be changed, so that the real line or its higher-dimensional equivalent, R^n, is being mapped into the manifold, at the expense of summing over all the topologically disjoint equivalence classes, π_1, of the original manifold. This idea will reappear for field theory models that possess nontrivial topological structure.

Difficulties with bound states in the path integral formalism were mentioned in passing in Sec. 2.3. However, while the general problem of bound states remains problematic in the path integral, analyses similar to this section can be developed for some systems with binding potentials. Because the potential is capable of binding there are a discrete set of states available, characterized by at least one quantum number. It is possible, in some instances, to apply resummation techniques to these quantum numbers to regain the standard form (2.50) of the path integral. Demonstrating an example of this is left as Exercise 3.9.

Exercise 3.9: Consider a point mass constrained to remain in an infinitely deep well of width L. Show that the measure of the path integral representation of the transition amplitude is characterized by integers, and these represent the number of bounces off the wall the particle experienced before arriving at its final position [10].

Perhaps the most important binding potential in all of quantum mechanics is the Coulomb potential. While the standard $1/r$ potential is well understood in terms of the differential form of the Schrödinger equation, the same cannot be said of the path integral form. It is not easy to recapture the bound state behavior of the system by considering some subset of trajectories.

References

[1] The concept of the generating functional was first introduced by W. Heisenberg and H. Euler, Z. Physik **98**, 714 (1936). The modern formulation stems primarily from the work of Schwinger; see J. Schwinger, Proc. Natl. Acad. Sci. (USA) **37**, 452 (1951); J. Schwinger, Phys. Rev. **82**, 664 (1951); J. Schwinger, *Particles and Sources*, Gordon and

Breach, New York, 1969; see also K. Symanzik, Z. Natürforschung **9A**, 10 (1954).

[2] R.P. Feynman and A.R. Hibbs, *Quantum Mechanics and Path Integrals*, McGraw-Hill, New York, 1965.

[3] This definition is adapted from G. Jona-Lasinio, Nuovo Cimento **34**, 1790 (1964). See also [4].

[4] B. Zumino in *Lectures on Elementary Particles and Quantum Field Theory, 1970 Brandeis University Summer Institute in Theoretical Physics*, edited by S. Deser, M. Grisaru, and H. Pendleton, MIT Press, Cambridge, Massachusetts, 1970.

[5] See, for example, the text by L. Schiff, *Quantum Mechanics, Third Edition*, McGraw-Hill, New York, 1968.

[6] L.S. Schulman, Phys. Rev. **176**, 1558 (1968).

[7] T. Eguchi, P.B. Gilkey, and A.J. Hanson, Phys. Rep. **66**, 213 (1980).

[8] C. Nash and S. Sen, *Topology and Geometry for Physicists*, Academic Press, New York, 1983.

[9] M. Nakahara, *Geometry, Topology, and Physics*, Adam Hilger, Bristol, 1990.

[10] M. Carreau, E. Farhi, and S. Gutmann, Phys. Rev. **D42**, 1194 (1990).

Chapter 4

Further Applications

In this chapter extensions of quantum mechanical path integrals will be presented. The concepts to be discussed will be applicable to the analysis of field theoretical path integrals, and they will reappear in the relevant chapters. However, the presentation of these ideas in this chapter will be mathematically less complicated and therefore simpler.

The chapter begins by introducing natural units in Sec. 4.1. In Sec. 4.2 the relationship of the quantum mechanical partition function to the path integral is demonstrated, and several applications to simple systems are made. In Sec. 4.3 the concept of symmetry in quantum mechanical systems is developed, and the implications of symmetries for various formulations of the generating functionals, derived from path integrals, are presented. In Sec. 4.4 the harmonic oscillator is reformulated in terms of coherent states, and a path integral representation of the vacuum persistence amplitude is derived using these coherent states. In Sec. 4.5 the mechanism of spontaneous breakdown of symmetry for a simple quantum mechanical system is discussed, and the effective potential is derived from the form of the free energy defined in Sec. 4.2. The ground state is understood in the harmonic oscillator approximation as a coherent state. The chapter closes in Sec. 4.6 with a discussion of implementation of constraints in the path integral formalism.

4.1 Natural Units

Quantum mechanical and quantum field theory calculations can be rife with powers of Planck's constant, $\hbar = 1.05 \times 10^{-27}$ erg-sec, and the speed of light, $c = 2.99 \times 10^{10}$ cm/sec. The manipulations can be greatly simplified in appearance by using *natural units*, in which $\hbar$ and c are set to unity. The suppression of these two constants is equivalent to suppressing two of the three fundamental units of measure: the centimeter, the second, and the gram. As a result, in natural units all physical quantities may be measured in powers of only one of these fundamental units, or any one of the units derived from them, such as energy.

For example, a typical mass appearing in formulas is that of the electron, where $m_e = 9.1 \times 10^{-28}$ gm in the SI system. In natural units m_e may be measured in electron-Volts, $m_e = m_e c^2 = 8.19 \times 10^{-7}$ erg $= 0.511$ MeV; inverse length, $m_e = (m_e c)/\hbar = 2.59 \times 10^{10}$ cm^{-1}; or inverse time, $m_e = (m_e c^2)/\hbar = 7.75 \times 10^{20}$ sec^{-1}. Conversely, the appearance of $1/m_e$ in a formula may represent a time, $\hbar/(m_e c^2) = 1.29 \times 10^{-21}$ sec, or a length, $\hbar/(m_e c) = 3.86 \times 10^{-11}$ cm.

Throughout the remainder of this book natural units will be employed, save for a brief reappearance by $\hbar$ in Chapter 8, where it will serve as a counting parameter. The choice of basic unit will be that of length, so that all physical quantities will be represented in powers of length. The sole exception to this will be the occasional mention of a mass in terms of MeV, where 1 MeV $= 5.07 \times 10^{20}$ cm^{-1} in natural units.

When using natural units the calculation of a physical quantity is first performed, and then the appropriate factors of $\hbar$ and c are inserted to give the quantity in SI units. For example, if the lifetime of a particle were calculated, and the formula derived for it took the form md^2, where m is some mass and d some fundamental length, then the proper factors are inserted of $\hbar$ and c are inserted into the hypothetical formula in the following way. The mass is converted to inverse length by multiplying it by $c/\hbar$, the distance d is unchanged, and the resulting length obtained from the formula $(mcd^2)/\hbar$ is converted to a time by dividing by c, so that the final result is $(md^2)/\hbar$. It is important to remember that masses, lengths, times, etc., appearing in the final result must be measured in the same system of units in which $\hbar$ and c are expressed. In the previous formula, this means that the mass must be measured in gm if $\hbar$ is measured in erg-sec. It is not difficult to see that, if $\hbar$ is measured in eV-sec and if the mass is to be measured in eV, the corresponding formula would be $(md^2)/(\hbar c^2)$. For this reason, masses are often stipulated in units of eV/c^2.

The choice of generic unit of length will be denoted L. It is noted in passing that the natural units associated with various mechanical properties of a system are time $= L$, energy $= L^{-1}$, mass $= L^{-1}$, momentum $= L^{-1}$, angular momentum $= L^0$, velocity $= L^0$, and force $= L^{-2}$.

In problems where electric charge is present Heavyside–Lorentz (rationalized) units [1] will be combined with natural units. It is recalled that

Coulomb's law takes the form, in rationalized units,

$$F = \frac{e^2}{4\pi r^2}, \tag{4.1}$$

where the unit of charge, e, is measured in terms of statcoulomb. In terms of the SI unit equivalents, the fundamental charge of the electron is given by

$$\frac{e^2}{4\pi} = 2.30 \times 10^{-28} \frac{\text{kg m}^3}{\text{s}^2}, \tag{4.2}$$

and this may be converted to a dimensionless constant in the natural unit system by dividing by $\hbar c$, with the result that

$$\frac{e^2}{4\pi} \approx \frac{1}{137.04}. \tag{4.3}$$

From the fact that the electric charge is a dimensionless constant, it follows that the electric and magnetic fields, $\mathbf{E}$ and $\mathbf{B}$, have the units $1/L^2$ in order that the force $\mathbf{F} = e\mathbf{E} + e\mathbf{v} \times \mathbf{B}$. From the definition of the vector potential, $\mathbf{B} = \nabla \times \mathbf{A}$, and the scalar potential, $\mathbf{E} = -\dot{\mathbf{A}} - \nabla\Phi$, it also follows that $\mathbf{A}$ and Φ have the units $1/L$.

Exercise 4.1: Show that the gravitational constant G appearing in Newton's law of gravitation,

$$F = -\frac{Gm_1 m_2}{r^2}, \tag{4.4}$$

is associated with the constant 2.59×10^{-66} cm^2 in natural units.

4.2 Statistical Mechanics

There is a deep relationship between quantum mechanical transition amplitudes, written as path integrals [2, 3], and the quantum mechanical partition function familiar from statistical mechanics [4]. Exploiting this relationship allows the thermal behavior of a system to be derived from the quantum mechanical transition amplitude, or propagator, in a very easy manner.

The partition function is defined for a system at thermal equilibrium, and so it is assumed that the Hamiltonian of the system, H, is time independent. The Hamiltonian is assumed to possess a complete set of Heisenberg picture eigenstates, denoted $|E_n\rangle$. The restriction to a discrete set,

78   **Further Applications**

indexed by some collection of integers n, is for notational simplicity. The partition function is given by

$$Z_\beta = \sum_n e^{-\beta E_n} = \sum_n \langle E_n | e^{-\beta H} | E_n \rangle = \text{Tr}\, e^{-\beta H}\,, \qquad (4.5)$$

where the sum runs over the complete set of states available to the system, and β is the inverse temperature, $1/(kT)$, where k is the Boltzmann constant. In natural units β is measured in length. The partition function can then be used to define thermal averages for observable quantities. The average energy of the system is given by

$$\langle E \rangle_\beta \equiv Z_\beta^{-1} \sum_n \langle E_n | H e^{-\beta H} | E_n \rangle = -\frac{\partial}{\partial \beta} \ln Z_\beta\,. \qquad (4.6)$$

The Helmholtz free energy F_β is defined as

$$F_\beta = -\frac{1}{\beta} \ln Z_\beta\,. \qquad (4.7)$$

Using (4.7) and denoting the volume of the system as V, the entropy S and pressure P are defined by

$$S = \beta^2 \frac{\partial F_\beta}{\partial \beta}\,,\quad P = -\frac{\partial F_\beta}{\partial V}\,. \qquad (4.8)$$

Knowledge of the partition function is therefore paramount to understanding the thermal behavior of the system. Because of the calculational advantages of path integrals, a relationship between the two is highly desirable. From the definition of the quantum mechanical transition amplitude,

$$Z_{ab} = \langle q_b, t_b | q_a, t_a \rangle = \langle q_b | e^{-iH(t_b - t_a)} | q_a \rangle\,, \qquad (4.9)$$

it is apparent that the Wick rotation $t_b - t_a \to -i\beta$ and the identification $q_a = q_b$ gives

$$\bar{Z}_{aa} = \langle q_a | e^{-\beta H} | q_a \rangle\,. \qquad (4.10)$$

Under the assumption that the $|q\rangle$ and $|E_n\rangle$ states are complete, integrating (4.10) gives

$$\int dq_a\, \bar{Z}_{aa} = \int dq_a\, \langle q_a | e^{-\beta H} | q_a \rangle = \sum_n \langle E_n | e^{-\beta H} | E_n \rangle = Z_\beta\,. \qquad (4.11)$$

The final step is to combine (4.11) with the path integral representation for $\bar{Z}_{aa}$, given by (2.47), continued to imaginary times. The variable $\tau = it$

is introduced, so that $\dot{q}$ denotes the derivative of q with respect to τ, and for simplicity, t_a is chosen to be 0. The result is

$$Z_\beta = \int dq_a \int_{q_a}^{q_a} \mathcal{D}p\,\mathcal{D}q \, \exp\left\{-\int_0^\beta d\tau \left[H(p,q) - ip\dot{q}\right]\right\}, \qquad (4.12)$$

For the special case that

$$H(p,q) = \frac{p^2}{2m} + V(q), \qquad (4.13)$$

the partition function reduces to

$$Z_\beta = \int dq_a \int_{q_a}^{q_a} \overline{\mathcal{D}}q \, \exp\left\{-\int_0^\beta d\tau \left[\tfrac{1}{2}m\dot{q}^2 + V(q)\right]\right\}, \qquad (4.14)$$

where the measure $\overline{\mathcal{D}}q$ is defined by (2.62). The relationship of (4.14) to the Euclidean time path integral, defined through the Wick rotation introduced in Sec. 2.3, is obvious. Very loosely speaking, the Wick-rotated path integral is as well behaved as the statistical mechanical version of the same problem. The relationship of the Euclidean path integral to the partition function thus serves to justify the Wick rotation.

It follows from the form of the path integral appearing in (4.14) that only periodic trajectories, i.e., those starting and ending at q_a, contribute to the partition function, and that the period of these trajectories must be β. This leads to the Kubo–Martin–Schwinger (KMS) condition [5], namely that all expectation values of Wick-rotated observables are periodic with period β. This can be verified from the definition of the thermal average of a Heisenberg operator, $O_H(t)$, given by

$$\langle O_H(t) \rangle_\beta = \sum_n \langle E_n | O_H(t) e^{-\beta H} | E_n \rangle. \qquad (4.15)$$

Using result (2.24) for the time development of Heisenberg picture operators, it follows that

$$O_H(t - i\beta) = e^{-i(i\beta)H} O_H(t) e^{i(i\beta)H}. \qquad (4.16)$$

Using (4.16) in (4.15) gives

$$\langle O_H(t) \rangle_\beta = \sum_n \langle E_n | e^{-\beta H} e^{-i(i\beta)H} O_H(t) e^{i(i\beta)H} | E_n \rangle$$

$$= \sum_n \langle E_n | e^{-\beta H} O_H(t - i\beta) | E_n \rangle = \langle O_H(t - i\beta) \rangle_\beta, \qquad (4.17)$$

Further Applications

thereby demonstrating the periodicity of the expectation value. If the observable itself undergoes a Wick rotation to imaginary time, $O_H(t) \to O_H(\tau)$, it follows from result (4.17) that

$$\langle O_H(\tau) \rangle_\beta = \langle O_H(\tau + \beta) \rangle_\beta . \tag{4.18}$$

Similar periodicity conditions exist for the thermal average of time-ordered products.

Exercise 4.2: Demonstrate result (4.18) by using the path integral form of the partition function (4.14).

The free energy, F_β, is readily expressed in terms of the path integral, giving

$$F_\beta = -\frac{1}{\beta} \ln \int dq_a \int_{q_a}^{q_a} \mathcal{D}p \mathcal{D}q \, \exp\left\{ -\int_0^\beta d\tau \left[H(p,q) - ip\dot{q} \right] \right\} . \tag{4.19}$$

For example, using the result of evaluating the path integral for the free particle, (3.4), gives the standard dimensionless result

$$F_\beta = -\frac{1}{\beta} \ln \left[\left(\frac{m}{2\pi\beta}\right)^{\frac{1}{2}} \int dq_a \right] \equiv -\frac{1}{\beta} \ln \left[\left(\frac{m}{2\pi\beta}\right)^{\frac{1}{2}} V \right] . \tag{4.20}$$

Combining (4.8) with result (4.20) immediately yields the perfect gas law, $PV = kT$.

The statistical mechanics relationship gives a method for calculating the ground state energy of a system. Form (4.5) shows that, in the limit $\beta \to \infty$, the partition function is dominated by the lowest lying energy eigenvalue for H, denoted E_g. This is written

$$\lim_{\beta \to \infty} e^{\beta E_g} Z_\beta = 1 , \tag{4.21}$$

so that

$$E_g = -\lim_{\beta \to \infty} \frac{1}{\beta} \ln Z_\beta . \tag{4.22}$$

Result (4.22) allows the ground state energy of a system to be derived from the Euclidean form of the path integral.

To demonstrate this for a nontrivial case the harmonic oscillator of Sec. 3.3 is considered. It follows from (3.54) and (3.60), and the prescription developed in this section, that the partition function is given by

$$Z_\beta = \int dq \left[\frac{m\omega}{2\pi \sinh(\omega\beta)}\right]^{\frac{1}{2}} \exp\left[-q^2 \frac{m\omega}{\sinh(\omega\beta)}(\cosh(\omega\beta) - 1)\right]$$
$$= \frac{1}{\sqrt{2\cosh(\omega\beta) - 2}} = \frac{1}{2\sinh(\frac{1}{2}\omega\beta)} . \qquad (4.23)$$

In the large β limit expression (4.23) has the leading behavior

$$\lim_{\beta \to \infty} Z_\beta \approx e^{-\frac{1}{2}\beta\omega} , \qquad (4.24)$$

from which it immediately follows that $E_g = \frac{1}{2}\omega$, the well-known result for the harmonic oscillator in one dimension.

In order to analyze the thermal behavior of time-ordered products of Heisenberg picture position operators, the path integral form for the free energy with a source, J, will be used. This is written, for imaginary times, as

$$F_\beta[J] =$$
$$-\frac{1}{\beta}\ln \int dq_a \int_{q_a}^{q_a} \mathcal{D}p\,\mathcal{D}q \,\exp\left\{-\int_0^\beta d\tau \left[H(p,q) - ip\dot{q} - Jq\right]\right\} . \qquad (4.25)$$

While such a system cannot be in thermal equilibrium, in the limit that the source term vanishes it will be assumed that equilibrium is attained. The source term is therefore merely an artifice for generating the time-ordered products. However, (4.25) is a generating functional whose derivatives give imaginary-time-ordered products. In order to obtain the real-time-ordered products, it is necessary to continue the time arguments back to real time by the replacement $\tau \to it$. The formal similarity of (4.25) to the generating functional $Z_{ab}[J]$ introduced in Sec. 3.2 is obvious.

For example, the thermal average of the position operator at the imaginary time τ is given by

$$\langle Q(\tau) \rangle_\beta = -\beta \frac{\delta F_\beta[J]}{\delta J(\tau)} \equiv Q_\beta(\tau) . \qquad (4.26)$$

Result (4.26) gives an expression for $Q_\beta(\tau)$ that is an implicit functional of J.

The analysis of time-ordered products yields an interesting result. In the limit that $\beta \to \infty$, thermal averages of time-ordered products are dominated by the ground state. This can be written

$$\lim_{\beta \to \infty} e^{\beta E_g} \langle T\{Q(\tau_1)\cdots\} \rangle_\beta = \langle E_g | T\{Q(\tau_1)\cdots\} | E_g \rangle . \qquad (4.27)$$

82   Further Applications

This shows that the Wick-rotated path integral, when integrated over all periodic paths whose interval of periodicity approaches infinity, serves as the generator for ordered products between the ground states of the theory.

> **Exercise 4.3:** Using the results of (3.62), show that the harmonic oscillator gives
> $$\lim_{\beta \to \infty} \langle Q^2 \rangle_\beta = \frac{1}{2m\omega} e^{-\frac{1}{2}\beta\omega}, \qquad (4.28)$$
> where Q is the Heisenberg picture operator evaluated at $t = 0$.

There is an important extension of the generating functional defined by (4.25). The generating functional $\Gamma_\beta[Q_\beta]$, known as the *effective action*, is defined by a Legendre transformation as

$$\Gamma_\beta[Q_\beta] = F_\beta[J] + \frac{1}{\beta} \int_0^\beta d\tau\, J(\tau) Q_\beta(\tau). \qquad (4.29)$$

Application of the chain rule to (4.29) gives

$$\begin{aligned} \frac{\delta \Gamma_\beta[Q_\beta]}{\delta Q_\beta(\tau)} &= \int_0^\beta d\tau'\, \frac{\delta F_\beta[J]}{\delta J(\tau')} \frac{\delta J(\tau')}{\delta Q_\beta(\tau)} \\ &\quad + \frac{1}{\beta} J(\tau) + \frac{1}{\beta} \int_0^\beta d\tau'\, \frac{\delta J(\tau')}{\delta Q_\beta(\tau)} Q_\beta(\tau'). \end{aligned} \qquad (4.30)$$

The definition (4.26) reduces expression (4.30) to

$$\frac{\delta \Gamma_\beta[Q_\beta]}{\delta Q_\beta(\tau)} = \frac{1}{\beta} J(\tau), \qquad (4.31)$$

In the limit that $J \to 0$ the effective action reduces to the free energy of the equilibrium system. Result (4.31) shows that this free energy, viewed as a functional of the equilibrium position, must satisfy

$$\left. \frac{\delta \Gamma_\beta[Q_\beta]}{\delta Q_\beta(\tau)} \right|_{J=0} = 0. \qquad (4.32)$$

Result (4.32) shows that the expectation value of the position extremizes the free energy of the equilibrium ($J = 0$) system. By recalling result (4.27) it is seen that, in the limit $\beta \to \infty$, Q_β becomes, up to the factor $e^{-\beta E_g}$, the expectation value of the position in the ground state of the system. Therefore, the extrema of the effective action, or free energy, are

Sec. 4.3 Symmetry and Generating Functionals   83

intimately related to the nature of the ground state (or states) available to the quantized system.

Returning to the $J \neq 0$ effective action, it follows from (4.31) that

$$\frac{\delta^2 \Gamma_\beta[Q_\beta]}{\delta Q_\beta(\tau)\, \delta Q_\beta(\tau_1)} = \frac{1}{\beta} \frac{\delta J(\tau_1)}{\delta Q_\beta(\tau)} \equiv P_\beta(\tau_1, \tau) \,. \tag{4.33}$$

Result (4.26) gives

$$\frac{\delta Q_\beta(\tau)}{\delta J(\tau_2)} = \langle T\{Q(\tau)Q(\tau_2)\} \rangle_\beta \equiv G_\beta(\tau, \tau_2) \,, \tag{4.34}$$

where the imaginary time-ordering is given by

$$T\{Q(\tau_1)Q(\tau_2)\} = \theta(\tau_1 - \tau_2)Q(\tau_1)Q(\tau_2) + \theta(\tau_2 - \tau_1)Q(\tau_2)Q(\tau_1) \,. \tag{4.35}$$

Results (4.33) and (4.34) are related to the Dirac delta in the manner

$$\begin{aligned}\int_0^\beta d\tau\, P_\beta(\tau_1, \tau) G_\beta(\tau, \tau_2) &= \int_0^\beta d\tau\, \frac{\delta J(\tau_1)}{\delta Q_\beta(\tau)} \frac{\delta Q_\beta(\tau)}{\delta J(\tau_2)} \\ &= \frac{\delta J(\tau_1)}{\delta J(\tau_2)} = \delta(\tau_1 - \tau_2) \,,\end{aligned} \tag{4.36}$$

It follows that $P_\beta(\tau, \tau')$ is the functional inverse, in the sense of (1.74), of $G_\beta(\tau, \tau')$. Higher-order derivatives of Γ_β have interesting properties, but their analysis will be deferred until the field theoretic extensions of these ideas are developed in Chapter 8. However, in Sec. 4.5 the effective action will be evaluated to demonstrate the spontaneous breakdown of symmetry in a quantum mechanical system.

4.3 Symmetry and Generating Functionals

The concept of symmetry plays such a central role in understanding quantum processes that its significance can scarcely be overestimated. The two great impulses of physical science, the classification of objects and the discovery of the dynamical laws governing them, have in particle and field theory become intertwined by considerations of symmetry. Up to this point the forms chosen for the action have been a legacy from systems important to classical mechanics. It will be seen that symmetry, and its implications for quantum dynamics, is a powerful tool for both classifying particles and constructing the action describing their dynamics. The presentation of

Further Applications

symmetry is most transparent in the path integral formulation of generating functionals. In this section the formulation of quantum mechanical symmetry at the classical and the quantum level will be presented.

The central feature of symmetry in physics is that each symmetry is associated with a conservation law, or conserved quantity. At the classical level [6] this is seen very easily from the Euler–Lagrange equation. If it is true that the Lagrangian density satisfies

$$\frac{\partial \mathcal{L}}{\partial q_j} = 0, \qquad (4.37)$$

then the action is said to be *cyclic* in the coordinate q_j. Using the definition (1.42) of the canonical momentum, it follows from the Euler–Lagrange equation that

$$\frac{dp_j}{dt} = 0, \qquad (4.38)$$

and the momentum is conserved. Expression (4.37) is equivalent to the following property of the Lagrangian density: it is invariant in form and content under a translation of the coordinate q_j by a time-independent constant c_j, i.e.,

$$\mathcal{L}(q_j, \dot{q}_j) = \mathcal{L}(q_j + c_j, \dot{q}_j). \qquad (4.39)$$

Property (4.39) is referred to as *translational invariance* of the action, and it immediately implies the conservation law (4.38).

By analogy with this simple example, in a classical Newtonian system symmetries are understood as variations or *transformations* of the coordinates that leave the action invariant [7]. For the moment, only *continuous* transformations will be considered. This means that each transformation is connected smoothly to the identity transformation, which is defined as the transformation that leaves the coordinates unchanged. Because the transformations are continuous, it suffices to consider only infinitesimal transformations, since finite transformations may be built up by repeated applications of them. An infinitesimal transformation of the coordinate q_j will be written $q_j \to q_j + \delta q_j$, where δq_j is an arbitrary function of t satisfying $(\delta q_j)^2 \approx 0$. The variation of the action under this transformation gives

$$\begin{aligned} \delta S &= \int_{t_a}^{t_b} dt \left(\frac{\partial \mathcal{L}}{\partial q_j} \delta q_j + \frac{\partial \mathcal{L}}{\partial \dot{q}_j} \delta \dot{q}_j \right) \\ &= \int_{t_a}^{t_b} dt \left[\frac{d}{dt} \left(\frac{\partial \mathcal{L}}{\partial \dot{q}_j} \delta q_j \right) + \left(\frac{\partial \mathcal{L}}{\partial q_j} - \frac{d}{dt} \frac{\partial \mathcal{L}}{\partial \dot{q}_j} \right) \delta q_j \right]. \end{aligned} \qquad (4.40)$$

Sec. 4.3 Symmetry and Generating Functionals

There is nothing new in (4.40); it is a tautology. However, if δS is evaluated for a set of classical trajectories, q_c, which must satisfy the *original* Euler–Lagrange equations of motion, then it follows that

$$\delta S = \int_{t_a}^{t_b} dt \, \frac{d}{dt} \left(\frac{\partial \mathcal{L}}{\partial \dot{q}_j} \delta q_j \right) \bigg|_{q=q_c} . \tag{4.41}$$

If $\delta q_j = 0$ at $t = t_a$ and $t = t_b$, then $\delta S = 0$, and these manipulations have simply reproduced the derivation of the Euler–Lagrange equations. However, if $\delta q_j \neq 0$ at the endpoint times *and* if $\delta S = 0$ under this transformation, then (4.41) gives

$$\frac{dG}{dt} = 0 , \tag{4.42}$$

where

$$G = \frac{\partial \mathcal{L}}{\partial \dot{q}_j} \delta q_j \bigg|_{q=q_c} . \tag{4.43}$$

If $\delta S = 0$ under this nontrivial variation, then this transformation is said to be a *symmetry* of the action, and to this symmetry there is associated a *charge* G, given by (4.43), which is conserved by virtue of (4.42).

Result (4.41) can be reversed. If the action is *not* invariant under the transformation, then δS does not vanish. Instead, it will have the general form

$$\delta S = \int_{t_a}^{t_b} dt \, F(t) , \tag{4.44}$$

which, when combined with (4.41), gives

$$\frac{dG}{dt} = F(t) . \tag{4.45}$$

It is still possible to define the charge G associated with the transformation, but it is no longer conserved.

The transformations on the coordinates do not exhaust the symmetries that might be present in the system. There is also the possibility of varying the domain of integration appearing in the action:

$$S = \int_{t_a}^{t_b} dt \, \mathcal{L} \quad \rightarrow \quad S' = \int_{t_a+\delta t}^{t_b+\delta t} dt \, \mathcal{L} . \tag{4.46}$$

Exercise 4.4: If the action is invariant under the transformation (4.46), show that the Hamiltonian is conserved along a classical trajectory.

86   Further Applications

Obviously, the most general form of transformation involves both the coordinates and the domain of integration. Such a transformation assumes greater meaning in the field theory context.

An important property of classical mechanical systems is the freedom to add the total time derivative of some function F of the canonical variables to the Lagrangian density,

$$\mathcal{L} \to \mathcal{L} + \frac{dF}{dt} \;\Rightarrow\; S \to S + F(t_b) - F(t_a) \,. \tag{4.47}$$

At the classical level (4.47) does not affect the dynamics of the system during the time interval since the function F cannot affect the variations in that interval. Because of this property the transformation (4.47) has been well studied in classical mechanics where it is used to define *canonical transformations* of the coordinates, i.e., those transformations that preserve the Poisson bracket structure of the theory [6]. This is equivalent to considering only those transformations of p and q whose Jacobian is unity. The volume of phase space is therefore invariant under a canonical transformation.

The consideration of symmetry in quantum mechanics is complicated by the Hilbert space structure that underlies the process of quantization [8]. A quantum mechanical symmetry transformation may involve a unitary or antiunitary transformation of the states of the theory as well as the operators that act upon them. For example, in quantum mechanical systems the appearance of a total time derivative in the action may signal nontrivial transformation properties for the states of the system. A *quantum symmetry* is therefore defined as a set of transformations on the operators *and* states of a theory that leaves a physically relevant transition element invariant. It may be that certain transition elements develop an overall phase, but this represents transformation properties of the physically relevant states of the theory. The presence of a phase factor will be discussed at the end of the section when gauge transformations are introduced.

In the path integral formulation there is a clear relationship between classical symmetries and their quantum counterparts. The classical component of the system is apparent in the presence of the classical action in the argument of the exponential. The quantum mechanical structure is present in the measure used to integrate the path integral. The initial and final states are present in the limits on the path integral. Inspection of the path integral shows that invariance of the action under a transformation is insufficient to guarantee that the transition amplitude itself is invariant, and this is because this same transformation may generate a nontrivial Jacobian in the measure. If a nontrivial Jacobian occurs, the classical symmetry does not survive the quantization procedure, and the theory is said

Sec. 4.3 Symmetry and Generating Functionals 87

to possess an *anomaly*. An anomaly may or may not be desirable, and this must be determined on a case by case basis.

For example, the classical Lagrangian given by

$$\mathcal{L} = \frac{1}{2\omega} \frac{\dot{q}^2}{q^2} \qquad (4.48)$$

is invariant under the time-independent rescaling $q \to e^\lambda q$. A naive insertion of the action (4.48) into the path integral of the form (2.50) with the measure (2.52) would be incorrect, since the measure (2.52) is not scale invariant. If the action (4.41) is to be quantized consistently with scale invariance, the measure of the corresponding path integral must differ from that of (2.52). The construction of this measure is left as an exercise.

Exercise 4.5: Starting with the action (4.48), find the action corresponding to (2.48), determine the corresponding scale transformation on p, and from a path integral of the form (2.47) construct the path integral equivalent to (2.50), verifying that it does retain scale invariance [9].

The structure of the measure reveals many of the quantum mechanical properties of the system. For example, if the path integral measure of (2.50) has the limits $\pm\infty$, then it is possible to prove a form of Ehrenfest's theorem. Ehrenfest's theorem states that the expectation values of the observables obey the classical form of Hamilton's equations of motion. In order to prove this theorem using the path integral formalism, each of the variables of integration is translated by an arbitrary constant, i.e., $q_j \to q_j + f_j$. Because of the limits on the integrations it is obvious that the path integral, in terms of the new variables, must yield the same result. In the large N limit this is equivalent to translating the $q(t)$ appearing in the action by a function $f(t)$ that vanishes at the times t_b and t_a. The latter property allows an integration by parts that drops the endpoint terms. Since $f(t)$ is arbitrary, it may be chosen to be infinitesimal, so that $[f(t)]^2 \approx 0$. The final result of this translation therefore gives the equality

$$\int_{q_a}^{q_b} \overline{\mathcal{D}q} \, \exp\left[i \int_{t_a}^{t_b} dt \, \mathcal{L}(q, \dot{q}) \right]$$
$$= \int_{q_a}^{q_b} \overline{\mathcal{D}q} \, \exp\left\{ i \int_{t_a}^{t_b} dt \, \left[\mathcal{L}(q, \dot{q}) + f(t) \left(\frac{\partial \mathcal{L}}{\partial q} - \frac{d}{dt}\frac{\partial \mathcal{L}}{\partial \dot{q}} \right) \right] \right\} . \qquad (4.49)$$

By choosing $f(t) = \epsilon \delta(t - t')$, where $\epsilon^2 \approx 0$ and t' is arbitrary, (4.49)

88   Further Applications

immediately reduces to

$$\int_{q_a}^{q_b} \overline{\mathcal{D}}q \left.\left(\frac{\partial \mathcal{L}}{\partial q} - \frac{d}{dt}\frac{\partial \mathcal{L}}{\partial \dot{q}}\right)\right|_{t=t'} \exp\left\{i\int_{t_a}^{t_b} dt \; [\mathcal{L}(q,\dot{q})]\right\} = 0 \; . \qquad (4.50)$$

Result (4.50) shows that the expectation value of the observable q must satisfy the classical Euler–Lagrange equation, a variant of Ehrenfest's theorem.

Another very important symmetry phenomenon can occur. It may be possible to quantize a *nonlinear*, i.e., nonquadratic, theory by writing down a formal perturbation series in such a way that the symmetry appears to be preserved. However, it may be discovered that the true ground state of the system is such that, due to the nonlinearity of the interactions, the fluctuations of the system around the ground state violate the symmetry. This is known as the *spontaneous breakdown of symmetry*, and it is one of the central theoretical ideas in many areas of physics, in particular particle and condensed matter physics. In general, determining the nature of the ground state in a nonlinear quantum system is a difficult task, but one in which path integrals have played a significant role. This will be discussed again in Sec. 4.5 for a simple quantum mechanical model.

The presence of a quantum symmetry has immediate ramifications for the transition elements and the time-ordered products. In the path integral formalism this is expressed as a set of identities that the generating functional must satisfy [10]. To demonstrate this, the results of rotational invariance will be analyzed. First, the basic properties of a rotation are reviewed.

Since a rotation must preserve the norm of an arbitrary vector, the transformation representing the rotation must be linear in the coordinates. The most general form such a transformation may then have is $x'_j = R_{jk} x_k$, where the R_{jk} are the elements of some matrix $\mathbf{R}$. It follows that the restriction of norm invariance gives

$$x'_i x'_i = R_{ij} R_{ik} x_j x_k = x_k x_k \;\Rightarrow\; R_{ij} R_{ik} = \delta_{jk} \; , \qquad (4.51)$$

so that $\mathbf{R}$ is an orthogonal matrix. An infinitesimal form of the transformation is achieved by expanding around the identity, so that $R_{jk} \approx \delta_{jk} + \delta R_{jk}$. The requirement that this transformation be orthogonal immediately yields

$$\delta_{jk} \approx (\delta_{ij} + \delta R_{ij})(\delta_{ik} + \delta R_{ik}) \approx \delta_{jk} + \delta R_{kj} + \delta R_{jk} \; , \qquad (4.52)$$

from which it follows that δR_{jk}, the infinitesimal form of the transformation, is given by an antisymmetric matrix. The 3×3 antisymmetric matrix

Sec. 4.3 Symmetry and Generating Functionals

representing the infinitesimal rotation can therefore have only three independent entries, and these are denoted $\delta\theta_j$ and correspond to the three independent angles of rotation. The matrix δR_{jk} can then be written $\delta R_{jk} = \delta\theta_i \varepsilon_{ijk}$, where the Levi–Civita symbol of (1.54) has been used to ensure that δR_{jk} is an antisymmetric matrix. The effect of an infinitesimal rotation upon a Cartesian coordinate system is to induce the change

$$x_j \to x_j + \delta R_{jk} x_k \,. \tag{4.53}$$

The inverse transformation is given, for the infinitesimal case, by

$$x_j \to x_j + \delta R_{kj} x_k = x_j - \delta R_{jk} x_k \,. \tag{4.54}$$

The transition amplitude describing the three-dimensional motion of a particle in the presence of both a central potential and a time-dependent source is given in Cartesian coordinates by the path integral

$$Z_{ab}[J] = \int_{\mathbf{x}_a}^{\mathbf{x}_b} \overline{\mathcal{D}} x \exp\left\{ i \int_{t_a}^{t_b} dt \left[\tfrac{1}{2} m \dot{x}_j \dot{x}_j - V(r) - J_j x_j \right] \right\} \,. \tag{4.55}$$

Similarly to (2.52), the measure is given by

$$\overline{\mathcal{D}} x = \lim_{N \to \infty} \left(\frac{m}{2\pi i \epsilon} \right)^{3N/2} \prod_{k=1}^{N-1} d^3 x_k \,, \tag{4.56}$$

where the subscript k refers to the time of the variable. In what follows the special case that $\mathbf{x}_a = \mathbf{x}_b = 0$ will be considered, so that the amplitude for origin to origin transition, $Z_{00}[J]$ is being considered. This restriction simplifies the results.

The path integral $Z_{00}[J = 0]$ is invariant under rotations of the coordinates of integration since both the measure and the action are invariant. It follows that a rotation gives

$$Z_{00}[J=0] \to \int_0^0 \overline{\mathcal{D}} x' \exp\left\{ i \int_{t_a}^{t_b} dt \left[\tfrac{1}{2} m \dot{x}'_j \dot{x}'_j - V'(r') \right] \right\}$$
$$= \int_0^0 \overline{\mathcal{D}} x \exp\left\{ i \int_{t_a}^{t_b} dt \left[\tfrac{1}{2} m \dot{x}_j \dot{x}_j - V(r) \right] \right\} = Z_{00}[J=0] \,, \tag{4.57}$$

where the rotational invariance of both V and the origin, $\mathbf{x} = 0$, has been used. For the $J \neq 0$ case, the infinitesimal rotation of the source, given by $J_j \to J_j + \delta R_{jk} J_k$, can be compensated by the inverse rotation of the variables of integration, given by $x_j \to x_j - \delta R_{jk} x_k$, since $J'_j x'_j = J_j x_j$

90   Further Applications

to $O(\delta R)$. This means that the transition element for the $J \neq 0$ case is invariant under rotations of the source, so that

$$Z_{00}[J] = Z_{00}[J'] . \tag{4.58}$$

By using the functional Taylor series technique of (1.28) for the infinitesimal case, (4.58) immediately gives the result

$$Z_{00}[J'] - Z_{00}[J] = 0 = \delta\theta_i \int dt\, \varepsilon_{ijk} J_k(t) \frac{\delta Z_{00}[J]}{\delta J_j(t)} . \tag{4.59}$$

Because $J \neq 0$ while the $\delta\theta_i$ are arbitrary, result (4.59) serves a generator for relationships between time-ordered products. For example, a single functional derivative of (4.59) with respect to J_k gives

$$\varepsilon_{ijk} \frac{\delta Z_{00}[J]}{\delta J_j(t_1)} + \varepsilon_{ijl} \int dt\, J_l(t) \frac{\delta^2 Z_{00}[J]}{\delta J_k(t_1)\,\delta J_j(t)} = 0 . \tag{4.60}$$

Result (4.60) relates different time-ordered products, assuming they exist, although they are evaluated between the particular $\mathbf{x} = 0$ states. Similar, somewhat more complicated identities can be derived for the case of nonzero initial and final coordinates. This is left as an exercise.

Exercise 4.6: Show that the general transition amplitude (4.55) satisfies the identity

$$\varepsilon_{ijk} \int dt\, J_k(t) \frac{\delta Z_{ab}[J]}{\delta J_j(t)} + \varepsilon_{ijk} \left(x_k^a \frac{\partial}{\partial x_j^a} + x_k^b \frac{\partial}{\partial x_j^b} \right) Z_{ab}[J] = 0 . \tag{4.61}$$

As another example of symmetry structure in quantum mechanics, the motion of a charged particle in an arbitrary electromagnetic potential is considered. The Lagrangian density for this system is given, in Cartesian coordinates, by [6]

$$\mathcal{L}(x, \dot{x}) = \tfrac{1}{2} m \dot{x}_j \dot{x}_j - V(x) - e\phi(x) + eA^j(x)\dot{x}^j , \tag{4.62}$$

where ϕ and $\mathbf{A}$ are the scalar and vector potential. Electromagnetic potentials are defined so that the equations of motion in which they appear are invariant under a *gauge transformation of the second kind*[1], which is defined by

$$A^j \to A^j + \frac{\partial \Lambda}{\partial x^j} , \quad \phi \to \phi - \frac{\partial \Lambda}{\partial t} , \tag{4.63}$$

---

[1] Gauge transformations of the first kind will be defined in Chapter 6.

Sec. 4.3 Symmetry and Generating Functionals

where Λ is an arbitrary function. Using the chain rule shows that (4.62) changes by the total time derivative of Λ under a gauge transformation of the second kind. The action therefore changes to

$$S \rightarrow S + e\Lambda(x_b, t_b) - e\Lambda(x_a, t_a) . \tag{4.64}$$

At the classical level the equations of motion for the particle are then unaffected by the gauge transformation. However, the quantum mechanical transition element is not invariant. Instead, the form of the path integral immediately yields

$$Z_{ab} \rightarrow e^{ie\Lambda(x_b, t_b)} Z_{ab} e^{-ie\Lambda(x_a, t_a)} . \tag{4.65}$$

Result (4.65) shows that the transition element develops an overall phase under the gauge transformation. Such a possibility was mentioned earlier in this section. In order to compensate for the phases generated by the propagator and to ensure that the quantum mechanical mechanical transition amplitude is invariant under gauge transformations of the second kind, the wave functions of the charged particles must simultaneously undergo a gauge transformation of the form

$$\langle x | \psi, t \rangle \rightarrow e^{ie\Lambda(x,t)} \langle x | \psi, t \rangle , \tag{4.66}$$

With the prescription (4.66), the transition element of the form (2.31) is invariant.

As a final note, it is interesting to examine the quantum mechanical form of the charge G, defined by (4.43). Using the definition of the canonical momentum, the charge G becomes the quantum mechanical operator

$$G = P_j \, \delta Q_j . \tag{4.67}$$

Assuming that $[\delta Q_j, Q_k] = 0$ gives

$$-i[Q_k, G] = \delta Q_k . \tag{4.68}$$

Therefore, in the quantum mechanical version of the system, the charge G *generates* the infinitesimal transformation of the coordinates, and for that reason G is known as the *generator* of the symmetry. If G is conserved, then it commutes with the Hamiltonian of the system. Therefore, G can be added to the set of simultaneous observables that includes the energy. It may be that a symmetry is associated with more than one generator. The three-dimensional rotations discussed earlier are associated with the the three generators

$$G_j = \varepsilon_{jkm} Q_k P_m , \tag{4.69}$$

92   Further Applications

and these generators form an *algebra* [11]. Loosely speaking, an algebra is a set of objects with a set of rules for multiplying and adding them. It is straightforward to show that the generators of (4.69) give the familiar angular momentum commutation relations

$$[G_j, G_k] = i\varepsilon_{jkm} G_m \ . \tag{4.70}$$

Result (4.70) is an example of a *Lie algebra*. These will be discussed in more detail in Chapter 7.

4.4 Harmonic Oscillator Coherent States

In this section the simple harmonic oscillator is revisited, and the path integral representation of the transition amplitude between coherent states is derived. This reformulation gives the ground state to ground state transition amplitude, sometimes referred to as the *vacuum persistence functional*, as a special case. In the process, the harmonic oscillator is reformulated in terms of creation and annihilation operators, and these, with some adaptations, give a natural expression of canonically quantized field theory.

The starting point is to rewrite the harmonic oscillator Hamiltonian, $H = P^2/2m + \frac{1}{2}m\omega^2 Q^2$, by introducing the following definitions due to Dirac:

$$\begin{aligned} a &= (\tfrac{1}{2}m\omega)^{\frac{1}{2}} \left[Q + \frac{iP}{m\omega} \right] , \\ a^\dagger &= (\tfrac{1}{2}m\omega)^{\frac{1}{2}} \left[Q - \frac{iP}{m\omega} \right] . \end{aligned} \tag{4.71}$$

These definitions give

$$[a, a^\dagger] = 1 \ , \tag{4.72}$$

$$H = \omega a^\dagger a + \tfrac{1}{2}\omega \ , \tag{4.73}$$

$$[H, a^\dagger] = \omega a^\dagger \ , \tag{4.74}$$

$$[H, a] = -\omega a \ . \tag{4.75}$$

Exercise 4.7: Verify relations (4.72) through (4.75).

The ground state, or *vacuum*, $|0\rangle$, is defined as the state that satisfies

$$a|0\rangle = 0 \ . \tag{4.76}$$

Sec. 4.4 Harmonic Oscillator Coherent States

In other words, the vacuum is annihilated by the operator a, and so a is known as an *annihilation operator*.

Exercise 4.8: Prove that the state $|0\rangle$ must exist.

The ground state is an eigenstate of H, $H|0\rangle = \frac{1}{2}\omega|0\rangle$. The state $|1\rangle$ is defined as $|1\rangle = a^\dagger|0\rangle$. It is also an eigenstate of H, since

$$H|1\rangle = [H, a^\dagger]|0\rangle + a^\dagger H|0\rangle = \omega a^\dagger|0\rangle + \frac{1}{2}\omega a^\dagger|0\rangle = \frac{3}{2}\omega|1\rangle. \quad (4.77)$$

Because $|1\rangle$ contains an additional unit of energy, ω, the operator $a^\dagger$ is referred to as a *creation operator*.

Exercise 4.9: Show that the state $|n\rangle = (n!)^{-\frac{1}{2}}(a^\dagger)^n|0\rangle$ satisfies

$$H|n\rangle = (n + \tfrac{1}{2})\omega|n\rangle, \quad (4.78)$$
$$\langle n|m\rangle = \delta_{nm}, \quad (4.79)$$
$$a^\dagger a|n\rangle = n|n\rangle. \quad (4.80)$$

Result (4.80) shows that $N = a^\dagger a$ counts the number of units of ω in the state $|n\rangle$, and is therefore referred to as the *number operator*.

Given a complex number λ, the *coherent state* $|\lambda\rangle$ is defined as [12]

$$|\lambda\rangle = e^{\lambda a^\dagger - \lambda^* a}|0\rangle. \quad (4.81)$$

The coherent states have many useful properties, and some of these are

$$|\lambda\rangle = \sum_{n=0}^{\infty} \frac{\lambda^n}{\sqrt{n!}}|n\rangle e^{-\frac{1}{2}\lambda^*\lambda}, \quad (4.82)$$

$$\frac{1}{2\pi i}\int d\lambda^* d\lambda |\lambda\rangle\langle\lambda| = \sum_{n=0}^{\infty} |n\rangle\langle n|, \quad (4.83)$$

$$\langle\sigma|\lambda\rangle = \exp(-\tfrac{1}{2}\sigma^*\sigma - \tfrac{1}{2}\lambda^*\lambda + \sigma^*\lambda), \quad (4.84)$$
$$\langle\sigma|a^\dagger|\lambda\rangle = \sigma^*\langle\sigma|\lambda\rangle, \quad (4.85)$$
$$\langle\sigma|N|\lambda\rangle = \sigma^*\lambda\langle\sigma|\lambda\rangle. \quad (4.86)$$

Exercise 4.10: Using the Baker–Campbell–Hausdorff formula (1.66), verify results (4.82) through (4.86).

94   Further Applications

The coherent states are not eigenstates of H, Q, or P. Their utility lies in the fact that they allow the *expectation values* of Q and P to be nonzero and well defined simultaneously, so that, in their general form, they lie somewhere between a purely position eigenstate and a purely momentum eigenstate. This is an important property, and one that allows their adaptation to more complicated field theory systems. The transition amplitude between coherent states then describes a quantum process between states with a more classical quality in the sense that it is possible to discuss both a position and momentum expectation value. Since the path integral strives to maintain the connection between quantum mechanics and classical mechanics it is natural to formulate this amplitude as a path integral and examine it.

By virtue of (4.83) the coherent states are complete, and this allows a construction of the path integral representation of the transition amplitude between an initial and final coherent state, $Z_{ab} = \langle \lambda_b, t_b | \lambda_a, t_a \rangle$, where

$$|\lambda, t\rangle \equiv e^{iHt}|\lambda, t=0\rangle. \qquad (4.87)$$

The derivation [13] is accomplished by using a repetition of the steps in Sec. 2.2, so that once again the time interval is partitioned into $N-1$ subintervals of duration ϵ, and complete sets of coherent states are inserted at the respective times. The result is

$$Z_{ab} = \int \mathcal{D}\lambda \prod_{j=0}^{N-1} \langle \lambda_{j+1}, t_{j+1} | \lambda_j, t_j \rangle, \qquad (4.88)$$

where the measure in (4.88) is given by

$$\mathcal{D}\lambda = \prod_{j=1}^{N-1} \frac{1}{2\pi i} d\lambda_j^* \, d\lambda_j, \qquad (4.89)$$

and the identifications $\lambda_N = \lambda_b$ and $\lambda_0 = \lambda_a$ have been made.

The individual infinitesimal elements appearing in (4.88) can be evaluated to find

$$\langle \lambda_{j+1}, t_{j+1} | \lambda_j, t_j \rangle$$
$$\approx \exp(-i\epsilon \omega \lambda_{j+1}^* \lambda_j) \exp(-\tfrac{1}{2}\lambda_{j+1}^* \lambda_{j+1} - \tfrac{1}{2}\lambda_j^* \lambda_j + \lambda_{j+1}^* \lambda_j). \qquad (4.90)$$

Using (4.90) in (4.88) gives the result

$$Z_{ab} = e^{-\tfrac{1}{2}(\lambda_b^* \lambda_b - \lambda_a^* \lambda_a)}$$
$$\times \int_{\lambda_a}^{\lambda_b} \mathcal{D}\lambda \, \exp\left\{i \sum_{j=0}^{N-1} \left[-i\lambda_{j+1}^* \lambda_j + i\lambda_j^* \lambda_j - \epsilon \omega \lambda_{j+1}^* \lambda_j\right]\right\}. \qquad (4.91)$$

Sec. 4.4 Harmonic Oscillator Coherent States

Taking the continuum limit of (4.91), $\lambda_{j+1} \approx \lambda_j + \epsilon\dot{\lambda}_j$, and integrating by parts, taking into account the boundary conditions, gives

$$Z_{ab} = e^{\frac{1}{2}(\lambda_b^*\lambda_b - \lambda_a^*\lambda_a)} \int_{\lambda_a}^{\lambda_b} \mathcal{D}\lambda \, \exp\left\{i \int_{t_a}^{t_b} dt \, \left[i\lambda^*\dot{\lambda} - \omega\lambda^*\lambda\right]\right\} . \quad (4.92)$$

Result (4.92) is reduced to the vacuum persistence functional by setting $\lambda_a = \lambda_b = 0$, since for that case the two coherent states reduce to $|0\rangle$.

While result (4.92) looks somewhat different than the form (2.47), it is possible to put it into a similar form with a small effort. First, the overall factor $\exp[\frac{1}{2}(\lambda_b^*\lambda_b - \lambda_a^*\lambda_a)]$ is absorbed into the path integral by using the fact that the limits on the path integral allow the argument to be written

$$\tfrac{1}{2}(\lambda_b^*\lambda_b - \lambda_a^*\lambda_a) = \int_{t_a}^{t_b} dt \, \frac{1}{2}\frac{\partial}{\partial t}(\lambda^*\lambda) . \quad (4.93)$$

The path integral becomes

$$Z_{ab} = \int_{\lambda_a}^{\lambda_b} \mathcal{D}\lambda \, \exp\left\{i \int_{t_a}^{t_b} dt \, \left[i\tfrac{1}{2}\lambda^*\dot{\lambda} - i\tfrac{1}{2}\dot{\lambda}^*\lambda - \omega\lambda^*\lambda\right]\right\} . \quad (4.94)$$

By making the change of variables

$$\lambda = (\tfrac{1}{2}m\omega)^{\frac{1}{2}}\left[q + \frac{ip}{m\omega}\right] , \quad (4.95)$$

along with its complex conjugate, the path integral becomes

$$Z_{ab} = \int_{q_a,p_a}^{q_b,p_b} \mathcal{D}p\,\mathcal{D}q \, \exp\left\{i \int_{t_a}^{t_b} dt \, \left[\tfrac{1}{2}p\dot{q} - \tfrac{1}{2}q\dot{p} - \frac{p^2}{2m} - \tfrac{1}{2}m\omega^2 q^2\right]\right\} , \quad (4.96)$$

where the measure is now given by

$$\mathcal{D}p\,\mathcal{D}q = \prod_{j=1}^{N-1} \frac{dp_j}{2\pi} \, dq_j . \quad (4.97)$$

Exercise 4.11: Verify (4.96).

The action appearing in the path integral (4.96) is, up to an integration by parts, the same as the one that would appear in (2.47). The measure

Further Applications

is slightly different from (2.48) since it is missing the initial integration over dp_0, but for that reason this measure is manifestly invariant under a canonical transformation since it is symmetric in the number of factors of dp and dq. The integration by parts may be done to create the presence of the usual form of the action, but it is critical to realize that the surface terms may not be discarded due to the presence of the boundary conditions on q and p. Form (4.96) of the path integral allows incorporation of more complicated boundary conditions on the particle's motion, but the price to be paid is an overall phase:

$$Z_{ab} = e^{i\frac{1}{2}(p_a q_a - p_b q_b)} \int_{q_a, p_a}^{q_b, p_b} \mathcal{D}p \mathcal{D}q \, \exp\left\{ i \int_{t_a}^{t_b} dt \, \mathcal{L}(p, q) \right\} . \tag{4.98}$$

Exercise 4.12: Add complex source terms of the form $J^*\lambda + J\lambda^*$ to the action appearing in (4.92) and evaluate the vacuum persistence functional using result (1.86) and the continuum version of (1.110). Hint: invert the operator $\partial/\partial t$ using the step function $\theta(t - t')$.

4.5 Spontaneously Broken Symmetry

In this section a simple quantum mechanical model of spontaneously broken symmetry will be analyzed. Very similar techniques will be developed for field theory models in Chapter 9, and the results presented in this section will demonstrate many of the ideas that are used in the standard model of particle physics to break symmetry and generate mass.

The model is one-dimensional, and the potential is nonlinear, given by

$$V(q) = -\tfrac{1}{2}\alpha q^2 + \tfrac{1}{4}\gamma q^4, \quad \alpha, \gamma > 0 . \tag{4.99}$$

Depicted in Fig. 4.1, this potential possesses two absolute minima, located at $q_m = \pm\sqrt{\alpha/\gamma}$. From energy considerations the classical motion of a particle moving in the potential (4.99) can be described qualitatively. For positive values of the total mechanical energy the particle will perform oscillations that are symmetric about the origin. However, for the range of total mechanical energy $0 > E \geq -\alpha^2/4\gamma$, the particle is trapped in one of the two wells and is constrained to perform oscillations that are *not* symmetric about the origin. While the potential has the symmetry $V(q) = V(-q)$, there exists a range of energies for the classical oscillations such that the domain of motion is not symmetric. When the energies are small enough, the bottom of each well can be given a quadratic approximation, so that the low-lying classical oscillations are harmonic in nature.

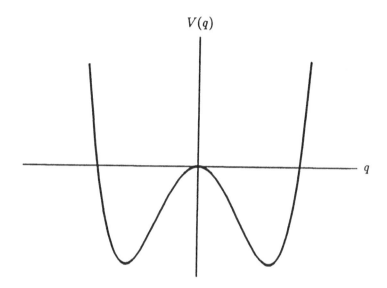

Fig. 4.1. The potential $V(q) = -\frac{1}{2}\alpha q^2 + \frac{1}{4}\lambda q^4$.

The basic classical behavior just outlined has a quantum mechanical counterpart, but it is somewhat more subtle. In effect, the potential (4.99) gives rise to the existence of two energetically degenerate ground states for the system: one whose low-energy excitations are oscillations about the left well, the other whose low-energy excitations are oscillations about the right well. It is expected, in analogy with the classical situation, that the low-energy oscillations can be approximated as those of a harmonic oscillator. The frequency of this approximate harmonic oscillation is the quantum mechanical analog of mass in the field theoretic extension of this model to be discussed in Chapter 9.

Demonstrating the existence of these two approximate harmonic oscillator ground states, or vacuums,[2] is best accomplished by using the path

---

[2] There are two acceptable plurals of the word vacuum: vacuums and vacua. The former is used here.

Further Applications

integral formulation of the free energy, given by

$$F_\beta = -\frac{1}{\beta} \ln \int dx \int_x^x \overline{\mathcal{D}} q \, \exp\left\{-\int_0^\beta d\tau \left[\tfrac{1}{2} m \dot{q}^2 + V(q)\right]\right\} . \tag{4.100}$$

Because of the nonlinearity of $V(q)$ it is not possible to obtain an exact evaluation of (4.100). However, it is possible to extract enough of the nonlinear behavior to demonstrate the existence of the two ground states. In so doing, a general method will be developed that is applicable to other systems.

For the moment the potential $V(q)$ will be considered to be general, and the specific case of (4.99) will considered again later. The analysis begins by noting that (4.100) has the form

$$F_\beta = -\frac{1}{\beta} \ln \int dx \, P_\beta(x) , \tag{4.101}$$

where $P_\beta(x)$ is a probability density given by the Euclidean path integral appearing in (4.100). In fact, $P_\beta(x)$ is the partition function density,

$$P_\beta(x) = \langle x | e^{-\beta H} | x \rangle . \tag{4.102}$$

From (4.101) it is apparent that the value of the free energy is determined predominately from the regions of integration for which the partition function density, $P_\beta(x)$, is a maximum. These are given by the values of x that satisfy

$$\frac{dP_\beta(x)}{dx} = 0, \quad \frac{d^2 P_\beta(x)}{dx^2} < 0 . \tag{4.103}$$

For this reason, it is convenient to introduce the *effective potential*, V_{eff}, defined by

$$V_{\text{eff}}(x) = -\frac{1}{\beta} \ln P_\beta(x) . \tag{4.104}$$

By virtue of (4.103) the effective potential determines the values of x that maximize the probability density. However, from the definition of the effective potential, at these values it must satisfy

$$\frac{dV_{\text{eff}}}{dx} = 0, \quad \frac{d^2 V_{\text{eff}}}{dx^2} > 0 , \tag{4.105}$$

so that the effective potential's *minima* determine the positions of maximum probability for observing the particle, i.e., those positions around which the particle is most probably oscillating.

Sec. 4.5 Spontaneously Broken Symmetry

From the definition of the effective action made in Sec. 4.3, F_β is to be viewed as a functional of the expectation value of the position operator, for notational simplicity denoted $\bar{x}$, so that $F_\beta = F_\beta(\bar{x})$. It is assumed that $\bar{x}$ is time independent, and this will certainly be true in the case $J = 0$. The free energy can be related to the effective potential by writing

$$F_\beta(\bar{x}) \approx -\frac{1}{\beta} \ln \left(P_\beta(\bar{x}) \int dx \right) = V_{\text{eff}}(\bar{x}) - \frac{1}{\beta} \ln \int dx \ . \qquad (4.106)$$

Form (4.106) can be incorporated into the lowest-order approximation to the exact effective action. From the fact that only derivatives of the effective action are important, the second term in (4.106) proportional to the logarithm of the volume can be discarded. The resulting approximation to the effective action is no longer a functional, but rather the function

$$\Gamma_\beta(\bar{x}) = V_{\text{eff}}(\bar{x}) + \frac{1}{\beta} \int_0^\beta d\tau \, J(\tau)\bar{x} \ . \qquad (4.107)$$

The analysis of $P_\beta(x)$ is facilitated by the techniques previously developed for path integrals. The form for $P_\beta(x)$,

$$P_\beta(x) = \int_x^x \mathcal{D}q \exp\left\{ -\int_0^\beta d\tau \left[\tfrac{1}{2} m\dot{q}^2 + V(q) + Jq \right] \right\} , \qquad (4.108)$$

can be analyzed by expanding the action around x by writing $q = x + y$. This will give the desirable simplification that $y = 0$ at both $\tau = 0$ and $\tau = \beta$. This results in

$$\begin{aligned}P_\beta(x) = {}& \exp\left[-\beta V(x) - x \int_0^\beta d\tau \, J(\tau) \right] \\ & \times \int_0^0 \mathcal{D}y \exp\left\{ -\int_0^\beta d\tau \left[\tfrac{1}{2} m\dot{y}^2 \right.\right. \\ & \left.\left. + \left(\frac{\partial V(x)}{\partial x} + J \right) y + \tfrac{1}{2} \frac{\partial^2 V(x)}{\partial x^2} y^2 + \cdots \right] \right\} , \end{aligned} \qquad (4.109)$$

where the ellipsis refers to the terms cubic and higher in y. If the terms represented by the ellipsis could be ignored as small, then the remaining path integral in (4.109) could easily be evaluated, since it would be nothing more than the Wick-rotated version of the Gaussian harmonic oscillator transition amplitude. Such an approximation is referred to as the *method*

Further Applications

of steepest descent [14], and it is commonly used in analyses of the effective potential.

However, it carries an inherent flaw. Ignoring the higher-order terms will not pose an immediate threat, apart from inaccuracy, unless there are values of x for which the coefficient of the quadratic term in (4.109) becomes negative. If the coefficent of the y^2 term becomes negative, then the Gaussian approximation breaks down, and ignoring the higher-order terms results in a path integral that is not defined. Because the function $P_\beta(x)$ is to be integrated over all values of x, the coefficient must remain positive for all real values of x in order to avoid this calamity.

Clearly, if the coefficient becomes negative, then the path integral might possibly be stabilized by the higher-order terms appearing in the expansion, but an analytic evaluation of the path integral to verify this becomes difficult, if not impossible. This can be demonstrated by the following simple analog [15]. The real-valued function $\exp(\frac{1}{2}\alpha x^2 - \frac{1}{4}\gamma x^4)$ is bounded and integrable for $\alpha, \gamma > 0$, but an attempt to integrate it by using an expansion in γ,

$$\int_{-\infty}^{\infty} dx\, e^{\frac{1}{2}\alpha x^2 - \frac{1}{4}\gamma x^4} = \sum_{n=0}^{\infty} \int_{-\infty}^{\infty} dx\, \frac{1}{n!}(-\tfrac{1}{4}\gamma x^4)^n e^{\frac{1}{2}\alpha x^2}, \qquad (4.110)$$

results in nonsense at every order. This drawback will become evident when (4.99) is analyzed.

Employing the results of (3.60), the method of steepest descent yields

$$P_\beta(x) \approx \left(\frac{m\omega}{2\pi \sinh \beta\omega}\right)^{\frac{1}{2}} \exp\left[-\beta V(x) - x \int_0^\beta d\tau\, J(\tau)\right], \qquad (4.111)$$

where the effects of the term linear in y in (4.109) have been suppressed as small. The ω appearing in (4.111) is x dependent, given by

$$\omega = \left(\frac{1}{m}\frac{\partial^2 V(x)}{\partial x^2}\right)^{\frac{1}{2}}. \qquad (4.112)$$

Exercise 4.13: Verify (4.111) and justify ignoring the term proportional to y originally in (4.109).

Sec. 4.5 Spontaneously Broken Symmetry   101

Exercise 4.14: Show that the prefactor appearing in (4.111) is a function of the determinant of the differential operator

$$D = \frac{1}{m^2}\frac{d^2}{d\tau^2} - \frac{\omega^2}{m^2}, \qquad (4.113)$$

evaluated with periodic boundary conditions.

The previously discussed weakness of the method of steepest descent appears here as an imaginary value for ω when the second derivative of the potential is negative. However, the expression (4.111) remains real even when ω is imaginary, although it cannot be well defined for all values of β due to the zeros of the sine function that appears when ω is imaginary. Nevertheless, it will be assumed that expression (4.111) is a good approximation of the path integral *in the vicinity of its minima* for all values of β.

Result (4.111) gives the effective potential

$$V_{\text{eff}} = V(x) + \frac{1}{\beta}x\int_0^\beta d\tau\, J(\tau) - \frac{1}{2\beta}\ln\left(\frac{m\omega}{2\pi\sinh\beta\omega}\right). \qquad (4.114)$$

The zero temperature form of the effective potential is found by taking the limit $\beta \to \infty$, so that, for the case that $J = 0$,

$$V_{\text{eff}}^\infty(x) = V(x) + \tfrac{1}{2}\omega. \qquad (4.115)$$

Exercise 4.15: Verify (4.115).

Result (4.115) is identical in content to result (4.24) if the value $V(\bar{x})$ is ignored. The value of ω at the minimum will describe the separation of low-lying energy levels in the harmonic oscillator approximation for the case that the system is found in that ground state.

At this point attention is returned to the specific case of (4.99). The zero temperature effective potential for this case is given by

$$V_{\text{eff}}^\infty(x) = -\tfrac{1}{2}\alpha x^2 + \tfrac{1}{4}\gamma x^4 + \tfrac{1}{2}\omega, \quad \omega = \sqrt{\frac{1}{m}(3\gamma x^2 - \alpha)}. \qquad (4.116)$$

The absolute minima of (4.116) are easily found for the case that γ is small. The result is

$$\bar{x} \approx \pm\left(\sqrt{\frac{\alpha}{\gamma}} - \frac{3}{4\alpha}\sqrt{\frac{\gamma}{2m}}\right), \qquad (4.117)$$

while $\bar{x} = 0$ corresponds to a local maximum. Result (4.117) shows the existence of two energetically degenerate ground states for the system, each of which is associated with the frequency

$$\omega \approx \sqrt{\frac{2\alpha}{m}}. \qquad (4.118)$$

Result (4.117) is *nonperturbative* since the inverse of the coupling strength γ appears in the expression for $\bar{x}$. Perturbation theory yields expressions that are not singular as $\gamma \to 0$ due to the assumption that the actual solution can be expanded in a power series in γ.

It is interesting to examine the effective potential for the case that $x \approx 0$ to determine the behavior in β. For a wide range of values for the constants the effective potential of (4.114) becomes

$$V_{\text{eff}}(x) \approx \left(\frac{3\gamma}{4\beta\alpha} - \frac{\alpha}{2}\right)x^2 + \tfrac{1}{4}\gamma x^4. \qquad (4.119)$$

Exercise 4.16: Find the range of values for the constants for which (4.119) is valid.

Result (4.119) shows that $\bar{x} = 0$ becomes the absolute minimum of the theory for the *critical temperature* $T_c = 2\alpha^2/3\gamma$. For temperatures greater than T_c the average thermal energy is sufficient to allow the particle to overcome the barrier at $x = 0$, and the oscillations become symmetric about the origin.

The fact that the system has two ground states available to it, representing oscillations about the left or right well, can be examined in the context of the coherent state formalism of Sec. 4.4, and this will provide considerable insight into the problem. The results of Sec. 4.4 give the harmonic oscillator transition element between coherent states as

$$Z_{ab} = e^{i\frac{1}{2}(q_a p_a - q_b p_b)} \int_{p_a}^{p_b} \mathcal{D}p \int_{q_a}^{q_b} \mathcal{D}q \, \exp\left\{i \int_{t_a}^{t_b} dt \left[p\dot{q} - \frac{p^2}{2m} - \tfrac{1}{2}m\omega^2 q^2\right]\right\}, \qquad (4.120)$$

where the measure is given by

$$\mathcal{D}q = \lim_{N\to\infty} \prod_{j=1}^{N-1} dq_j, \quad \mathcal{D}p = \lim_{N\to\infty} \prod_{j=1}^{N-1} \frac{dp_j}{2\pi}. \qquad (4.121)$$

Sec. 4.5 Spontaneously Broken Symmetry 103

It will now be *assumed* that a form similar to (4.120) is valid for the potential (4.99). In order for this to have a chance of being legitimate, it must be possible to choose an initial and final coherent state such that stable oscillations are defined by the path integral. Such a constraint on the initial and final coherent states is known as a *self-consistency condition*, and it will determine the structure of the ground state in terms of a harmonic oscillator approximation. For the moment it will be assumed that $p_a = p_b = 0$. After denoting $q_a = q_b = q_0$ and integrating the momenta, the transition amplitude becomes

$$Z_{00} = \int_{q_0}^{q_0} \overline{\mathcal{D}}q \, \exp\left\{ i \int_{t_a}^{t_b} dt \, \left[\tfrac{1}{2} m \dot{q}^2 + \tfrac{1}{2} \alpha q^2 - \tfrac{1}{4} \gamma q^4 \right] \right\} . \qquad (4.122)$$

The limits on this transition amplitude may be incorporated into the action by translating the variables of integration by q_0. The result is

$$\begin{aligned} Z_{00} &= \int_0^0 \overline{\mathcal{D}}q \, \exp\left\{ i \int_{t_a}^{t_b} dt \, \left[\tfrac{1}{2} m \dot{q}^2 + \tfrac{1}{2} q_0^2 - \tfrac{1}{4} \gamma q_0^4 \right. \right. \\ &\quad \left. \left. + \left(\gamma q_0^3 - \alpha q_0 \right) q - \tfrac{1}{2} \left(3 \gamma q_0^2 - \alpha \right) q^2 - \gamma q_0 q^3 - \tfrac{1}{4} \gamma q^4 \right] \right\} . \qquad (4.123) \end{aligned}$$

As a minimal set of criteria for stable oscillations the term linear in q must vanish and the term quadratic must have a negative coefficient. The first condition reflects the fact that a linear term represents a source and therefore cannot be present for a system in a state of equilibrium. The second condition avoids the calamity of imaginary frequencies that represent exponentially damped or unstable behavior. These two conditions immediately yield $q_0 = \pm\sqrt{\alpha/\gamma}$, reproducing the dominant part of (4.117). It is also obvious that the action appearing in (4.123) no longer possesses the symmetry $q \to -q$ by virtue of the presence of a cubic term. Thus, this system has undergone *spontaneous breakdown of symmetry*. At the critical temperature determined earlier the oscillations become symmetric about the origin and the symmetry is restored.

In the approach that led to (4.123), it is evident that the mechanism for breakdown of symmetry is the fact that the ground state of the system causes the operator Q to develop a nontrivial expectation value q_0, even though the potential is symmetric. In fact, it is even possible to write down explicitly a first-order approximation to this ground state in terms of the harmonic oscillator coherent states of the previous section. This approximate ground state can be used to derive the form (4.123) from a perturbative representation of the transition element.

Exercise 4.17: Starting from a coherent state and using standard quantum mechanical perturbation theory, derive the form (4.123) and the criteria for stable oscillations.

This result demonstrates a very important idea. The values of q_0 determined by this latter method are nothing more than the constant solutions to the *classical* equations of motion obtained by demanding that

$$\left.\frac{\partial V(q)}{\partial q}\right|_{q=q_0} = 0, \quad \left.\frac{\partial^2 V(q)}{\partial q^2}\right|_{q=q_0} > 0, \qquad (4.124)$$

where $V(q)$ is the potential in the original untranslated action. These solutions give a reasonably good first-order approximation to the ground state structure derived by a more careful analysis of the effective potential, which takes into account quantum corrections. This is an extension of the *saddle-point* evaluation technique first introduced in Sec. 1.5 to the nonlinear and non-Gaussian case. However, nonlinearity precludes making the claim that the classical solutions to the equations of motion contain all the quantum behavior, as was the case for quadratic actions. Nevertheless, the results obtained here strongly indicate that, as a first-order approximation, translating the variables of the action by a classical solution with the appropriate boundary conditions continues to be a valuable technique, even in a nonlinear quantum system.

An important demonstration of this idea lies in the following problem. As the system is cooled to its critical temperature, the nontrivial ground states become available, and the system must transit from the $q_0 = 0$ vacuum to one of the $q_0 \neq 0$ vacuum. Because of the degeneracy in energy, either vacuum is satisfactory. The system may be pushed into one or the other of the available ground states by applying an external force in the form of a source term in the action. This is similar to the alignment of a ferromagnet by applying a small external magnetic field as the system is cooled through its critical temperature. There are subtleties in this argument that involve positive definiteness of the free energy, or *convexity* of the effective potential and that induce singular behavior in the effective action at $J = 0$. A discussion of these problems lies outside the scope of this section [15, 16]. However, it is natural to question if the system may transit between the two different possible vacuums. The answer is easily formulated using the path integral. Such a transition element, describing a transition from the left vacuum to the right vacuum, is given by

$$Z_{L \to R} = \int_{-q_0}^{q_0} \overline{\mathcal{D}}q \, \exp\left\{i \int_{t_a}^{t_b} dt \, \left[\tfrac{1}{2}m\dot{q}^2 + \tfrac{1}{2}\alpha q^2 - \tfrac{1}{4}\gamma q^4\right]\right\}. \qquad (4.125)$$

> **Exercise 4.18**: Evaluate (4.125) and relate the result to double well oscillations familiar from standard quantum mechanics.

This problem is also related to the transition element from an unstable ground state to a stable ground state, and this is discussed in Chapter 9.

4.6 Constraints

In this section the implementation of constraints within the path integral formalism is discussed. The study of constraints in quantum mechanics is subtle and significant. It will be seen that, with some important adjustments to the measure [17], the method of Lagrange multipliers familiar from classical mechanics can be used in the path integral.

In a classical mechanical system whose phase space consists of $2n$ degrees of freedom, $\{p_1, q_1, \ldots, p_n, q_n\}$, a constraint consists of some relation between a subset of the coordinates. For simplicity, it will be assumed that this constraint has no explicit time dependence, i.e., it is a *scleronomous* constraint. Such a set of r constraints can be written in terms of a set of functions of the coordinates,

$$f_i(p,q) = 0, \quad \frac{\partial f^i(p,q)}{\partial t} = 0, \quad i = 1, \ldots, r, \qquad (4.126)$$

or, in variational form,

$$\delta f_i(p,q) = \frac{\partial f_i(p,q)}{\partial q_j} \delta q_j + \frac{\partial f_i(p,q)}{\partial p_j} \delta p_j = 0. \qquad (4.127)$$

Each constraint defines a hypersurface in the $2n$-dimensional phase space $\{p, q\}$. It is assumed that the r hypersurfaces possess a nontrivial intersection that constitutes the reduced phase space available to the constrained system; otherwise, the system of constraints cannot be realized.

Because there are constraints, the $2n$ variables are no longer independent, and there must be forces present that are not *a priori* known in order to create the constraint. The Euler–Lagrange equations of the form (1.41) or Hamilton's equations (1.47) are no longer valid since they were derived under the assumption that all coordinates have independent variations, and this is no longer true by virtue of (4.127).

At the classical level the solution [6] is to introduce a Lagrange multiplier λ_i for each constraint, and the Lagrangian density is modified to read

$$\mathcal{L}(p,q) \to \mathcal{L}(p,q) + \sum_{i=1}^{r} \lambda_i f_i(p,q) \,. \tag{4.128}$$

The modified action is entirely equivalent to the original action for the case that the constraints are satisfied. The presence of the Lagrange multipliers allows the modified action to be varied, and the multipliers are chosen so that the variations of the $2n$ original coordinates are once again independent. This results in $2n$ modified Hamilton equations:

$$\begin{aligned}\dot{q}_j &= \frac{\partial H(p,q)}{\partial p_j} - \sum_{i=1}^{r} \lambda_i \frac{\partial f_i(p,q)}{\partial p_j} \,, \\ \dot{p}_j &= -\frac{\partial H(p,q)}{\partial q_j} + \sum_{i=1}^{r} \lambda_i \frac{\partial f_i(p,q)}{\partial q_j} \,.\end{aligned} \tag{4.129}$$

The $2n$ dynamical equations of (4.129), coupled with the r constraint equations (4.126), allow a solution for all $2n + r$ unknowns.

The constraint surface $f_i(p,q) = 0$ must also be consistent with the time development of the system. This means that time derivatives of the constraint must also vanish when evaluated on the hypersurface defined by the original or *primary* constraints. Anticipating the need to define a quantum mechanical form for constraints, this will be cast into the language of Poisson brackets. The Poisson bracket of each primary constraint with the Hamiltonian is evaluated, and this yields a function of the p and q, denoted

$$\{f_i, H\}_{p,q} = g_i(p,q) \,. \tag{4.130}$$

If g_i fails to vanish on the hypersurface defined by the initial constraints, then it constitutes an additional or *secondary* constraint. The Poisson brackets of H with these secondary constraints are now found and evaluated on the constraint hypersurface defined by both the primary constraints, f_i, and the secondary constraints, g_i, possibly yielding yet more secondary constraints. It will be assumed that this process eventually terminates, and that it is possible to find a hypersurface in phase space where all constraints are satisfied. At the classical level it is standard practice to enforce the secondary constraints by hand when necessary, so that there is usually no distinction made between primary and secondary constraints. However, the quantum mechanical picture is quite different.

Even at the classical level there is an additional subtlety, and that is the possibility that the Poisson brackets involving pairs of constraints may not vanish on the constraint hypersurface so far determined. It is very often the case that these Poisson brackets yield constants and therefore cannot simply be added to the list of secondary constraints. The solution to this dilemma is to use Dirac's implementation of constraints [18]. The constraints, including both original and secondary, are broken into two groups: first-class constraints, ϕ_i, and second-class constraints, θ_j. The first-class constraints have vanishing Poisson brackets with all other constraints, so that

$$\{\phi_i, \phi_j\}_{p,q} = \{\phi_i, \theta_j\}_{p,q} = 0, \quad \forall i,j, \tag{4.131}$$

while the second-class constraints are all those constraints that do not have this property. The Poisson brackets of the second-class constraints are denoted

$$\{\theta_i, \theta_j\}_{p,q} = t_{ij} \tag{4.132}$$

and may be viewed as the elements of a matrix t that is antisymmetric by virtue of the antisymmetry of the Poisson brackets. It will be assumed that this matrix possesses an inverse, denoted c, which will also be antisymmetric. The elements of this matrix are denoted c_{ij}. The Dirac brackets of two quantities A and B are defined in terms of the Poisson brackets as

$$\{A, B\}_D = \{A, B\}_{p,q} - c_{ij}\{A, \theta_i\}_{p,q}\{\theta_j, B\}_{p,q}. \tag{4.133}$$

Exercise 4.19: Prove that all Dirac brackets between pairs of constraints vanish, as well as all those between the constraints and the Hamiltonian.

The Dirac bracket possesses the same general properties as the Poisson bracket. It is therefore understood that *all* Poisson brackets are to be replaced by Dirac brackets in the Poisson bracket formulation of mechanics, and this yields a mechanics consistent with the reduced phase space of the constrained system.

Constraints are more difficult to implement consistently in quantum mechanical systems. This is because the $2n$ coordinates become operators, and these operators possess nontrivial commutation relations among each other. By substituting operators for coordinates in the constraint it becomes apparent that ordering ambiguities will be present unless all the operators in the constraint equation commute among themselves. For that reason this section will restrict attention at the quantum mechanical level to operator constraints of the form

$$f_i(P_1, \ldots, P_n) \equiv g_i(P) = 0, \quad f_i(Q_1, \ldots, Q_n) \equiv f_i(Q) = 0, \tag{4.134}$$

and for simplicity, it will be assumed that there is only one type of initial constraint, either a Q-type or a P-type constraint.

The statement (4.134), as written, cannot be understood in the same manner that classical constraints are. This is because operators are defined only by their action on some Hilbert space. As a result, the constraints of (4.134) can be true only on a subspace of the total Hilbert space, referred to as the *physical subspace*, which is the analog of the constraint hypersurface determined for the classical system. If $|\psi\rangle_P$ is a state entirely contained in the physical subspace, then the constraints are realized if

$$f_i(Q)|\psi\rangle_P = 0 \,. \tag{4.135}$$

The orthogonal complement of the physical subspace is, for obvious reasons, called the *unphysical subspace*. A realization of the constraints in the manner (4.135) is said to be a *strong* implementation. It is possible to realize the constraints in a weaker manner by finding the set of states such that

$$_P\langle\psi'|f_i(Q)|\psi\rangle_P = 0 \,, \tag{4.136}$$

and therefore (4.136) is known as the *weak* implementation. The weak implementation is possible since only the matrix elements of observables are physically relevant.

Exercise 4.20: Show that the physical subspace determined by the weak implementation is always larger or equal to the physical subspace determined by the strong implementation.

However, once the physical subspace is determined, a second difficulty may arise, and that is the quantum mechanical analog of the Poisson bracket problem. The time derivatives of the constraints are generated by the Hamiltonian, H, so that

$$\frac{\partial f_j(Q)}{\partial t} = i[H, f_j(Q)] \equiv g_j(P,Q) \tag{4.137}$$

becomes another constraint. If it is true that

$$_P\langle\psi'|g_j(P,Q)|\psi\rangle_P \neq 0 \,, \tag{4.138}$$

then the original physical subspace is not compatible with the dynamics of the system. In effect, a nonzero form for (4.138) indicates that the dynamics of the system are such that a state initially in the physical subspace will evolve in time so that it develops a component in the unphysical subspace, thereby violating the constraint. If this is the case, then the physical subspace must be amended to accommodate this secondary constraint. Higher-order commutators need to be considered as well. Therefore, as in the classical case, the primary constraints of (4.135) may give rise to a hierarchy of secondary constraints, and the question of their compatibility can be answered only by a detailed study of the specific case. In what follows it will be assumed that the hierarchy of constraints terminates in a finite number of secondary constraints, and that a physical subspace can be found such that all constraints are satisfied.

It may be that the commutators of the constraints do not vanish in the physical subspace, and this is the quantum mechanical analog of the Poisson bracket problem solved by Dirac. As before, the constraints are broken into two groups: first-class constraints, ϕ_i, and second-class constraints, θ_j. The first-class constraints possess vanishing commutators with all constraints, so that

$$[\phi_i, \phi_j] = [\phi_i, \theta_j] = 0 , \qquad (4.139)$$

while the second-class constraints are all the remaining constraints. The commutators of the second-class constraints are denoted

$$[\theta_i, \theta_j] = t_{ij} , \qquad (4.140)$$

and the inverse c of the antisymmetric matrix t is assumed to exist. The Dirac commutator [18] of two observables A and B is then defined in terms of the standard commutators as

$$[A, B]_D = [A, B] - c_{ij}[A, \theta_i][\theta_j, B] . \qquad (4.141)$$

Exercise 4.21: Show that the Dirac commutators of all constraints vanish, as well as those of the constraints and the Hamiltonian, when evaluated in the physical subspace.

The quantum mechanics of the constrained system is obtained by restricting consideration to the physical subspace and by replacing all commutators with Dirac commutators.

As an example, the quantum mechanical motion of a free particle in a plane is described in polar coordinates by the Hamiltonian

$$H = \frac{P_r^2}{2m} + \frac{P_\theta^2}{2mR^2} , \qquad (4.142)$$

110   Further Applications

along with the commutators

$$[R, P_r] = i, \quad [\theta, P_\theta] = i. \tag{4.143}$$

The constraint $R - a = 0$, forcing the particle to stay on a circle, does not commute with H, rather $[R - a, H] = iP_r/m$. The hierarchy terminates at this point since the secondary constraint commutes with the Hamiltonian. The physical subspace is identified as the one where the constraints $\phi_1 = P_r$ and $\phi_2 = R - a$ are satisfied. However, both of these constraints are second class since

$$[\phi_1, \phi_2] = -i. \tag{4.144}$$

The t matrix therefore takes the form

$$t = \begin{pmatrix} 0 & -i \\ i & 0 \end{pmatrix}, \tag{4.145}$$

so that its inverse is given by

$$c = \begin{pmatrix} 0 & -i \\ i & 0 \end{pmatrix}. \tag{4.146}$$

Therefore, for this constrained system the Dirac commutator is given by

$$[A, B]_D = [A, B] + i[A, P_r][R, B] - i[A, R][P_r, B]. \tag{4.147}$$

Constraints manifest themselves in the measure of the path integral for the constrained system. For the moment attention will be restricted to the case of one initial constraint of the form $f(Q)$. The general technique for implementing the constraint in the path integral is to construct a projection operator onto the physical subspace of the $|q\rangle$ states. Denoted P_P^q, the projection operator must be *idempotent*, so that $(P_P^q)^2 = P_P^q$. Its action on an arbitrary state $|\psi\rangle$ of the total Hilbert space is such that the state $|\psi\rangle_P \equiv P_P|\psi\rangle$ satisfies

$$_P\langle \psi' |f(Q)|\psi\rangle_P = 0. \tag{4.148}$$

The projection operator can be constructed from the eigenstates of the Q_j, which in the unconstrained form are written $|q_1, \ldots, q_n\rangle \equiv |q\rangle$. In terms of the eigenvalues the constraint equation can be written

$$f(q_1, \ldots, q_n) = 0, \tag{4.149}$$

and it is assumed to possess a root of the form

$$q_n = r(q_1, \ldots, q_{n-1}) \equiv r(q_\perp), \tag{4.150}$$

where, for notational simplicity, it is assumed that it is q_n that can be found, and $q_\perp$ denotes that the solution lies in the subspace of the q_i that is orthogonal to the constrained variable q_n. The projection operator is then defined as

$$P_P^q = N_q \int d^n q \, \delta(q_n - r(q_1, \ldots, q_{n-1})) \, |q\rangle\langle q| \,. \tag{4.151}$$

Exercise 4.22: Verify that (4.151) is idempotent if $N_q^{-1} = \delta(0)$, where $\delta(0)$ is the Dirac delta $\delta(q)$ evaluated at $q = 0$.

From the definition of the Dirac delta (1.16), it is apparent that the normalization factor N_q^{-1} is $V_p/2\pi$, where V_p is the volume of the momentum space canonically conjugate to the constrained coordinate.

Assuming that q_n possesses a canonical momentum p_n, it is possible to write the projection operator in terms of momentum eigenstates as well. It is not difficult to show by direct evaluation that the projection operator, given by

$$P_P^p = N_p \int \frac{d^n p}{(2\pi)^n} \, \delta(p_n) |p\rangle\langle p| \,, \tag{4.152}$$

has the property

$$P_P^p P_P^q P_P^p = P_P^p \,, \tag{4.153}$$

if the normalization $N_p = 2\pi\delta(0)$ is chosen. The operator P_P^p is *not* idempotent. Instead, it satisfies $P_P^p P_P^p = c P_P^p$, where c is proportional to the total volume of the phase space associated with q_n and p_n. However, it follows from (4.153) that the projection operator $P_P^q P_P^p$ is idempotent and can therefore be used to partition the time interval of the transition element. The presence of the delta function, $\delta(p_n)$, in (4.152) is directly related to the necessity of satisfying the secondary constraint generated by the Hamiltonian commutator. In order that the system stay in the physical subspace that is orthogonal to q_n, it is necessary for the canonical momentum p_n to vanish.

It is straightforward to repeat the steps of Sec. 2.2 and find the form of the path integral representation of the constrained system. This is accomplished by using the product of the two projection operators, (4.151) and (4.152), to define the infinitesimal intermediate matrix elements. A repetition of the steps used in Sec. 2.2 gives

$$Z_{ab} = \int_{q_a}^{q_b} \mathcal{D}^c p \, \mathcal{D}^c q \, \exp\left\{ i \int_{t_a}^{t_b} dt \, \mathcal{L}(p, q) \right\} \,, \tag{4.154}$$

Further Applications

where the constrained measure is given by

$$\mathcal{D}^c p \, \mathcal{D}^c q = 2\pi\delta(0) \lim_{N\to\infty} \prod_{j=0}^{N-1} \frac{d^n p_j}{(2\pi)^{n-1}} \delta(p_{nj})$$

$$\times \prod_{i=1}^{N-1} d^n q_i \, \delta\big(q_{ni} - r(q_{1i}, \ldots, q_{(n-1)i})\big) . \quad (4.155)$$

All but one of the $\delta(0)$ factors have been cancelled in the product. The final step is to note that the Dirac delta in q_n may be rewritten as

$$\delta\big(q_{nj} - r(q_{1j}, \ldots, q_{(n-1)j})\big) =$$
$$\frac{\partial f(q_j)}{\partial q_{nj}} \delta(f(q_j)) = \frac{\epsilon}{2\pi} \frac{\partial f(q_j)}{\partial q_{nj}} \int_{-\infty}^{\infty} d\lambda_j \, \exp\left[i\epsilon\lambda_j f(q_j)\right] , \quad (4.156)$$

where $\epsilon = (t_b - t_a)/N$, and properties (1.3) and (1.16) of the Dirac delta have been used. Assembling all the results back into the path integral gives the form

$$Z_{ab} = \int_{q_a}^{q_b} \mathcal{D}p \, \mathcal{D}q \, \mathcal{D}\lambda \, \exp\left\{ i \int_{t_b}^{t_a} dt \, \left[\mathcal{L}(p,q) + \lambda f(q)\right]\right\} , \quad (4.157)$$

where $\mathcal{D}p$ and $\mathcal{D}q$ constitute the original unconstrained measure defined by (2.46), and the constraint measure $\mathcal{D}\lambda$ is given by

$$\mathcal{D}\lambda = 2\pi\delta(0) \lim_{N\to\infty} \delta(p_{n0}) \prod_{j=1}^{N-1} \frac{\epsilon}{2\pi} \delta(p_{nj}) \frac{\partial f(q_j)}{\partial q_{nj}} \, d\lambda_j . \quad (4.158)$$

It is apparent from (4.157) that the modified form of the Lagrangian (4.128) has reappeared in the action. The Lagrange multipliers now appear in the path integral measure much as if they were additional coordinates with no conjugate momenta.

It is crucial to generalize this result to the case of many constraints. In particular, it is important to deal with a special type of constraint, one where the system of equations that determines the $\dot{q}_i$ in terms of the p_i and q_i is underdetermined. Such a case leads immediately to a constraint on the phase space, and this generic type of constraint is known as a *gauge* constraint for reasons that will become clear in Chapter 7. In its simplest form this occurs if the Lagrangian is independent of the subset of velocities $\{\dot{q}_i\}$, where $i = 1, \ldots, M$. If this simple version is not true, it may still be possible to perform a canonical transformation to a set of variables such that the transformed Lagrangian is independent of some subset of the new velocities. If so, it follows immediately that

$$p_i = \frac{\partial \mathcal{L}}{\partial \dot{q}_i} = 0, \quad i = 1, \ldots, M, \qquad (4.159)$$

and this is understood as a set of M primary constraints on the phase space of the system.

Since the action is assumed to contain some reference to the q_i involved in determining the constraints of (4.159), it follows that there will be M secondary constraints. These are determined at the classical level by the Poisson bracket, which gives

$$\{p_i, H\}_{p,q} = -\frac{\partial H}{\partial q_i} = G_i(p, q). \qquad (4.160)$$

It will be assumed that all the constraints so determined are first class, and that there are no further secondary constraints.

In order to accommodate these constraints in the measure of the path integral [19] it is necessary to solve the classical constraint equations, since these determine the physical subspace of the path integral measure. It is assumed, therefore, that the G_i are solvable functions of the q_i in the sense that each possesses a *unique* root for the q_i that is canonically conjugate to the constrained p_i. If this is true, then the solutions of the $G_i = 0$ constraints will form a set of M roots, denoted $q_i = r_{i\perp}$, since each root must lie in the part of phase space orthogonal to the constrained set $\{q_i\}$. The result is simply M copies of the single constraint problem. If there is not a unique solution to the constraint equation, an ambiguity is induced into the quantized system, and it is assumed that the path integral represents a perturbative expansion around one of the possible solutions [17]. For the unique case, the constraint measure is therefore given by

$$\mathcal{D}^c p \mathcal{D}^c q = [2\pi\delta(0)]^M \lim_{N \to \infty} \prod_{j=0}^{N-1} \prod_{i=1}^{N-1} \frac{d^n p_j}{(2\pi)^{n-M}} d^n q_i$$

$$\times \prod_{m=1}^{M} \delta(p_{mj}) \delta(q_{mi} - r_{mi\perp}). \qquad (4.161)$$

The subscripts i and j appearing on the variables in (4.161) refer to the index associated with the time of the variables of integration. The product of Dirac deltas appearing in (4.161) can be rewritten by using the multidimensional extension of result (1.4),

$$\prod_{i=1}^{N-1} \prod_{m=1}^{M} \delta(q_{mi} - r_{mi\perp}) = \det\left|\frac{\delta G}{\delta q}\right| \prod_{i=1}^{N-1} \prod_{m=1}^{M} \delta(G_{mi}), \qquad (4.162)$$

Further Applications

where $\delta G_{im}/\delta q_{jk}$ is the $(im)(jk)$th element of the $M(N-1) \times M(N-1)$ matrix appearing in the determinant.

> **Exercise 4.23**: Prove (4.162) by using an integral form for the product of Dirac deltas and performing a suitable change of variables that generates the determinant as the Jacobian of the transformation.

Result (4.162) may be reexpressed by using the Poisson bracket to write

$$\frac{\partial G_{im}}{\partial q_{jk}} = \{p_{jk}, G_{im}\}_{p_j,q_j} . \quad (4.163)$$

This form allows a canonical transformation from the p_j to a new set of coordinates $\chi_j(q,p)$ and their canonical conjugate coordinates, $Q_j(q,p)$. The total volume of phase space is, by definition, invariant under this transformation, so that

$$\prod_{i=1}^{N-1} \prod_{m=1}^{M} d\chi_{im} \, dQ_{im} = \prod_{i=1}^{N-1} \prod_{m=1}^{M} dp_{im} \, dq_{im} . \quad (4.164)$$

The Poisson bracket is invariant as well, so that

$$\{p_{im}, G_{jk}\}_{p_i,q_i} = \{\chi_{im}, G_{jk}\}_{p_i,q_i} . \quad (4.165)$$

It is necessary to add the time derivative of the generating functional of the canonical transformation to the action, but from the discussion of (4.47) it follows that this merely contributes a phase to the incoming and outgoing states since it may be integrated exactly. Putting these results together gives the final constraint measure,

$$\mathcal{D}^c p \, \mathcal{D}^c q = [2\pi\delta(0)]^M \lim_{N \to \infty} \prod_{j=0}^{N-1} \prod_{i=1}^{N-1} \frac{d^n p_j}{(2\pi)^{n-M}} \, d^n q_i$$

$$\times \det\{\chi, G\} \prod_{m=1}^{M} \delta(\chi_{jm}) \, \delta(G_{jm}) . \quad (4.166)$$

The constraints can be incorporated into the action by introducing suitable Lagrange multipliers, but this will not be written explicitly. The canonically transformed momentum χ, which must vanish, is referred to as a *gauge condition*. This nomenclature has its origin in gauge field theory, where the absence of a canonical momentum for some coordinates is related to the gauge symmetry of the Lagrangian. The constraint measure for this system is characterized by the appearance of the determinant of the Poisson bracket for the primary and secondary constraints. The presence of two second-class constraints, θ_1 and θ_2, requires an additional modification of the measure proportional to $\|\det\{\theta_1, \theta_2\}\|^{1/2}$, but this will not be derived here [20].

> **Exercise 4.24**: Derive the topological measure of Sec. 3.4 for the case of a free particle moving in a circle by implementing the constraint $r = a$ on a free particle moving in a plane.

References

[1] See, for example, J.D. Jackson, *Classical Electrodynamics, Second Edition*, Wiley, New York, 1975.

[2] R.P. Feynman, *Statistical Mechanics*, Benjamin, Reading, Mass., 1972.

[3] F.W. Wiegel, Phys. Rep. **16**, 57 (1975).

[4] K. Huang, *Statistical Mechanics*, Wiley, New York, 1963.

[5] R. Kubo, J. Phys. Soc. Japan **12**, 570 (1957); P.C. Martin and J. Schwinger, Phys. Rev. **115**, 1342 (1959).

[6] See, for example, H. Goldstein, *Classical Mechanics, Second Edition*, Wiley, New York, 1983.

[7] This presentation follows that of C. Itzykson and J.-B. Zuber, *Quantum Field Theory*, McGraw-Hill, New York, 1980.

[8] A discussion of quantum mechanical symmetries may be found in A. Messiah, *Quantum Mechanics*, Vol. 2, Wiley, New York, 1966.

[9] T.D. Lee and C.N. Yang, Phys. Rev. **128**, 885 (1962).

[10] The derivation of the symmetry-induced identities presented here is adapted from B. Zumino in *Lectures on Elementary Particles and Quantum Field Theory, 1970 Brandeis University Summer Institute in Theoretical Physics*, ed. S. Deser, M. Grisaru, and H. Pendleton, MIT Press, Cambridge, Massachusetts, 1970.

[11] The abstract mathematical properties of algebras can be found in T. Hungerford, *Algebra*, Holt, Rinehart, and Winston, New York, 1973.

[12] The use of coherent states was introduced into field theory and quantum optics in R.J. Glauber, Phys. Rev. **130**, 2529 (1963), although their use in quantum mechanics predates that. A general review of coherent state properties is found in W.-M. Zhang, D.H. Feng, and R. Gilmore, Rev. Mod. Phys. **62**, 867 (1991). See also [13].

[13] The use of coherent states to formulate the path integral, sometimes referred to as the *holomorphic representation*, is predominately the work of Klauder; see for example J.R. Klauder, Ann. Phys. **11**, 123 (1960); J.R. Klauder, Phys. Rev. D**19**, 2349 (1979); J.R. Klauder and B.S. Skagerstam, *Coherent States, Applications in Physics and Mathematical Physics*, World Scientific, Singapore, 1985.

[14] For an early review of this technique see J. Iliopoulos, C. Itzykson, and A. Martin, Rev. Mod. Phys. **47**, 165 (1975).

[15] R.J. Rivers, *Path Integral Methods in Quantum Field Theory*, Cambridge University Press, New York, 1987.

[16] The relation of the degeneracy of the ground states to the singular nature of the effective potential at $J = 0$ is discussed in S. Norimatsu, K. Yamamoto, and A. Tanaka, Phys. Rev. D**35**, 2009 (1987).

[17] An excellent review of the implementation of constraints in both classical and quantum systems is given by K. Sundermeyer, *Constrained Dynamics*, Springer-Verlag, Berlin, 1983.

[18] P.A.M. Dirac, Proc. Roy. Soc. A**246**, 326 (1958); P.A.M. Dirac, *Lectures on Quantum Mechanics*, Belfer Graduate School of Science, Yeshiva University, New York, 1964.

[19] The implementation of first-class constraints in the path integral measure was first derived by L.D. Faddeev, Teoret. i Mat. Fiz. **1**, 3 (1969), translated in Theoret. Math. Phys. **1**, 1 (1970).

[20] The measure consistent with second-class constraints is derived in P. Senjanovic, Ann. Phys. **100**, 227 (1976).

Chapter 5

Grassmann Variables

In all the results obtained so far standard analytic techniques of real and complex variables have been employed. It has been assumed without discussion that these variables commute with each other, or that, as in the case of (2.1), it is the *commutator* of two operators that is nonzero. Numbers of this kind will hereafter be referred to as c-number variables, while such operators will be classified as *bosonic*. However, there are naturally occurring systems that require a different kind of variable. These variables anticommute[1] and are referred to as Grassmann [1] or a-number variables, while the associated operators are classified as *fermionic*. In this chapter the general properties of these variables will be presented. Since many of these properties are counterintuitive, it is important to define and implement the formal manipulations of these variables carefully. These a-number variables will be used to construct an anticommuting form of classical and quantum mechanics, and eventually to evaluate the path integral representation of a transition amplitude in Grassmann quantum mechanics as presented in the book by DeWitt [2].

The need for Grassmann variables is related, in the long run, to the mathematical problem of writing a second-order differential operator as the product of two first-order differential operators. Grassmann variables become necessary when such an operator is to be represented in a path integral or a functional, and it was in the context of a functional formulation of field theory that they were first introduced by Schwinger [3] and used later by Berezin [4]. This will be discussed again in the next chapter when the path integral associated with fermionic fields is developed,

---

[1] It is possible to have variables that neither commute nor anticommute, but these will not be considered here.

but simple examples of this will be given later in this section for several quantum mechanical Grassmann systems. Grassmann variables also play a central role in BRST invariance in gauge field theories and in formulations of supersymmetry, both of which are discussed in later sections.

In Sec. 5.1 the basic properties of real and complex Grassmann variables are introduced, and differentiation and integration are defined. In Sec. 5.2 Gaussian integrals of Grassmann variables are evaluated for both the real and complex case. In Sec. 5.3 classical mechanics is developed for the case of Grassmann variables, and the equivalent forms for the Euler-Lagrange equation and Hamilton's equations are presented. In Sec. 5.4 quantum mechanics for operators with a-number eigenvalues or anticommutation relations is developed. In Sec. 5.5 path integral representations of several Grassmann quantum systems are developed and evaluated. Section 5.5 also introduces the fermionic coherent state, which will be of great importance in Chapter 6, where the path integral for quantum fields is derived. The chapter closes with Sec. 5.6, where the rudiments of supersymmetric quantum mechanics are briefly discussed.

5.1 Basic Definitions

If $\{\eta_a\}$, $a = 1, \ldots, N$, denotes a set of real Grassmann variables, then these variables satisfy the anticommutation relation

$$\eta_a \eta_b + \eta_b \eta_a \equiv \{\eta_a, \eta_b\} = 0 , \qquad (5.1)$$

for all possible a and b. It immediately follows from the anticommutation relation that $\eta_a^2 = 0$. However, a real or complex c-number variable x is assumed to commute with the Grassmann variables, giving $x\eta_a = \eta_a x$. These Grassmann variables are real in the sense that the operation of complex conjugation leaves them invariant, so that $\eta_a^* = \eta_a$. It is not possible to visualize the Grassmann variables as performing either the ordinal or quantifying role associated with standard real numbers. Instead, the Grassmann variables should be viewed as a means to identify a particular algebra, the concept of which was introduced in Sec. 4.3. In this sense it should be noted that the real number zero is both a c-number and a Grassmann number, since it anticommutes with all other Grassmann variables as well as itself. The definition (5.1) of Grassmann variables can therefore be viewed as the multiplication rules of the Grassmann algebra.

For a system of N real Grassmann variables it is possible to write an arbitrary product of N Grassmann variables as

$$\eta_{a_1} \eta_{a_2} \cdots \eta_{a_N} = \varepsilon^{a_1 a_2 \cdots a_N} \eta_1 \eta_2 \cdots \eta_N , \qquad (5.2)$$

Sec. 5.1 Basic Definitions

where $\varepsilon^{abc\cdots}$ is the Levi–Civita symbol of order N defined in (1.54).

> **Exercise 5.1:** Verify (5.2).

It is possible to consider functions of Grassmann variables, in particular the polynomials. It is straightforward to see that any polynomial in Grassmann variables must be of finite order for a finite number of Grassmann variables. This follows directly from the fact that $\eta_a^2 = 0$. For example, if $f(x,\eta)$ is a function of the c-number variable x and a polynomial in the single real Grassmann variable η, then it must have the form $f(x,\eta) = f_1(x) + f_2(x)\eta$. Because of the connection such functions have to supersymmetry, they are sometimes referred to as superfunctions, and the combination of x and η is known as superspace.

> **Exercise 5.2:** Show that a general function $f(x,\eta_a)$ for N real Grassmann variables has 2^N terms in its polynomial expansion.

A general polynomial function of N real Grassmann variables will have 2^{N-1} terms that are a product of an even number of Grassmann variables, and 2^{N-1} terms that are a product of an odd number of Grassmann variables.

> **Exercise 5.3:** Show that a product of an odd number of Grassmann variables is another Grassmann variable, i.e., satisfies the properties (5.1) of a Grassmann variable, while a product of an even number of Grassmann variables commutes with all other variables.

These results show that a general polynomial of Grassmann variables has equal degrees of freedom in both Grassmann and standard variables. This is a central feature of supersymmetry, where the number of anticommuting degrees of freedom is constrained to match the number of commuting degrees of freedom.

Any c-number function that possesses a Taylor series expansion around zero can be adapted to a-number variables by substituting the a-number variable into the power series. For example, the exponential of a Grassmann variable η is given by

$$e^\eta = 1 + \eta , \qquad (5.3)$$

since all other higher-order terms vanish. This power series representation

Grassmann Variables

of the exponential may be used to prove the Baker–Campbell–Hausdorff rule for Grassmann variables. If η and ξ are two Grassmann variables, then it follows that

$$e^{\eta+\xi} = e^{\eta} e^{\xi} e^{-\frac{1}{2}[\eta,\xi]} \,. \tag{5.4}$$

Exercise 5.4: Verify (5.4).

A complex Grassmann variable, ζ, may be defined by combining two real Grassmann variables, η_R and η_I, in the form $\zeta = \eta_R + i\eta_I$, so that $\zeta^* = \eta_R - i\eta_I$. It follows directly from the Grassmann nature of η_R and η_I that $\zeta^2 = \zeta^{*2} = 0$. Since η_R and η_I are real Grassmann variables, the complex conjugate of ζ is given by $\zeta^* = \eta_R - i\eta_I$. The product of a complex Grassmann variable with its complex conjugate is therefore given by

$$\zeta^*\zeta = i(\eta_R\eta_I - \eta_I\eta_R) = i[\eta_R, \eta_I] \,. \tag{5.5}$$

If the product (5.5) is to be real, so that $(\zeta^*\zeta)^* = \zeta^*\zeta$, then the product of any two real Grassmann variables must satisfy

$$(\eta_a\eta_b)^* = \eta_b\eta_a = -\eta_a\eta_b \,. \tag{5.6}$$

Exercise 5.5: Extend result (5.6) to the case of a product of N complex Grassmann variables ζ_a to show that

$$(\zeta_a\zeta_b\cdots\zeta_c)^* = \zeta_c^*\cdots\zeta_b^*\zeta_a^* \,. \tag{5.7}$$

The operation of complex conjugation for Grassmann variables is then, in form, similar to the Hermitian adjoint operation on matrices.

Differentiation of products of Grassmann variables requires care since the order of the variables determines the sign of the product. Differentiation may act from the left or the right of a product, and the direction will determine the sign of the result. The left and right derivative are defined as

$$\frac{\partial \eta_b}{\partial \eta_a} = -\eta_a \frac{\overleftarrow{\partial}}{\partial \eta_b} = \delta_{ab} \,, \tag{5.8}$$

and it follows that

$$\frac{\partial}{\partial \eta_a}(\eta_b\eta_c) = \frac{\partial \eta_b}{\partial \eta_a}\eta_c - \frac{\partial \eta_c}{\partial \eta_a}\eta_b = \delta_{ab}\eta_c - \delta_{ac}\eta_b \,, \tag{5.9}$$

Sec. 5.1 Basic Definitions

yielding an anti-Leibniz property. The right derivative yields

$$(\eta_b \eta_c) \frac{\overleftarrow{\partial}}{\partial \eta_a} = \delta_{ac} \eta_b - \delta_{ab} \eta_c . \tag{5.10}$$

If p denotes the left position of the variable η_b in the product $\eta_a \cdots \eta_b \cdots \eta_c$, where $p = 0$ is the leftmost position, then the extension of (5.9) is

$$\frac{\partial}{\partial \eta_b}(\eta_a \cdots \eta_b \cdots \eta_c) = (-1)^p (\eta_a \cdots \eta_c) . \tag{5.11}$$

In what follows the left derivative is used exclusively, but the reader is reminded that other conventions are possible.

In order to define integration for Grassmann variables the properties of the differential of a Grassmann variable must be determined. For a system of N Grassmann variables the differential operator will be assumed to take the same form as its c-number counterpart,

$$d = d\eta_a \frac{\partial}{\partial \eta_a} . \tag{5.12}$$

This definition has the immediate consequence that

$$d(\eta_a \eta_b) = d\eta_a \, \eta_b - d\eta_b \, \eta_a . \tag{5.13}$$

The differential of a product is constrained to satisfy the Leibniz property,

$$d(\eta_a \eta_b) = d\eta_a \, \eta_b + \eta_a \, d\eta_b , \tag{5.14}$$

so that a comparison of (5.13) with (5.14) gives

$$\{\eta_a, d\eta_b\} = 0 , \tag{5.15}$$

demonstrating that $d\eta_b$ is itself a Grassmann variable.

For a single real Grassmann variable there are only two integrals of interest, $\int d\eta$ and $\int d\eta \, \eta$, since the Grassmann nature of η and $d\eta$ forces all others vanish. The Grassmann property of $d\eta$ shows that $\int d\eta$, assumed to be a definite integral, must result in a Grassmann variable. The integral $\int d\eta \, \eta$ commutes with all other Grassmann variables and therefore must result in a c-number. Because Grassmann numbers are neither ordinal nor quantal, the definition of a Grassmann measure similar to Jordan measure is unavailable and it is impossible to understand either of these integrals as a Riemann sum. Some additional criterion is therefore needed in order to define these integrals. It is recalled that the property of translational invariance is critical to evaluation of Gaussian integrals for c-number variables. Such a property will be valuable later and therefore the Grassmann integrals will be defined so that they are invariant under Grassmann translation. This means that the change of variable $\eta = \eta' + \xi$, where ξ is a constant a-number translation of the original variable of integration satisfying $d\xi = 0$, is assumed to leave the two integrals invariant. This assumption gives

$$\int d\eta' = \int d\eta ,$$
$$\int d\eta' \, \eta' = \int d\eta \, (\eta - \xi) = \xi \int d\eta + \int d\eta \, \eta . \qquad (5.16)$$

It follows from (5.16) that

$$\int d\eta = 0 . \qquad (5.17)$$

Since zero is a Grassmann number, result (5.17) is consistent with the algebra. Since $\int d\eta \, \eta$ commutes with all other Grassmann numbers it must be given by a c-number. The normalization of η will be chosen so that

$$\int d\eta \, \eta = -\int \eta \, d\eta = 1 . \qquad (5.18)$$

The actions of left and right differentiation yield identical results to left and right integration when applied to Grassmann variables.

For a real Grassmann variable the choice of the normalization in (5.18) has the consequence that

$$\left(\int d\eta \, \eta\right)^* = \int \eta^* \, d\eta^* = -\int d\eta^* \eta = 1 , \qquad (5.19)$$

where (5.18) and the Grassmann nature of $d\eta^*$ have been used. It follows that

$$d\eta^* = -d\eta \qquad (5.20)$$

for a real Grassmann variable.

The rules of integration may be used to find the Grassmann form of the Jacobian for a change of integration variables. The Jacobian is determined by demanding that any integration involving N Grassmann variables, η_a, is invariant under the change to a new set of Grassmann variables, θ_a. For simplicity, the change of variables will be assumed to be of the form

$$\eta_a = M_{ab} \theta_b , \qquad (5.21)$$

where the M_{ab} are elements of the invertible c-number matrix $\mathbf{M}$. The linearity and c-number nature of this transformation ensure that the θ

variables also satisfy a Grassmann algebra. The Jacobian, J_G, of the Grassmann measure associated with this transformation is determined by demanding that the integration result is invariant, so that

$$\int d\eta_1 \cdots d\eta_N \, f(\eta_1,\ldots,\eta_N) = \int d\theta_1 \cdots d\theta_N \, J_G \, f'(\theta_1,\ldots,\theta_N) , \qquad (5.22)$$

where f' is the function obtained by simply substituting the transformation (5.21) to the variables θ_a into f. The rules of integration (5.17) and (5.18) show that only the term in f proportional to $\eta_1 \cdots \eta_N$, i.e., the term containing all N of the Grassmann variables, contributes to the integral on the left of (5.22). As a result, the Jacobian of the measure must be such that it cancels the factors generated by the transformation (5.21). This means that

$$\eta_1 \cdots \eta_N = J_G^{-1} \theta_1 \cdots \theta_N \qquad (5.23)$$

under this transformation. A direct substitution yields

$$\begin{aligned} \eta_1 \cdots \eta_N &= M_{1a_1} M_{2a_2} \cdots M_{N a_N} \theta_{a_1} \cdots \theta_{a_N} \\ &= \varepsilon^{a_1 a_2 \cdots a_N} M_{1a_1} M_{2a_2} \cdots M_{N a_N} \theta_1 \cdots \theta_N , \end{aligned} \qquad (5.24)$$

where the result of Exercise 5.1 has been used. Recalling the definition (1.54) of the determinant and comparing (5.23) to (5.24), shows that the Jacobian J_G is given by $J_G = (\det M)^{-1}$. From the fact that $M_{ab} = \partial \eta_a / \partial \theta_b$ it follows that

$$J_G = \left[\det \left(\frac{\partial \eta}{\partial \theta} \right) \right]^{-1} , \qquad (5.25)$$

which is the inverse of result (1.101).

The extension of integration to complex Grassmann variables is not without subtlety. The complex variable η may be written in terms of two real Grassmann variables, θ_1 and θ_2, in the form $\eta = \theta_1 + i\theta_2$, so that

$$d\eta = d\theta_1 + i d\theta_2 . \qquad (5.26)$$

From the rules (5.7) for complex conjugation it follows that

$$(d\eta)^* = -d\theta_1 + i d\theta_2 = -d(\eta^*) \equiv -d\eta^* . \qquad (5.27)$$

Therefore, for η complex,

$$\left(\int d\eta \, \eta \right)^* = \int \eta^* (d\eta)^* = -\int \eta^* d\eta^* = \int d\eta^* \, \eta^* . \qquad (5.28)$$

124   Grassmann Variables

Exercise 5.6: Using the form of the Jacobian for Grassmann variables and defining $\eta = \theta_1 + i\theta_2$, where θ_1 and θ_2 are real Grassmann variables, show that

$$\int d\eta^* \, \eta^* \int d\eta \, \eta = -\int d\eta^* d\eta \, \eta^* \eta = 1 \,. \tag{5.29}$$

Therefore, for consistency, it must be that the integrals over complex Grassmann variables have the form

$$\int d\eta \, \eta = \int d\eta^* \, \eta^* = 1 \,,$$

$$\int d\eta = \int d\eta^* = 0 \,. \tag{5.30}$$

This is consistent with both results (5.28) and (5.29).

The Dirac delta can be defined for Grassmann variables in an extremely simple manner. By extension of the c-number case, the Dirac delta for a single Grassmann variable should satisfy

$$\int d\eta \, \delta(\eta - \eta') f(\eta) = f(\eta') \,, \tag{5.31}$$

where η and η' are both Grassmann variables. It is straightforward, using the rules of integration and the polynomial representation of f, to show that

$$\delta(\eta - \eta') = \eta - \eta' \,. \tag{5.32}$$

Exercise 5.7: Verify that (5.32) satisfies (5.31).

It is worth noting that the Grassmann Dirac delta satisfies

$$\delta(0) = 0 \,,$$
$$\delta(\alpha\eta) = \alpha\delta(\eta) \,,$$
$$\delta(\eta - \eta') = -\delta(\eta' - \eta) \,,$$
$$\frac{\partial}{\partial \eta}\delta(\eta - \eta') = 1 \,. \tag{5.33}$$

The generalization of the Dirac delta to many Grassmann variables is constructed from the product of single Dirac deltas. However, it requires that the sequence of Grassmann variables in the definition of the Dirac delta is an odd permutation of the order of the Grassmann variables appearing in the measure. An integral representation of the Dirac delta is simple to construct. For a single Grassmann variable it follows that

$$\delta(\eta - \eta') = \int d\xi \, e^{\xi(\eta - \eta')} , \qquad (5.34)$$

where ξ is also a Grassmann variable. The generalization to many Grassmann variables is ambiguous since the order of the variables of integration in the measure determines the order of the factors of the form (5.32) that make up the Dirac delta. This is illustrated in Exercise 5.8.

Exercise 5.8: For the case of N real Grassmann variables verify that

$$\int d\xi_{a_1} \cdots d\xi_{a_N} \exp\{\xi_j(\eta_j - \eta'_j)\} = (\eta_{a_N} - \eta'_{a_N}) \cdots (\eta_{a_1} - \eta'_{a_1}) . \qquad (5.35)$$

Exercise 5.9: Show that any function, $f(\eta)$, of the Grassmann variable η satisfies

$$\left(\eta \frac{\partial}{\partial \eta} + \frac{\partial}{\partial \eta} \eta\right) f(\eta) = f(\eta) , \qquad (5.36)$$

so that

$$\left(\eta \frac{\partial}{\partial \eta} + \frac{\partial}{\partial \eta} \eta\right) = 1 . \qquad (5.37)$$

5.2 Gaussian Grassmann Integrals

In this section the Grassmann variable integrals similar in form to (1.97) will be evaluated, in expectation that they will be useful for later path integrals. A key ingredient in this procedure is the result that, for a complex Grassmann variable η,

$$\int d\eta^* \, d\eta \, e^{-\eta^* \eta} = 1 , \qquad (5.38)$$

which is obtained by using a power series expansion of the exponential and the rules of Grassmann integration.

For the case of N complex variables, the extension of (5.38) is given by

$$I = \int d\eta_1^* d\eta_1 \cdots d\eta_N^* d\eta_N \, \exp(-\eta_j^* M_{jk} \eta_k) \,, \tag{5.39}$$

where the M_{jk} can be viewed as the elements of an $n \times n$ c-number matrix **M**. The matrix **M** will be assumed to be Hermitian, so that I must be a real c-number. Since **M** is Hermitian it can be diagonalized, as in Sec. 1.5, with some unitary matrix **U**, so that the matrix $\mathbf{D} = \mathbf{UMU}^\dagger$ has the elements

$$D_{jk} = \lambda^{(j)} \delta_{jk} \,, \tag{5.40}$$

where $\lambda^{(j)}$ is the jth eigenvalue of **M**. The Grassmann variables are transformed to $\eta_j = U_{jk} \xi_k$, and, using the unitarity of **U**, the Jacobian of this transformation is $\det \mathbf{U} \det \mathbf{U}^\dagger = \det \mathbf{UU}^\dagger = 1$. Using the properties of Grassmann variables, the transformed integral I can be written

$$I = \prod_{j=1}^{N} \int d\xi_j^* d\xi_j \, \exp(-\lambda^{(j)} \xi_j^* \xi_j) \,, \tag{5.41}$$

where there is no sum on j. Using (5.38) the total integral reduces to

$$I = \prod_{j=1}^{N} \lambda^{(j)} = \det \mathbf{M} \,, \tag{5.42}$$

and, once again, the inverse of the c-number result is obtained.

The case of real Grassmann Gaussian integrals can now be analyzed. For $\{\theta_j\}$ a set of N real Grassmann variables, the integral of interest is

$$I = \int d\theta_1 \cdots d\theta_N \, \exp(-\theta_j M_{jk} \theta_k) \,. \tag{5.43}$$

Attention will be restricted to the case that M_{jk} is a real c-number. As before, the M_{jk} may be viewed as the elements of a real matrix **M**. While there are no other restrictions on **M**, it can be shown that only the antisymmetric part of **M** contributes to I.

Exercise 5.10: By defining the antisymmetric part of **M** as the matrix $\mathbf{M}^A = \frac{1}{2}(\mathbf{M} - \mathbf{M}^T)$, show that

$$\theta_j M_{jk} \theta_k = \theta_j M_{jk}^A \theta_k \,. \tag{5.44}$$

Sec. 5.2 Gaussian Grassmann Integrals

In what follows it will be assumed that **M** is a real antisymmetric matrix.

The analysis of I is greatly facilitated by first squaring it and writing the square as

$$I^2 = \int d\theta_1 \cdots d\theta_N d\xi_1 \cdots d\xi_N \; \exp(-\theta_j M_{jk}\theta_k - \xi_j M_{jk}\xi_k), \qquad (5.45)$$

where the $\{\xi_j\}$ are also real Grassmann variables, and the commutativity of bilinear products of Grassmann variables has been used to combine the arguments of the exponentials. This integral can now be transformed to a complex form by defining the new variables $\eta_j = \frac{1}{2}(\theta_j + i\xi_j)$. By virtue of the antisymmetry of **M** the argument of the exponential becomes

$$-\theta_j M_{jk}\theta_k - \xi_j M_{jk}\xi_k = -\eta_j^* M_{jk}\eta_k. \qquad (5.46)$$

The analysis of the Jacobian is left as an exercise.

> **Exercise 5.11:** Show that the Jacobian of the above transformation is given by
>
> $$J_G = \begin{cases} (-1)^{3N/2}(2i)^N & \text{if } N \text{ is even}, \\ (-1)^{3(N-1)/2}(2i)^N & \text{if } N \text{ is odd}. \end{cases} \qquad (5.47)$$

Combining these results gives

$$I^2 = J_G \int d\eta_1^* d\eta_1 \cdots d\eta_N^* d\eta_N \; \exp(-\eta_j^* M_{jk}\eta_k). \qquad (5.48)$$

Using the fact that $i\mathbf{M}$ is a Hermitian matrix shows that the integral is given by

$$I^2 = J_G \det \mathbf{M}. \qquad (5.49)$$

Combining this with the results of Exercise (1.11) shows that the integral vanishes if N is odd. This must be the case, since, if N is odd, the integral must result in a Grassmann number. For N even, the results of the Jacobian give

$$I = 2^{N/2}\sqrt{\det \mathbf{M}}. \qquad (5.50)$$

The extension of these results to the case where terms linear in the Grassmann variables are present is left as a pair of exercises.

128 Grassmann Variables

Exercise 5.12: For the case of N real Grassmann variables, $\{\theta_j\}$, and N real Grassmann constants, $\{K_j\}$, show that, if N is even and $\mathbf{M}$ is a real antisymmetric matrix,

$$\int d\theta_1 \cdots d\theta_N \exp(-\theta^T \mathbf{M}\theta + K^T\theta)$$
$$= 2^{(N/2)}\sqrt{\det \mathbf{M}} \exp(-\tfrac{1}{4} K^T \mathbf{M}^{-1} K), \qquad (5.51)$$

where an obvious matrix notation has been employed.

Exercise 5.13: For the case of N complex Grassmann variables, $\{\eta_j\}$, and N complex Grassmann constants, $\{K_j\}$, show that

$$\int d\eta_1^* d\eta_1 \cdots d\eta_N^* d\eta_N \exp(-\eta^\dagger \mathbf{M}\eta + K^\dagger \eta + \eta^\dagger K)$$
$$= \det \mathbf{M} \exp(K^\dagger \mathbf{M}^{-1} K) \quad , \qquad (5.52)$$

where $\mathbf{M}$ is a Hermitian matrix.

5.3 Classical Grassmann Mechanics

In this section classical mechanics will be formulated for the case of Grassmann variables. Of course, results that are quite different from the usual form of classical mechanics will be obtained. However, the familiar concepts of the action and the Hamiltonian will survive the transition. Equally important, the Poisson bracket formalism can be adapted to Grassmann variables, thereby enabling the definition of quantum Grassmann systems.

The simple case of a single Grassmann particle will be considered. The first step is to parameterize the Grassmann variable η, representing the "position" of the particle, with the time t, so that $\eta = \eta(t)$. It is assumed that η is a real, dimensionless Grassmann variable, although this is not necessary. It follows from the Grassmann nature of η that $\{\eta(t), \eta(t')\} = 0$. Because η is a Grassmann-valued function of t, it possesses a time derivative, denoted $\dot{\eta}$, and it is not difficult to show that $\dot{\eta}$ is also a Grassmann variable.

The next step is to construct an action. It is obvious that a term of

Sec. 5.3 Classical Grassmann Mechanics

the form $\dot{\eta}^2$ vanishes, and so cannot be present in the action. A possible Lagrangian density for such a system is

$$S = \int_{t_a}^{t_b} dt\, \mathcal{L} = i \int_{t_a}^{t_b} dt\, \eta\, \dot{\eta} \,. \tag{5.53}$$

There are two points of interest regarding this action. First, if η were a c-number variable, then the action (5.53) would, after an integration by parts, reduce to the difference of η^2 at the endpoints of the integral. A key property in such a demonstration is the fact that, for c-number variables, $\eta\dot{\eta} = \dot{\eta}\eta$. Since this is not the case for Grassmann variables, the action (5.53) cannot be shown to vanish by this argument. Second, because of property (5.2), the occurrence of the factor of i is necessary to make the action real.

In order to derive the equation of motion for this system η is varied, $\eta \to \eta + \delta\eta$, where $\delta\eta$ is another Grassmann varible, assumed to satisfy $\delta\eta(t_b) = \delta\eta(t_a) = 0$. The variation of the action is set to zero for the classical trajectory, and this results in

$$\delta S = \int dt\, (i\eta\, \delta\dot{\eta} + i\delta\eta\, \dot{\eta}) = -2i \int dt\, \dot{\eta}\, \delta\eta = 0\,, \tag{5.54}$$

where an integration by parts has been performed. This results in the *first-order* differential equation of motion

$$\dot{\eta} = 0\,. \tag{5.55}$$

A slightly more complicated action demonstrates an ambiguity that is not present in (5.53). If $K(t)$ is an a-number source, then a possible action is given by

$$\mathcal{L} = i\eta\,\dot{\eta} - i\eta K \,. \tag{5.56}$$

Varying the action (5.56) yields the equation of motion

$$\dot{\eta} = \tfrac{1}{2} K \,. \tag{5.57}$$

However, if the order of K and η is reversed in the second term of (5.56), the equation of motion becomes

$$\dot{\eta} = -\tfrac{1}{2} K \,. \tag{5.58}$$

Because of such ambiguities a set of conventions must be adopted. First, it will be assumed that all actions are constructed to give c-numbers, so that only bilinear Grassmann terms, or products of bilinear Grassmann terms,

130   Grassmann Variables

appear in the action. Second, it will be assumed that all derivatives of Grassmann variables will be placed at the right side of the term it appears in. Third, all occurrences of the Grassmann variables will be placed to the left of any a-number or c-number sources appearing in the term.

In order to systematize the equation of motion a definition of functional differentiation compatible with Grassmann variables must be made. If $\mathcal{F}[\eta]$ is a functional of $\eta(t)$, then the functional derivative of $\mathcal{F}$ is defined as

$$\frac{\delta \mathcal{F}}{\delta \eta(t)} \equiv \frac{\partial}{\partial \varepsilon} \left(\mathcal{F}[\eta(\tau) + \varepsilon \delta(\tau - t)] - \mathcal{F}[\eta] \right), \qquad (5.59)$$

where ε is a constant (time-independent) Grassmann number. For consistency with the differential of (5.1), the functional differential is then given by

$$\delta \mathcal{F} = \int dt \, \delta \eta_i(t) \, \frac{\delta \mathcal{F}}{\delta \eta_i(t)} . \qquad (5.60)$$

Hamilton's equations of motion for Grassmann systems can now be derived by constructing a Lagrangian density as similar to (1.45) as possible. Using ρ to denote the momentum canonically conjugate to η, a general Lagrangian density may be written

$$\mathcal{L}(\eta, \dot{\eta}) = \rho \dot{\eta} - H(\rho, \eta), \qquad (5.61)$$

where $H(\rho, \eta)$ is the Hamiltonian of the system. The ordering in (5.61) coincides with the previous conventions. Using the definition of the functional differential, it follows that

$$\delta S = \int dt \left[\delta \rho \left(\dot{\eta} - \frac{\partial H}{\partial \rho} \right) + \delta \eta \left(\dot{\rho} - \frac{\partial H}{\partial \eta} \right) \right]. \qquad (5.62)$$

From this Hamilton's equations of motion are deduced to be

$$\dot{\rho} = \frac{\partial H}{\partial \eta}, \quad \dot{\eta} = \frac{\partial H}{\partial \rho}. \qquad (5.63)$$

The sign difference between Hamilton's equations for a Grassmann system and those for a c-number system is an inevitable outgrowth of the fact that ρ and $\dot{\eta}$ anticommute. It follows from (5.61) that

$$\rho = -\frac{\partial \mathcal{L}}{\partial \dot{\eta}} . \qquad (5.64)$$

Sec. 5.3 Classical Grassmann Mechanics

Exercise 5.14: Show that Hamilton's equations of motion and (5.64) imply the usual form of the Euler–Lagrange equation for Grassmann systems,

$$\frac{d}{dt}\frac{\partial \mathcal{L}}{\partial \dot{\eta}} - \frac{\partial \mathcal{L}}{\partial \eta} = 0, \tag{5.65}$$

and that the Hamiltonian satisfies

$$\frac{dH}{dt} = -\frac{\partial \mathcal{L}}{\partial t}. \tag{5.66}$$

The Poisson bracket for a Grassmann system is defined as

$$\{A, B\}_{\rho,\eta} \equiv \frac{\partial A}{\partial \eta_j}\frac{\partial B}{\partial \rho_j} + \frac{\partial A}{\partial \rho_j}\frac{\partial B}{\partial \eta_j}, \tag{5.67}$$

and will be referred to as an *antibracket*. This definition has been chosen so that

$$\frac{dA}{dt} = \frac{\partial A}{\partial t} + \dot{\eta}\frac{\partial A}{\partial \eta} + \dot{\rho}\frac{\partial A}{\partial \rho} = \frac{\partial A}{\partial t} + \{H, A\}_{\rho,\eta}, \tag{5.68}$$

a result identical to the c-number form of classical mechanics. The antibracket is unusual since its symmetry or antisymmetry depends on the two objects, A and B, that form its arguments. If A and B are both c-number products of Grassmann variables, then their derivatives with respect to ρ or η yield a-numbers, and therefore

$$\{A, B\}_{\rho,\eta} = -\{B, A\}_{\rho,\eta}, \tag{5.69}$$

while, for all other cases,

$$\{A, B\}_{\rho,\eta} = \{B, A\}_{\rho,\eta}. \tag{5.70}$$

This has the ramification that, since H is assumed to be c-number, the antibracket $\{A, H\}$ is antisymmetric if A is a c-number and is symmetric if A is an a-number. It is easy to see that the Poisson antibracket of η_j and ρ_k is given by

$$\{\eta_j, \rho_k\}_{\rho,\eta} = \delta_{jk} \tag{5.71}$$

and is symmetric.

As an example of an important system, the action for the Grassmann harmonic oscillator can be written in terms of two real Grassmann variables, η_1 and η_2, as

$$\mathcal{L} = i\,\eta_1\dot{\eta}_1 + i\,\eta_2\dot{\eta}_2 - 2i\omega\eta_1\eta_2, \tag{5.72}$$

where ω is a c-number constant. The Euler–Lagrange equation gives the two equations of motion

$$\dot{\eta}_1 = \omega \eta_2 ,$$
$$\dot{\eta}_2 = -\omega \eta_1 . \qquad (5.73)$$

The two first-order equations of (5.73) can be combined to find

$$\ddot{\eta}_j = -\omega^2 \eta_j , \quad j = (1,2) , \qquad (5.74)$$

which is the harmonic oscillator equation of motion. This is a simple example of factorizing a second-order differential operator into the product of two first-order differential operators by using Grassmann variables. By using (5.64), the momenta are identified as

$$p_j = i\eta_j , \qquad (5.75)$$

so that the Hamiltonian is given by

$$H = i\omega \eta_1 \eta_2 . \qquad (5.76)$$

Given the identification of the momenta in (5.75), there is an ambiguity in how the Hamiltonian should be written, i.e., whether or not the η_j in (5.76) should be replaced by $-ip_j$. This difficulty has arisen because η_j and p_j are not independent variables, and could not be in any system that gives rise to first-order differential equations. In order that the Poisson antibracket formalism work for this case, the relation between p_j and η_j must be taken into account, so that

$$\frac{\partial H}{\partial p_1} = \omega \eta_2 , \quad \frac{\partial H}{\partial p_2} = -\omega \eta_1 . \qquad (5.77)$$

As a final note, the action for this system can be written very compactly by introducing the complex variable $\eta = \eta_1 + i\eta_2$. The action then takes the form

$$\mathcal{L} = i\eta^* \dot{\eta} - \omega \eta^* \eta . \qquad (5.78)$$

The momentum canonically conjugate to η is then given by $p = i\eta^*$. An added bonus is the absence of any ambiguity in the Hamiltonian in this basis, given by $H = -ip\eta$. In the next section this action will be elevated to a quantum mechanical system.

5.4 Grassmann Quantum Mechanics

In this section the quantum mechanics of Grassmann systems will be presented. While fermionic field theories will be discussed in detail in Chapter 6, this section and the next will present many of the ideas necessary to construct them. In Sec. 5.5 the results of this section will be used to construct path integral representations of transition amplitudes for simple anticommuting systems, and many of these results will be carried over to fermionic field theories.

The quantum mechanical version of Grassmann systems consists of realizing the classical mechanical formalism of Sec. 5.3 as a theory of operators defined over a Hilbert space. This will result in associating an operator $\mathcal{N}_j$ to the Grassmann coordinates η_j. After identifying the canonical momentum, $i\mathcal{P}_j$, the system is quantized by assuming the anticommutator

$$\{\mathcal{N}_j, \mathcal{P}_k\} = \delta_{jk} \, , \tag{5.79}$$

where an anticommutator has been chosen in accordance with the symmetry of the classical Poisson bracket.

Relation (5.79) creates an algebra that the operators of the system must satisfy. It will be seen that this typically leads to a statement of the form $\mathcal{N}_j^2 = \alpha_j^2$, where α_j is a c-number variable. It is straightforward to show that, if $\alpha_j \neq 0$, the operator $\mathcal{N}_j$ possesses only two eigenvalues, $\pm \alpha_j$. If $\alpha_j = 0$, all nonzero eigenvalues of $\mathcal{N}_j$ must be pure Grassmann numbers, and the operator $\mathcal{N}$ is said to be *nilpotent* since $\mathcal{N}^2 = 0$.

The case of $\alpha_j = 0$ will be considered first, and for simplicity, only a single degree of freedom will be considered initially. The case of nonzero α_j is treated in an example at the end of this section. Denoting the operator by $\mathcal{N}$, the previous argument shows that its eigenvalues must be pure Grassmann. Its eigenstate is denoted $|\eta\rangle$, and it is assumed that

$$\mathcal{N}|\eta\rangle = \eta|\eta\rangle \, , \tag{5.80}$$

where η is a real Grassmann number. It is assumed that the eigenstates $|\eta\rangle$ possess an inner product, $\langle \eta'|\eta\rangle$. For the moment, the form of this inner product and its behavior under complex conjugation are unknown; however, certain of its properties can be deduced. Even in more complicated systems, it must take the form of a superfunction of the arguments of the states. If K is another Grassmann variable, it then follows that

$$K\langle \eta|\eta'\rangle = \langle -\eta|-\eta'\rangle K \, , \tag{5.81}$$

and this immediately gives the fact that Grassmann inner products possibly may not commute with other Grassmann variables.

Grassmann Variables

For the case of a single variable η, it is assumed that the dual space is the set of all linear functionals f_η that map $|\eta\rangle$ into an a-number or a superfunction, $f_\eta : |\eta'\rangle \mapsto f(\eta, \eta')$. A technicality occurs because of the presence of Grassmann variables, so that the mapping may be antilinear,

$$f_\eta : \eta_1 |\eta_2\rangle \mapsto \eta_1 f(\eta, \eta_2) = f(-\eta, -\eta_2)\eta_1 \,. \tag{5.82}$$

As in the c-number case, these linear functionals may be understood as the action of the inner product, so that

$$f_\eta : \eta_1 |\eta_2\rangle \mapsto \eta_1 \langle \eta | \eta_2 \rangle = \langle -\eta | -\eta_2 \rangle \eta_1 \,, \tag{5.83}$$

where, because of the antilinear nature of the functional, it has been necessary to choose a convention for moving the dual states through any Grassmann variables that may be present. In what follows, the left-handed convention will be employed. For the case that η is real it is assumed that $\mathcal{N}$ is Hermitian, so that it acts on the dual space in the manner

$$\langle \eta | \eta = (\mathcal{N}|\eta\rangle)^\dagger = \langle \eta | \mathcal{N}^\dagger = \langle \eta | \mathcal{N} \,. \tag{5.84}$$

Using the Hermiticity of $\mathcal{N}$, expressed in (5.84), and the left-handed convention gives

$$\langle \eta' | \mathcal{N} | \eta \rangle = \eta \langle \eta' | \eta \rangle = \eta' \langle \eta' | \eta \rangle \,, \tag{5.85}$$

so that

$$(\eta' - \eta)\langle \eta' | \eta \rangle = 0 \,. \tag{5.86}$$

Recalling the properties of the Grassmann algebra, relation (5.86) shows that the inner product must be proportional to $(\eta - \eta')$. For simplicity, the constant of proportionality is fixed to be unity, so that

$$\langle \eta' | \eta \rangle = \eta - \eta' \,. \tag{5.87}$$

Comparing (5.87) with the Grassmann Dirac delta (5.32) shows that the Grassmann states are orthonormal in the same manner as the c-number states of Sec. 2.1, as expressed in (2.4). However, this gives the result

$$\langle \eta' | \eta \rangle^* = -\langle \eta | \eta' \rangle = \langle -\eta | -\eta' \rangle \,. \tag{5.88}$$

In addition, it shows that the state $|\eta\rangle$ has zero norm since $\langle \eta | \eta \rangle = 0$. Any state with a norm less than or equal to zero is referred to as a *ghost*. The presence of ghosts in the Hilbert space threatens the probabilistic interpretation of quantum mechanics unless they can be excluded from

Sec. 5.4 Grassmann Quantum Mechanics

contributing to quantum processes. If they cannot, the theory must be discarded as pathological. This will be discussed in more detail in Chapter 7, when the unitarity of gauge theories is analyzed.

It is assumed that the states $|\eta\rangle$ are complete. This means that the operator P_η, defined by

$$P_\eta = \int d\eta \, |\eta\rangle\langle\eta| = 1 , \qquad (5.89)$$

is a unit projection operator. Using results obtained so far, the straightforward series of steps

$$\begin{aligned} P_\eta^\dagger &= \int |\eta\rangle\langle\eta| (d\eta)^* = -\int |\eta\rangle\langle\eta| d\eta \\ &= -\int d\eta \, |-\eta\rangle\langle-\eta| = \int d\eta \, |\eta\rangle\langle\eta| = P_\eta , \qquad (5.90) \end{aligned}$$

demonstrates that P_η is a Hermitian operator.

Exercise 5.15: Using the form of the inner product (5.87), verify the completeness relation (5.87) by showing that

$$\langle \eta_1 | \eta_2 \rangle = \int d\eta \, \langle \eta_1 | \eta \rangle\langle \eta | \eta_2 \rangle . \qquad (5.91)$$

The generalization of these results to systems with more degrees of freedom is not straightforward. If $\mathcal{N}_1$ and $\mathcal{N}_2$ denote two nilpotent operators, and $|\eta_1, \eta_2\rangle$ is a *simultaneous* eigenstate of both these operators, then it follows that

$$\{\mathcal{N}_1, \mathcal{N}_2\} = 0 , \qquad (5.92)$$

so that these operators must have a vanishing anticommutator. It is not possible to deduce the form of the inner product for such a system; however, a method of constructing the inner product will be given in the second example of this section.

If K is an arbitrary Grassmann variable, then the anticommutator of $\mathcal{N}$ with K satisfies

$$\langle \eta' | \{K, \mathcal{N}\} | \eta \rangle = K(\eta - \eta')\langle \eta' | \eta \rangle = 0 , \qquad (5.93)$$

where the form (5.87) has been used. This yields the result that $\{K, \mathcal{N}\} = 0$. In addition, it follows that all c-number variables commute with both the inner products and the operators.

136 Grassmann Variables

In a Grassmann system where η and ρ are distinct variables at the classical level it is possible to construct a form of Grassmann quantum mechanics that is very similar to standard c-number quantum mechanics. Again, for simplicity only one degree of freedom will be considered. If $\mathcal{N}$ is to be the operator associated with the Grassmann position, then $i\mathcal{P}$ denotes the operator associated with the canonically conjugate momentum, where $\mathcal{P}$ is assumed to be a Hermitian operator. From the fact that the Poisson antibracket (5.71) is symmetric for η and ρ, and that the Poisson brackets of classical mechanics are elevated to operator algebras in quantum mechanics, it is assumed that $\mathcal{N}$ and $i\mathcal{P}$ satisfy the symmetric *anticommutator*

$$\{\mathcal{N}, i\mathcal{P}\} \equiv i\mathcal{N}\mathcal{P} + i\mathcal{P}\mathcal{N} = i \;\Rightarrow\; \{\mathcal{N}, \mathcal{P}\} = 1 \,. \tag{5.94}$$

The eigenstates of $\mathcal{P}$ are denoted $|\rho\rangle$, and they obey

$$\mathcal{P}|\rho\rangle = \rho|\rho\rangle \,, \tag{5.95}$$

possess an inner product similar to (5.87),

$$\langle \rho'|\rho\rangle = \rho - \rho' \,, \tag{5.96}$$

and are assumed to be complete,

$$\int d\rho \, |\rho\rangle\langle\rho| = 1 \,. \tag{5.97}$$

Using the rules derived so far, it follows from the defining algebra (5.94) that

$$\eta' - \eta = \langle \eta | \{\mathcal{N}, \mathcal{P}\} | \eta' \rangle = (\eta - \eta')\langle \eta | \mathcal{P} | \eta' \rangle \,, \tag{5.98}$$

so that

$$\langle \eta | \mathcal{P} | \eta' \rangle = -1 = \frac{\partial}{\partial \eta}\langle \eta | \eta' \rangle \,. \tag{5.99}$$

From (5.99), a straightforward series of steps yields

$$\rho\langle\eta|\rho\rangle = \langle\eta|\mathcal{P}|\rho\rangle = \int d\eta'\, \langle\eta|\mathcal{P}|\eta'\rangle\langle\eta'|\rho\rangle$$
$$= -\int d\eta \, \langle\eta|\rho\rangle = -\frac{\partial}{\partial \eta}\langle\eta|\rho\rangle \,, \tag{5.100}$$

where the formal identity of integration and differentiation for Grassmann variables has been exploited. The differential equation of (5.100) is similar to that of standard quantum mechanics, *sans* the factor of i. For this system, the position representation of Grassmann quantum mechanics yields

$$\mathcal{N} \to \eta \,, \quad \mathcal{P} \to \frac{\partial}{\partial \eta} \,, \tag{5.101}$$

Sec. 5.4 Grassmann Quantum Mechanics

which is identical to result (5.37). The solution of (5.100) gives

$$\langle \eta | \rho \rangle = e^{\eta \rho}, \tag{5.102}$$

which is the Grassmann equivalent of a plane wave.

> **Exercise 5.16**: Use result (5.102) to verify the completeness relation (5.97).

It is possible to define states that project onto superfunctions. The adjoint of such a state, denoted $\langle f |$, is a linear functional on the single degree of freedom eigenstates, $| \eta \rangle$, so that

$$\langle f | \eta \rangle = f_c^* + f_a^* \eta, \tag{5.103}$$

where f_c and f_a are c-numbers. It follows that the inner product of two such states is given by

$$\langle f | h \rangle = \int d\eta \, \langle f | \eta \rangle \langle \eta | h \rangle = f_a^* h_c + f_c^* h_a. \tag{5.104}$$

From simple inspection it is apparent that values for f_a and f_c exist such that $\langle f | f \rangle \leq 0$, so that these states are potential ghosts.

The quantum mechanical version of Grassmann mechanics consists of the replacements

$$\eta \to \mathcal{N}, \quad \rho \to i\mathcal{P}, \tag{5.105}$$

in the classical Hamiltonian, H, but only in those cases where η and ρ are distinct from each other. In the Schrödinger picture the time development of an arbitrary state $| \psi \rangle_S$ is assumed to be given by

$$H(\mathcal{N}, \mathcal{P}) | \psi, t \rangle_S = i \frac{\partial}{\partial t} | \psi, t \rangle_S. \tag{5.106}$$

If the classical Hamiltonian is a real c-number, then the quantum mechanical operator H, obtained by substituting operators for η and ρ, will be Hermitian, so that $H^\dagger = H$. For such a case, the Schrödinger equation gives

$$i \frac{\partial}{\partial t} \,_S\langle \psi, t | = -_S\langle \psi, t | H^\dagger(\mathcal{N}, \mathcal{P}) = -_S\langle \psi, t | H(\mathcal{N}, \mathcal{P}). \tag{5.107}$$

Under these circumstances the time dependence of Heisenberg picture observables may be derived. The Heisenberg picture observable O_H obeys the relation

$$_H\langle \psi | O_H(t) | \psi \rangle_H = \,_S\langle \psi, t | O_S | \psi, t \rangle_S, \tag{5.108}$$

138 Grassmann Variables

Assuming O_s has no explicit time-dependence, it follows, by differentiating and using the Schrödinger equation, that

$$i\frac{\partial}{\partial t}O_H = O_H H - H O_H = [O_H, H]. \qquad (5.109)$$

As examples of quantum mechanical Grassmann systems, two simple actions will be analyzed. These examples are drawn from the book by DeWitt [2].

5.4.1 The Free Particle

The first action is given by $\mathcal{L} = i\eta\dot{\eta}$. In this case η is real, and the canonical momentum is given by $p = i\eta$. The anticommutation relation (5.94) becomes

$$\{\mathcal{N}, \mathcal{N}\} = 1 \Rightarrow \mathcal{N}^2 = \frac{1}{2}. \qquad (5.110)$$

This action is therefore an example of a theory where, in the nomenclature introduced at the beginning of this section, $\alpha_j \neq 0$. The fact that $\mathcal{N}$ and $\mathcal{P}$ are not independent has led to result (5.110). Equation (5.110) can be understood only as a *constraint*; the physically allowed states, $|P\rangle$, of the quantum system must satisfy

$$\mathcal{N}^2|P\rangle = \tfrac{1}{2}|P\rangle. \qquad (5.111)$$

Paradoxically, this action has forced $\mathcal{N}$ to possess eigenstates with real c-number eigenvalues. The state $|\eta\rangle$, discussed earlier in this section, is therefore *not* an eigenstate of $\mathcal{N}$. If $|\alpha\rangle$ is an eigenstate of $\mathcal{N}$, with the real c-number eigenvalue α, then

$$\mathcal{N}^2|\alpha\rangle = \alpha^2|\alpha\rangle \Rightarrow \alpha = \pm\frac{1}{\sqrt{2}}. \qquad (5.112)$$

There are, therefore, two eigenstates of $\mathcal{N}$, and these are denoted $|+\rangle$ and $|-\rangle$, for the eigenvalues $+\sqrt{2}/2$ and $-\sqrt{2}/2$, respectively. Both these states must have projections onto the Grassmann state $|\eta\rangle$ that result in superfunctions. In such a case, the constraint $\mathcal{N}^2 = 1/2$ can be represented, using the result of Exercise 5.9, as

$$\langle\eta|\mathcal{N}^2|\pm\rangle = \frac{1}{2}\left(\eta\frac{\partial}{\partial\eta} + \frac{\partial}{\partial\eta}\eta\right)f_\pm(\eta). \qquad (5.113)$$

From this equation it follows that, in position representation,

$$\mathcal{N}^2 = \frac{1}{2}\left(\eta\frac{\partial}{\partial\eta} + \frac{\partial}{\partial\eta}\eta\right) = \left(\eta + \frac{1}{2}\frac{\partial}{\partial\eta}\right)^2, \qquad (5.114)$$

so that, in position representation,

$$\mathcal{N} = \eta + \frac{1}{2}\frac{\partial}{\partial \eta} . \qquad (5.115)$$

It is now straightforward to find the eigenfunctions of $\mathcal{N}$; these must satisfy

$$\left(\eta + \frac{1}{2}\frac{\partial}{\partial \eta}\right) f_{\pm}(\eta) = \pm \frac{1}{\sqrt{2}} f_{\pm}(\eta) . \qquad (5.116)$$

A simple calculation yields

$$f_{+}(\eta) = \langle \eta | + \rangle = \frac{1 + \sqrt{2}\,\eta}{\sqrt{2\sqrt{2}}}, \quad f_{-}(\eta) = \langle \eta | - \rangle = \frac{1 - \sqrt{2}\,\eta}{\sqrt{2\sqrt{2}}} . \qquad (5.117)$$

It is not difficult to show that the two states are orthogonal. However, in the normalizations chosen for (5.117) it follows that

$$\langle \pm | \pm \rangle = \int d\eta\, f_{\pm}^{*}(\eta) f_{\pm}(\eta) = \pm 1 , \qquad (5.118)$$

so that $|-\rangle$ corresponds to a negative norm ghost state. The presence of the negative norm state creates difficulties whose resolution will not be considered here.

5.4.2 The Harmonic Oscillator

The second action is that of the harmonic oscillator, given in complex notation by $\mathcal{L} = i\mathcal{N}^{\dagger}\dot{\mathcal{N}} + \omega \mathcal{N}^{\dagger}\mathcal{N}$, where $\mathcal{N}^{\dagger}$ is the operator extension of η^{*}. Since the momentum canonically conjugate to $\mathcal{N}$ is given by $i\mathcal{N}^{\dagger}$, the two coordinates must satisfy the anticommutation relation

$$\{\mathcal{N}, \mathcal{N}^{\dagger}\} = 1 , \qquad (5.119)$$

along with the additional constraints that $\mathcal{N}^2 = \mathcal{N}^{\dagger 2} = 0$. Because $\mathcal{N}$ and $\mathcal{N}^{\dagger}$ have a nonzero anticommutator, it is not possible to find a simultaneous eigenstate for both of these operators. It is also not possible to associate the position representation (5.101) with these operators, since that would violate $\mathcal{N}^{\dagger} = \mathcal{P}$.

Therefore, rather than the eigenstates of $\mathcal{N}$, the eigenstates of the Hamiltonian,

$$H = \omega \mathcal{N}^{\dagger}\mathcal{N} , \qquad (5.120)$$

will be found, and this will demonstrate a method of great importance in quantum field theory. In the Heisenberg picture the coordinate $\mathcal{N}$ must satisfy the equation of motion

$$\dot{\mathcal{N}} = -i[\mathcal{N}, H] = -i\omega\mathcal{N} \,. \tag{5.121}$$

The solution to this equation and the anticommutation relations is accomplished by the *mode expansion*. In the mode expansion, the operator is written as a combination of the c-number solutions to the equation of motion and secondary operators that satisfy the (anti)commutation relations. There is only one c-number solution to (5.121), given by $\exp(-i\omega t)$. Therefore, the mode expansions of $\mathcal{N}$ and $\mathcal{N}^\dagger$ are given by

$$\mathcal{N} = a e^{-i\omega t}, \quad \mathcal{N}^\dagger = a^\dagger e^{i\omega t} \,, \tag{5.122}$$

where a and $a^\dagger$ are time-independent operators satisfying

$$a^2 = a^{\dagger 2} = 0, \quad \{a, a^\dagger\} = 1 \,. \tag{5.123}$$

The combination of (5.122) with (5.123) satisfies both the equation of motion and the canonical anticommutation relation. The Hamiltonian then takes the time-independent form, $H = \omega a^\dagger a$, identical to that of the harmonic oscillator of Sec. 4.4. While the algebra of (5.123) is different, a method similar to that of Sec. 4.4 may be used to solve for the energy eigenvalues. The ground state $|0\rangle$ is introduced, with the properties that

$$a|0\rangle = 0, \quad \langle 0|0\rangle = 1 \,, \tag{5.124}$$

so that, once again, a plays the role of an annihilation operator. The state $|1\rangle$ is constructed by the formula

$$|1\rangle = a^\dagger |0\rangle \,. \tag{5.125}$$

The anticommutation relations (5.123) show that $\langle 1|1\rangle = 1$ and $\langle 1|0\rangle = 0$, and that there are no other states in the theory, since $a^{\dagger 2} = 0$. The energy eigenvalues of these states are given by

$$H|0\rangle = 0, \quad H|1\rangle = \omega|1\rangle \,. \tag{5.126}$$

The Grassmann harmonic oscillator is fundamentally different from its c-number counterpart since only one excited state exists. The energy occupation of the system is therefore either zero or one unit of ω. In this respect the Grassmann oscillator is *fermionic*. The harmonic oscillator of Sec. 4.4 is *bosonic* since an arbitrary number of units of ω may occupy a state. In position representation, the forms for the ground state and excited state may be realized as superfunctions. The operators $\mathcal{N}$ and $\mathcal{N}^\dagger$ may be expressed in terms of a complex Grassmann variable η as

$$\mathcal{N} = \eta + \frac{1}{2}\frac{\partial}{\partial \eta^*}, \quad \mathcal{N}^\dagger = \eta^* + \frac{1}{2}\frac{\partial}{\partial \eta}, \quad (5.127)$$

and the two energy eigenfunctions of the theory are given by the superfunctions

$$f_0(\eta^*, \eta) = \langle \eta, \eta^* | 0 \rangle = \frac{1}{\sqrt{2}}\left(1 + \sqrt{2}\eta^* + 2\eta^*\eta\right),$$

$$f_1(\eta^*, \eta) = \langle \eta, \eta^* | 1 \rangle = \frac{1}{2}\left(1 + 2\sqrt{2}\eta - 2\eta^*\eta\right), \quad (5.128)$$

normalized to the Grassmann measure $d\eta\, d\eta^*$.

Exercise 5.17: Verify (5.127) and (5.128).

Result (5.128) allows a determination of the inner product, giving

$$\begin{aligned}\langle \eta, \eta^* | \eta', \eta'^* \rangle &= \frac{3}{4} + \frac{\sqrt{2}}{2}(\eta + \eta^* + \eta' + \eta'^*) + \eta^*\eta' + 2\eta\eta'^* \\ &\quad + \sqrt{2}(\eta^*\eta'\eta'^* + \eta\eta^*\eta') - \sqrt{2}(\eta\eta'\eta'^* + \eta\eta^*\eta'^*) \\ &\quad + 3\eta\eta^*\eta'\eta'^*,\end{aligned} \quad (5.129)$$

and

$$\int d\eta\, d\eta^* |\eta, \eta^*\rangle\langle \eta, \eta^*| = 1. \quad (5.130)$$

Exercise 5.18: Verify (5.129) and (5.130).

5.5 Grassmann Path Integrals

In this section the quantum transition amplitude between two Grassmann coherent states, associated with the fermionic harmonic oscillator analyzed in the Sec. 5.4, will be derived and evaluated using the results of previous sections. In the next section the fermionic harmonic oscillator will be combined with the bosonic harmonic oscillator to obtain a *supersymmetric* quantum mechanical system.

Throughout this section attention will be restricted to the fermionic harmonic oscillator of Sec. 5.4. The starting point is the definition of fermionic coherent states. These are analogous in form to the bosonic coherent states of Sec. 4.4, but employ Grassmann variables. In terms of the creation and annihilation operators of (5.123) and the ground state $|0\rangle$ of (5.124), the coherent state is written, using the complex Grassmann variable η, as

$$|\eta\rangle = \exp\left(\eta a^\dagger + \eta^* a\right)|0\rangle. \qquad (5.131)$$

From the definition (5.131) it is straightforward to find the inner product of two coherent states, as well as the expectation value of the creation and annihilation operators. The Grassmann coherent states share many similar properties to the c-number coherent states of Sec. 4.4. Among these are the following.

$$\langle\eta|\xi\rangle = \eta^*\xi + (1 + \tfrac{1}{2}\eta\eta^*)(1 + \tfrac{1}{2}\xi\xi^*), \qquad (5.132)$$

$$\int d\eta\, d\eta^*\, |\eta\rangle\langle\eta| = |0\rangle\langle 0| + |1\rangle\langle 1| \equiv 1, \qquad (5.133)$$

$$\langle\eta|a^\dagger|\xi\rangle = \eta^*(1 + \tfrac{1}{2}\xi\xi^*) = \eta^*\langle\eta|\xi\rangle, \qquad (5.134)$$

$$\sum_{n=0}^{1}\langle n|\eta\rangle\langle\xi|n\rangle = \exp(\tfrac{1}{2}\eta\eta^* + \tfrac{1}{2}\xi\xi^* + \eta\xi^*), \qquad (5.135)$$

$$\langle\eta|a^\dagger a|\xi\rangle = \eta^*\xi = \eta^*\xi\langle\eta|\xi\rangle. \qquad (5.136)$$

Exercise 5.19: Verify (5.132) through (5.136).

The results of Exercise 5.19 allow the derivation of a path integral representation of the transition amplitude between two instantaneous fermionic coherent states. Combining (5.133) with a repetition of the steps in Sec. 2.1 yields

$$\langle\eta_b, t_b | \eta_a, t_a\rangle =$$

$$\int \mathcal{D}\eta\, \langle\eta_b, t_b | \eta_{N-1}, t_{N-1}\rangle\langle\eta_{N-1}, t_{N-1}| \cdots |\eta_1, t_1\rangle\langle\eta_1, t_1 | \eta_a, t_a\rangle, \quad (5.137)$$

where the measure appearing in (5.137) is given by

$$\mathcal{D}\eta = \prod_{j=1}^{N-1} d\eta_j\, d\eta_j^*, \qquad (5.138)$$

and $t_{j+1} - t_j = \epsilon$. Each of the N infinitesimal matrix elements is further reduced by using the form of the Hamiltonian, $H = \omega a^\dagger a$, and result (5.136) to obtain

$$\langle \eta_{j+1}, t_{j+1} | \eta_j, t_j \rangle =$$
$$\langle \eta_{j+1} | e^{-i\epsilon H} | \eta_j \rangle \approx \exp(-i\epsilon\omega\eta^*_{j+1}\eta_j)\langle \eta_{j+1} | \eta_j \rangle . \quad (5.139)$$

Finally, making the formal identification $\eta_{j+1} - \eta_j = \epsilon\dot\eta_j$, and dropping all terms $O(\epsilon^2)$, allows the matrix element to be written

$$\langle \eta_{j+1} | \eta_j \rangle \approx 1 - \tfrac{1}{2}\epsilon\eta^*_j\dot\eta_j + \tfrac{1}{2}\epsilon\dot\eta^*_j\eta_j = \exp\left[-\epsilon\tfrac{1}{2}(\eta^*_j\dot\eta_j - \dot\eta^*_j\eta_j)\right] . \quad (5.140)$$

Exercise 5.20: Verify (5.140).

Therefore, the entire transition element becomes, in the limit $N \to \infty$,

$$\langle \eta_b, t_b | \eta_a, t_a \rangle =$$
$$\int_{\eta_a}^{\eta_b} \mathcal{D}\eta \exp\left\{i\int_{t_a}^{t_b} dt \left[\tfrac{1}{2}i\eta^*\dot\eta - \tfrac{1}{2}i\dot\eta^*\eta - \omega\eta^*\eta\right]\right\} , \quad (5.141)$$

where the limits indicate that $\eta_0 = \eta_a$ and $\eta_N = \eta_b$. Once again, the familiar form of the path integral has emerged, with the slight, but important, difference that the time derivatives in the action appear in a symmetrized manner.

It is instructive to evaluate the path integral (5.141) in a manner as similar as possible to the c-number harmonic oscillator of Sec. 3.3. There, the action was expanded about a c-number solution to the classical equation of motion, $q = x + q_c$, and the boundary conditions on q_c were chosen to match the initial and final states of the system. It is not possible to apply this technique to the Grassmann amplitude (5.141), where, for simplicity, the limit $\omega \to 0$ will be taken. Because the classical equation of motion is first order, it is not possible to match *both* the initial and final states, and still satisfy the equation of motion.

However, one of the key aspects of evaluating the c-number harmonic oscillator by this method was the result that the fluctuations, x, had the property that $x = 0$ at the initial and final times. It is still possible to define fluctuations for the Grassmann system with this property, and this will allow integrations by parts in the action. This is accomplished by using the translation

$$\eta = \chi + \chi_c , \quad (5.142)$$

where χ_c is given by

$$\chi_c = \theta(t-\tau)\eta_b + \theta(\tau-t)\eta_a ,\qquad (5.143)$$

where θ is the step function defined in (2.17), and τ is an arbitrary time in the interval (t_a, t_b). The function χ_c has the property that it matches both the initial and final states at the respective times, so that χ must vanish at the initial and final times. However, by virtue of (2.18), its derivative is given by

$$\frac{\partial}{\partial t}\chi_c = (\eta_b - \eta_a)\delta(t-\tau) .\qquad (5.144)$$

Rewriting (5.141) by substituting (5.142) gives

$$\langle \eta_b, t_b | \eta_a, t_a \rangle = \exp(\tfrac{1}{2}\eta_b^*\eta_a - \tfrac{1}{2}\eta_a^*\eta_b)$$
$$\times \int_0^0 \mathcal{D}\chi \exp\left\{i\int_{t_a}^{t_b} dt \left[\tfrac{1}{2}i\chi^*\dot\chi - \tfrac{1}{2}i\dot\chi^*\chi \right.\right.$$
$$\left.\left. + i\chi^*(\eta_b - \eta_a)\delta(t-\tau) - i(\eta_b^* - \eta_a^*)\chi\delta(t-\tau)\right]\right\}.\qquad (5.145)$$

The next step in the evaluation is to write the action appearing in (5.145) in a matrix form, allowing result (5.52) to be used. This is accomplished by using the identity

$$\int dt\, \chi^*\dot\chi = \int dt\, dt'\, \chi^*(t)D(t-t')\chi(t') ,\qquad (5.146)$$

where $D(t-t') = \frac{d}{dt}\delta(t-t')$. It is straightforward to show that

$$\int dt\, \theta(t_1-t)D(t-t_2) = \delta(t_1-t_2) ,\qquad (5.147)$$

so that $\theta(t'-t)$ is the inverse of $D(t-t')$ in the functional matrix formalism developed in Sec. 1.4. This allows the remaining path integral of (5.145) to be evaluated, and this is left as an exercise.

Exercise 5.21: By using (5.52) and the result that $\theta(0) = \tfrac{1}{2}$, show that the $\omega = 0$ path integral of (5.145) can be evaluated to obtain

$$\langle \eta_b, t_b | \eta_a, t_a \rangle = \exp(\eta_b^*\eta_a - \tfrac{1}{2}\eta_b^*\eta_b - \tfrac{1}{2}\eta_a^*\eta_a) .\qquad (5.148)$$

Sec. 5.5 Grassmann Path Integrals

Exercise 5.22: Using similar arguments, show that the $\omega \neq 0$ case may be evaluated to obtain

$$\langle \eta_b, t_b | \eta_a, t_a \rangle = \exp(\eta_b^* \eta_a e^{-i\omega(t_b - t_a)} - \tfrac{1}{2}\eta_b^* \eta_b - \tfrac{1}{2}\eta_a^* \eta_a) \,. \tag{5.149}$$

The partition function associated with the Grassmann modes is given by the standard definition,

$$Z_\beta = \sum_n \langle n | e^{-\beta H} | n \rangle \,. \tag{5.150}$$

and this may be related to the Grassmann transition amplitude in a manner similar to the bosonic case of Sec. 4.2. Using the completeness of the Grassmann coherent states, the partition function (5.150) can be written

$$Z_\beta = \sum_n \int d\eta \, d\eta^* \, \langle n | \eta \rangle \langle \eta | e^{-\beta H} | n \rangle \,. \tag{5.151}$$

Because of the anticommutativity of the Grassmann variables, it follows that

$$Z_\beta = \int d\eta \, d\eta^* \sum_n \langle -\eta | e^{-\beta H} | n \rangle \langle n | \eta \rangle = \int d\eta \, d\eta^* \, \langle -\eta | e^{-\beta H} | \eta \rangle \,. \tag{5.152}$$

The shows that the partition function is obtained from the real time Grassmann transition amplitude

$$Z_{ab} = \langle -\eta | e^{-iH(t_b - t_a)} | \eta \rangle \,, \tag{5.153}$$

by replacing $t_b - t_a \to -i\beta$ and summing over all final and initial values that are *antiperiodic*.

Using (5.149) and result (5.153), it is straightforward to show that the partition function of the fermionic oscillator is given by

$$Z_\beta = 1 + e^{-\beta\omega} \,. \tag{5.154}$$

Exercise 5.23: Verify (5.154).

5.6 Supersymmetric Quantum Mechanics

In this section a simple quantum mechanical model exhibiting Bose–Fermi symmetry, or *supersymmetry* [5, 6], will be presented. While there is no current experimental evidence to indicate any manifestation of supersymmetry in nature, models possessing supersymmetry have many compelling and desirable theoretical properties [6]. Supersymmetry has been described as a solution in search of a problem; however, since its popularity remains undiminished, a discussion of its simpler properties, within the context of supersymmetric quantum mechanics [2, 7], is included.

The starting point of the model is to combine the bosonic oscillator of Sec. 4.4 with the fermionic oscillator discussed in the previous two sections. The action is given by

$$S = \int dt \left(\tfrac{1}{2} m\dot{x}^2 - \tfrac{1}{2} m\omega^2 x^2 + \tfrac{1}{2} i\eta^*\dot{\eta} - \tfrac{1}{2} i\dot{\eta}^*\eta - \omega\eta^*\eta \right) , \qquad (5.155)$$

so that the fermionic and the bosonic oscillator possess the same fundamental frequency ω. This degeneracy in energy between the fermionic and the bosonic sectors is the *sine qua non* of supersymmetry. The action (5.155) is invariant under the following simultaneous transformations [2],

$$\begin{aligned} x &\to x + \delta x \equiv x - i\eta\delta\alpha^* - i\eta^*\delta\alpha , \\ \eta &\to \eta + \delta\eta \equiv \eta + (m\dot{x} + im\omega x)\delta\alpha , \\ \eta^* &\to \eta^* + \delta\eta^* \equiv \eta^* + (m\dot{x} - im\omega x)\delta\alpha^* , \end{aligned} \qquad (5.156)$$

where $\delta\alpha$ is an infinitesimal time-independent Grassmann number. It is straightforward to verify that, to the lowest order in $\delta\alpha$, the transformations of (5.156) leave the action (5.155) invariant, but only if x and η are constrained to satisfy their respective equations of motion, $\ddot{x} = -\omega^2 x$ and $\dot{\eta} = -i\omega\eta$.

This invariance, when combined with the results of Sec. 4.3, leads to the presence of two conserved *charges* in the system. This follows directly from the fact that, for coordinates satisfying the equation of motion,

$$\delta S = \int dt \, \frac{d}{dt} \left(\delta\eta_j \frac{\partial \mathcal{L}}{\partial \dot{\eta}_i} + \delta x \frac{\partial \mathcal{L}}{\partial \dot{x}} \right) = 0 , \qquad (5.157)$$

from which the two charges can be obtained,

$$\begin{aligned} G &= (m\dot{x} + im\omega x)\,\eta , \\ G^\dagger &= (m\dot{x} - im\omega x)\,\eta^* . \end{aligned} \qquad (5.158)$$

Sec. 5.6 Supersymmetric Quantum Mechanics

It is straightforward to verify that $\dot{\mathcal{G}} = \dot{\mathcal{G}}^\dagger = 0$ when combined with the equations of motion.

While the charges of (5.158) are derived from the classical Lagrangian, they can be elevated to operators by rewriting them in terms of the Bose operators, Q and P, defined in Sec. 2.1, and the Fermi operators $\mathcal{N}$ and $\mathcal{P}$, defined in Sec. 5.4. This gives

$$\mathcal{G} = (P + im\omega Q)\mathcal{N},$$
$$\mathcal{G}^\dagger = (P - im\omega Q)\mathcal{P}. \tag{5.159}$$

The operators Q and P are assumed to commute with $\mathcal{N}$ and $\mathcal{P}$. It can be verified that the two operators of (5.159) are nilpotent, $\mathcal{G}^2 = \mathcal{G}^{\dagger 2} = 0$. These two operators *generate* the supersymmetry transformations of (5.156) in the sense that

$$[Q, \mathcal{G}] = i\mathcal{N},$$
$$[Q, \mathcal{G}^\dagger] = i\mathcal{P},$$
$$\{\mathcal{N}, \mathcal{G}^\dagger\} = P - im\omega Q, \tag{5.160}$$

and these are the operator forms for (5.156).

The anticommutator of the operators of (5.159) is given by

$$\{\mathcal{G}, \mathcal{G}^\dagger\} = P^2 + m^2\omega^2 Q^2 + 2m\omega\mathcal{P}\mathcal{N} = 2mH. \tag{5.161}$$

Result (5.161) is a general property of supersymmetry: the algebra of the generators of the supersymmetric transformations closes on the generators of time (and, in the field theoretic case, spatial) translations, i.e., the Hamiltonian (and the momentum).

The restriction to the solutions of the equations of motion can eliminated by introducing the c-number *auxiliary* coordinate z into the action,

$$\mathcal{L} = \tfrac{1}{2}m\dot{x}^2 - \tfrac{1}{2}m\omega^2 x^2 + \tfrac{1}{2}i\eta^*\dot{\eta} - \tfrac{1}{2}i\dot{\eta}^*\eta - \omega\eta^*\eta - \tfrac{1}{2}z^2. \tag{5.162}$$

Under the variations

$$\delta x = i\eta\,\delta\alpha^* + i\eta^*\,\delta\alpha$$
$$\delta\eta = (m\dot{x} + im\omega x + z)\delta\alpha,$$
$$\delta z = -(i\dot{\eta}^* - \omega\eta^*)\delta\alpha - (i\dot{\eta} + \omega\eta)\delta\alpha^*, \tag{5.163}$$

it can be shown that the Lagrangian develops the total derivative

$$\mathcal{L} \to \mathcal{L} + \frac{dF}{dt},$$
$$F = \tfrac{1}{2}i(m\dot{x} + im\omega x + z)\eta\,\delta\alpha^* + \tfrac{1}{2}i(m\dot{x} - im\omega x + z)\eta^*\,\delta\alpha. \tag{5.164}$$

Exercise 5.24: Verify (5.164) and show that the generators for this transformation are conserved when evaluated using a solution of the equations of motion, and that their anticommutator is proportional to the Hamiltonian of the system.

The presence of the auxiliary coordinate z raises the number of bosonic degrees of freedom to two, thereby matching them in number with the fermionic degrees of freedom. For that reason, the equations of motion are no longer necessary to make the symmetry exact. This is a general property of all supersymmetric models, where the number of fermionic degrees of freedom must match the number of bosonic degrees of freedom in order to avoid the necessity of invoking the equations of motion.

Of course, the dynamics of the theory (5.155) is rather uninteresting since the fermionic and bosonic sectors do not interact. It is possible to introduce nonlinear terms that allow the two sectors of the theory to interact and to maintain supersymmetry. An example of such an action is given by

$$\mathcal{L} = \tfrac{1}{2}m\dot{x}^2 - \tfrac{1}{2}(\gamma x^2 - \mu^2)^2 + \tfrac{1}{2}i\eta^*\dot{\eta} - \tfrac{1}{2}i\dot{\eta}^*\eta - \gamma x \eta^*\eta \,. \tag{5.165}$$

Exercise 5.25: Find the supersymmetry transformations that leave (5.165) invariant, and construct the generators of these transformations. Describe the implications of the conservation of the supersymmetry charge for the states of the theory [2].

References

[1] H. Grassmann, *Hermann Grassmanns Gesammelte Mathematische und Physikalische Werke*, Vols. I1 and I2, Teubner, Leipzig, 1894.

[2] B. DeWitt, *Supermanifolds*, Cambridge University Press, Cambridge, 1985.

[3] See, for example, J. Schwinger, *Particles and Sources*, Gordon and Breach, New York, 1969.

[4] F.A. Berezin, *The Method of Second Quantization*, Academic Press, New York, 1966.

[5] Supersymmetry was first formulated by Y.A. Golfand and E.S. Likhtman, JETP Lett. **13**, 323 (1971). It was discovered in the context of the spinning string model by P. Ramond, Phys. Rev. **D3**, 2415 (1971).

[6] Applications of supersymmetry to field theory models are discussed extensively in P. West, *Introduction to Supersymmetry and Supergravity, Second Edition*, World Scientific, Singapore, 1990.

[7] Relativistic supersymmetric quantum mechanics has been used to prove index theorems (to be discussed in Chapter 9) in L. Alvarez-Gaumé, Comm. Math. Phys. 90, 161 (1983), and D. Friedan and P. Windey, Nucl. Phys. B235, 395 (1984).

Chapter 6

Field Theory

While standard quantum mechanics is extremely useful in describing systems where the potential energy is known, it has limitations that prevent it from being a complete description of nature. The two outstanding shortcomings are its incompatibility with the special theory of relativity [1] and its inability to describe systems where the number of particles is not constant. It was to remedy these two faults that relativistic quantum field theory was developed.

Historically, the first goal pursued was the formulation of a relativistic version of the Schrödinger equation, and the first candidates were the Klein–Gordon equation [2] and the Dirac equation [3]. Both of these were originally formulated as linear differential equations with potentials put in by hand much like the Schrödinger equation. However, in addition to describing the behavior of the wave functions, these theories may also be quantized by imposing an algebra on the wave functions, or *fields*, that solve them, thereby giving rise to quantum field theory [4]. This leads directly to the Fock space representation of the particles associated with the theory which reveals that both of these equations are useful since they describe fundamentally different kinds of particles. It is found that the Klein–Gordon or scalar field describes *spinless bosons*, while the Dirac field is associated with *spin-$\frac{1}{2}$ fermions*. In addition to these fields, it was found that the electromagnetic or Maxwell field and its extension, the Yang–Mills field, could also be treated in a similar manner.

In order to describe processes where the number of particles may change it is necessary to introduce terms nonlinear in the fields, and these terms are constrained to be relativistically invariant. Formulating such theories is easiest in an action functional approach, in which relativistic invariance is treated as an *a priori* symmetry of the action. The need for relativistic invariance places profound restrictions on the types of fields that may appear, as well as the nonlinear terms that can be present. Demands for additional symmetries, such as gauge invariance, further constrain the types of actions allowable.

The problem of quantum field theory then becomes one of solving the equations of motion generated by varying the action while simultaneously imposing the quantization conditions that describe the particle content of the theory. Unfortunately, very few nonlinear quantum field theories have ever been solved exactly, although many rigorous results have been established. As a result, the content of the theory is often analyzed by perturbative techniques, and there is good cause to believe that these methods are even less reliable in the field theory case than in the quantum mechanical case. Indeed, order by order, the application of perturbation theory results in divergent quantities, and it is necessary to institute a *renormalization* procedure to render the results meaningful. In some cases, such as quantum electrodynamics (QED), the perturbatively obtained results have been outstanding; in others, such as quantum chromodynamics (QCD), such results have been murky and are, at best, only an indication of the true dynamical behavior. Results based on nonperturbative techniques are more scant, but many of these have come from the path integral formulation of quantum field theory.

Quantum field theory is, to say the least, a much-studied area of physics and numerous outstanding texts exist [5, 6, 7, 8, 9, 10]. However, the purpose of this chapter is to develop the path integral representation of quantized field processes. In order to do this in a meaningful way a connection to the Fock space used to model the asymptotic scattering states must be made. Without such a connection the meaning of the path integral becomes unclear, and its application to measurable processes, such as scattering and particle production, becomes an *auto da fé* rather than a straightforward method. In Sec. 6.1 a simple mechanical model is used to derive the path integral measure of a free field theory. This is used as a heuristic device to motivate later techniques. In Sec. 6.2 relativistic invariance is formulated in terms of a symmetry transformation of the fields, and scalar, spinor, and vector representations are presented. In Sec. 6.3 the action functionals consistent with free or linear theories are constructed for scalar and spinor fields, and the Euler–Lagrange equation is derived. In Sec. 6.4 Noether's theorem is presented and applications are made to demonstrate the implications of relativistic invariance for conservation laws at the classical level. In Sec. 6.5 the free fields are quantized and the Fock space representation is developed. In Sec. 6.6 the S-matrix is defined and S-matrix elements are reduced to give an expression to transition amplitudes in terms of time-ordered products of field operators. In Sec. 6.7 these transition amplitudes are given a perturbative representation in terms of time-ordered products of interaction picture fields. In Sec. 6.8 coherent states for the field theoretic case are presented. The results of the chapter are then combined to derive the path integral representation of the vacuum persistence functional in a nonlinear field theory, as well as a list of the assumptions necessary to derive it by this approach.

6.1 A Mechanical Model

In this section a simple mechanical model of a free field theory will be presented [11]. The purpose is to motivate certain aspects of field theory, as well as giving a heuristic definition of the path integral measure.

The model is a one-dimensional string of length L, fixed at each end, with N point particles spaced uniformly along it, so that the separation of each of the masses is $\Delta x = L/(N+1)$. Each of the point particles has the mass m, so that the mass per length of the string is $\rho = Nm/L$, and the position along the string of the nth particle is given by $x_n = n\Delta x$. The point masses are constrained to vibrate only in one fixed vertical direction, and the vertical displacement of the nth particle is denoted y_n. The kinetic energy of the point particles is given by

$$T = \sum_{n=1}^{N} \tfrac{1}{2} m \dot{y}_n^2 . \tag{6.1}$$

It is assumed that there is a potential energy associated with distortions of the string and that this, to the lowest order in the y_n, can be written

$$V = \sum_{n=1}^{N+1} \tfrac{1}{2}(\kappa N/L)(y_n - y_{n-1})^2 , \tag{6.2}$$

where κ is a constant with the units of length^{-2}, therefore identified as the string tension, while $y_0 = y_{N+1} = 0$. The classical action for this system is therefore given by

$$S = \int_{t_a}^{t_b} dt \sum_{n=1}^{N+1} \left[\tfrac{1}{2} m \dot{y}_n^2 - \tfrac{1}{2}(\kappa N/L)(y_n - y_{n-1})^2 \right] . \tag{6.3}$$

A Lagrangian density equivalent to (6.3) is given by

$$S = \int_{t_a}^{t_b} dt \sum_{n=1}^{N+1} \left[p_n \dot{y}_n - \frac{p_n^2}{2m} - \tfrac{1}{2}(\kappa N/L)(y_n - y_{n-1})^2 \right] . \tag{6.4}$$

Field Theory

Varying (6.3) or (6.4) results in a set of N coupled harmonic oscillator differential equations.

The *continuum limit* of (6.4) can be obtained by the following replacements. The displacement from equilibrium for the nth particle is written $y_n(t) = \phi(x_n, t)$, where ϕ is a function of both x and t, while the momentum of the nth particle is given by $p_n(t) = \pi(x_n, t)\Delta x$. The function $\pi(x, t)$ is referred to as the *momentum density*. Assuming that $\Delta x \to 0$ as ρ is held fixed, y_n is replaced with

$$\dot{y}_n(t) \to \frac{\partial \phi(x_n, t)}{\partial t} \equiv \dot{\phi}(x_n, t),$$

$$y_n(t) - y_{n-1}(t) \to \frac{\partial \phi(x_n, t)}{\partial x_n} \Delta x \equiv \partial \phi(x_n, t) \Delta x, \quad (6.5)$$

so that in the limit $\Delta x \to 0$ the action becomes

$$S =$$
$$\int_{t_a}^{t_b} dt \sum_{n=1}^{N+1} \Delta x \left[\pi(x_n, t)\dot{\phi}(x_n, t) - \frac{\pi^2(x_n, t)}{2\rho} - \frac{1}{2}\kappa \left(\frac{\partial \phi(x_n, t)}{\partial x_n} \right)^2 \right]$$
$$\to \int_{t_a}^{t_b} dt \int_0^L dx \left[\pi\dot{\phi} - \frac{\pi^2}{2\rho} - \frac{1}{2}\kappa(\partial\phi)^2 \right] \quad (6.6)$$

The function ϕ is a *field* variable, since it describes the configuration of the string throughout space.

The equation of motion for ϕ is found by first varying (6.3) and then taking the continuum limit. The result is

$$\left(\frac{1}{v^2} \frac{\partial^2}{\partial t^2} - \frac{\partial^2}{\partial x^2} \right) \phi(x, t) = 0, \quad (6.7)$$

where $v = \sqrt{k/\rho}$. The equation (6.7) describes *transverse waves* travelling with the velocity of propagation v along the string, and any function of the form $f(x \pm vt)$ is a solution. In the remainder of this book it will be assumed that the velocity of propagation is $v = c = 1$ in natural units, and for that reason it is convenient to introduce the d'Alembertian operator, $\Box$, which in one spatial dimension stands for

$$\Box_x \equiv \left(\frac{\partial^2}{\partial t^2} - \frac{\partial^2}{\partial x^2} \right). \quad (6.8)$$

When there is no chance of confusion the subscript will be dropped, so that (6.7) can be written $\Box\phi = 0$. The generalization to n Cartesian spatial dimensions is

$$\Box_x \equiv \frac{\partial^2}{\partial t^2} - \sum_{i=1}^{n} \frac{\partial^2}{\partial x_i^2} . \qquad (6.9)$$

The quantized version of this system is obtained by imposing the standard canonical commutation relations on the y_n and p_n and taking the continuum limit. Since y_n and p_n are assumed to be time dependent, they must be understood as Heisenberg picture operators. Therefore, the commutation relation must be enforced at *equal times*, since at different times the commutation relations of the operators depend upon the dynamics. The equal-time commutation relation (ETCR) gives

$$[y_j(t), p_k(t)] = i\,\delta_{jk} \;\Rightarrow\; [\phi(x_j,t), \pi(x_k,t)] = i\frac{\delta_{jk}}{\Delta x} . \qquad (6.10)$$

In the continuum limit (6.10) gives the ETCR

$$[\phi(x,t), \pi(y,t)] = i\delta(x-y) . \qquad (6.11)$$

Exercise 6.1: Demonstrate the validity of identifying the $\Delta x \to 0$ limit of the right-hand side of (6.10) with the continuum Dirac delta.

From (6.6) it is apparent that

$$\pi(x,t) = \frac{\delta S}{\delta \dot\phi(x,t)} . \qquad (6.12)$$

Relation (6.12) is an extension of the mechanical result to a continuum system and is assumed to be valid in every approach to quantum field theory based on the canonical quantization condition (6.11) [4]. Result (6.12) allows the identification of the canonical momentum in more complicated systems, and from that the canonical commutation relations may be imposed to quantize the system. For the mechanical model of this section it is not difficult to see that the continuum limit of (6.3) gives $\pi = \dot\phi$.

At this point there are two approaches to analyzing the behavior of this system. The first is to construct an operator formalism built around the equation of motion (6.7) and the commutation relation (6.11). This will be pursued in Sec. 6.5 for relativistic fields. The second approach is to use the results of previous chapters in conjunction with the continuum methods illustrated here to write down a path integral representation of a quantum field theory transition element. Since this has great heuristic value, the second approach will be taken.

156   Field Theory

Because the basic model is a set of coupled harmonic oscillators the results of Sec. 4.4 will be used as the generalized transition element for the string. The N particle extension of (4.96) gives the transition element

$$\langle q_b, p_b, t_b | q_a, p_a, t_a \rangle = \exp \tfrac{1}{2} i \sum_{n=1}^{N} (p_{na} q_{na} - p_{nb} q_{nb}) \int_{q_a, p_a}^{q_b, p_b} \mathcal{D}p\, \mathcal{D}q$$

$$\times \exp \left\{ i \int_{t_a}^{t_b} dt \sum_{n=1}^{N} \left[p_n \dot{q}_n - \frac{p_n^2}{2m} - \frac{\kappa N}{2L} (q_n - q_{n-1})^2 \right] \right\}, \quad (6.13)$$

where the measure is given by

$$\mathcal{D}p\, \mathcal{D}q = \prod_{n=1}^{N} \prod_{j=1}^{N_t - 1} \frac{dp_{nj}}{2\pi} dq_{nj}. \quad (6.14)$$

and N_t specifies the number of partitions of $t_b - t_a$ that have been made. The initial and final configurations for the string states are given by specifying the arbitrary real functions, $\phi_a(x)$, $\phi_b(x)$, $\pi_a(x)$, and $\pi_b(x)$. These functions determine the initial and final values of the q and p variables through the identifications

$$q_{n(a,b)} = \phi_{(a,b)}(x_n), \quad p_{n(a,b)} = \pi_{(a,b)}(x_n) \Delta x. \quad (6.15)$$

Because of the relationship (6.15), the initial and final states will be indexed by the ϕ and π functions. Applying the continuum limit to (6.13) and (6.14) gives the transition element

$$\langle \phi_b, \pi_b, t_b | \phi_a, \pi_a, t_a \rangle = \int_{\phi_a, \pi_a}^{\phi_b, \pi_b} \mathcal{D}\pi\, \mathcal{D}\phi \exp \left\{ i \int_{t_a}^{t_b} dt \int_0^L dx \left[\tfrac{1}{2} \pi \dot{\phi} \right. \right.$$

$$\left. \left. - \tfrac{1}{2} \dot{\pi} \phi - \frac{\pi^2}{2\rho} - \tfrac{1}{2} \kappa (\partial \phi)^2 \right] \right\}, \quad (6.16)$$

where the measure has become

$$\mathcal{D}\pi\, \mathcal{D}\phi = \lim_{N, N_t \to \infty} \prod_{n=1}^{N} \prod_{j=1}^{N_t - 1} \Delta x \frac{d\pi(x_n, t_j)}{2\pi} d\phi(x_n, t_j), \quad (6.17)$$

and the phase factor present in (6.13) has been absorbed into the argument of the exponential by the techniques of Sec. 4.4.

A final simplification is achieved by scaling to natural units. Choosing $v = 1$, or $\kappa = \rho$, and making the replacements $\pi \to \sqrt{\kappa} \pi$ and $\phi \to \phi/\sqrt{\kappa}$

Sec. 6.1 A Mechanical Model

gives the form

$$\langle \phi_b, \pi_b, t_b | \phi_a, \pi_a, t_a \rangle = \int_{\phi_a, \pi_a}^{\phi_b, \pi_b} \mathcal{D}\pi \mathcal{D}\phi \exp\left\{ i \int_{t_a}^{t_b} dt \int_0^L dx \left[\tfrac{1}{2}\pi\dot{\phi} \right. \right.$$

$$\left. \left. - \tfrac{1}{2}\dot{\pi}\phi - \tfrac{1}{2}\pi^2 - \tfrac{1}{2}(\partial\phi)^2 \right] \right\}, \quad (6.18)$$

where the measure now reads

$$\mathcal{D}\pi \mathcal{D}\phi = \lim_{N, N_t \to \infty} \prod_{n=1}^{N} \prod_{j=1}^{N_t - 1} \Delta x \, \frac{d\pi(x_n, t_j)}{2\pi} \, d\phi(x_n, t_j). \quad (6.19)$$

In these units ϕ is dimensionless, while π has units of (length)$^{-1}$. These units for ϕ and π are specific to the case of one spatial dimension, so that ϕ and π will have different units for a string free to vibrate in more than one spatial dimension. For the special case that $\pi_a = \pi_b = 0$ it is straightforward to integrate the π variables by writing the action as a Riemann sum to obtain

$$\langle \phi_b, t_b | \phi_a, t_a \rangle =$$

$$\int_{\phi_a}^{\phi_b} \overline{\mathcal{D}}\phi \exp\left\{ i \int_{t_a}^{t_b} dt \int_0^L dx \left[\frac{1}{2}\left(\frac{\partial\phi}{\partial t}\right)^2 - \frac{1}{2}\left(\frac{\partial\phi}{\partial x}\right)^2 \right] \right\}, \quad (6.20)$$

where the measure in (6.20) is given by

$$\overline{\mathcal{D}}\phi = \lim_{N, N_t \to \infty} \prod_{n=1}^{N} \prod_{j=1}^{N_t - 1} \left(\frac{\Delta x}{2\pi i \Delta t}\right)^{\frac{1}{2}} d\phi(x_n, t_j). \quad (6.21)$$

Exercise 6.2: Verify (6.20) and (6.21).

A number of important points have emerged. The first is that continuous systems can be analyzed by action techniques. The details of the action approach to field theory will be presented in Sec. 6.3. The second point is that canonical commutation relations can be defined for a field theory in a manner very similar to that of standard quantum mechanics. The third point is that the path integral describing transitions between arbitrary initial and final spatial configurations of the system could be defined by analogy with the N-body quantum mechanical system, and several forms for the measure were found.

158 Field Theory

The content of these path integrals gives the following intuitive picture. The transition between two quantum field theory states that describe possible configurations for the system is found by summing over all possible paths through *function space* between these two configurations, and each of these paths is weighted by the exponential of the value of the action along the path. This summation is represented by treating the field at each space and time point as an independent variable and integrating over all possible valuables. In order to perform this integration the action must be broken into a Riemann sum, and this complicates the evaluation if derivatives are present, since these relate the field at separate points. This approach to defining the path integral will be referred to as *configuration measure*. Later in this chapter the path integral transition amplitude between coherent states will be derived from the operator formalism. The result will be equivalent, save that the measure will be written as an integration over Fourier components of the fields. The path integral (6.20) with a source term will be evaluated in Chapter 8 to generate time-ordered products of field operators, as well as to identify the perturbative representation of nonlinear field theories.

6.2 Relativity and Group Theory

In order for physics to be meaningful, it must be possible for two different observers to compare their observations. This means that measurements of fundamental observables must be *related* from one observer to another by a mutually agreed upon set of rules. This requires identifying both the rules for relating observables and the set of observables that can be related by these rules. For the case that two observers are in relative motion in the absence of a gravitational field the method is that of *special relativity* [1].

There are two postulates of special relativity: first, that the speed of light is invariant for all observers regardless of relative motion, and second, that the fundamental equations of physics must take the same form for all inertial observers. The first postulate identifies the mathematical relationship between the measurements of different observers, while the second identifies those mathematical objects from which theories may be built.

The first postulate can be understood in the following way. If $(\mathbf{x}, t)$ and $(\mathbf{x}', t')$ represent the spatial coordinates and the time of a light pulse according to two observers observing its motion, then in natural units ($c = 1$) it must be that

$$t^2 - x^j x^j = t'^2 - x'^j x'^j . \tag{6.22}$$

It is convenient to introduce relativistic notation. The variable t is denoted x^0, and an event in space and time is then described by the four-vector $x = (x^0, \mathbf{x})$, whose four components are denoted x^μ, $\mu = 0, \ldots, 3$. Repeated Greek indices will be summed over all four values, while repeated Latin indices will represent a sum over only the spatial indices $(1, 2, 3)$. The equation (6.22) can be viewed as relation between dot products of vectors in a four-dimensional space-time, and this is best represented by introducing the *metric tensor* $g_{\mu\nu}$, $(\mu, \nu = 0, 1, 2, 3)$. The metric tensor can be visualized as a 4×4 symmetric matrix g that, in the absence of a gravitational field, has the elements $g_{00} = -g_{11} = -g_{22} = -g_{33} = 1$, with all other elements vanishing. Equation (6.22) can then be written

$$g_{\mu\nu} x^\mu x^\nu \equiv x_\mu x^\mu = g_{\mu\nu} x'^\mu x'^\nu \equiv x'_\mu x'^\mu , \qquad (6.23)$$

where $x_\mu \equiv g_{\mu\nu} x^\nu$. Where there is no possibility of confusion it will be assumed that x^2 represents $x \cdot x \equiv x_\mu x^\mu$. The elements x^μ form the *contravariant* components of x, while the x_μ are the *covariant* components, and x is said to be a vector in Minkowski space. It is obvious that x^μ and x_μ are not the same, and that it is the *contraction* of x^μ against x_μ, written $x^\mu x_\mu$, that is the physically meaningful product. The inverse of $g_{\mu\nu}$ is denoted $g^{\mu\nu}$, so that $g^{\mu\nu} g_{\nu\rho} = \delta^\mu{}_\rho$. The two matrices $g_{\mu\nu}$ and $g^{\mu\nu}$ can therefore be used to "raise" and "lower" the space-time indices.

The relativistic form of the gradient operator is given by $\partial = (\partial/\partial t, \vec{\nabla})$, and the components of the gradient are written $\partial_\mu = \partial/\partial x^\mu$. It follows that the d'Alembertian introduced in Sec. 6.1 can be written $\Box = g^{\mu\nu} \partial_\mu \partial_\nu = \partial_\mu \partial^\mu$. The differential operator d for a function of x can then be written $d = dx^\mu \partial_\mu$.

The value of x^2 gives the square of the length of the vector x. Because of the metric of Minkowski space, this value can be positive, zero, or negative, and these respective cases describe vectors that are *timelike*, *lightlike*, or *spacelike*. This will also be true for the infinitesimal line element, ds, found from dx^μ by

$$ds^2 = dx_\mu dx^\mu = g_{\mu\nu} dx^\mu dx^\nu . \qquad (6.24)$$

The physical constraint (6.22) can be used to determine the allowed transformations between coordinate systems, which, with no loss of generality, can be written $x \to x' = x'(x)$. According to (6.22), under such a transformation the line element (6.24) must be invariant, so that

$$ds^2 = g_{\mu\nu} dx'^\mu dx'^\nu = g_{\mu\nu} \frac{\partial x'^\mu}{\partial x^\alpha} \frac{\partial x'^\nu}{\partial x^\beta} dx^\alpha dx^\beta = g_{\alpha\beta} dx^\alpha dx^\beta , \qquad (6.25)$$

and it immediately follows that

$$g_{\alpha\beta} = g_{\mu\nu} \frac{\partial x'^\mu}{\partial x^\alpha} \frac{\partial x'^\nu}{\partial x^\beta} . \qquad (6.26)$$

160   Field Theory

Using the fact that g is independent of x, differentiation of (6.26) and use of the symmetry of $g_{\mu\nu}$ shows that the change of coordinates must satisfy

$$\frac{\partial^2 x'^{\mu}}{\partial x^{\alpha} \partial x^{\beta}} = 0 \,. \tag{6.27}$$

The general solution to (6.27) is given by

$$x'^{\mu} = \Lambda^{\mu}{}_{\nu} x^{\nu} + b^{\mu} \,, \tag{6.28}$$

where $\Lambda^{\mu}{}_{\nu}$ and b^{μ} are independent of x. For the case that $b^{\mu} = 0$ (6.28) is known as a *Lorentz transformation*. For the general case that $b^{\mu} \neq 0$ it is known as a *Poincaré transformation*. For the moment it is useful to concentrate on the properties of Lorentz transformations, so that $\Lambda^{\mu}{}_{\nu} = \partial x'^{\mu}/\partial x^{\nu}$ and

$$g_{\mu\nu} = g_{\alpha\beta} \Lambda^{\alpha}{}_{\mu} \Lambda^{\beta}{}_{\nu} \,. \tag{6.29}$$

If the values $\Lambda^{\mu}{}_{\nu}$ are viewed as the elements of a 4×4 matrix Λ, then the form of g and relation (6.29) give

$$\det g = \det(\Lambda^T g \Lambda) \;\Rightarrow\; \det \Lambda = \pm 1 \,. \tag{6.30}$$

The case $\det \Lambda = +1$ is known as a *proper* Lorentz transformation, while $\det \Lambda = -1$ corresponds to an *improper* Lorentz transformation. Examples of improper transformations include *time reversal*, $(t, \mathbf{x}) \to (-t, \mathbf{x})$, and *spatial inversion*, $(t, \mathbf{x}) \to (t, -\mathbf{x})$. The latter transformation changes a right-handed spatial coordinate system to a left-handed coordinate system, and it is therefore referred to as a *parity* transformation. The behavior of quantum systems under parity transformations is very important, and the action of this special transformation on the fields of a theory will be discussed.

Exercise 6.3: Show that

$$(\Lambda^0{}_0)^2 = 1 + \Lambda^j{}_0 \Lambda^j{}_0 \,. \tag{6.31}$$

Result (6.31) shows that $|\Lambda^0{}_0| \geq 1$, so that the transformations may be characterized by the range of values for $\Lambda^0{}_0$. If $\Lambda^0{}_0 \geq +1$, the transformation is said to be *orthochronous*, while the case $\Lambda^0{}_0 \leq -1$ is called *nonorthochronous*. It is obvious that nonorthochronous transformations correspond to the generalization of time reversal. The Lorentz transformations can now be placed into four categories: proper orthochronous

($L_+^\uparrow$), improper orthochronous ($L_-^\uparrow$), proper nonorthochronous ($L_+^\downarrow$), and improper nonorthochronous ($L_-^\downarrow$).

Result (6.22) is very similar to the restriction that rotations, familiar from Sec. 4.3, must preserve the length of spatial vectors. It should be apparent that the set of rotations, i.e., those transformations that act solely upon the spatial coordinates, forms a subset of all possible Lorentz transformations, since they also leave x^2 invariant. In matrix form this subset can be written

$$\Lambda_R = \begin{pmatrix} 1 & 0 \\ 0 & \mathbf{R} \end{pmatrix}, \qquad (6.32)$$

where $\mathbf{R}$ is the orthogonal 3 × 3 matrix representing the rotation. From the analysis of Sec. 4.3 it is recalled that $\mathbf{R}$ is characterized by three free parameters, and these are the three angles θ_i necessary to describe the rotation. In addition, it is assumed that transformations of the form (6.32) include spatial inversions. Those transformations of the form (6.32) that preserve parity therefore belong to $L_+^\uparrow$, while those that change parity belong to $L_-^\uparrow$.

An infinitesimal Lorentz transformation takes the form

$$\Lambda^\mu{}_\nu \approx \delta^\mu{}_\nu + \lambda^\mu{}_\nu, \qquad (6.33)$$

and substituting (6.33) into (6.29) gives

$$0 = g_{\nu\alpha}\lambda^\alpha{}_\mu + g_{\mu\alpha}\lambda^\alpha{}_\nu = \lambda_{\mu\nu} + \lambda_{\nu\mu}, \qquad (6.34)$$

where $g_{\mu\nu}\lambda^\nu{}_\beta \equiv \lambda_{\mu\beta}$. It follows from (6.34) that $\lambda_{\mu\nu}$ must be a real, antisymmetric 4 × 4 matrix, and therefore it has six arbitrary elements, or *parameters*. Three of these have already been identified as the angles of rotation associated with mixing the spatial coordinates. The other three parameters characterize the mixing of time and spatial coordinates, and are identified as the three components of the relative velocity of the two coordinate systems. These three parameters characterize the "boosts," i.e., transformations of the form

$$\begin{pmatrix} x'^0 \\ x'^j \end{pmatrix} = \frac{1}{\sqrt{1-v^2}} \begin{pmatrix} 1 & -v \\ -v & 1 \end{pmatrix} \begin{pmatrix} x^0 \\ x^j \end{pmatrix} \qquad (6.35)$$

where v is the relative velocity of the two coordinate frames assumed to lie along the j axis. The boosts give rise to the more popularly known effects of special relativity, those of time dilation and length contraction. Since $\Lambda^0{}_0 \geq +1$, it follows that the boosts belong to $L_+^\uparrow$. Likewise, time reversal must belong to $L_-^\downarrow$, and a full inversion, $x^\mu \to -x^\mu$, belongs to $L_+^\downarrow$.

162   Field Theory

There are other important four-vectors. From the fact that the energy E and time t are canonically conjugate, while the spatial momentum $\mathbf{p}$ and the spatial coordinates $\mathbf{x}$ are canonically conjugate, it follows that the energy-momentum vector $p = (E, \mathbf{p})$ should transform like the space-time vector x. For a particle with the rest mass m, the energy-momentum vector is assumed to satisfy

$$p^2 = p \cdot p = p_\mu p^\mu = E^2 - \mathbf{p} \cdot \mathbf{p} = m^2 \ . \qquad (6.36)$$

Only the positive root for E is considered physically meaningful. However, all observers agree upon the inertial mass of a particle. It also follows that $p_\mu x^\mu = Et - \mathbf{p} \cdot \mathbf{x}$ is a Lorentz invariant. Another important four-vector is formed by the scalar potential ϕ and the vector potential $\mathbf{A}$ of electrodynamics. The four-vector $A = (\phi, \mathbf{A})$ must transform like x in order that Maxwell's equations take the same form in every Lorentz frame.

From the discussion so far it is apparent that an arbitrary Lorentz transformation can be thought of as a 4×4 matrix Λ that is completely specified by giving the values of the six free parameters and to which of the four general categories it belongs. If $\Lambda(v, \theta)$ and $\Lambda(v', \theta')$ are two members of $L_+^\uparrow$, then, in order for special relativity to be a sensible method of connecting various observers, it must be possible to find a third transformation, i.e., a set of six parameters v'' and θ'', such that

$$\Lambda(v'', \theta'') = \Lambda(v', \theta') \Lambda(v, \theta) \ , \qquad (6.37)$$

and this transformation must also belong to $L_+^\uparrow$. Property (6.37) means that the set of all Lorentz transformations belonging to $L_+^\uparrow$ forms a *group* under matrix multiplication [12, 13]. A group G is a collection of objects $\{g\}$, along with a definition of multiplication of these objects, such that the product of any two objects in the group is also a member of the group, $g_1 g_2 \in G$, and such that the product is associative, $(g_1 g_2) g_3 = g_1(g_2 g_3)$. In addition, the group must possess an identity element, $I \in G$, such that $Ig = g$, and each member g of the group must possess an inverse, $g^{-1} \in G$, so that $gg^{-1} = I$.

Groups fall into two general categories, *discrete* and *continuous*. A discrete group has a finite number of elements. A simple example is the set of integers $\{1, -1\}$. The members of a continuous group are characterized by a set of parameters that can take any values in some set of intervals. Although not necessary, it is standard to identify the identity element with the origin $\{0\}$ of the space of parameters. By inspection, (6.37) reveals that the set of all Lorentz transformations comprising $L_+^\uparrow$ forms a continuous six parameter group, and this is usually referred to as the *restricted Lorentz group* [14]. A subtlety enters when the other categories of Lorentz transformation are considered, and this is because they are not smoothly connected to the identity element. This can be seen intuitively in three dimensions, where it is obvious that there is no way to connect a right-handed coordinate system to a left-handed coordinate system by a rotation. Therefore, the spatial inversions cannot be reached by starting at the identity element and *smoothly* varying the six parameters. This means that the set of *all* Lorentz transformations is not *connected*.

This brief discussion of the specific case of the Lorentz group has only hinted at the rich structures available to groups, and the study of group theory has been a centerpiece of modern mathematics. More details of group structure will be presented in Chapter 7 when Lie algebras are discussed, but, for the moment, some basic ideas need to be presented. Attention will be restricted to continuous groups.

A continuous group is classified by the number and range of parameters and by the way the parameters *compose* when members of the group are multiplied. If α and β are two sets of parameters that describe two members of the group, then the product of the two members gives a third member, described by the set of parameters γ. It must be that γ is a function of α and β, and this function $\gamma(\alpha, \beta)$ gives the composition rules for the group. These properties transcend a specific matrix form for the group. A specific matrix form for the members of a group, such as the one just derived for the Lorentz group, is therefore referred to as a *representation* of the group, and there may be many of these. The set of objects upon which the specific representation acts is called the *representation space*. For the Lorentz group representation in terms of a 4×4 matrix, the representation space used to derive it was the set of space-time vectors x^μ. Each specific representation of a group is determined by the representation space chosen for it. This is quite similar to the case of quantum mechanics, where a position space representation or a momentum space representation may be chosen, and these determine the form of the operators appearing in the eigenvalue problems.

Once a representation space is defined, it is then possible to form *tensor* or *Kronecker products* of members of the space, and these constitute representation spaces for *direct product representations* of the group. For example, the tensor product $A^{\mu\nu} = x^\mu x^\nu$ of space-time vectors transforms as $A'^{\mu\nu} = \Lambda^\mu{}_\alpha \Lambda^\nu{}_\beta A^{\alpha\beta}$. A particular representation is called *irreducible* if its action on the representation space leaves no nontrivial element unchanged. If this criterion is not met, the representation is said to be *reducible*. Determining the irreducible representations of groups and their representation spaces is of great importance in physics. An example of this no doubt famil-

164 Field Theory

iar to the reader is the set of all rotations in three-dimensional space. The angular momentum operators of quantum mechanics serve as *generators* for the rotations, since their action on a function serves to rotate its argument. Generators associated with groups and algebras will be discussed in more detail in Chapter 7. Tensor products of the representation space of the rotations correspond to quantum mechanical systems where there are several types of angular momentum present. The addition of angular momentum through the Clebsch–Gordan procedure is precisely the task of finding the irreducible representations of a direct product representation of the rotation group.

If all the members of a group commute, then the group is said to be *abelian*. For example, identifying the number zero as the identity and using addition, the set of integers forms an abelian group. Groups whose members have nontrivial commutation properties are referred to as *nonabelian*. It is possible that a group possesses a subcollection H of elements which itself forms a group. Such a subcollection is referred to as a *subgroup*. In the case of the Lorentz group, it is apparent that the spatial rotations form a subgroup. The Poincaré transformation (6.28) adds the translations, $x^\mu \to x^\mu + a^\mu$, to the Lorentz group, and these translations constitute an abelian subgroup of the Poincaré group.

Exercise 6.4: Show that the Poincaré transformations form a group.

There are many cases where different groups possess similar structure, and it is possible to set up maps between them. Such is the case of the Lorentz group and the group SL(2,C). The group SL(2,C) is defined as the set of all complex 2×2 matrices with a unit determinant, so that all members are unimodular and invertible. A 2×2 complex matrix **U** possesses four complex elements for a total of eight arbitrary numbers, or parameters. The constraint that det **U** $= +1$ gives two equations that the parameters must satisfy, i.e., that the real part of det **U** is $+1$, while the imaginary part vanishes. Therefore, two of the original eight parameters may be replaced by the solutions to these equations, and a member of SL(2,C) is then specified by six independent parameters.

Exercise 6.5: Prove that the set of all complex 2×2 matrices of determinant $+1$ forms a group under matrix multiplication.

Since the Lorentz group and SL(2,C) possess an identical number of pa-

Sec. 6.2 Relativity and Group Theory   165

rameters, it is natural to inquire if there is a mapping of SL(2,C) onto the Lorentz group. The mapping must associate each element $\mathbf{U} \in SL(2,C)$ with an element Λ of the Lorentz group. This is accomplished in the following way. The Pauli spin matrices are extended to a four-vector notation by the definitions

$$\sigma_0 = \begin{pmatrix} 1 & 0 \\ 0 & 1 \end{pmatrix} = I, \quad \sigma_1 = \begin{pmatrix} 0 & 1 \\ 1 & 0 \end{pmatrix},$$

$$\sigma_2 = \begin{pmatrix} 0 & -i \\ i & 0 \end{pmatrix}, \quad \sigma_3 = \begin{pmatrix} 1 & 0 \\ 0 & -1 \end{pmatrix}. \quad (6.38)$$

Exercise 6.6: Verify that the matrices of (6.38) satisfy

$$\{\sigma_i, \sigma_j\} = \delta_{ij} I,$$
$$\sigma_i \sigma_j = i\varepsilon^{ijk} \sigma_k,$$
$$\sigma_\mu \sigma^\mu = g^{\mu\nu} \sigma_\mu \sigma_\nu = -2\sigma_0,$$
$$\mathrm{Tr}(\sigma_\mu \sigma_\nu) = 2\delta_{\mu\nu}. \quad (6.39)$$

The 2×2 matrix $\rlap{/}x$ is formed from the four-vector x and the matrices of (6.38) by the definition $\rlap{/}x = \sigma_\mu x^\mu = \sigma_0 x_0 + \sigma_j x^j$, and has the property that $\det \rlap{/}x = x_\mu x^\mu$. Although $\rlap{/}x$ has been defined using the coordinates x^μ, it is clear that any four-vector v can be associated with $\rlap{/}v$ without altering the following results. If $\mathbf{U}$ is a member of SL(2,C), then it can be used to transform $\rlap{/}x$ according to

$$\rlap{/}x = \mathbf{U}^\dagger \rlap{/}x' \mathbf{U} \Rightarrow \rlap{/}x' = (\mathbf{U}^\dagger)^{-1} \rlap{/}x \mathbf{U}^{-1}. \quad (6.40)$$

Because $\det \mathbf{U} = 1$, the determinant of (6.40) is invariant under this transformation, and it follows that $x'_\mu x'^\mu = \det \rlap{/}x' = \det \rlap{/}x = x_\mu x^\mu$. Therefore, the action of $\mathbf{U}$ on $\rlap{/}x$ defines a Lorentz transformation since the length $x_\mu x^\mu$ is preserved. In general, it is not not possible for $\mathbf{U}$ to be a unitary matrix, for then the trace of $\rlap{/}x$, $\mathrm{Tr}\,\rlap{/}x = 2t$, would be invariant under the transformation, implying that t is unchanged. Such a transformation would therefore correspond to a spatial rotation, and could not represent the effects of a boost. There is a subgroup of SL(2,C) corresponding to the unitary transformations, and since these have three free parameters they correspond to the spatial rotation subgroup of the Lorentz group.

Because all observers can agree on the form of the matrices (6.38), the transformed coordinates are given by

$$\rlap{/}x = \sigma_\mu x^\mu = \sigma_\mu x'^\nu (\Lambda^{-1})^\mu{}_\nu = \mathbf{U}^\dagger \rlap{/}x' \mathbf{U}, \quad (6.41)$$

where (6.41) simply means that for every Λ^{-1} there is at least one corresponding **U**. The particular form for (6.41) is for later convenience. Using (6.39) allows the exact form of the Lorentz transformation associated with **U** to be obtained, since (6.41) yields

$$x'^\nu (\Lambda^{-1})^\mu{}_\nu = \tfrac{1}{2} \mathrm{Tr}\,(\sigma_\mu \mathbf{U}^\dagger \not{x}' \mathbf{U}) \,. \tag{6.42}$$

Exercise 6.7: Construct the SL(2,C) transformation associated with the infinitesimal Lorentz transformation (6.33).

There are two unusual aspects to the representation (6.41). The first is that there is an ambiguity in the mapping. If the matrix **U** belongs to SL(2,C), then it is easy to see that the matrix $-\mathbf{U}$ also belongs to SL(2,C). However, the relationship (6.41) shows that *both* of these matrices are associated to the *same* Lorentz transformation Λ^{-1}. This representation of SL(2,C) is therefore not single valued, and it is in fact two to one. Such a representation is not *faithful*, and this has profound consequences for physical systems modelled with this representation. It is the source of the fermionic nature of the particles modelled with this representation, since the wave function or field must change sign under a rotation of 2π to accommodate the double-valued nature of the representation. The mapping of SL(2,C) is *onto* the Lorentz group, so that for every Lorentz transformation there is a **U**. However, it is not an *isomorphism* (one to one).

The second important aspect of this representation of SL(2,C) constitutes a drawback in that there is no way to represent improper Lorentz transformations.

Exercise 6.8: Prove that (6.41) maps SL(2,C) onto only the proper Lorentz group by proving that it is impossible to associate any $\mathbf{U} \in$ SL(2,C) with the parity operator.

This is not necessarily a calamity, but it presents a limitation for the types of particles and fields that can be modelled with this representation. The physical system being modelled by this representation must be such that it remains either left- or right-handed, for otherwise a parity operation must exist that allows the parity of the system to be changed. However, if no parity operator is required, then this representation is adequate.

It is possible to choose the parity of the representation by hand. The original choice $\not{x} = \sigma_0 x^0 + \sigma_j x^j$ could have been replaced by the parity reversed form $\tilde{\rlap{/}x} = \sigma_0 x^0 - \sigma_j x^j \equiv \tilde{\sigma}_\mu x^\mu$ without sacrificing the general structure of the representation. The matrix associated with the Lorentz transformation of x determined by acting upon $\tilde{\rlap{/}x}$ is labelled $\tilde{U}$, so that

$$\tilde{\rlap{/}x} = \tilde{\sigma}_\mu x'^\nu (\Lambda^{-1})^\mu{}_\nu = \tilde{U}^\dagger \tilde{\rlap{/}x}' \tilde{U} . \tag{6.43}$$

This ambiguity in the choice of the representation of the Lorentz group has a physical significance that will be made clear in the next section when the Weyl spinor field is given an action.

The absence of a parity operator can be remedied very easily by using a *bispinor* representation of SL(2,C). A bispinor representation is obtained by using the Pauli spin matrices to construct a set of four 4×4 matrices, denoted γ_μ,

$$\gamma_0 = \begin{pmatrix} I & 0 \\ 0 & -I \end{pmatrix}, \quad \gamma_i = \begin{pmatrix} 0 & -\sigma_i \\ \sigma_i & 0 \end{pmatrix}. \tag{6.44}$$

The indices of the γ matrices are raised and lowered with the metric tensor. The matrices of (6.44) have the property that

$$\{\gamma_\mu, \gamma_\nu\} = \gamma_\mu \gamma_\nu + \gamma_\nu \gamma_\mu = 2g_{\mu\nu} \mathbf{1} , \tag{6.45}$$

where $\mathbf{1}$ is the 4×4 identity matrix. The γ matrices may be alternately defined as any set of four matrices that satisfy the algebra of (6.45). The matrices of (6.44) are not the only set of 4×4 matrices that satisfy (6.45), and in fact any unitary transformation of (6.44) gives another representation of (6.45). The specific matrices of (6.44) are referred to as the Dirac representation of the algebra (6.45).

In the bispinor representation (6.44) $\rlap{/}x = \gamma_\mu x^\mu$ has the property that $\det \rlap{/}x = (x_\mu x^\mu)^2$, and so transformations that preserve the determinant again represent a Lorentz transformation. If U is a 4×4 matrix, the transformation of $\rlap{/}x$ is defined as

$$\rlap{/}x = U^{-1} \rlap{/}x' U = \gamma_\mu x^\nu (\Lambda^{-1})^\mu{}_\nu . \tag{6.46}$$

It is noted that it is $\mathbf{U}^{-1}$, rather than $\mathbf{U}^\dagger$, that appears in (6.46). This is because the trace of $\rlap{/}x$ is zero in the bispinor representation, and zero is a Lorentz invariant. The form of the transformation (6.46) therefore must be chosen to preserve the trace, and for this reason the inverse of $\mathbf{U}$ appears. While this representation of the Lorentz group is still unfaithful, it is now easy to find a $\mathbf{U}$ that represents a parity change because of the increased size of the matrix $\mathbf{U}$.

Exercise 6.9: Verify that $U = \gamma_0$ generates a spatial inversion, and that

$$\gamma^0 \gamma^\mu \gamma^0 = \gamma^{\mu\dagger} . \tag{6.47}$$

The matrix $\gamma^5 = i\gamma^0\gamma^1\gamma^2\gamma^3 = -\frac{1}{4}i\varepsilon_{\mu\nu\rho\sigma}\gamma^\mu\gamma^\nu\gamma^\rho\gamma^\sigma$ has many important properties. In the Dirac representation γ^5 has the form

$$\gamma_5 = \gamma^5 = \begin{pmatrix} 0 & I \\ I & 0 \end{pmatrix} . \tag{6.48}$$

Defining the matrix $\sigma_{\mu\nu} = \frac{1}{2}i[\gamma_\mu, \gamma_\nu]$, it is easy to verify that the following properties of the γ matrices are *independent* of the representation chosen.

$$\{\gamma^5, \gamma^\mu\} = [\gamma^5, \sigma_{\mu\nu}] = 0 , \tag{6.49}$$

$$\not{a}\not{b} = a \cdot b\,\mathbf{1} - i\sigma_{\mu\nu}a^\mu b^\nu , \tag{6.50}$$

$$[\sigma^{\mu\nu}, \gamma^\rho] = 4g^{\nu\rho}\gamma^\mu - 4g^{\mu\rho}\gamma^\nu , \tag{6.51}$$

$$\mathrm{Tr}(\gamma^5\gamma^\mu\gamma^\nu\gamma^\rho\gamma^\sigma) = -4i\varepsilon^{\mu\nu\rho\sigma} = 4i\varepsilon_{\mu\nu\rho\sigma} . \tag{6.52}$$

Exercise 6.10: Verify relations (6.49) through (6.52).

Exercise 6.11: Using the infinitesimal form of a Lorentz transformation defined by (6.33), show that the infinitesimal form of the bispinor transformation associated with the Lorentz transformation is given by

$$U(\Lambda) \approx 1 + \tfrac{1}{4}i\lambda_{\mu\nu}\sigma^{\mu\nu} . \tag{6.53}$$

Exercise 6.12: Suppose that the world were three-dimensional, i.e., one time and two spatial degrees of freedom. Determine the properties of this lower-dimensional world's Lorentz group.

It is now possible to classify the types of fields whose transformation properties are given by special relativity. A general field, $\Psi(x)$, is a function of space-time. If a point P in space-time is described by the coordinates x in one observer's frame and x' in another, then the two coordinates are related by a Lorentz transformation, $x'^\mu = \Lambda^\mu{}_\nu x^\nu$. The two observers must be able to relate their observations of the same field at the point P, and this means that $\Psi'(x') = D(\Lambda)\Psi(x)$, where the prime on Ψ allows for the fact that Ψ may have components that are transformed by some irreducible representation $D(\Lambda)$ of the Lorentz group. The situation is analogous to the behavior of vectors in a three-dimensional space when a rotation is performed. For such a case, not only do the coordinates of the point where the vector is attached change, the components of the vector along the new coordinates change as well, although it is the same vector that is being described in both coordinate systems. The requirement that two observers must be able to compare measurements determines the types of fields that are fundamental. They must form a representation space for the Lorentz group or the group SL(2,C), since SL(2,C) transformations may be mapped into the Lorentz group.

A real *scalar* field $\phi(x)$ is simply a single degree of freedom at the point P, and therefore it transforms as

$$\phi'(x') = \phi(x) . \tag{6.54}$$

A real *vector* field $\mathbf{A}(x)$ transforms as do the coordinates, so that it possesses four real components or degrees of freedom, denoted $A^\mu(x)$ in their contravariant form. The transformation properties are therefore

$$A'^\mu(x') = \Lambda^\mu{}_\nu A^\nu(x) . \tag{6.55}$$

A *spinor* field transforms according to the matrix $\mathbf{U}$ determined from (6.41) in the 2×2 case, or from (6.46) in the 4×4 case. In the first possibility, the field $\psi_a(x)$ is a two-component complex object referred to as a *Weyl spinor*, while in the second case $\psi_a(x)$ is a four-component complex object referred to as a *Dirac spinor*. Obviously, the Weyl spinor possesses four degrees of freedom, while the Dirac spinor has eight. The transformation of the spinor is then given by

$$\psi'_a(x') = U_{ab}\psi_b(x) . \tag{6.56}$$

A *Rarita-Schwinger* field is a combination of a Weyl or Dirac spinor and a vector, so that it is written $\psi^\mu_a(x)$, and possesses either 16 or 32 degrees of freedom. Its transformation property is simply a combination of the two, so that

$$\psi'^\mu_a(x') = \Lambda^\mu{}_\nu U_{ab}\psi^\nu_b(x) . \tag{6.57}$$

While it is also possible to define a *tensor* field $h_{\mu\nu}$, such a field is intimately related to the structure of gravitation and general relativity, and its discussion therefore lies outside the scope of this book.

The definitions of the scalar and vector fields may be enlarged to include *pseudoscalar* and *pseudovector* fields. These fields transform as scalar and vector fields with the exception that they change sign under a spatial inversion. For example, the pseudoscalar field ϕ_P has the property that

$$\phi_P(-\mathbf{x},t) = -\phi_P(\mathbf{x},t) \ . \tag{6.58}$$

As a few final remarks, it is easy to see that the volume element of space-time is invariant under proper Lorentz transformations, while it develops a sign change under an improper Lorentz transformation. The Levi–Civita symbol transforms similarly. This is because they are examples of a more general object, a *tensor density*. Under changes of coordinates, tensor densities are also multiplied by factors of the Jacobian of the transformation.

Exercise 6.13: Verify that $\varepsilon^{\mu\nu\rho\sigma} \to -\varepsilon^{\mu\nu\rho\sigma}$ under an improper Lorentz transformation.

Now that the types of fields allowed by special relativity have been identified, it is necessary to develop the equations of motion for each of them, and this is the subject of the next section.

6.3 Classical Free Fields

In this section the basic *classical* dynamics of the fields identified in the Sec. 6.2 will be developed. These dynamics will be formulated in an action functional approach in such a way that the variation of the action gives a first- or second-order partial differential equation of motion that is *covariant*, i.e., it takes the same form in all coordinate systems related by Lorentz transformations. The particle content of these field theories will be determined in Sec. 6.5 when they are quantized.

The classical action functional is assumed to be a real, dimensionless number of the form

$$S = \int_{\mathcal{M}} d^4x \, \mathcal{L}(\Psi_\alpha, \partial_\mu \Psi_\alpha) \ , \tag{6.59}$$

where the Lagrangian density $\mathcal{L}$ is some local function of the fields Ψ_α and their gradients, α being a label for all the different types of fields in the action, while $\mathcal{M}$ represents an integration over the entire space-time manifold available to the fields. Manifest covariance of the equations of motion is assured if the Lagrangian density takes the same form and value in all frames, and this simply means that the Lagrangian density must be a Lorentz scalar. Therefore, under a Lorentz transformation it must be that $\mathcal{L}'(x') = \mathcal{L}(x)$, and this is true only if all terms appearing in $\mathcal{L}$ are Lorentz scalars. This will guarantee that the action density is *form invariant* under a Lorentz transformation, and thus, using the terminology of Sec. 4.3, the Lorentz group is a *symmetry* of the action. It is typically assumed that the action is an integral over all space-time, and for that case the full symmetry is actually the Poincaré group. The implications of this symmetry will be explored in the next section.

The equations of motion are determined by demanding that the action functional is stationary against variations of the field around the classical solutions [15], $\Psi_\alpha \to \Psi_\alpha + \delta\Psi_\alpha$, where $\delta\Psi_\alpha$ is chosen to vanish at the boundaries of the manifold, or at infinity if there are no boundaries. It follows that

$$\delta S = \int_{\mathcal{M}} d^4x \left[\frac{\partial \mathcal{L}}{\partial \Psi_\alpha} \delta\Psi_\alpha + \frac{\partial \mathcal{L}}{\partial(\partial_\mu \Psi_\alpha)} \delta(\partial_\mu \Psi_\alpha) \right] = 0 . \qquad (6.60)$$

Using $\delta(\partial_\mu \Psi_\alpha) = \partial_\mu \delta\Psi_\alpha$ and integrating by parts gives

$$\delta S = \int_{\mathcal{M}} d^4x \left[\frac{\partial \mathcal{L}}{\partial \Psi_\alpha} - \partial_\mu \frac{\partial \mathcal{L}}{\partial(\partial_\mu \Psi_\alpha)} \right] \delta\Psi_\alpha = 0 . \qquad (6.61)$$

Because $\delta\Psi_\alpha$ is arbitrary, the variation of the action vanishes only if the Euler–Lagrange equations of motion,

$$\frac{\partial \mathcal{L}}{\partial \Psi_\alpha} - \partial_\mu \frac{\partial \mathcal{L}}{\partial(\partial_\mu \Psi_\alpha)} = 0 , \qquad (6.62)$$

are satisfied. The Euler–Lagrange equations (6.62) will generate linear partial differential equations if the action is chosen to be quadratic in the fields. Linear partial differential equations have the useful property that superpositions of solutions are also solutions, and this allows a general wave-packet solution to the equation of motion. For that reason the quadratic actions form the basis of most field and particle analysis. Quadratic actions for the scalar and spinor fields will now be presented, and the space-time manifold will be assumed to be infinite Minkowski space. Because of the additional problems associated with gauge invariance, the analysis of vector fields will be presented in Chapter 7.

6.3.1 The Scalar Field

If $\phi(x)$ is a scalar field, then any power of ϕ transforms as a scalar. A possible quadratic Lagrangian is then given by

$$\mathcal{L} = \tfrac{1}{2} \partial_\mu \phi \, \partial^\mu \phi - \tfrac{1}{2} m^2 \phi^2 , \qquad (6.63)$$

where m is some constant. In order that the action be dimensionless, the field ϕ must have the units of inverse length, and that implies that the constant m also has the units of inverse length.

The Euler–Lagrange equation for (6.63) immediately gives the Klein–Gordon equation [2]

$$(\partial_\mu \partial^\mu + m^2)\phi = (\Box + m^2)\phi = 0 , \qquad (6.64)$$

where the definition of the d'Alembertian (6.8), and the equivalence $\Box = \partial_\mu \partial^\mu$ has been used. Solutions to (6.64) are given by plane waves of the form

$$\phi^{(+)}(p) = \exp(ipx), \quad \phi^{(-)}(p) = \exp(-ipx) , \qquad (6.65)$$

where $px = p_\mu x^\mu$ and the four-vector p must satisfy $p^2 = m^2$. It is assumed that p_0 appearing in (6.65) has been chosen to be the positive value $p_0 = +\sqrt{m^2 + \mathbf{p}\cdot\mathbf{p}}$. Because of the obvious similarity to (6.36) the constant m will be identified as the mass of the particles associated with the field, but a proof of this must wait until the field is quantized in Sec. 6.5.

A general solution to (6.64) is found as a *wave packet* of the form

$$\phi(x) = \int \frac{d^3p}{(2\pi)^{3/2}} \frac{1}{2p_0} \left[f(\mathbf{p}) e^{ipx} + f^*(\mathbf{p}) e^{-ipx} \right] , \qquad (6.66)$$

where $\mathbf{p}$ denotes the spatial part of p, and the function $f(\mathbf{p})$ is arbitrary. Any initially localized wave packet of the form (6.66) will disperse, and this may be seen by taking the nonrelativistic limit of the *dispersion relation* $p_0 = +\sqrt{m^2 + \mathbf{p}^2}$.

Exercise 6.14: Starting with the Gaussian wave packet solution to (6.64) described by $f(\mathbf{p}) = C \exp(-\alpha \mathbf{p}^2)$, show that, as t increases beyond $t = 0$, the wave packet disperses. Show that the *group velocity* v_g, i.e., the speed of the peak of the wave packet, must satisfy $v_g < 1$.

Exercise 6.15: Determine the behavior of the function $f(\mathbf{p})$ under a Lorentz transformation.

A complex scalar field ϕ may be defined by using two real scalar fields, ϕ_1 and ϕ_2, associated with the same value of m. The Lagrangian takes the form

$$\mathcal{L} = \tfrac{1}{2}\partial_\mu \phi_1 \partial^\mu \phi_1 + \tfrac{1}{2}\partial_\mu \phi_2 \partial^\mu \phi_2 - \tfrac{1}{2}m^2(\phi_1^2 + \phi_2^2) = \partial_\mu \phi^* \partial^\mu \phi - m^2 \phi^* \phi , \qquad (6.67)$$

where $\phi = (\phi_1 + i\phi_2)/\sqrt{2}$. The action defined by (6.67) is invariant under complex conjugation, as well as phase changes of the form $\phi \to e^{i\lambda}\phi$. This symmetry will be analyzed in Sec. 6.4.

> **Exercise 6.16:** Construct a general wave packet solution to the equations of motion for the complex field.

6.3.2 Spinor Fields

The action for a two-component Weyl spinor [16] is constructed first. The first-order differential matrix operator $\not{\partial} = \sigma^\mu \partial_\mu = \sigma^0 \partial_0 + \sigma^j \partial_j = \sigma_0 \partial_0 - \sigma \cdot \nabla$ is introduced, so that the Weyl spinor transforms under the representation **U** determined by (6.41). Under a Lorentz transformation, the bilinear form $\psi^\dagger \not{\partial} \psi$, which stands for $\psi_a^*(\not{\partial})_{ab}\psi_b$, becomes

$$\psi^\dagger(x)\not{\partial}\psi(x) \to \psi'^\dagger(x')\not{\partial}'\psi'(x')$$
$$= \psi^\dagger(x)\mathbf{U}^\dagger \not{\partial}' \mathbf{U} \psi(x) = \psi^\dagger(x)\not{\partial}\psi(x) \,, \qquad (6.68)$$

so that this bilinear form is invariant. However, a term of the form $\psi^\dagger \psi \equiv \psi_a^* \psi_a$ is *not* invariant, since $\mathbf{U}^\dagger \neq \mathbf{U}^{-1}$. Therefore, the only possible bilinear action for a Weyl spinor that is first order in $\not{\partial}$ is given by

$$S = \int_{\mathcal{M}} d^4x \, i\psi^\dagger \not{\partial} \psi \,. \qquad (6.69)$$

so that ψ has dimensions of length$^{-3/2}$.

> **Exercise 6.17:** Prove that S is real and a Poincaré invariant.

Applying the Euler–Lagrange equation to the action (6.69) gives the first-order equation of motion for the Weyl spinor,

$$\not{\partial}\psi = 0 \,, \qquad (6.70)$$

along with its Hermitian conjugate. Iterating this equation and using the fact that σ^0 commutes with the Pauli spin matrices gives

$$\sigma_i \partial_i \sigma_j \partial_j \psi = \sigma_0 \partial_0 \sigma_i \partial_i \psi = (\sigma_0)^2 \partial_0^2 \psi = \partial_0^2 \psi \,. \qquad (6.71)$$

The property (6.39) of the Pauli spin matrices gives

$$\sigma^j_{ab}\sigma^k_{bc}\partial_j \partial_k = \delta^{jk}\delta_{ac}\partial_j \partial_k + i\varepsilon^{jkl}\sigma^l_{ac}\partial_j \partial_k = \delta_{ac}\nabla^2 \,, \qquad (6.72)$$

thereby reducing (6.71) to

$$\nabla^2 \psi_a = \frac{\partial^2}{\partial t^2}\psi_a \Rightarrow \Box \psi_a = 0, \ a = (1,2). \tag{6.73}$$

Each component of a solution to (6.70) must automatically satisfy the *massless* wave equation.

Any solution to (6.70) must then take the form

$$\psi_p(x) = \begin{bmatrix} f_1(p) \\ f_2(p) \end{bmatrix} e^{\pm ipx}, \tag{6.74}$$

where $p^2 = 0$. In the limit that $p_\mu = 0$, there are two independent solutions to (6.70)

$$u_o = \begin{pmatrix} 1 \\ 0 \end{pmatrix}, \ v_o = \begin{pmatrix} 0 \\ 1 \end{pmatrix}, \tag{6.75}$$

with the obvious properties that $u_o^\dagger u_o = v_o^\dagger v_o = 1$ and $u_o^\dagger v_o = 0$. The construction of the solutions for $p_\mu \neq 0$ is facilitated by noting that, if $p^2 = 0$, then

$$(\sigma_o p_o + \sigma_j p^j)(\sigma_o p_o - \sigma_i p^i) = (p_o^2 - \mathbf{p}^2)I = 0, \tag{6.76}$$

and that

$$\not{p} e^{\pm ipx} = \pm i(\sigma_o p_o + \sigma_j p^j)e^{\pm ipx}. \tag{6.77}$$

Denoting ϵ_p as the positive value for p_0, $\epsilon_p = +|\mathbf{p}|$, a possible solution is obtained by

$$\begin{aligned} u(p)e^{\pm ipx} &= N_p(\sigma_o \epsilon_p - \sigma_j p^j)u_o e^{\pm ipx} \\ &= \frac{1}{\sqrt{2\epsilon_p(\epsilon_p - p^3)}} \begin{pmatrix} -\epsilon_p + p^3 \\ p^1 + ip^2 \end{pmatrix} e^{\pm ipx}, \end{aligned} \tag{6.78}$$

where N_p is the normalization factor, and where $px = \epsilon_p t - \mathbf{p} \cdot \mathbf{x}$ in (6.78) and throughout all that follows. The other solution is obtained similarly, and is given by

$$v(p)e^{\pm ipx} = \frac{1}{\sqrt{2\epsilon_p(\epsilon_p + p^3)}} \begin{pmatrix} -p^1 + ip^2 \\ \epsilon_p + p^3 \end{pmatrix} e^{\pm ipx}. \tag{6.79}$$

The normalizations of the spinor solutions have been chosen so that the inner products are given by

$$u^\dagger(p)u(p) = v^\dagger(p)v(p) = 1, \tag{6.80}$$

while the solutions possess a type of orthonormality,

$$u^\dagger(p)u(\tilde{p}) = v^\dagger(p)v(\tilde{p}) = u^\dagger(p)v(\tilde{p}) = 0 , \qquad (6.81)$$

where $\tilde{p} = (\epsilon_p, -\mathbf{p})$. A general wave packet solution to (6.70) can then be written

$$\psi = \int \frac{d^3p}{(2\pi)^3} \left[d^*(\mathbf{p})u(p)e^{-ipx} + f(\mathbf{p})v(p)e^{ipx} \right] . \qquad (6.82)$$

The choice of the specific solution to associate with e^{ipx} and e^{-ipx} is arbitrary. The packet (6.82) describes a nondispersing form travelling at the speed of light.

In order to understand the physical meaning of the spinor solution it is recalled that in nonrelativistic quantum mechanics the Pauli spin matrices are used to represent the spin operator $\vec{S}$ for the electron by writing $\vec{S} = \frac{1}{2}\vec{\sigma}$. In the particular case being studied here they play a similar role, although the proof of that must wait until Sec. 6.5 where angular momentum is analyzed. It will be shown that in relativistic field theory only the projection of the spin angular momentum along the direction of motion, referred to as the *helicity*, is a meaningful operator, and this is defined for a solution of momentum $\mathbf{p}$ as

$$\Sigma_p = \frac{1}{2}\frac{\sigma_j p^j}{\epsilon_p} . \qquad (6.83)$$

Exercise 6.18: Prove that Σ_p has the eigenvalues $\pm\frac{1}{2}$, and that the spinors $u(p)$ and $v(p)$ are eigenvectors with eigenvalue $-\frac{1}{2}$.

Both solutions belong to the same helicity, and the absence of a parity operation for the Weyl spinor is the ultimate source of this. This will be discussed at the end of this section. In this respect it is important to note that the two solutions can be related throught the transformation

$$u(\tilde{p}) = \sigma_2 v^*(p) . \qquad (6.84)$$

It would have been possible to begin with the other representation of the Lorentz group (6.43) by starting with the differential operator $\tilde{\partial} = \tilde{\sigma}^\mu \partial_\mu$. It is straightforward to repeat the steps of this subsection and construct the solutions to the parity reversed equation

$$\tilde{\partial}\psi = 0 . \qquad (6.85)$$

176   Field Theory

> **Exercise 6.19**: Find the normalized solutions to equation (6.85) and show that they have helicity $+\frac{1}{2}$.

The construction of bilinear Lorentz scalars for Dirac spinor fields [3] is slightly more complicated. From the form (6.53) of an infinitesimal transformation, it follows that an arbitrary Lorentz transformation of the bispinor field must be represented by

$$\mathbf{U} = \exp\left(\tfrac{1}{4} i \omega_{\mu\nu} \sigma^{\mu\nu}\right), \tag{6.86}$$

where $\omega_{\mu\nu}$ are the elements of a real, antisymmetric 4×4 matrix, and are functions of the six parameters of the Lorentz transformation.

> **Exercise 6.20**: Find the exact relationship of the $\omega_{\mu\nu}$ to the elements $\Lambda^{\mu}{}_{\nu}$ of the Lorentz transformation that $\mathbf{U}$ represents.

It follows from (6.86) that

$$\mathbf{U}^\dagger = \exp\left(-\tfrac{1}{4} i \omega_{\mu\nu} \sigma^{\mu\nu\dagger}\right) \neq \mathbf{U}^{-1} . \tag{6.87}$$

However, from property (6.47) $\mathbf{U}^\dagger$ may be converted to an inverse by using the fact that $\gamma^0 \sigma^{\mu\nu\dagger} \gamma^0 = \sigma^{\mu\nu}$, so that

$$\gamma^0 \mathbf{U}^\dagger \gamma^0 = \mathbf{U}^{-1} . \tag{6.88}$$

Defining the field variable $\bar{\psi} \equiv \psi^\dagger \gamma^0$, it follows that under a Lorentz transformation

$$\bar{\psi}'(x') = \psi^\dagger(x) \mathbf{U}^\dagger \gamma^0 = \psi^\dagger(x) \gamma^0 \mathbf{U}^{-1} = \bar{\psi}(x) \mathbf{U}^{-1} , \tag{6.89}$$

and therefore the quantity $\bar{\psi}\psi \equiv \bar{\psi}_a \psi_a$ is a Lorentz scalar. In addition, combining (6.46) with the reasoning of (6.68) shows that the form $\slashed{\partial} = \gamma^\mu \partial_\mu$ can be used to define the Lorentz scalar $\bar{\psi} \slashed{\partial} \psi$. Combining this with the previous result gives the Dirac spinor action

$$S = \int_{\mathcal{M}} d^4x\, i\bar{\psi}(\slashed{\partial} + im)\psi . \tag{6.90}$$

Since ψ must have units of (length)$^{-3/2}$ in order that the first term in the action is dimensionless, it follows that the constant m must have units of (length)$^{-1}$, and can be understood as a mass.

Sec. 6.3 Classical Free Fields

The Euler-Lagrange equations associated with the action (6.90) give two forms: the Dirac equation,

$$(\not\partial + im)\psi = 0, \tag{6.91}$$

and its conjugate

$$\bar\psi(\overleftarrow{\not\partial} - im) = 0, \tag{6.92}$$

where the differential operator in the conjugate equation is understood to act to the left. Using the algebra (6.45) of the Dirac matrices, it follows that

$$\not\partial\not\partial = \tfrac{1}{2}\{\gamma^\mu,\gamma^\nu\}\partial_\mu\partial_\nu + \tfrac{1}{2}[\gamma^\mu,\gamma^\nu]\partial_\mu\partial_\nu = 1g^{\mu\nu}\partial_\mu\partial_\nu = 1\Box, \tag{6.93}$$

so that the Dirac equation gives

$$(\not\partial - im)(\not\partial + im)\psi = (\not\partial\not\partial + m^2)\psi = (\Box + m^2)\psi = 0. \tag{6.94}$$

Therefore, each component of the bispinor solves the *massive* wave equation familiar from the scalar field case.

The solutions to the Dirac equation must therefore take the form $\psi = w(p)e^{\pm ipx}$, where $w(p)$ is a four-component spinor, and $p^2 = m^2$. In the rest frame of the particle where $\mathbf{p} = 0$, px becomes mt, and the Dirac equation reduces to

$$(\not\partial + im)\psi = (\gamma^0 \frac{\partial}{\partial t} + im)w(0)e^{\pm imt}. \tag{6.95}$$

Using the form (6.44) for γ^0, it is easy to find the solutions to (6.95). For the choice of the negative sign they are given by

$$u_1(0)e^{-imt} = \begin{pmatrix} 1 \\ 0 \\ 0 \\ 0 \end{pmatrix} e^{-imt}, \quad u_2(0)e^{-imt} = \begin{pmatrix} 0 \\ 1 \\ 0 \\ 0 \end{pmatrix} e^{-imt}, \tag{6.96}$$

while for the choice of the positive sign they are given by

$$v_1(0)e^{imt} = \begin{pmatrix} 0 \\ 0 \\ 1 \\ 0 \end{pmatrix} e^{imt}, \quad v_2(0)e^{imt} = \begin{pmatrix} 0 \\ 0 \\ 0 \\ 1 \end{pmatrix} e^{imt}. \tag{6.97}$$

The solutions for $\mathbf{p} \neq 0$ can be found by boosting these solutions from the rest frame by a Lorentz transformation. However, a far easier method is to use the identity similar to (6.76),

$$(\not p + m\mathbf{1})(\not p - m\mathbf{1}) = \not p\not p - m^2\mathbf{1} = (p^2 - m^2)\mathbf{1} = 0, \tag{6.98}$$

178   Field Theory

and to note that the action of the Dirac operator $\not{\partial} + im$ on a plane-wave is given by

$$(\not{\partial} + im)e^{\pm ipx} = \pm i(\not{p} \pm m)e^{\pm ipx}. \tag{6.99}$$

Therefore, the solutions to the Dirac equation can be found by applying the appropriate factor from (6.98) to the respective spinors. The solutions to the Dirac equation obtained from (6.96) and (6.97) then take the form, for $s = (1, 2)$,

$$u_s(p)e^{-ipx} = N_p(m + \not{p})u_s(0)e^{-ipx}, \tag{6.100}$$
$$v_s(p)e^{ipx} = N_p(m - \not{p})v_s(0)e^{ipx}, \tag{6.101}$$

where N_p is a normalization factor.

Exercise 6.21: Show that the solutions to the conjugate Dirac equation (6.92) are given by

$$\bar{u}_s(p)e^{ipx} = N_p \bar{u}_s(0)(m + \not{p})e^{ipx}, \tag{6.102}$$
$$\bar{v}_s(p)e^{-ipx} = N_p \bar{v}_s(0)(m - \not{p})e^{-ipx}. \tag{6.103}$$

If the normalization factor is chosen to be $N_p = 1/\sqrt{2m(m + \epsilon_p)}$, where $\epsilon_p = +\sqrt{m^2 + \mathbf{p}^2}$, it can be shown that the inner products of the spinors satisfy

$$\bar{u}_s(p)u_{s'}(p) = -\bar{v}_s(p)v_{s'}(p) = \delta_{ss'}, \tag{6.104}$$
$$\bar{u}_s(p)v_{s'}(p) = 0, \tag{6.105}$$
$$u_s^\dagger(p)u_{s'}(p) = v_s^\dagger(p)v_{s'}(p) = \frac{\epsilon_p}{m}\delta_{ss'}, \tag{6.106}$$
$$\bar{u}_s(\tilde{p})u_{s'}(p) = -\bar{v}_s(\tilde{p})v_{s'}(p) = \frac{\epsilon_p}{m}\delta_{ss'}, \tag{6.107}$$
$$\sum_{s=1}^{2}[\bar{u}_s^a(p)u_s^b(p) - \bar{v}_s^a(p)v_s^b(p)] = \delta_{ab}, \tag{6.108}$$

where, again, $\tilde{p} = (\epsilon_p, -\mathbf{p})$, and there is an implicit sum over the spinor indices in all but (6.108).

Exercise 6.22: Verify relations (6.104) through (6.108).

Sec. 6.3 Classical Free Fields   179

Using these spinors, it is possible to build a general solution to the Dirac equation in terms of a wave packet. Such a solution will be written in a form that will be convenient later and is given by

$$\psi = \int \frac{d^3p}{(2\pi)^{3/2}} \sqrt{\frac{m}{\epsilon_p}} \left[b_s(\mathbf{p}) u_s(p) e^{-ipx} + d_s^*(\mathbf{p}) v_s(p) e^{ipx} \right], \qquad (6.109)$$

where there is an implicit sum over the s index.

It is also possible to define other terms bilinear in the Dirac fields. For example, the term $\bar{\psi}\gamma^\mu\psi$ transforms as the contravariant components of a vector. Others are left as an exercise.

Exercise 6.23: Show that the terms $\bar{\psi}\gamma_5\psi$ and $\bar{\psi}\gamma_5\gamma^\mu\psi$ transform respectively as a pseudoscalar and pseudovector.

The physical content of the solutions to the Weyl and Dirac equations is best examined in the quantized form, but some aspects can be seen at the classical stage. A Weyl spinor can never be massive, and this is a consequence of the absence of a parity operator. The demonstration of this is based upon the relation of helicity to *handedness*. Helicity is the projection of the spin angular momentum onto the vector **p** that defines the direction of motion for the solution. If the solution's helicity is positive, then such an angular momentum is represented by a *right-handed* rotation about the direction of motion. In the other case, a negative helicity represents a *left-handed* rotation about the direction of motion. If a Weyl spinor were massive, then it would be possible to boost into a frame in which its value of **p** underwent a sign change, and such a transformation would automatically change a right-handed rotation into a left-handed rotation and *vice versa*. Such a boost would therefore be accompanied by a change in parity for the solution. However, it follows from the analysis of SL(2,C) that no such transformation exists in the two-component representation. Therefore, the Weyl spinor is constrained to be massless, thereby preventing the existence of a boost that changes the sign of **p**. A set of naturally occurring particles, the neutrinos, are candidates for being modelled as Weyl spinors. Within experimental error they appear to be massless; however, all observed neutrinos are left-handed, while the antineutrinos, to be defined in Sec. 6.5, are all right-handed. This is easily represented using a wave packet of Weyl spinors.

It is the existence of a parity transformation in the Dirac representation that allows the bispinor to be massive. Because of the Weyl spinor's relation to handedness, it is instructive to consider the handedness of the solutions to the Dirac equation, and this is accomplished by using γ_s to define a projection operator for different handedness, or *chirality*. The right (R) and left (L) pieces of the Dirac spinor are defined as

$$\psi^{(R)} = \tfrac{1}{2}(1+\gamma_s)\psi, \quad \psi^{(L)} = \tfrac{1}{2}(1-\gamma_s)\psi, \quad (6.110)$$

so that $\psi = \psi^{(R)} + \psi^{(L)}$. The left and right pieces are eigenfunctions of γ_s,

$$\gamma_s\psi^{(R)} = \tfrac{1}{2}(\gamma_s + (\gamma_s)^2)\psi = \tfrac{1}{2}(1+\gamma_s)\psi = +\psi^{(R)},$$
$$\gamma_s\psi^{(L)} = \tfrac{1}{2}(\gamma_s - (\gamma_s)^2)\psi = -\tfrac{1}{2}(1-\gamma_s)\psi = -\psi^{(L)}, \quad (6.111)$$

and for this reason γ_s is called the *chirality* operator.

Of course, all the statements derived so far are independent of the representation chosen for the γ matrices. However, there is a particular form of the γ matrices, known as the Weyl or chiral representation, that best demonstrates the chiral content of the Dirac equation. In the Weyl representation the γ matrices are given by

$$\gamma^0 = \begin{pmatrix} 0 & -I \\ -I & 0 \end{pmatrix}, \quad \gamma^j = \begin{pmatrix} 0 & \sigma^j \\ -\sigma^j & 0 \end{pmatrix}, \quad (6.112)$$

so that γ_s is given by

$$\gamma_s = \begin{pmatrix} I & 0 \\ 0 & -I \end{pmatrix}. \quad (6.113)$$

Therefore, if ψ is a solution to the Dirac equation in the chiral representation, then the bispinor breaks into the sum of two two-component spinors χ and ξ,

$$\psi^{(R)} = \begin{pmatrix} \xi \\ 0 \end{pmatrix}, \quad \psi^{(L)} = \begin{pmatrix} 0 \\ \chi \end{pmatrix}. \quad (6.114)$$

The mass term in the Dirac action becomes

$$m\bar{\psi}\psi = -m\xi^\dagger\chi - m\chi^\dagger\xi, \quad (6.115)$$

while the so-called kinetic term reduces to

$$i\bar{\psi}\slashed{\partial}\psi = i\xi^\dagger(\partial_0 + \sigma_j\partial_j)\xi + i\chi^\dagger(\partial_0 - \sigma_j\partial_j)\chi. \quad (6.116)$$

Varying the $m \neq 0$ action in terms of ξ and χ leads to a set of coupled first-order differential equations. However, it is obvious by inspection of (6.116) that the $m = 0$ form of the Dirac action breaks into the sum of the action for two Weyl spinors of opposite parity, and that the two Weyl actions transform into each other under the action of a spatial inversion, which is represented by γ_0.

Chirality is most easily pictured for the massless case of a Dirac bispinor. By comparing (6.116) with (6.68) it is apparent that χ corresponds to the case of a negative helicity Weyl spinor, while ξ is the positive helicity Weyl spinor of (6.85). It follows from the definitions of ξ and χ and the form of γ_s, given in the chiral representation by (6.113), that ξ is the right-handed chiral mode while χ is the left-handed mode. Therefore, for a massless Dirac spinor chirality and helicity are exactly correlated.

However, it is possible to construct a Lorentz invariant mass term using solely the left or the right chirality spinors of (6.110). For the left-handed case this mass term takes the form

$$\tfrac{1}{4} im \psi^T (1+\gamma_s) C \gamma_0 (1+\gamma_s) \psi - \tfrac{1}{4} im \psi^\dagger (1+\gamma_s) C \gamma_0 (1+\gamma_s) \psi^* \;, \qquad (6.117)$$

where C is referred to as the *charge conjugation* matrix and is given in the Weyl representation by

$$C = \begin{pmatrix} -i\sigma_2 & 0 \\ 0 & i\sigma_2 \end{pmatrix} \;. \qquad (6.118)$$

In the Weyl representation this can be written in terms of the two component spinor χ of (6.114) as

$$im\,\chi^T \sigma^2 \chi - im\,\chi^\dagger \sigma^2 \chi^* \;. \qquad (6.119)$$

Exercise 6.24: Show that the term (6.117) is a Lorentz scalar by using the fact that $\sigma_2 \sigma_j \sigma_2 = -\sigma_j^*$.

This result can be related to a standard Dirac bispinor mass term by introducing a constraint. The *Majorana* spinor ψ^M is defined as a Dirac bispinor ψ that satisfies the constraint

$$\psi = C \bar\psi^T \;. \qquad (6.120)$$

In the Weyl representation this gives a constraint between the two two-component spinors ξ and χ,

$$\xi = i\sigma_2 \chi^* \;, \quad \chi = -i\sigma_2 \xi^* \;. \qquad (6.121)$$

The two constraints of (6.121) are equivalent. Comparison of (6.121) to (6.84) immediately reveals that the constraint is identical to the relationship between the two solutions of the Weyl equation with the same chirality.

Exercise 6.25: Find the unitary transformation from the Dirac representation to the Weyl representation. Use this to express the two Weyl spinors of (6.114) in terms of the Dirac bispinors. Find the form of the Dirac spinors that satisfy the Majorana constraint.

It is then straightforward to show that the Dirac mass term $\bar{\psi}\psi$ reduces to the mass term (6.119) when $\psi = \psi^M$. In effect, the constraint reduces the eight degrees of freedom of a Dirac spinor to the four of a Weyl spinor in a Lorentz covariant way.

While the discussion has centered on the behavior of the fields under a parity transformation P, there is also the possibility of time reversal T. Time reversal in quantum mechanics and quantum field theory is associated with the antiunitary transformation on the states

$$\langle T\psi | T\phi \rangle = \langle \psi | \phi \rangle^* = \langle \phi | \psi \rangle . \qquad (6.122)$$

The action of time-reversal on the general field $\psi_\alpha(x)$ is given by

$$\psi_\alpha(\mathbf{x},t) \xrightarrow{T} \eta T_{\alpha\beta} \psi_\beta(\mathbf{x},-t) , \qquad (6.123)$$

where η is an overall phase factor, $|\eta|^2 = 1$, chosen to reflect the content of the field, and $T_{\alpha\beta}$ is matrix constructed to leave the (anti)commutation relations invariant. The details of this construction will not be presented here [5]. There is also the possibility of charge conjugation C, already defined as the matrix (6.118) for the spinor field, but understanding the content of charge conjugation requires quantizing the fields.

6.4 Symmetry and Noether's Theorem

In this section the properties of a classical theory that are induced by a symmetry of the action will be discussed. In Sec. 4.3 it was shown that a symmetry of the action is accompanied by a conservation law at the classical level and a set of identities for matrix elements at the quantum level. It is Noether's theorem [17] that summarizes the extension of these ideas to the case of classical fields [15].

In the field theoretic case, a symmetry of the action is some transformation of the fields and their space-time arguments that leaves the action invariant and the Lagrangian density form invariant. The Lorentz and Poincaré transformations discussed in Sec. 6.2 are examples of such simultaneous transformations of the forms of the fields and their arguments that,

Sec. 6.4 Symmetry and Noether's Theorem

by construction, leave the action invariant. In this section only continuous transformations will be considered, and this gives the advantage of being able to work with an infinitesimal form. Such an infinitesimal transformation is written

$$\psi_\alpha(x) \to \psi'_\alpha(x') = \psi_\alpha(x) + \delta\psi_\alpha(x) . \tag{6.124}$$

It is important to note that $\delta\psi_\alpha(x)$ includes both the infinitesimal change in the form of ψ_α as well as the change in the coordinate, which is assumed to take the form

$$x^\mu \to x^{\mu\prime} = x^\mu + \delta x^\mu . \tag{6.125}$$

Both δx and $\delta\psi_\alpha$ are infinitesimal, and any power of either beyond linear can be ignored, and this includes cross terms.

Because the transformation may simultaneously involve the form of the fields as well as the space-time variables, including any derivatives present, it should be clear that the variation does not necessarily commute with the partial derivatives occurring in the action, i.e., $\delta(\partial_\mu\psi_\alpha) \neq \partial_\mu\delta\psi_\alpha$. For that reason it is very important to separate out the part of the transformation that involves solely the functional form of the field and not the variation of its argument. This is denoted as a second kind of variation of the field,

$$\bar\delta\psi_\alpha(x) = \psi'_\alpha(x) - \psi_\alpha(x) , \tag{6.126}$$

and this variation has the desirable property that

$$\bar\delta(\partial_\mu\psi_\alpha) = \partial_\mu\bar\delta\psi_\alpha . \tag{6.127}$$

The two types of variation are related, and it follows that

$$\begin{aligned}\psi'_\alpha(x') &= \psi'_\alpha(x) + \delta x^\mu \partial_\mu \psi_\alpha(x) \\ \Rightarrow \bar\delta\psi_\alpha(x) &= \delta\psi_\alpha(x) - \delta x^\mu \partial_\mu\psi_\alpha(x) .\end{aligned} \tag{6.128}$$

Under an arbitrary infinitesimal transformation the Lagrangian density becomes $\mathcal{L}(x) \to \mathcal{L}'(x')$, and this can be expressed, using (6.128), as

$$\delta\mathcal{L}(x) = \mathcal{L}'(x') - \mathcal{L}(x) = \bar\delta\mathcal{L}(x) + \delta x^\mu \partial_\mu \mathcal{L}(x) . \tag{6.129}$$

The definition of the second kind of variation gives

$$\bar\delta\mathcal{L} = \frac{\partial \mathcal{L}}{\partial \psi_\alpha}\bar\delta\psi_\alpha + \frac{\partial \mathcal{L}}{\partial(\partial_\nu\psi_\alpha)}\bar\delta(\partial_\nu\psi_\alpha) . \tag{6.130}$$

Using (6.127) the order of variation and derivative may be interchanged, and using (6.128) allows the entire variation of $\mathcal{L}$ under the transformation

Field Theory

to be written

$$\delta\mathcal{L} = \frac{\partial\mathcal{L}}{\partial\psi_\alpha}\delta\psi_\alpha - \frac{\partial\mathcal{L}}{\partial\psi_\alpha}(\partial_\mu\psi_\alpha)\delta x^\mu + \frac{\partial\mathcal{L}}{\partial(\partial_\nu\psi_\alpha)}\partial_\nu\delta\psi_\alpha$$
$$- \frac{\partial\mathcal{L}}{\partial(\partial_\nu\psi_\alpha)}\partial_\nu(\partial_\mu\psi_\alpha\delta x^\mu) + \frac{\partial\mathcal{L}}{\partial x^\mu}\delta x^\mu . \quad (6.131)$$

Under this transformation the action S becomes

$$S = \int d^4x\, \mathcal{L}(x) \to S' = \int d^4x'\, \mathcal{L}'(x') , \quad (6.132)$$

where it is assumed that the domain of integration for the action is all of Minkowski space. Otherwise, the domain of integration must be altered as well to reflect the change in the limits. If this transformation is a symmetry, then $S = S'$. In the infinitesimal case it follows from the continuity of the transformation that

$$S' = \int [d^4x + \delta(d^4x)]\, [\mathcal{L}(x) + \delta\mathcal{L}(x)] . \quad (6.133)$$

The variation of the measure is determined from the Jacobian of the transformation, and in the infinitesimal case it is given by

$$\delta(d^4x) = d^4x\, \partial_\mu \delta x^\mu . \quad (6.134)$$

Exercise 6.26: Verify result (6.134).

Therefore, the entire change in the action is given by

$$\delta S = \int d^4x\, (\mathcal{L}\partial_\mu \delta x^\mu + \delta\mathcal{L}) . \quad (6.135)$$

By substituting (6.131) into (6.135) and rearranging the terms, the variation of the action can be written

$$\delta S = \int d^4x\, \partial_\mu \left[\left(g^\mu{}_\nu \mathcal{L} - \frac{\partial\mathcal{L}}{\partial(\partial_\mu\psi_\alpha)}\partial_\nu\psi_\alpha \right) \delta x^\nu + \frac{\partial\mathcal{L}}{\partial(\partial_\mu\psi_\alpha)}\delta\psi_\alpha \right]$$
$$+ \int d^4x \left[\frac{\partial\mathcal{L}}{\partial\psi_\alpha} - \partial_\nu \frac{\partial\mathcal{L}}{\partial(\partial_\nu\psi_\alpha)} \right] (\delta\psi_\alpha - \delta x^\mu \partial_\mu\psi_\alpha) . \quad (6.136)$$

Clearly, the second integral will vanish if ψ_α is a solution of the Euler–Lagrange equation. The first integral takes the form of a total divergence,

Sec. 6.4 Symmetry and Noether's Theorem

and the term in the parentheses is identified as a *current*,

$$J^\mu = \left(g^\mu{}_\nu \mathcal{L} - \frac{\partial \mathcal{L}}{\partial(\partial_\mu \psi_\alpha)} \partial_\nu \psi_\alpha\right) \delta x^\nu + \frac{\partial \mathcal{L}}{\partial(\partial_\mu \psi_\alpha)} \delta \psi_\alpha \,. \tag{6.137}$$

If the transformation is a symmetry, then $\delta S = 0$. If the current associated with a symmetry is evaluated using a solution to the Euler–Lagrange equation, then it follows from (6.136) that this current must be conserved,

$$\partial_\mu J^\mu = 0 \,. \tag{6.138}$$

Results (6.137) and (6.138) are the extensions of the results of Sec. 4.3 for mechanical systems, usually referred to as *Noether's theorem* [17].

Result (6.138) allows the definition of the *charge* Q by integrating J^0 over a *spatial* volume. It follows from the conservation law (6.138) and Gauss's theorem that

$$\frac{\partial Q}{\partial t} = \int_V d^3x \, \frac{\partial}{\partial t} J^0(\mathbf{x}, t) = -\int_V d^3x \, \nabla \cdot \mathbf{J}(\mathbf{x}, t) = -\int_S d\mathbf{S} \cdot \mathbf{J}(\mathbf{x}, t) \,, \tag{6.139}$$

where S is the surface bounding the volume $\mathcal{V}$. Assuming that the fields vanish at the surface of the volume, or that the fields vanish at infinity and the volume is all of space, it follows that the charge Q is conserved in time. Otherwise, (6.139) describes the flow of charge out of or into the volume. The reader knowledgeable in special relativity may be worried about the covariance of the procedure of defining the charge Q as a spatial volume integral, since such a definition is frame dependent. It is clear that Q, as defined by (6.139) is *not* a Lorentz scalar. This may be remedied by introducing the four-vector S_μ that represents the covariant form of three-surfaces. In the rest frame, this four-vector has the infinitesimal elements $dS = (d^3x, 0, 0, 0)$. The form of dS in another frame may be found by performing a Lorentz transformation on the rest frame form. The manifest Lorentz scalar form for the charge is then given by

$$Q = \int dS_\mu \, J^\mu \,. \tag{6.140}$$

It is clear that this definition reduces to (6.139) in the rest frame.

Noether's theorem shows that, for every distinct symmetry transformation available to the action, there is a corresponding conserved quantity. This means that, if the transformations form a continuous group, the number of conserved quantities will match the number of parameters necessary to completely specify the group member. In what follows several examples of the Noether charge Q associated with important symmetries will be presented.

6.4.1 Translational Invariance

In many systems the action is *translationally invariant*. The action is unchanged if all the fields $\psi_\alpha(x)$ appearing in the action are replaced by $\psi_\alpha(x-a)$, where a is a constant vector. This is true since the change in the fields can be compensated by a change in the variable x that leaves the domain of integration unchanged, assuming there are no functions appearing in the action that are not translationally invariant. For the infinitesimal case of this transformation, the field changes *form* according to

$$\psi'_\alpha(x) = \psi_\alpha(x) + a^\mu \partial_\mu \psi_\alpha(x) , \qquad (6.141)$$

and therefore

$$\psi'_\alpha(x-a) \equiv \psi'_\alpha(x') = \psi_\alpha(x) . \qquad (6.142)$$

From (6.142) it follows that $\delta\psi_\alpha = 0$ and $\delta x^\mu = -a^\mu$ for this case.

The construction of the conserved charges is obtained by simple substitution of these results into (6.137). The conserved current takes the form

$$J^\mu = \left[\frac{\partial \mathcal{L}}{\partial(\partial_\mu \psi_\alpha)} \partial_\nu \psi_\alpha - g^\mu{}_\nu \mathcal{L}\right] \delta a^\nu . \qquad (6.143)$$

The expression in the brackets is usually denoted $T^\mu{}_\nu$ and is referred to as the *stress-energy tensor*. It describes the flow of energy and momentum created by space-time variations in the field configuration. Because a^μ is a constant and $J^\mu = T^\mu{}_\nu a^\nu$, a^μ plays no role in the conservation of the current J^μ, and therefore the stress-energy tensor itself must obey $\partial_\mu T^\mu{}_\nu = 0$. Every attempt is made to construct the action so that the stress-energy tensor is symmetric, i.e., $T_{\mu\nu} = g_{\mu\rho} T^\rho{}_\nu = T_{\nu\mu}$, for otherwise there are difficulties in a physical interpretation of the theory, in particular for the definition of angular momentum [5].

The case of time translation invariance is obtained by setting $a^\nu = \epsilon \delta_{\nu 0}$. For such a case the quantity Q/ϵ is denoted H and is given by

$$H = \int d^3x \left(\pi_\alpha \dot\psi_\alpha - \mathcal{L}\right) , \qquad (6.144)$$

where the definition (6.12) of the canonical momentum density π_α has been used. Because of the complete analogy to the mechanical case, the quantity H is identified as the Hamiltonian of the system, and the quantity appearing in parentheses in (6.144) is referred to as the *Hamiltonian density*. The case of spatial translations is obtained similarly.

Sec. 6.4 Symmetry and Noether's Theorem

Exercise 6.27: Show that the three-vector **P**, given by

$$\mathbf{P} = \int d^3x\, \pi_\alpha \nabla \psi_\alpha \,, \tag{6.145}$$

is conserved in a translationally invariant system, and that the combination $(H, \mathbf{P})$, referred to as the energy-momentum P^μ, transforms as a four-vector.

The physical interpretation of P^μ is based on an analogy with mechanical systems, where the energy is the invariant quantity associated with time-translations, while spatial momentum is associated with spatial translations. Thus, the four parameters a^μ necessary to specify the translation have given rise to a conserved four-vector.

6.4.2 Lorentz Invariance and Angular Momentum

In this example the Lorentz invariance of the action will be used to determine the angular momentum of a classical field configuration. This is possible because the rotations form a subgroup of the Lorentz group, and therefore the Lorentz transformation properties of the fields play a critical role in this analysis.

The action of a Lorentz transformation on a general field can be written

$$\Psi'_\alpha(x') = D_{\alpha\beta}(\Lambda)\Psi_\beta(x) \,, \tag{6.146}$$

where D is an element of an irreducible representation of the Lorentz group. Noting that $x' = \Lambda x \Rightarrow x = \Lambda^{-1} x'$, form (6.146) may be written

$$\Psi'_\alpha(x) = D_{\alpha\beta}(\Lambda)\Psi_\beta(\Lambda^{-1}x) \,. \tag{6.147}$$

In the infinitesimal case $D_{\alpha\beta} = \delta_{\alpha\beta} + \delta D_{\alpha\beta}$ and $(\Lambda^{-1})^\mu{}_\nu x^\nu = x^\mu - \lambda^\mu{}_\nu x^\nu \equiv x^\mu - \delta x^\mu$, where the $\lambda^\mu{}_\nu$ are defined in (6.33). This gives the variation of the form of the fields

$$\delta\Psi_\alpha(x) = \overline{\delta}\Psi_\alpha(x) = \delta D_{\alpha\beta}\Psi_\beta(x) - \lambda^\mu{}_\nu x^\nu \partial_\mu \Psi_\alpha(x) \,. \tag{6.148}$$

These variations, combined with Noether's theorem, give the current associated with the Lorentz invariance of the action,

$$J^\mu = \frac{\partial \mathcal{L}}{\partial(\partial_\mu \Psi_\alpha)}(\delta D_{\alpha\beta}\Psi_\beta - \lambda^\nu{}_\rho x^\rho \partial_\nu \Psi_\alpha) \,. \tag{6.149}$$

The conserved quantities associated with restricting D to the spatial rotations can be extracted from this current, and these will be identified as the components of the angular momentum of the field configuration. It is recalled from the analysis of Sec. 4.3 that the spatial rotations form a continuous group whose actions on the spatial coordinates are given by $\delta x^i = \delta\theta^k \varepsilon^{kij} x^j = \lambda_{ij} x^j$, where the $\delta\theta^k$ are the components of the infinitesimal angle of rotation. The relation of the spatial rotations to the Lorentz transformation Λ, given by (6.32), shows that the parameters of the infinitesimal Lorentz transformation coincide with the rotation parameters. Therefore, restricting the Lorentz transformation parameters to this subset gives the conserved current associated with rotations. To proceed any further the exact type of field must be specified, since this determines the form of δD.

A scalar field ϕ is such that $\delta D = 0$, and therefore the second part of (6.149) reduces to

$$J^\mu = \delta\theta^i \frac{\partial \mathcal{L}}{\partial(\partial_\mu \phi)} \varepsilon^{ijk} x^j \partial_k \phi . \tag{6.150}$$

Because the $\delta\theta^i$ constitute three arbitrary constants, it is clear that (6.150) gives three conserved currents, and the charges associated with them are given by

$$L^i = \int d^3x\, \varepsilon^{ijk} \pi x^j \partial_k \phi . \tag{6.151}$$

Because these quantities are dimensionless and associated with rotations, and because of the occurrence of the orbital angular momentum operators familiar from quantum mechanics, (6.151) is referred to as the *orbital angular momentum* of the scalar field configuration. The scalar field possesses no other form of angular momentum apart from the orbital contribution, and therefore the scalar field is said to be *spinless*, or spin zero.

The angular momentum of the Dirac field may be analyzed by noting that the infinitesimal Lorentz transformation on the form of the field, given by (6.53), reduces to

$$\delta U_{ab} = \tfrac{1}{4} i\, \delta\theta^j \varepsilon^{jkm} \sigma^{km}_{ab} , \tag{6.152}$$

when restricted to a spatial rotation. From the definition $\sigma^{jk} = \tfrac{1}{2} i [\gamma^j, \gamma^k]$, and using the Dirac representation, it follows that

$$\sigma^{ij} = -\varepsilon^{ijk} S^k, \quad S^k = \begin{pmatrix} \sigma^k & 0 \\ 0 & \sigma^k \end{pmatrix} . \tag{6.153}$$

Using (6.153) in (6.152) and employing the identity $\varepsilon^{jkl}\varepsilon^{jkm} = 2\delta_{lm}$, (6.152) becomes

$$\delta U_{ab} = -\tfrac{1}{2} i\, \delta\theta^k S^k . \tag{6.154}$$

Sec. 6.4 Symmetry and Noether's Theorem

From the Dirac action (6.90), it follows that $\pi = i\bar{\psi}\gamma^0 = i\psi^\dagger$. Using this in the definition of the Noether charge shows that the *total* angular momentum of the Dirac field is given by

$$J^k = \int d^3x \, \psi^\dagger (\tfrac{1}{2} S^k + i\varepsilon^{kjm} x^m \partial_j)\psi \, , \tag{6.155}$$

so that $\tfrac{1}{2} S^k$ is identified as the intrinsic or *spin* angular momentum of the Dirac field. The total angular momentum is therefore the combination of the orbital and spin angular momentum. It is not difficult to see that the static solutions of the Dirac equation given by (6.96) and (6.97) correspond to $\pm \tfrac{1}{2}$ for the eigenvalues of $\tfrac{1}{2} S^3$. The relation of $\tfrac{1}{2} S^k$ to the helicity will be presented in Sec. 6.5, while the discussion of the intrinsic angular momentum of the vector field will be deferred to Chapter 7.

6.4.3 Phase and Chiral Invariance

Many times the action is invariant under a *gauge transformation of the first kind*, which consists of changing the phase of a complex-valued field by an arbitrary *constant* phase. For example, the action (6.67) is invariant under the change $\phi \to e^{i\lambda}\phi$, while the spinor actions (6.69) and (6.90) possess a similar symmetry. For λ infinitesimal, this transformation gives the infinitesimal change in the form of the field

$$\bar{\delta}\Psi_\alpha = i\lambda \Psi_\alpha = \delta \Psi_\alpha \, , \tag{6.156}$$

which gives rise to the current

$$J^\mu = i\lambda \frac{\partial \mathcal{L}}{\partial(\partial_\mu \Psi_\alpha)} \Psi_\alpha \, . \tag{6.157}$$

Dropping the factor $i\lambda$ as irrelevant and using the Dirac action (6.90), the current (6.157) is given by $J^\mu = \bar{\psi}\gamma^\mu\psi$, and will be identified in Chapter 7 as the electromagnetic current associated with the bispinor field.

The Dirac action may also possess another interesting invariance. In the massless limit, $m = 0$, the theory is invariant under a *chiral transformation* of the form

$$\psi \to e^{i\alpha\gamma_5}\psi \, . \tag{6.158}$$

Exercise 6.28: Prove that the massless Dirac action is invariant under chiral transformations, and find the conserved current associated with this invariance.

Field Theory

The presence of chiral invariance in a bispinor field theory is therefore associated with masslessness of the theory. On the other hand, the absence of chiral invariance is symptomatic of the presence of mass in the bispinor sector.

6.4.4 Charges as Symmetry Generators

All of the conservation laws presented in this section are of course true only at the classical level. Their verity at the quantum level must be ascertained, and this requires further developments. However, one point can be made without detailed knowledge of the quantized theories.

Each of the charges associated with a symmetry serves, at the quantum level, as the *generator* of that symmetry just as it did in the quantum mechanical case discussed in Sec. 4.4. This is easily demonstrated for a symmetry associated with the change in form $\psi_\alpha \to \psi_\alpha + \delta\psi_\alpha$. For such a symmetry the associated charge is given, by Noether's theorem, to be

$$Q = \int d^3x \, \delta\psi_\alpha \frac{\partial \mathcal{L}}{\partial(\partial_0 \psi_\alpha)} \,. \tag{6.159}$$

However, from Sec. 6.1 the momentum canonically conjugate to ψ_α is given by

$$\pi_\alpha = \frac{\partial \mathcal{L}}{\partial(\partial_0 \psi_\alpha)} \,, \tag{6.160}$$

so that the charge becomes

$$Q = \int d^3x \, \delta\psi_\alpha \pi_\alpha \,. \tag{6.161}$$

Using the postulated canonical commutation relation (6.11), generalized to read

$$[\psi_\alpha(\mathbf{x},t), \pi_\beta(\mathbf{y},t)] = i\delta_{\alpha\beta}\delta^3(\mathbf{x}-\mathbf{y}) \,, \tag{6.162}$$

it follows that

$$i[Q, \psi_\alpha(\mathbf{y},t)] = \int d^3x \, \delta\psi_\alpha(\mathbf{x},t)\delta^3(\mathbf{x}-\mathbf{y}) = \delta\psi_\alpha(\mathbf{y},t) \,, \tag{6.163}$$

where it has been assumed that $\delta\psi_\alpha$ commutes with ψ_α. The conservation of the charge allows the time argument of the fields in the integral that defines the charge to be shifted to match the time of the field with which it is being commuted. Within these limits it follows that Q generates the symmetry transformation.

> **Exercise 6.29**: Find the conditions for which the equal-time commutation relations may be replaced with equal-time anticommutation relations such that (6.163) still holds.

6.5 Canonical Quantization

While this is a book on path integral methods in quantum processes, and such an approach is often viewed as the antithesis of "canonical" field theory, the former is actually dependent upon the latter for much of its physical interpretation. This is true because of the necessity to understand the relationship between fields and particles, and this is clear only within the framework of canonical quantization techniques. The path integral is a powerful tool for analyzing field theories, but the ultimate interpretation of the results relies heavily upon the intuitions and methods of canonical quantization.

Canonical quantization consists of finding solutions of the equations of motion that also satisfy the generalized version of the ETCR quantization condition (6.11),

$$[\psi_\alpha(\mathbf{x},t), \pi_\beta(\mathbf{y},t)]_\pm = i\delta_{\alpha\beta}\delta^3(\mathbf{x}-\mathbf{y}) \,, \tag{6.164}$$

where the possibility of choosing *anticommutation* relations has been made available. Of course, implementing this procedure must be done on a case by case basis. However, the basic approach is the mode expansion introduced in Sec. 5.5. The two types of theories developed so far, scalar and spinor, will now be analyzed.

6.5.1 Scalar Field Quantization

For a real scalar field with the action (6.63) the momentum canonically conjugate to the field is given by $\pi = \dot\phi$. The modes of the field must satisfy $(\Box + m^2)\phi = 0$, and assuming that the field is defined over all of space-time, these are the plane waves $e^{\pm ipx}$, with $p_0 = \epsilon_p = +\sqrt{\mathbf{p}^2 + m^2}$. A general mode expansion of ϕ is written

$$\phi(x) = \int \frac{d^3p}{(2\pi)^{3/2}} \frac{1}{\sqrt{2\epsilon_p}} \left(a_p e^{-ipx} + a_p^\dagger e^{ipx}\right) \,. \tag{6.165}$$

192   Field Theory

From the relationship $\dot{\phi} = \pi$, it follows that

$$\pi(x) = i \int \frac{d^3p}{(2\pi)^{3/2}} \sqrt{\frac{\epsilon_p}{2}} \left(a_p^\dagger e^{ipx} - a_p e^{-ipx} \right) . \tag{6.166}$$

Assuming the commutation relations

$$[a_p, a_k] = [a_p^\dagger, a_k^\dagger] = 0 , \quad [a_p, a_k^\dagger] = \delta^3(\mathbf{k} - \mathbf{p}) , \tag{6.167}$$

the ETCR becomes

$$[\phi(\mathbf{x}, t), \pi(\mathbf{y}, t)] = i \int \frac{d^3p}{(2\pi)^3} \exp i\mathbf{p}\cdot(\mathbf{x} - \mathbf{y}) = i\delta^3(\mathbf{x} - \mathbf{y}) , \tag{6.168}$$

thereby satisfying the quantization condition.

The operators a_p and $a_p^\dagger$ are interpreted as annihilation and creation operators, respectively. The vacuum state $|0\rangle$ is introduced, and it satisfies

$$\langle 0 | a_p^\dagger = 0, \quad a_p | 0 \rangle = 0 . \tag{6.169}$$

The state $|\mathbf{p}\rangle = a_p^\dagger |0\rangle$ satisfies

$$\langle 0 | \mathbf{p} \rangle = 0, \quad \langle \mathbf{p} | \mathbf{k} \rangle = \langle 0 | a_p a_k^\dagger | 0 \rangle = \delta^3(\mathbf{p} - \mathbf{k}) . \tag{6.170}$$

To understand the physical content of these states, the result of applying the energy-momentum operator (6.144) and (6.145) is calculated. The Hamiltonian of the system, defined by (6.144), may be written in terms of ϕ and π as

$$H = \int d^3x \left(\tfrac{1}{2}\pi^2 + \tfrac{1}{2}\nabla\phi\cdot\nabla\phi + \tfrac{1}{2}m^2\phi^2 \right) . \tag{6.171}$$

Exercise 6.30: Show that the mode expansions of ϕ and π give

$$H = \tfrac{1}{2} \int d^3p \, \epsilon_p \left(a_p^\dagger a_p + a_p a_p^\dagger \right) . \tag{6.172}$$

At this point the concept of *normal ordering* is introduced. This simply means moving all annihilation operators to the right in all expressions in which they occur. The normal ordered Hamiltonian is denoted $:H:$, and it follows that

$$:H: = \int d^3p \, \epsilon_p \, a_p^\dagger a_p , \tag{6.173}$$

Sec. 6.5 Canonical Quantization

so that the full Hamiltonian is given by

$$H = :H: + E_o, \quad E_o = \tfrac{1}{2}\delta^3(0) \int d^3p \, \epsilon_p \, . \tag{6.174}$$

It is apparent that E_o is a divergent quantity, proportional to the spatial volume $\delta^3(0)$, representing the energy of the vacuum state, i.e., $H|0\rangle = E_o|0\rangle$. Such a divergent term could have been expected from the arguments of Sec. 6.1 where the mechanical model that led to the field theory was an infinite assembly of harmonic oscillators, each of which contributes a nonzero ground state energy when quantized. In general, only differences in energy concern physical processes, and for that reason E_o will be discarded. There is a notable exception to this, and that is general relativity [18], where the ground state energy affects the overall behavior of space-time. This is but one of the many reasons that the quantization of gravitation, as represented by Einstein's version of general relativity, has been a formidable problem.

The spatial momentum operator **P** is found similarly and is given by

$$\mathbf{P} = \int d^3x \, \pi \nabla \phi = \int d^3p \, \mathbf{p} \, a_\mathbf{p}^\dagger a_\mathbf{p} \tag{6.175}$$

where a term of the form $\delta^3(0) \int d^3p \, \mathbf{p}$ has been evaluated to zero in the finite volume limit.

It is now easy to see that the state $|\mathbf{p}\rangle$ satisfies

$$:H:|\mathbf{p}\rangle = \epsilon_p|\mathbf{p}\rangle, \quad \mathbf{P}|\mathbf{p}\rangle = \mathbf{p}|\mathbf{p}\rangle, \tag{6.176}$$

so that the state $|\mathbf{p}\rangle$ has the natural interpretation as a particle of energy ϵ_p and spatial momentum **p**. The extension of this to many-particle states is straightforward, and they are obtained as tensor product of the single particle states. A state with n particles is given by

$$|\mathbf{p}_1, \ldots, \mathbf{p}_n\rangle = \frac{1}{\sqrt{n!}} a_{\mathbf{p}_1}^\dagger \cdots a_{\mathbf{p}_n}^\dagger |0\rangle \, . \tag{6.177}$$

The set of all the states of the form (6.177) is referred to as a *Fock space* [19] representation of the theory. It is straightforward to show that the many-particle states are eigenstates of the energy-momentum operator P_μ introduced in Exercise 6.27, and that the eigenvalues are simply the additive sum of the respective single-particle eigenvalues. Once again, as in Sec. 4.4, the combination $a_\mathbf{p}^\dagger a_\mathbf{p}$ is functioning as a number operator, counting the number of particles in the state with the four-momentum p.

194 Field Theory

Using the normalization of (6.177), it can be shown that the Fock space possesses the unit projection operator

$$\sum_{j=0}^{\infty} |j\rangle\langle j| = 1 , \qquad (6.178)$$

where the nth term is given by

$$|n\rangle\langle n| = \int d^3p_1 \cdots d^3p_n |p_1,\ldots,p_n\rangle\langle p_1,\ldots,p_n| . \qquad (6.179)$$

Exercise 6.31: Verify (6.178).

Exercise 6.32: Show that

$$[P_\mu, \phi] = -i\partial_\mu \phi . \qquad (6.180)$$

Result (6.180) shows that the formalism familiar from quantum mechanics has generalized in a covariant fashion to quantum field theory.

6.5.2 Dirac Field Quantization

The same procedure may be followed for the Dirac field. The mode expansion of (6.109), given by

$$\psi = \int \frac{d^3p}{(2\pi)^{3/2}} \sqrt{\frac{m}{\epsilon_p}} [b_p^s u_s(p) e^{-ipx} + d_p^{s\dagger} v_s(p) e^{ipx}] , \qquad (6.181)$$

is inserted into the Hamiltonian, which is given by

$$H = -i \int d^3x \, \bar{\psi}(\gamma\cdot\nabla + im)\psi . \qquad (6.182)$$

Exercise 6.33: Show that the expansion (6.181), combined with results (6.104) through (6.107), reduces the Hamiltonian to

$$H = \int d^3p \, \epsilon_p \left(b_p^{s\dagger} b_p^s - d_p^s d_p^{s\dagger} \right) . \qquad (6.183)$$

Sec. 6.5 Canonical Quantization

In order that the expansion (6.181) allows the interpretation of d_p as a destruction operator while simultaneously allowing the normal ordered Hamiltonian to have positive eigenvalues, the operators b and d must *anticommute*. Assuming the algebra

$$\{b_p^{s\dagger}, b_k^{s'(\dagger)}\} = \{b_p^{s\dagger}, d_k^{s'\dagger}\} = \{d_p^s(\dagger), d_k^{s'(\dagger)}\} = 0,$$
$$\{b_p^{s\dagger}, b_k^{s'}\} = \{d_p^{s\dagger}, d_k^{s'}\} = \delta_{ss'}\delta^3(\mathbf{p}-\mathbf{k}),\qquad(6.184)$$

allows a definition of the states $|\mathbf{p},s\rangle = b_p^{s\dagger}|0\rangle$ and $|\overline{\mathbf{p}},s'\rangle = d_p^{s'\dagger}|0\rangle$, which are eigenstates of the normal ordered Hamiltonian with the positive eigenvalues ϵ_p. A divergent negative energy that must be subtracted from the Hamiltonian has appeared as a result of normal ordering. The so-called Dirac sea of negative energy appears because of the need to include negative energy solutions to obtain completeness. An immediate implication of the anticommutation relations is that the fields satisfy the equal-time anticommutation relation (ETAR),

$$\{\psi_a^\dagger(\mathbf{x},t), \psi_b(\mathbf{y},t)\} = i\delta_{ab}\delta^3(\mathbf{x}-\mathbf{y}),\qquad(6.185)$$

where a,b label the bispinor indices.

Exercise 6.34: Verify (6.185).

The index s on the particle states may now be understood in terms of the helicity Σ, which can be constructed from the definition (6.153) of the spin operator $\frac{1}{2}S^j$. It is given by

$$\Sigma = i\frac{1}{m}\int d^3x\,\psi^\dagger \tfrac{1}{2}S^j\partial_j\psi = \int d^3p\,\tfrac{1}{2}\frac{p^j}{\epsilon_p}\left(b_p^{s\dagger}\sigma^j_{ss'}b_p^{s'} - d_p^{s\dagger}\sigma^j_{ss'}d_p^{s'}\right).\qquad(6.186)$$

Exercise 6.35: Verify the form of the second integral appearing in (6.186), and show that $[\Sigma, H] = 0$.

For the case that the particle's momentum is directed along the x^3 axis, it is easy to verify that Σ has the particle states as eigenstates with the eigenvalues $\pm\frac{1}{2}$, corresponding to $s = (1,2)$. Thus s corresponds to the helicity or, speaking loosely, the spin states of the particles associated with this field. Because Σ commutes with the Hamiltonian, the helicity can be a member of the set of observables that includes the energy and momentum.

196   Field Theory

> **Exercise 6.36**: Complete the construction of the energy-momentum operator for a Dirac field and show that the commutation relation $[P_\mu, \Psi] = -i\partial_\mu \Psi$ holds despite the anticommutative nature of the spinor fields.

Many-particle states are constructed similarly to the case of the scalar field. However, another immediate result of the anticommutation relations is that $(b_p^{s\dagger})^2|0\rangle = 0$, so that there are no states where two particles share identical values of energy, momentum, and spin. This property is referred to as the *exclusion principle*, and it profoundly affects the statistical behavior of these particles. Particles that have such a behavior are said to be *fermionic* and obey Fermi-Dirac statistics. No such restriction exists for particles that have commutation relations, e.g., the scalar particles discussed earlier. For this reason, the scalar particles obey Bose-Einstein statistics and are said to be *bosonic*. These results have been elevated to the rigorous and general *spin and statistics* theorem [20, 21], which states that the statistical behavior of particles in three spatial dimensions is determined from their spin. All particles with integral spin must obey Bose-Einstein statistics, while all half-integral spin particles must obey Fermi-Dirac statistics. The spin and statistics theorem is proved from the very general assumptions of Lorentz invariance in four dimensions, locality of the equations of motion, and the presence of a positive-definite norm for the states in the theory. In lower-dimensional systems this result fails, and therefore exotic relations between spin and statistics can occur. The result also fails if the Fock space possesses an indefinite metric, so that zero and negative norm states may be present. States with this property occur in gauge theories, where great care must be taken in order to preserve a probabilistic interpretation. It will be seen in Chapter 7 that the presence of zero norm states in gauge theories leads to the presence of scalar fields that must be quantized using anticommutation relations.

In Dirac's original work [3] the states created by $b_s^\dagger$ were associated with the electron. The states created by $d_s^\dagger$ are then associated with its *antiparticle*, the positron. The positron has mass and spin identical to that of the electron; however, it will be seen in Chapter 7 that the electric charge of the positron is opposite that of the electron. The appearance of antiparticles in the quantized version of a field theory is a direct consequence of relativistic invariance. In some theories the particle and its antiparticle coincide, as in the case of the real scalar field of this section. The charge conjugation operator C is defined as the operator that interchanges particle with antiparticle in the field expansions. The charge conjugation matrix has already been defined in (6.118) for spinor fields, and the interchange of particle and antiparticle is given by the transformation

$$\psi \xrightarrow{C} \eta C \bar{\psi}^T ,\qquad (6.187)$$

where $\bar{\psi}^T$ is the transpose of $\bar{\psi}$, and η is a phase factor. It is this transformation that changes the left-handed neutrino into the right-handed antineutrino. It is a fundamental and general theorem that the transformation composed of simultaneous charge conjugation, parity, and time-reversal, CPT, is a fundamental symmetry of nature [21]. While individual symmetries may be violated, such as parity in the weak interactions or CP in kaon decays, no evidence for the violation of CPT has ever been observed.

Exercise 6.37: Quantize the complex scalar field (6.67) and construct the states. Evaluate the orbital angular momentum operator.

Exercise 6.38: Quantize the Weyl spinor action (6.69) and construct the states, determining their helicity.

6.6 The S-Matrix

In this section the dynamics of particle interaction will be developed within the framework of the S-matrix formalism [22]. In the previous sections the fields played the role of Heisenberg picture operators, and the Fock space states were therefore time independent. Since these states are exact eigenstates of the Hamiltonian, it is clear that their time development in the Schrödinger picture is very simple, $|\psi, t\rangle = \exp(-iE_\psi t)|\psi, 0\rangle$. As a result, the states in the Schrödinger picture that are orthonormal remain orthonormal, even when the inner product is taken between these states at different times. Therefore, the theories derived from quadratic actions and quantized have so far described noninteracting or *free* field theories.

In order to allow particle interaction it is necessary to introduce terms nonlinear, i.e., cubic and higher, in the fields. Of course, these terms must remain relativistically invariant. It should also be clear that the formal structures developed in the previous sections, such as the Hamiltonian, angular momentum, and the momentum, will retain their physical interpretation. However, these quantities are to be evaluated using fields that solve the equations of motion, and these fields must also satisfy the canonical equal-time (anti)commutation relations. Unfortunately, the combination of nonlinearity and the quantization condition usually leads to a problem that cannot be solved exactly, unlike the free or linear field theories. There are notable exceptions, but most of the exactly solvable theories are in lower-dimensional space-times. An example of an exactly solvable coupled theory, the Schwinger model, is discussed in Chapter 9.

The most commonly used technique to solve field theories is a perturbative method. In this approach the exact solution is constructed from solutions to the related free field theories. One of the difficulties encountered at the outset of such an approach is how to relate the exact fields to the experimentally observed particles, which seem to behave, within limits, much like the relativistic particles that populate the Fock space introduced in the previous section. This is resolved by postulating an *asymptotic* condition for the behavior of the exact fields. For simplicity, this condition will be presented for the scalar case; the generalization is straightforward.

In the beginning one postulates an action $\mathcal{L}(\Phi)$, and it is assumed that this action has a set of terms that are quadratic and linear in the fields, denoted $\mathcal{L}_o$. It is assumed that $\mathcal{L}_o$ defines a theory that can be solved exactly by using the techniques of the previous section, or some hybrid thereof. The field that solves this simpler problem is denoted $\phi(x)$, and associated to it is a Fock space of particle states that will be assumed, for now, to be massive. However, the real object of interest is $\Phi(x)$, and this field solves the nonlinear problem defined by the total action as well as the canonical commutation relation. Because of the nonlinearity it is reasonable to expect the particle states associated with the field Φ to possess nontrivial properties. Among these is the possibility that the Schrödinger picture particle states are no longer orthogonal at different times. Using the probability amplitude interpretation of the inner product, this situation therefore corresponds to measurable transitions such as scattering and particle production and decay. The problem then becomes that of calculating the overlap of Schrödinger picture particle states at different times.

The Lehmann–Symanzik–Zimmermann (LSZ) [23] formulation of scattering processes in field theory assumes that there exist *asymptotic fields*, denoted ϕ_{in} and ϕ_{out}. These asymptotic fields are associated with the well-defined Fock space constructed by solving the simpler theory defined by $\mathcal{L}_o$. These fields, typically free, represent the particles of the full, interacting field theory before and after they emerge from the scattering region, and therefore they correspond to the limits $t \to \pm\infty$ of Φ. In the LSZ formalism they are related to the interacting field Φ in the following way. If $f(x)$ represents a wave-packet solution to $(\Box + m^2)f = 0$, then the asymptotic limit of Φ is assumed to satisfy

$$\lim_{t\to+\infty} \left(\int d^3x\, f(x) \frac{\overleftrightarrow{\partial}}{\partial t} \Phi(x) - \sqrt{Z} \int d^3x\, f(x) \frac{\overleftrightarrow{\partial}}{\partial t} \phi_{\text{out}}(x) \right) = 0, \quad (6.188)$$

with a similar case for the limit $t \to -\infty$ and ϕ_{in}. In (6.188) the symbol $\overleftrightarrow{\partial}$ represents

$$f \frac{\overleftrightarrow{\partial}}{\partial t} g \equiv f \frac{\partial g}{\partial t} - \frac{\partial f}{\partial t} g, \quad (6.189)$$

and Z is a constant called the *wave-function renormalization*. The necessity for this constant will be discussed shortly. The limit in (6.188) is understood to be in the weak sense, so that the limit is true only for the matrix elements of the operators. The presence of the constant Z in (6.188) leads to the identification of Φ as the *unrenormalized field*.

Relation (6.188) serves to define the asymptotic fields, but makes no assumptions about their form or particle content. The connection to the particle content of the theory is made by assuming that the modes of the asymptotic in and out fields define a Fock space in a manner similar to the Fock space of the free field ϕ. The respective Fock spaces are labeled as $|\alpha\rangle_{\text{in}}$ and $|\beta\rangle_{\text{out}}$. They are constructed by choosing an appropriate form for f in (6.188) and assuming that this defines a creation or annihilation operator that acts on the in or out states. For example, if the theory is free of exotic behavior in the asymptotic region then the choice for f, given by

$$f(x) = f_p(x) = \frac{1}{\sqrt{(2\pi)^3}} \frac{e^{-ipx}}{\sqrt{2\epsilon_p}}, \quad (6.190)$$

is assumed to define a creation operator,

$$|p,\alpha\rangle_{\text{in}} = a^\dagger_{\text{in}}(p)|\alpha\rangle_{\text{in}}$$

$$= -\frac{i}{\sqrt{Z}} \lim_{t\to-\infty} \int d^3x\, f_p(x) \frac{\overleftrightarrow{\partial}}{\partial t} \Phi(x)|\alpha\rangle_{\text{in}}, \quad (6.191)$$

where α represents all the other particles in the state. Result (6.191) is certainly true for the case that the asymptotic field coincides with the free scalar field. Therefore, by analogy with the scalar field discussion of Sec. 6.4, it is possible to define an annihilation operator $a_{\text{in}}(p)$, as well as the respective annihilation and creation operators for the out states, and this is accomplished by choosing the complex conjugate of (6.191). It is assumed that both the in and out states are complete and, until gauge fields are considered, are assumed to possess a positive-definite norm.

A quantum process is defined as $_{\text{out}}\langle\beta|\alpha\rangle_{\text{in}}$, and this transition element is interpreted, up to a normalization constant, as the probability amplitude for a transition between the two states. From it such physically measurable quantities as cross sections and lifetimes may be derived. It is assumed that there exists an operator S [22] that maps the in states into the out state with the same respective properties, i.e., the same form for the function f in (6.191), so that $S|\alpha\rangle_{\text{in}} = |\alpha\rangle_{\text{out}}$. The S operator has the matrix elements

$$_{\text{out}}\langle\beta|\alpha\rangle_{\text{in}} = {}_{\text{in}}\langle\beta|S^\dagger|\alpha\rangle_{\text{in}} \equiv S^\dagger_{\beta\alpha}. \tag{6.192}$$

From the assumed completeness of the in and out states it follows that

$$_{\text{in}}\langle\alpha|\beta\rangle_{\text{in}} = \sum_{n(out)} {}_{\text{in}}\langle\alpha|n\rangle\langle n|\beta\rangle_{\text{in}}$$

$$= \sum_{n(in)} {}_{\text{in}}\langle\alpha|S|n\rangle\langle n|S^\dagger|\beta\rangle_{\text{in}} = {}_{\text{in}}\langle\alpha|SS^\dagger|\beta\rangle_{\text{in}}, \tag{6.193}$$

so that the S-operator must be unitary. Unitarity of the S-matrix also represents the fact that the total transition probability from some in state to the entire spectrum of out states must be unity, for otherwise probability is not conserved. Denoting the norm of the in state $|\alpha\rangle_{\text{in}}$ by $|N_\alpha|^2$, the unitarity of the S-matrix preserves probability according to

$$1 = \frac{1}{|N_\alpha|^2}{}_{\text{in}}\langle\alpha|\alpha\rangle_{\text{in}} = \frac{1}{|N_\alpha|^2}\sum_n |{}_{\text{out}}\langle n|\alpha\rangle_{\text{in}}|^2 = \frac{1}{|N_\alpha|^2}S_{\alpha n}S^\dagger_{n\alpha}. \tag{6.194}$$

From the fact the in and out states are constructed from the fields ϕ_{in} and ϕ_{out}, it follows that the S operator maps them into one another.

Exercise 6.39: Show that the S operator satisfies

$$S\phi_{\text{out}}S^{-1} = \phi_{\text{in}}. \tag{6.195}$$

Using the assumptions made so far, an expression for the S-matrix in terms of the interacting Heisenberg fields can be derived. A general formula will be derived and applied inductively to obtain this expression. The starting point is the general object

$$W = {}_{\text{out}}\langle\alpha|T\{\Phi(x_1),\ldots,\Phi(x_n)\}|\mathbf{p},\beta\rangle_{\text{in}}, \tag{6.196}$$

where T stands for time ordering of the fields and is defined identically to the quantum mechanical time-ordering of Heisenberg picture operators of

Sec. 2.1. For scalar fields the time-ordered product is given by

$$T\{\phi(x_1)\phi(x_2)\} = \theta(t_1 - t_2)\phi(x_1)\phi(x_2) + \theta(t_2 - t_1)\phi(x_2)\phi(x_1) \, . \qquad (6.197)$$

The in state contains a particle of momentum **p**, and this particle will now be *reduced* from the state. This is done by writing W as

$$\begin{aligned}
W = \;& -\frac{i}{\sqrt{Z}} \lim_{t \to -\infty} \int d^3x \, f_p(x) \overleftrightarrow{\frac{\partial}{\partial t}} {}_{\text{out}}\langle \alpha | T\{\Phi(x_1) \cdots \Phi(x_n)\} \Phi(x) | \beta \rangle_{\text{in}} \\
= \;& -\frac{i}{\sqrt{Z}} \lim_{t \to -\infty} \int d^3x \, f_p(x) \overleftrightarrow{\frac{\partial}{\partial t}} {}_{\text{out}}\langle \alpha | T\{\Phi(x_1) \cdots \Phi(x_n)\} \Phi(x) | \beta \rangle_{\text{in}} \\
& + \frac{i}{\sqrt{Z}} \lim_{t \to +\infty} \int d^3x \, f_p(x) \overleftrightarrow{\frac{\partial}{\partial t}} {}_{\text{out}}\langle \alpha | \Phi(x) T\{\Phi(x_1) \cdots \Phi(x_n)\} | \beta \rangle_{\text{in}} \\
& - \frac{i}{\sqrt{Z}} \lim_{t \to +\infty} \int d^3x \, f_p(x) \overleftrightarrow{\frac{\partial}{\partial t}} {}_{\text{out}}\langle \alpha | \Phi(x) T\{\Phi(x_1) \cdots \Phi(x_n)\} | \beta \rangle_{\text{in}} \, ,
\end{aligned}$$
$$(6.198)$$

where it is assumed that f_p has been chosen to give a localized wave packet. Using the definition of time ordering in conjunction with the limits allows the field $\phi(x)$ to be moved inside the time-ordered product:

$$\lim_{t \to +\infty} T\{\phi(x)\phi(x_1) \cdots \phi(x_n)\} = \lim_{t \to +\infty} \phi(x) T\{\phi(x_1) \cdots \phi(x_n)\} \, . \qquad (6.199)$$

As a result, the first two terms of (6.198) may be combined by noting that

$$\left(\lim_{t \to +\infty} - \lim_{t \to -\infty} \right) \int d^3x \, f_p(x) \overleftrightarrow{\frac{\partial}{\partial t}} {}_{\text{out}}\langle \alpha | T\{\Phi(x)\Phi(x_1) \cdots \Phi(x_n)\} | \beta \rangle_{\text{in}}$$
$$= \int_{-\infty}^{\infty} dt \, \frac{\partial}{\partial t} \int d^3x \, f_p(x) \overleftrightarrow{\frac{\partial}{\partial t}} {}_{\text{out}}\langle \alpha | T\{\Phi(x)\Phi(x_1) \cdots \Phi(x_n)\} | \beta \rangle_{\text{in}} \, .$$
$$(6.200)$$

Using definition (6.189), and the fact that

$$\frac{\partial^2}{\partial t^2} f_p(x) = (\nabla^2 - m^2) f_p(x) \, , \qquad (6.201)$$

allows an integration by parts in (6.200) to move ∇^2 from f_p to the field operator. Surface terms are dropped since f is localized, and this gives the final form for W:

$$W = \frac{i}{\sqrt{Z}} \int d^4x \, f_p(x) \, (\Box + m^2) \, {}_{\text{out}}\langle \alpha | T\{\Phi(x)\Phi(x_1)\cdots\Phi(x_n)\} | \beta \rangle_{\text{in}}$$
$$- {}_{\text{out}}\langle \alpha | a^\dagger_{\text{out}}(p) T\{\Phi(x_1)\cdots\Phi(x_n)\} | \beta \rangle_{\text{in}} \, . \qquad (6.202)$$

The second term in (6.202) will vanish unless there is a particle in the out state with the same momentum. Such a condition would correspond to a particle unaffected by the quantum process and will be referred to as *forward scattering*. Normally such terms are ignored; however, there are cases where the in and out states are coherent states, and for that case the forward scattering term does not necessarily vanish, in particular if the coherent states are not the same. Coherent states are discussed in Sec. 6.8. For now, the forward scattering terms will be ignored.

Result (6.202) can be iterated to reduce all the particles from the in state, and similar formulas, differing only by the complex conjugate, can be used to reduce all the particles from the out state. The process terminates when the in and out states are reduced to the vacuum state $|0\rangle$, and it is assumed that this vacuum state is the same for both the in and out states. The final result is

$${}_{\text{out}}\langle \mathbf{k}_1, \ldots, \mathbf{k}_r | \mathbf{p}_1, \ldots, \mathbf{p}_n \rangle_{\text{in}} =$$
$$\frac{i^{n+r}}{Z^{(n+r)/2}} \int d^4x_1 \cdots d^4x_{n+r} \, f^*_{k_1}(x_1) \cdots f_{p_n}(x_{n+r})$$
$$\times (\Box_1 + m^2) \cdots (\Box_{n+r} + m^2) \, \langle 0 | T\{\Phi(x_1)\cdots\Phi(x_{n+r})\} | 0 \rangle \, . \qquad (6.203)$$

Exercise 6.40: Verify result (6.203).

Similar reduction formulas for spinors may be obtained by almost identical arguments. For example, from the assumption that the unrenormalized spinor field satisfies $\Psi \to \sqrt{Z_2} \, \psi$ in the asymptotic limit, an incoming electron state is defined as

$$b^{s\dagger}_{\text{in}}(\mathbf{p}) = \lim_{t\to -\infty} \frac{1}{\sqrt{Z_2}} \int d^3x \, \frac{1}{(2\pi)^{3/2}} \sqrt{\frac{m}{\epsilon_p}} e^{-ipx} \Psi^\dagger(x) u_s(p)$$
$$\equiv \lim_{t\to -\infty} \frac{1}{\sqrt{Z_2}} \int d^3x \, \Psi^\dagger_a(x) U^a_s(p,x) \, , \qquad (6.204)$$

where $u_s(p)$ is the spinor part of the free solution defined in (6.100). Of course, a similar expression defines a positron creation operator by using

$v_s^\dagger$ and Ψ. Because of the anticommutativity, it is necessary to modify the definition of time ordering for Dirac fields, so that

$$T\{\psi_a(x_1)\psi_b(x_2)\} = \theta(t_1 - t_2)\psi_a(x_1)\psi_b(x_2) - \theta(t_2 - t_1)\psi_b(x_2)\psi_a(x_1) \,. \quad (6.205)$$

It is then possible to show that, ignoring forward scattering,

$$\sqrt{Z_2}\,_{\text{out}}\langle \alpha\,|T\{\Psi_{a_1}(x_1)\cdots\}|\,p,s,\beta\rangle_{\text{in}} =$$
$$-\int d^4x\,_{\text{out}}\langle \alpha\,|T\{\Psi_a^\dagger(x)\Psi_{a_1}(x_1)\cdots\}|\,\beta\rangle_{\text{in}} \left(\overleftarrow{\partial\!\!\!/} - im\right) U_s^a(p,x) \,. \quad (6.206)$$

Exercise 6.41: Verify (6.206) and derive the reduction formulas for positron states. Obtain the form for the completely reduced matrix element.

The S-matrix is now expressed in terms of the local field operators and, if the solutions for the fields are available, they may now be used to evaluate quantum processes. Unfortunately, very few exactly solvable systems are known, other than the free field theories already presented, and of the known solvable interacting theories, there are none that correspond to a four-dimensional Minkowski space problem. There are also conceptual problems with the LSZ reduction technique presented in this section. For example, it assumes that there is a one-to-one correspondence between measurable particle states and the fields of the underlying theory. It is widely believed that this feature must break down in quantum chromodynamics (QCD), where the underlying particle modes of the Fock space are conjectured to be confined in bound states that, speaking loosely, can never be broken. Unfortunately, the presence of *confinement* in 3+1 QCD has never been demonstrated analytically, although the combination of high speed computers and the path integral formalism has provided strong evidence for it.

A related problem in the LSZ formalism occurs when some or all of the fields are associated with massless particles. Such fields give rise to long-range, Coulomb-like potentials, and these persist out to infinity. For such a case it is not possible to argue that all interactions become insignificant at asymptotic times anymore than one could claim that a charge in uniform motion does not possess an electric and magnetic field. Insisting that all interactions can be ignored at asymptotic times leads ultimately to a problem in massless theories lumped under the rubric of *infrared divergences*. These are not a calamity in quantum electrodynamics, where they can be controlled by redefining the charged in and out states to possess electromagnetic fields. However, the infrared divergences of QCD may be the mechanism of confinement, and therefore the assumption that the asymptotic fields are free may be a very grave error in that theory. Unfortunately, the very interesting problem of infrared divergences lies, for the most part, outside the scope of this book. Some simple aspects are discussed in Chapter 9.

These problems notwithstanding, the central feature of (6.203) and the fully reduced spinor matrix element is that all the quantum dynamics is contained in the vacuum expectation value of the time-ordered product of the Heisenberg fields. It is the evaluation of these quantities that is the central problem of quantum field theory. The utility of the path integral representation of field theory lies in evaluating such time-ordered products of field operators. However, there is one last step to be taken before the path integral may be derived.

6.7 The Interaction Picture

Because solving field theory problems is so difficult, perturbation theory is often the only recourse. Covariant perturbation theory [24, 25, 26] as a calculational technique can be derived by different methods. In this section the *interaction picture* approach [27] will be used, and it is defined in the same way as its quantum mechanical counterpart. The interacting field Φ, which satisfies the full nonlinear problem, is constructed from the field ϕ, which satisfies a solvable linear problem.

The Hamiltonian of the system is split into two parts, $H = H_0 + H_I$. It is assumed that H_0 defines a solvable problem, while H_I represents the remainder of the Hamiltonian terms. The fields whose time developments are given by H_0 are denoted ϕ_α, along with their conjugate momenta π_α, so that these fields satisfy $[H_0, \phi_\alpha] = -i\dot\phi_\alpha$. The time dependence of the interacting field Φ_α is given by $[H, \Phi_\alpha] = -i\dot\Phi_\alpha$, where H is a functional of the Φ_α fields and their conjugate momenta Π_α.

The Φ_α fields are then constructed from the ϕ_α fields by making the assumption that they and their conjugate momenta are related by a time-dependent unitary transformation $U(t)$ of the form

$$\phi_\alpha(\mathbf{x},t) = U(t)\Phi_\alpha(\mathbf{x},t)U^{-1}(t), \quad \pi_\alpha(\mathbf{x},t) = U(t)\Pi_\alpha(\mathbf{x},t)U^{-1}(t). \quad (6.207)$$

Applying $\partial/\partial t$ immediately yields

$$\begin{aligned}\dot\phi_\alpha(x) &= \dot U(t)U(t)\phi_\alpha(x) + \phi_\alpha(x)U(t)\dot U^{-1}(t) + U(t)\dot\Phi_\alpha(x)U^{-1}\\ &= [\dot U(t)U^{-1}(t), \phi_\alpha(x)] + i[H[\phi_\alpha], \phi_\alpha(x)], \quad (6.208)\end{aligned}$$

Sec. 6.7 The Interaction Picture

where the identity $\dot{U}U^{-1} = -U\dot{U}^{-1}$ has been used, and the identification

$$H[\phi_\alpha] = U(t)H[\Phi_\alpha]U^{-1}(t) \tag{6.209}$$

has been made. The Hamiltonian of (6.209) is now in the interaction picture, and is no longer time independent since it does not commute with H_0.

Substituting $\dot{\phi}_\alpha = i[H_0, \phi_\alpha]$ into (6.208) gives

$$\begin{aligned}[\dot{U}(t)U^{-1}(t), \phi_\alpha(x)] &= -i[H[\phi_\alpha] - H_0[\phi_\alpha], \phi_\alpha(x)] \\ &\equiv -i[H_I(t), \phi_\alpha(x)],\end{aligned} \tag{6.210}$$

from which follows the identification

$$\dot{U}(t)U^{-1}(t) = -iH_I(t). \tag{6.211}$$

Introducing the *evolution operator* $E(t, t') = U(t)U^{-1}(t')$, relation (6.211) becomes

$$\frac{\partial}{\partial t}E(t, t') = -iH_I(t)E(t, t'). \tag{6.212}$$

From its definition the evolution operator must satisfy the initial condition $E(t, t) = 1$. Using this allows the first-order differential equation to be integrated, giving

$$E(t, t_0) = 1 - i\int_{t_0}^{t} d\tau\, H_I(\tau)E(\tau, t_0). \tag{6.213}$$

Iteration of this equation leads to the representation of the evolution operator in terms of a time-ordered operator,

$$E(t, t_0) = T\left\{\exp\left[-i\int_{t_0}^{t} d\tau\, H_I(\tau)\right]\right\}. \tag{6.214}$$

Exercise 6.42: Verify that (6.214) is the solution to (6.213).

The vacuum expectation value of the time-ordered product of Heisenberg fields can now be given a representation in terms of the evolution operator and the free fields. The form is rewritten as

$$\begin{aligned}G_{\alpha\cdots\beta}(x_1, \ldots, x_n) &= \langle 0|T\{\Phi_\alpha(x_1)\cdots\Phi_\beta(x_n)\}|0\rangle \\ &= \langle 0|T\{U^{-1}(t_1)\phi_\alpha(x_1)U(t_1)\cdots U^{-1}(t_n)\phi_\beta(x_n)U(t_n)\}|0\rangle \\ &= \langle 0|U^{-1}(t_+)T\{\phi_\alpha(x_1)\cdots\phi_\beta(x_n)E(t_+, t_-)\}U(t_-)|0\rangle,\end{aligned} \tag{6.215}$$

where $t_\pm$ represents the limits $t \to \pm\infty$. The last step in (6.215) is valid due to the nature of time ordering, which gives

$$T\{U(t_+)U^{-1}(t)\phi_\alpha(x)U(t)U^{-1}(t_-)\} = T\{\phi(x)U(t_+)U^{-1}(t_-)\}. \quad (6.216)$$

Now the assumed asymptotic behavior of the fields can be used to determine the action of $U(t_\pm)$ on the vacuum. From the fact that the in field can be used to define an annihilation operator, and that annihilation operator destroys the vacuum, in the spinor field case it follows that

$$0 = \sqrt{Z_2}\langle n|a_{\text{in}}(\mathbf{p})|0\rangle = \lim_{t\to-\infty}\int d^3x\, U_s^\dagger(p,x)\langle n|\Psi(x)|0\rangle$$

$$= \lim_{t\to-\infty}\int d^3x\, U_s^\dagger(p,x)\langle n|U^{-1}(t)\psi(x)U(t)|0\rangle. \quad (6.217)$$

This can be true only if $U(t_-)|0\rangle = \lambda_-|0\rangle$, where λ_- is some constant, for otherwise the annihilation operator present in the last step of (6.217) would not destroy the state created by $U(t_-)$. Likewise, it follows that $\langle 0|U^{-1}(t_+) = \lambda_+\langle 0|$. It follows immediately that

$$\langle 0|E(t_+,t_-)|0\rangle = \langle 0|U(t_+)U^{-1}(t_-)|0\rangle = \frac{1}{\lambda_+\lambda_-}. \quad (6.218)$$

Inserting (6.218) into (6.215) gives

$$G_{\alpha\cdots\beta}(x_1,\ldots,x_n) = \frac{\langle 0|T\{\phi_\alpha(x_1)\cdots\phi_\beta(x_n)E(t_+,t_-)\}|0\rangle}{\langle 0|E(t_+,t_-)|0\rangle}. \quad (6.219)$$

The factor appearing in the denominator of (6.219) is the vacuum-to-vacuum transition amplitude in the interaction picture, and its presence serves to cancel *disconnected* processes appearing in the expansion of the numerator. The explanation of this remark will be found in Chapter 8, where a graphical representation of the expansion of (6.219) is discussed. However, for this reason $G_{\alpha\cdots\beta}(x_1,\ldots,x_n)$ is referred to as the *connected Green's function* for the particular quantum process to which it is associated. The fact that this function is expressed as a ratio allows nonzero factors common to both the numerator and denominator to be dropped.

In principle there is no impediment to proceeding by using the expansions determined so far for the free fields. The exponential representing the evolution operator is expanded in a power series and the resulting time-ordered products are evaluated by commuting the annihilation operators to the right and the creation operators to the left, eventually arriving at the value of the Green's function at each order of the expansion. Such a procedure can be refined into a set of rules, referred to as the *Feynman rules* [26]

Sec. 6.8 The Path Integral for Field Theory   207

for the perturbation expansion, and these are based around a technique for simplifying the vacuum expectation value of time-ordered products of free fields known as *Wick's theorem* [28]. However, rather than defining the Feynman rules in the context of operator representations of the fields, the path integral representation of the vacuum-to-vacuum transition element in the presence of an external source will be developed. In Chapter 8 this will be used to define the Feynman rules in a very simple manner.

It should be pointed out that the unitary transformation between the interacting field and the free field does not exist as a well-defined mathematical object. It has been shown that, at best, in the infinite volume limit U is an *improper* transformation in that its matrix elements in the Fock space of the free theory vanish [29]. This is a strong indication of trouble ahead for the perturbative representation of the Green's function, and it should therefore come as no surprise that many terms in the expansion are divergent. Methods for controlling and removing these divergences exist for a subclass of theories that are said to be *renormalizable*, and this is the reason for the presence of the factor $\sqrt{Z}$ in the reduction formulas. These methods will be discussed briefly in Chapter 8.

6.8 The Path Integral for Field Theory

In this section the path integral representation of the vacuum-to-vacuum transition element will be constructed. By adding a source term, a path integral can serve as a generating functional for the connected Green's functions of the theory. The method is to use coherent states [30], already encountered for the quantum mechanical case in Chapters 4 and 5, to partition the evolution operator into infinitesimal time periods. This is essentially the field theoretic extension of the approach used in Chapters 2 and 4 to construct the quantum mechanical path integral. This approach [31, 32, 33] has the advantage that the measure for the path integral is clear at each step. In effect, it leads to a momentum space representation of the measure, rather than the configuration space measure derived heuristically in Sec. 6.1. The measure will be derived for a Minkowski space representation of the theory. However, the derived measure will be related to Euclidean measure in both momentum and configuration space. This is discussed later in the section.

6.8.1 The Scalar Field Case

The scalar field case will be considered first. The starting point is the idea that the interacting field, assumed to be massive, obeys some nonlinear equation, so that $(\Box + m^2)\Phi \neq 0$. If the path integral is to represent solutions to such an equation, then a class of functions larger than that of the plane-wave solutions to $(\Box + m^2)\phi = 0$ must be used. This class is defined in the following way. The relation between p_0 and $\mathbf{p}$, namely that $p_0 = +\sqrt{\mathbf{p}^2 + m^2}$, is relaxed, so that $p_0 \equiv \omega$ is now an independent degree of freedom, and p now denotes $(\omega, \mathbf{p})$. Such a set of functions now accommodates a much larger variety of behaviors.

A general real-valued function may be constructed from these eigenfunctions as a four-dimensional Fourier transform, and it is written

$$\psi(x) = \int \frac{d^4p}{(2\pi)^2} \left[\tilde{\psi}(p) e^{-ipx} + \tilde{\psi}^*(p) e^{ipx} \right] . \tag{6.220}$$

where $px = \omega t - \mathbf{p} \cdot \mathbf{x}$. Expression (6.220) may be rewritten by singling out the ω dependence and denoting

$$\int \frac{d\omega}{\sqrt{2\pi}} \tilde{\psi}(p) e^{-i\omega t} = \int \frac{d\omega}{\sqrt{2\pi}} \tilde{\psi}(\mathbf{p}, \omega) e^{-i\omega t} \equiv \frac{1}{\sqrt{2\epsilon_p}} e^{-i\epsilon_p t} \tilde{\psi}_\mathbf{p}(t) , \tag{6.221}$$

where the presence of ϵ_p is for later convenience. Using this notation, the function of (6.220) becomes

$$\psi(x) = \int \frac{d^3p}{(2\pi)^{3/2}} \frac{1}{\sqrt{2\epsilon_p}} \left[\tilde{\psi}_\mathbf{p}(t) e^{-ipx} + \tilde{\psi}^*_\mathbf{p}(t) e^{ipx} \right] , \tag{6.222}$$

where $px = \epsilon_p t - \mathbf{p} \cdot \mathbf{x}$ in (6.222). Result (6.222) has the same *form* as the wave-packet solution (6.66) to the free field equation of motion, but with coefficients that are now time dependent.

The coherent state $|\psi(t)\rangle$ is defined in terms of the *free field* Fock space operators $a_\mathbf{p}$ and $a^\dagger_\mathbf{p}$ of (6.167), associated with the free field ϕ,

$$|\psi(t)\rangle = \exp\left\{ \int d^3p \left[\tilde{\psi}_\mathbf{p}(t) a^\dagger_\mathbf{p} - \tilde{\psi}^*_\mathbf{p}(t) a_\mathbf{p} \right] \right\} |0\rangle . \tag{6.223}$$

Using the Baker–Campbell–Hausdorff formula shows that

$$\langle \psi(t) | \phi(x) | \psi(t) \rangle = \psi(x) , \tag{6.224}$$

$$\langle \varphi(t) | \psi(t) \rangle =$$
$$\exp\left\{ -\tfrac{1}{2} \int d^3p \left[\tilde{\psi}^*_\mathbf{p}(t) \tilde{\psi}_\mathbf{p}(t) + \tilde{\varphi}^*_\mathbf{p}(t) \tilde{\varphi}_\mathbf{p}(t) - 2\tilde{\varphi}^*_\mathbf{p}(t) \tilde{\psi}_\mathbf{p}(t) \right] \right\} . \tag{6.225}$$

Exercise 6.43: Verify (6.224) and (6.225).

Sec. 6.8 The Path Integral for Field Theory

It is stressed that a different Fock space representation of the particle spectrum, perhaps associated with the presence of interactions, would lead to quite different results.

The coherent states possess a form of completeness. The statement of this completeness is complicated by the presence of the measure d^3p in the definition (6.223) and the result (6.225). For that reason it is common practice to quantize the system in a box, so that there is a discrete set of eigenfunctions. Using such an approach allows the integrals over **p** in (6.223) and (6.225) to be replaced by a sum over the indices of the eigenvalues, and products over the discrete indices are well defined. In the end of the calculation the infinite volume or continuum limit is taken, resulting in the reappearance of factors of the volume. This is the multidimensional extension of the technique familiar from quantum mechanical path integrals, where the action was rewritten as a Fourier sum. At the end of the calculation the infinite volume limit reproduces the inverse factors of the measure in the manner

$$\sum_n = \frac{L}{2\pi} \sum_n \frac{2\pi}{L} \to \frac{L}{2\pi} \int dp . \qquad (6.226)$$

A variant of this technique will be used here. For the purposes of this section the element d^3p will stand for $(2\pi)^3/V$, where the limit $V \to \infty$ is understood. Integrals over **p** will be replaced by Riemann sums weighted by d^3p. This approach allows a definition of the measure required by the coherent states for completeness.

The completeness of the coherent states can now be written as

$$1 = \int \mathcal{D}\psi(t) \, |\psi(t)\rangle\langle\psi(t)| , \qquad (6.227)$$

where the measure is given by

$$\mathcal{D}\psi(t) = \prod_{\mathbf{p}} \frac{d^3p}{2\pi i} \, d\tilde{\psi}_{\mathbf{p}}(t) \, d\tilde{\psi}_{\mathbf{p}}^*(t) , \qquad (6.228)$$

so that $\tilde{\psi}_{\mathbf{p}}(t)$ is the variable of integration. The price paid for not quantizing the system in a box is now apparent in the rather poorly defined concept of a product over a continuum of values **p**. For that reason the form (6.228) of the measure must be understood as the continuum limit of the discrete case.

210   Field Theory

> **Exercise 6.44**: Verify that the projection operator defined by (6.227) is equivalent to the Fock space unit projection operator (6.178) by taking the continuum limit of the discrete form of the measure, i.e., the measure written for the case that the integral is replaced by a Riemann sum.

One last step needs to be taken before the path integral may be constructed, and that is simply to note that the free field $\phi(x)$ and its canonical momentum $\pi(x)$ have the matrix elements, for $\delta t \approx 0$,

$$\frac{\langle \psi(t+\delta t)|\phi(x)|\psi(t)\rangle}{\langle \psi(t+\delta t)|\psi(t)\rangle} = \int d^4p \left[\tilde{\psi}^*(p)\phi_p(\mathbf{x},t+\delta t) + \tilde{\psi}(p)\phi_p^*(\mathbf{x},t)\right]$$
$$= \psi(x) + O(\delta t), \qquad (6.229)$$

$$\frac{\langle \psi(t+\delta t)|\pi(x)|\psi(t)\rangle}{\langle \psi(t+\delta t)|\psi(t)\rangle} = i\int d^4p\, \epsilon_p \left[\tilde{\psi}^*(p)\phi_p(\mathbf{x},t+\delta t) - \tilde{\psi}(p)\phi_p^*(\mathbf{x},t)\right]$$
$$= \rho(x) + O(\delta t), \qquad (6.230)$$

where

$$\rho(x) \equiv i \int d^4p\, \epsilon_p \left[\tilde{\psi}^*(p)\phi_p(x) - \tilde{\psi}(p)\phi_p^*(x)\right]. \qquad (6.231)$$

The perturbative representation (6.219) of the coherent state transition element is now written as the product of N transition elements separated by infinitesimal times using the familiar procedure of inserting complete sets of states at the $N-1$ intermediate times. This gives

$$\langle \psi, t_+ | E(t_+, t_-) | \psi, t_- \rangle =$$
$$\prod_{j=0}^{N-1} \int \mathcal{D}\psi(t_j) \langle \psi(t_{j+1}) | E(t_{j+1}, t_j) | \psi(t_j) \rangle, \qquad (6.232)$$

where the identifications $t_N = t_+$ and $t_0 = t_-$ have been made.

Each of the infinitesimal elements can now be reduced by noting, as in Exercise 2.3, that time ordering becomes trivial in the limit $t_{j+1} - t_j = \delta t \to 0$. The result is

$$E(t_{j+1}, t_j) = T\left\{\exp\left(-i\int_{t_j}^{t_{j+1}} d\tau\, H_I[\phi,\pi]\right)\right\}$$
$$= \exp\left\{-i\delta t \int d^3x\, \mathcal{H}_I\left[\phi(\mathbf{x},t_j), \pi(\mathbf{x},t_j)\right]\right\}, \qquad (6.233)$$

where the interaction Hamiltonian H_I has been assumed to be a general, local, normal-ordered functional of ϕ and π. Using (6.229), (6.230), (6.233),

Sec. 6.8   The Path Integral for Field Theory   211

and suppressing terms of $O(\delta t^2)$ gives

$$\langle \psi(t_{j+1}) | E(t_{j+1}, t_j) | \psi(t_j) \rangle =$$
$$\langle \psi(t_{j+1}) | \psi(t_j) \rangle \exp\left(-i\delta t \int d^3x \, \mathcal{H}_I[\psi(\mathbf{x}, t_j), \rho(\mathbf{x}, t_j)]\right), \quad (6.234)$$

The remaining matrix element in (6.234) is given by

$$\langle \psi(t_{j+1}) | \psi(t_j) \rangle = \exp\left\{-\tfrac{1}{2} \int d^3p \left[\tilde{\psi}_\mathbf{p}^*(t_j + \delta t) \tilde{\psi}_\mathbf{p}(t_j + \delta t)\right.\right.$$
$$\left.\left. + \tilde{\psi}_\mathbf{p}^*(t_j) \tilde{\psi}_\mathbf{p}(t_j) - 2\tilde{\psi}_\mathbf{p}^*(t_j + \delta t) \tilde{\psi}_\mathbf{p}(t_j)\right]\right\}$$
$$\approx \exp\left\{-\tfrac{1}{2} \delta t \int d^3p \, \tilde{\psi}_\mathbf{p}^*(t_j) \overleftrightarrow{\frac{\partial}{\partial t_j}} \tilde{\psi}_\mathbf{p}(t_j)\right\}. \quad (6.235)$$

Result (6.235) may be written in terms of $\rho(x)$ and $\psi(x)$ by exploiting their relationship to $\tilde{\psi}_p$, given by (6.222) and (6.231), and noting that (6.221) allows $\partial \tilde{\psi}_\mathbf{p}(t)/\partial t$ to be expressed in terms of the $\tilde{\psi}_p$. It is tedious but straightforward to show that

$$\langle \psi(t_{j+1}) | \psi(t_j) \rangle = \exp\left\{-\tfrac{1}{2} \delta t \int d^3p \, \tilde{\psi}_\mathbf{p}^*(t_j) \overleftrightarrow{\frac{\partial}{\partial t_j}} \tilde{\psi}_\mathbf{p}(t_j)\right\} =$$
$$\exp\left\{i\delta t \int d^3x \, \tfrac{1}{2} \left[\rho\dot{\psi} - \dot{\rho}\psi - \rho^2 - \nabla\psi \cdot \nabla\psi - m^2\psi^2\right]\right\}\Bigg|_{t=t_j}. \quad (6.236)$$

Exercise 6.45: Verify (6.236).

The quadratic part of the action $\mathcal{L}_o$ has appeared, albeit in symmetrized form. Combining (6.236) with (6.234), substituting that result back into (6.232), and taking the $\delta t \to 0$ limit gives the the coherent state transition element:

$$\langle \psi, t_+ | \psi, t_- \rangle =$$
$$\int_{\psi_-}^{\psi_+} \mathcal{D}\rho \, \mathcal{D}\psi \exp\left\{i \int d^4x \left[\tfrac{1}{2}\rho\dot{\psi} - \tfrac{1}{2}\dot{\rho}\psi - \mathcal{H}(\psi, \rho)\right]\right\}, \quad (6.237)$$

where the limits reflect the fact that the functions ψ and ρ have the asymptotic forms $\psi_\pm = \psi(\mathbf{x}, t_\pm)$ and $\rho_\pm = \rho(\mathbf{x}, t_\pm)$, and $\mathcal{H} = \mathcal{H}_o + \mathcal{H}_I$ denotes the full Hamiltonian density. The measure is given by

$$\mathcal{D}\rho\mathcal{D}\psi = \lim_{N\to\infty} \prod_{j=1}^{N-1} \prod_{\mathbf{p}} \frac{d^3p}{2\pi i} \, d\tilde{\psi}_{\mathbf{p}}^*(t_j) \, d\tilde{\psi}_{\mathbf{p}}(t_j) \,. \tag{6.238}$$

Result (6.237) is the general field theoretic path integral in the scalar case.

The boundary conditions on ψ and ρ are extremely important, and they must be taken into account whenever an integration by parts involving the time is performed. It is easily seen from the form of the coherent state of (6.223) that the vacuum-to-vacuum transition element is obtained by setting $\psi_\pm = \rho_\pm = 0$. In this respect, the vacuum-to-vacuum transition element allows integration by parts with no change in form since ψ and ρ vanish at $t_\pm$. For the vacuum transition amplitude case it follows that

$$_+\langle 0|0\rangle_- = \int_0^0 \mathcal{D}\rho\mathcal{D}\psi \, \exp\left\{i\int d^4x \left[\rho\dot{\psi} - \mathcal{H}(\psi,\rho)\right]\right\}, \tag{6.239}$$

so that the vacuum transition is evaluated by using a sum over all functions that begin and end at zero.

A final adjustment to the measure is possible since the relationship (6.221) is identical, up to the factor $d\omega\sqrt{\epsilon_p/\pi}$ appearing in the definition, to the change of variables via a Fourier series for the continuum approach discussed in Sec. 3.3. Therefore, the Jacobian J for this change of variables is already known and is given by (3.50). The final result is

$$\mathcal{D}\rho\mathcal{D}\psi = J^2 \prod_p \frac{\epsilon_p \, d^3p \, d^2\omega}{2\pi^2 i} \, d\tilde{\psi}^*(p) \, d\tilde{\psi}(p) \,, \tag{6.240}$$

where the product now runs over all values of the four-vector p. The measure is now a dimensionless product of the modes of the "field" ψ and the "canonical momentum" ρ.

In the event that the basis Hamiltonian takes the simple form

$$H_0 = \int d^3x \, \mathcal{H}_0 = \int d^3x \, \left[\tfrac{1}{2}\rho^2 + \tfrac{1}{2}\nabla\psi\cdot\nabla\psi + \tfrac{1}{2}m^2\psi^2\right] \,, \tag{6.241}$$

and $\mathcal{H}_I$ is independent of ρ, it is possible to integrate the ρ variable from the theory. This is accomplished by using the expansions (6.220) and (6.231) in those terms in the Lagrangian that involve ρ. Using the notation

$$\rho(x) = i\int \frac{d^4p}{(2\pi)^2} \, \epsilon_p \left[\psi(p)e^{ipx} - \psi^*(p)e^{-ipx}\right] =$$
$$i\int \frac{d^4p}{(2\pi)^2} \, \epsilon_p \left[\psi(p) - \psi^*(-p)\right] e^{ipx} \equiv i\int \frac{d^4p}{(2\pi)^2} \, \epsilon_p \, \tilde{\rho}(p)e^{ipx} \,, \tag{6.242}$$

Sec. 6.8 The Path Integral for Field Theory   213

it follows that $\tilde{\rho}^*(p) = -\tilde{\rho}(-p)$. A similar definition, $\tilde{\phi}(p) = \tilde{\psi}(p) + \tilde{\psi}^*(-p)$, gives

$$\psi(x) = \int \frac{d^4p}{(2\pi)^2} \tilde{\phi}(p) e^{ipx}, \qquad (6.243)$$

which has the property $\tilde{\phi}^*(p) = \tilde{\phi}(-p)$. Using the expansions (6.242) and (6.243) gives

$$\int d^4x \left(\rho\dot{\psi} - \tfrac{1}{2}\rho^2\right) = \int d^4p \left[\tfrac{1}{2}\epsilon_p^2\, \tilde{\rho}(p)\tilde{\rho}(-p) + \epsilon_p \omega\, \tilde{\rho}(p)\tilde{\phi}(-p)\right]$$

$$= \int d^4p \, \tfrac{1}{2}\epsilon_p^2 \left[\tilde{\rho}(p) - \frac{\omega}{\epsilon_p}\tilde{\phi}(p)\right]\left[\tilde{\rho}(-p) + \frac{\omega}{\epsilon_p}\tilde{\phi}(-p)\right] + \int d^4x \, \tfrac{1}{2}\dot{\psi}^2. \quad (6.244)$$

The Jacobian for the change of variables to $\tilde{\rho}$ and $\tilde{\phi}$ is simply a factor of $\tfrac{1}{2}$ for each value of p. The $\tilde{\rho}$ variables are then translated according to

$$\tilde{\pi}(p) = \tilde{\rho}(p) - \frac{\omega}{\epsilon_p}\tilde{\phi}(p), \qquad (6.245)$$

and from the properties of $\rho(p)$ and $\phi(p)$, this automatically implies

$$-\tilde{\pi}^*(p) = \tilde{\rho}(-p) + \frac{\omega}{\epsilon_p}\tilde{\phi}(-p). \qquad (6.246)$$

In the measure it is possible to pair $d\tilde{\rho}(p)$ and $d\tilde{\rho}(-p)$ by breaking the space p into two pieces, one with $\omega > 0$, the other with $\omega < 0$, and making the product over only one of the pieces in the following way:

$$\prod_p d\tilde{\rho}(p) = \prod_{p(\omega>0)} d\tilde{\rho}(p)\, d\tilde{\rho}(-p) = \prod_{p(\omega>0)} -d\tilde{\pi}(p)\, d\tilde{\pi}^*(p). \qquad (6.247)$$

Using the same definition, the integral appearing in (6.244) can be written as the sum

$$\sum_p d^4p \, \tfrac{1}{2}\epsilon_p^2 \left[\tilde{\rho}(p) - \frac{\omega}{\epsilon_p}\tilde{\phi}(p)\right]\left[\tilde{\rho}(-p) + \frac{\omega}{\epsilon_p}\tilde{\phi}(-p)\right]$$

$$= -\sum_{p(\omega>0)} d^4p \, \epsilon_p^2 \, \tilde{\pi}^*(p)\tilde{\pi}(p). \qquad (6.248)$$

where the property $\tilde{\pi}^*(-p) = -\tilde{\pi}(p)$ has been used. The $\tilde{\pi}$ variables can now be easily integrated out of the path integral by using the result

$$\left[\prod_{p(\omega>0)} d\tilde{\pi}^*(p)\, d\tilde{\pi}(p)\right] \exp\left[-i \sum_{p(\omega>0)} d^4p\, \epsilon_p^2\, \tilde{\pi}^*(p)\tilde{\pi}(p)\right]$$

$$= \prod_{p(\omega>0)} \frac{2\pi}{\epsilon_p^2\, d^4p} = \prod_p \sqrt{\frac{2\pi}{\epsilon_p^2\, d^4p}}. \qquad (6.249)$$

Using (6.249) and (6.244) gives the more commonly used variant of the path integral

$$_+\langle 0|0\rangle_- = \int_0^0 \bar{\mathcal{D}}\psi \, \exp\left[i\int d^4x \, \mathcal{L}(\psi)\right] , \qquad (6.250)$$

where the Lagrangian density is given by

$$\mathcal{L}(\psi) = \tfrac{1}{2}\partial_\mu\psi\partial^\mu\psi - \mathcal{H}_I(\psi) , \qquad (6.251)$$

and the measure is now given by

$$\bar{\mathcal{D}}\psi = \prod_p \frac{i\, d^3p\, d^2\omega}{\sqrt{(2\pi)^3 d^4p}}\, d\tilde{\phi}(p) . \qquad (6.252)$$

The relation of ψ to $\tilde{\phi}(p)$ is given by (6.243).

While care has been taken to carry along all factors of volumes necessary to normalize the path integral and make the measure dimensionless, in actuality these factors are of no importance whatsoever to an evaluation of the time-ordered products. This is because the principal goal is an evaluation of the connected Green's function. Recalling the definition (6.219), it is apparent that only the normalized vacuum transition amplitude is meaningful. As a result, all normalization factors present in the measure of the path integral representation of this quantity will cancel when the connected Green's functions are calculated. For this reason, most discussions of the measure have little or no reference to the normalization factors. From this point forward their role in normalized transition amplitudes will be ignored. However, it is not always the case that a normalized amplitude is being calculated, and in those cases it is necessary to maintain these factors in the measure.

6.8.2 The Dirac Field Case

The construction of the path integral for bispinor fields is similar, but complicated by the spinor indices and by the anticommutation relations. The latter, when combined with the results of Sec. 5.5, show that Grassmann variables will be needed to construct coherent states consistent with anticommutativity. The starting point is the definition of spinor functions in the enlarged space of functions. This is done identically to the scalar case by extending the wave-packet solution (6.109) to the case that the coefficients are time dependent, so that

$$\eta(x) = \int \frac{d^3p}{(2\pi)^{3/2}} \sqrt{\frac{m}{\epsilon_p}} \left[\beta_{ps}(t) u_s(p) e^{-ipx} + \alpha^*_{ps}(t) v_s(p) e^{ipx}\right] , \qquad (6.253)$$

Sec. 6.8 The Path Integral for Field Theory

where the u_s and v_s are the standard free bispinor solutions, and $px = \epsilon_p t - \mathbf{p} \cdot \mathbf{x}$. The function $\bar{\eta}$ is defined similarly. As in the scalar case, these coefficients are written as Fourier transforms,

$$\beta_{\mathbf{p}s}(t) = e^{i\epsilon_p t} \int \frac{d\omega}{\sqrt{2\pi}} \beta_s(p) e^{-i\omega t},$$

$$\alpha^*_{\mathbf{p}s}(t) = e^{-i\epsilon_p t} \int \frac{d\omega}{\sqrt{2\pi}} \alpha^*_s(p) e^{i\omega t}. \tag{6.254}$$

However, unlike the scalar field case, the coefficients are Grassmann variables and therefore anticommute. The definitions of (6.254) give

$$\eta(x) = \int \frac{d^4p}{(2\pi)^2} \sqrt{\frac{m}{\epsilon_p}} \left[\beta_s(p) u_s(p) e^{-ipx} + \alpha^*_s(p) v_s(p) e^{ipx} \right], \tag{6.255}$$

where $px = \omega t - \mathbf{p} \cdot \mathbf{x}$. Using properties (6.106) and (6.107) allows the coefficients to be written

$$\beta_{\mathbf{p}s}(t) = \int \frac{d^3x}{(2\pi)^{3/2}} \sqrt{\frac{m}{\epsilon_p}} \eta^\dagger(x) u_s(p) e^{ipx}, \tag{6.256}$$

$$\alpha_{\mathbf{p}s}(t) = \int \frac{d^3x}{(2\pi)^{3/2}} \sqrt{\frac{m}{\epsilon_p}} v_s^\dagger(p) \eta(x) e^{ipx}. \tag{6.257}$$

where $px = \epsilon_p t - \mathbf{p} \cdot \mathbf{x}$.

Using the Fock space operators $b_{\mathbf{p}}^s$ and $d_{\mathbf{p}}^s$, associated with the free spinor field $\psi(x)$, the Grassmann coherent state $|\eta, t\rangle$ is defined, as in (5.131), by

$$|\eta(t)\rangle = \exp\left\{ \int d^3p \, [b_{\mathbf{p}}^{s\dagger} \beta^*_{\mathbf{p}s}(t) + b_{\mathbf{p}}^s \beta_{\mathbf{p}s}(t) \right.$$
$$\left. + d_{\mathbf{p}}^{s\dagger} \alpha^*_{\mathbf{p}s}(t) + d_{\mathbf{p}}^s \alpha_{\mathbf{p}s}(t)] \right\} |0\rangle. \tag{6.258}$$

These have many properties similar to the scalar case, and it is left as an exercise to show that

$$\langle \eta(t) | \psi(x) | \eta(t) \rangle = \eta(x), \tag{6.259}$$

$$\langle \eta(t) | \bar{\psi}(x) | \eta(t) \rangle = \bar{\eta}(x), \tag{6.260}$$

$$\langle \eta'(t) | \eta(t) \rangle = \exp\left\{ -\tfrac{1}{2} \int d^3p \, [\beta'^{*}_{\mathbf{p}s}(t) \beta'_{\mathbf{p}s}(t) + \beta^*_{\mathbf{p}s}(t) \beta_{\mathbf{p}s}(t) \right.$$
$$- 2\beta'^{*}_{\mathbf{p}s}(t) \beta_{\mathbf{p}s}(t) + \alpha'^{*}_{\mathbf{p}s}(t) \alpha'_{\mathbf{p}s}(t) + \alpha^*_{\mathbf{p}s}(t) \alpha_{\mathbf{p}s}(t)$$
$$\left. - 2\alpha'^{*}_{\mathbf{p}s}(t) \alpha_{\mathbf{p}s}(t)] \right\}. \tag{6.261}$$

216   Field Theory

Exercise 6.46: Verify relations (6.259) and (6.261).

The statement of completeness is very similar to that of (5.133), except for factors of d^3p. It is written

$$\int \mathcal{D}\eta(t) \, |\eta(t)\rangle\langle\eta(t)| = 1 \,, \tag{6.262}$$

where the measure is given by

$$\mathcal{D}\eta(t) = \prod_{\mathbf{p},s} \left(d^3p\right)^2 d\alpha^*_{\mathbf{p},s}(t) \, d\alpha_{\mathbf{p},s}(t) \, d\beta^*_{\mathbf{p},s}(t) \, d\beta_{\mathbf{p},s}(t) \,. \tag{6.263}$$

Exercise 6.47: Verify (6.263).

The results (6.229) and (6.230) also have their spinor counterparts, given by

$$\frac{\langle\eta(t+\delta t)|\psi(x)|\eta(t)\rangle}{\langle\eta(t+\delta t)|\eta(t)\rangle} = \eta(x) + O(\delta t) \,, \tag{6.264}$$

$$\frac{\langle\eta(t+\delta t)|\bar\psi(x)|\eta(t)\rangle}{\langle\eta(t+\delta t)|\eta(t)\rangle} = \bar\eta(x) + O(\delta t) \,. \tag{6.265}$$

The path integral for spinor fields may now be derived. The derivation is identical in spirit to that of the scalar field, so that the S-matrix between two Grassmann-valued coherent states is partitioned into the product of N infinitesimal elements as in (6.232). The entire content of the path integral is once again contained in the form of the infinitesimal matrix element. The jth factor reduces to

$$\langle \eta(t_j + \delta t) | E(t_j + \delta t, t_j) | \eta(t_j) \rangle =$$

$$\langle \eta(t_j + \delta t) | \eta(t_j) \rangle \exp\left\{-i\delta t \int d^3x \, \mathcal{H}_I[\eta(\mathbf{x},t_j), \bar\eta(\mathbf{x},t_j)]\right\} \,, \tag{6.266}$$

where terms $O(\delta t^2)$ have been discarded, and it has been assumed that $\mathcal{H}_I$ is a local, normal-ordered function of the free field operators ψ and $\bar\psi$.

Using (6.261) the remaining matrix element in (6.266) can be shown to reduce to

$$\langle \eta(t_j + \delta t) | \eta(t_j) \rangle =$$

$$\exp\left\{-\tfrac{1}{2}\delta t \int d^3p \left[\beta^*_{\mathbf{p},s}(t_j) \overset{\leftrightarrow}{\frac{\partial}{\partial t_j}} \beta_{\mathbf{p},s}(t_j) + \alpha^*_{\mathbf{p},s}(t_j) \overset{\leftrightarrow}{\frac{\partial}{\partial t_j}} \alpha_{\mathbf{p},s}(t_j) \right]\right\}. \tag{6.267}$$

Sec. 6.8 The Path Integral for Field Theory

Using the definitions (6.253), (6.256), and (6.257) it follows that (6.267) is given by

$$\langle \eta(t_j + \delta t) | \eta(t_j) \rangle =$$
$$\exp\left\{ i\delta t \int d^3x \left[\tfrac{1}{2}\eta^\dagger i \overleftrightarrow{\partial_0} \eta + \bar\eta(i\gamma\cdot\nabla - m)\eta \right] \right\}\bigg|_{t=t_j} . \qquad (6.268)$$

Exercise 6.48: Verify result (6.268). To do this requires proving that free spinors satisfy a variant of (6.108) given by

$$u_s^b(p) u_s^{\dagger a}(p) + v_s^b(\tilde p) v_s^{\dagger a}(\tilde p) = \delta_{ab} , \qquad (6.269)$$

where $\tilde p = (\epsilon_p, -\mathbf{p})$.

Inserting these results back into the product of infinitesimal matrix elements and taking the $\delta t \to 0$ limit gives the bispinor path integral:

$$_+\langle \eta(t_+) | \eta(t_-) \rangle_- =$$
$$\int_{\eta_-}^{\eta_+} \mathcal{D}\eta \, \exp\left\{ i \int d^4x \left[\tfrac{1}{2}\bar\eta i\gamma_0 \overleftrightarrow{\partial_0} \eta - \mathcal{H}(\eta,\bar\eta) \right] \right\} . \qquad (6.270)$$

where $\mathcal{H}$ is the full Hamiltonian density of the theory, and the measure is given by

$$\mathcal{D}\eta = \lim_{N\to\infty} \prod_{j=1}^{N-1} \prod_{\mathbf{p},s} (d^3p)^2 \, d\alpha_{\mathbf{p}s}^*(t_j) \, d\alpha_{\mathbf{p}s}(t_j) \, d\beta_{\mathbf{p}s}^*(t_j) \, d\beta_{\mathbf{p}s}(t_j) . \qquad (6.271)$$

If the vacuum persistence functional is considered, so that $\eta_\pm = 0$, the path integral allows an integration by parts to become

$$_+\langle 0 | 0 \rangle_- = \int_0^0 \mathcal{D}\eta \, \exp\left[i \int d^4x \, \mathcal{L}(\eta,\bar\eta) \right] . \qquad (6.272)$$

The relation of the variables of the measure to the fields is given by (6.253). Once again, the product over times in the measure may be replaced by a product over ω, so that in the limit $N \to \infty$ the measure becomes

$$\mathcal{D}\eta = J^{-8} \prod_{p,s} \left(\frac{d^3p \, d^2\omega}{2\pi} \right)^2 d\alpha_s^*(p) \, d\alpha_s(p) \, d\beta_s^*(p) \, d\beta_s(p) , \qquad (6.273)$$

where J is the Jacobian (3.50).

6.8.3 Euclidean Measure

In many analyses of path integrals they are first continued to Euclidean times, defined by the Wick rotation $t \to -it$. There are many reasons to do this. First, and most obvious, the exponential becomes a damping factor that suppresses large deviations from the classical trajectories through function space. As in the quantum mechanical case discussed in Sec. 2.3, the Wick rotation to Euclidean space means that using the sum over field configurations and the exponential of the action to define a sensible measure is a much more tractable mathematical problem. Second, and more subtle, the Euclidean formulation allows the differential operators appearing in the action to become Hermitian operators on a Euclidean four-dimensional manifold. This results in a well-defined eigenvalue problem for the Euclidean modes of the respective fields. Once the eigenvalue problem is solved, the Euclidean form for each field is written as an expansion over these eigenmodes, and the measure is written, up to a dimensionful factor, as an integration over the coefficients of the eigenmodes appearing in the expansion.

The question immediately arises as to the relationship of this Euclidean measure to the one presented in the previous two subsections. In order to discuss this question it is necessary to reflect upon the meaning of the measure. The measure represents a sum over all possible functions or field configurations belonging to a *specific* class of functions, and each field configuration is weighted by the exponential of the action evaluated for that function to give its contribution to the path integral. In the previous two subsections this class of functions was defined by using the eigenfunctions associated with the differential operators appearing in the free field actions for the scalar and spinor fields. This gives an explicit recipe for constructing the fields and functions appearing in the action, and the sum over this class of functions is given by an integration over the coefficients of the Fourier sum or transform that represents the function. It will be assumed that two formulations of the measure of a path integral are equivalent if the eigenmodes used to define each measure span the same function space. To answer the question of how the Euclidean and the Minkowski measure and field expansions are related then requires an examination of the two specific cases of this section and a comparison of the space of functions spanned by the measure. Such a comparison is relatively easy to do for the quadratic actions discussed so far, but remains an intractable problem for most nonlinear theories.

For the scalar field case, the Euclidean field is expanded in terms of the solutions to the Euclidean eigenvalue problem for the operator $\Box + m^2$. The Euclidean form for $\Box$, denoted $\Box_E$, is obtained by the Wick rotation

Sec. 6.8 The Path Integral for Field Theory

and is given explicitly by

$$\Box_E = -\frac{\partial^2}{\partial t^2} - \nabla^2 , \qquad (6.274)$$

The operator $\Box$ is hyperbolic, while the operator $\Box_E$ is elliptic. The solutions,

$$\psi_p(x) = \frac{1}{(2\pi)^2} \exp ipx \equiv \frac{1}{(2\pi)^2} \exp i(\omega t - \mathbf{p} \cdot \mathbf{x}) , \qquad (6.275)$$

to the massive eigenvalue problem,

$$(\Box_E + m^2)\psi_p(x) = \lambda_p \psi_p(x) \qquad (6.276)$$

have *positive-definite* eigenvalues $\lambda_p = m^2 + \omega^2 + \mathbf{p}^2 \equiv m^2 + p_E^2$, are orthogonal,

$$\int d^4x \, \psi_k^*(x)\psi_p(x) = \delta^4(k - p) , \qquad (6.277)$$

and are complete,

$$\int d^4p \, \psi_p^*(x)\psi_p(y) = \delta^4(x - y) , \qquad (6.278)$$

in the four-dimensional Euclidean manifold.

By inspection this set of functions coincides exactly with the set of functions used in (6.220) to expand the scalar field in the Minkowski space representation. Therefore, the two measures are identical since they span the same space of functions. For this case a prescription for continuing the Minkowski field variables into the Euclidean region is given by the generalized Wick rotation,

$$t \to -it , \quad \omega \to i\omega , \qquad (6.279)$$

where ω is the frequency occurring in the Fourier transforms and eigenvalues. The Euclidean field $\phi_E(x)$ is obtained from the Minkowski field $\phi_M(x)$ by the Wick rotation

$$\phi_E(x) = \phi_M(\mathbf{x}, -it) . \qquad (6.280)$$

Exercise 6.49: Show that the Fourier components of the Euclidean and Minkowski field are related by

$$\tilde{\phi}_E(p) = i\tilde{\phi}_M(\mathbf{p}, i\omega) . \qquad (6.281)$$

The standard method for evaluating the Minkowski space path integral is to continue it to the Euclidean region, evaluate it, and continue it back to the Minkowski region by the replacements $t \to it$ and $\omega \to -i\omega$. Due to the equivalence of the Euclidean and Minkowski measure for the free scalar field, it will come as no surprise in Chapter 8 that the Euclidean and Minkowski path integrals representing the vacuum transition element for the free field may be evaluated to find the same functional form. The difference, of course, will be in the presence of Euclidean momenta in the final form of the Euclidean path integral. However, when continued back to Minkowski space, the Euclidean space path integral is identical to the Minkowski space path integral developed in the previous sections, modulo regularization subtleties that will be discussed in Chapter 8. It is generally believed that this equivalence breaks down in the presence of nontrivial interactions, and that the Wick rotation is a necessity to define the general path integral. This point will be discussed again in Chapter 7 for the case of gauge fields, where the Euclidean space formulation avoids many of the ambiguities that beset the Minkowski space path integral.

The eigenvalue equation for the Euclidean form of the Dirac equation is

$$\left(i\gamma^0 \frac{\partial}{\partial t} + \gamma \cdot \nabla + im\right) \eta_p(x) \equiv (\gamma_E^\mu \partial_\mu^E + im)\eta_p(x) = \lambda_p \eta_p(x) \;. \quad (6.282)$$

The term im may be absorbed into λ_p, and so it is irrelevant to the problem. The solutions take the form $\eta_p = w(p)e^{-ipx}$, where $w(p)$ is a bispinor function of p. Substituting this form into the Dirac representation of (6.282) gives

$$D(p)w(p) = (\lambda_p - im)w(p) \equiv \lambda'_p w(p) \;, \quad (6.283)$$

where $D(p)$ is given by

$$D(p) = \begin{pmatrix} \omega & i\mathbf{p}\cdot\sigma \\ -i\mathbf{p}\cdot\sigma & -\omega \end{pmatrix} \;. \quad (6.284)$$

Exercise 6.50: Show that the solutions to (6.283) are given by two sets of degenerate eigenspinors corresponding to the eigenvalues $\pm\sqrt{\omega^2 + \mathbf{p}^2} = \pm p_E$.

The eigenspinor solutions of (6.283) can be constructed in the following way. If $\xi^{(j)}$, $j = (1,2)$, represents any two orthonormal 2×1 spinors, such

Sec. 6.8 The Path Integral for Field Theory   221

that $\xi^{(j)\dagger}\xi^{(k)} = \delta_{jk}$, then the eigenspinors $w_\pm(p)$, corresponding respectively to $\pm p_E$, are given by

$$w_+^{(j)}(p) = \frac{1}{\sqrt{2p_E(\omega + p_E)}} \begin{pmatrix} (\omega + p_E)\xi^{(j)} \\ -i\sigma\cdot\mathbf{p}\,\xi^{(j)} \end{pmatrix}, \quad (6.285)$$

$$w_-^{(j)}(p) = \frac{1}{\sqrt{2p_E(\omega + p_E)}} \begin{pmatrix} -i\sigma\cdot\mathbf{p}\,\xi^{(j)} \\ (\omega + p_E)\xi^{(j)} \end{pmatrix}. \quad (6.286)$$

These normalized solutions allow the Euclidean field to be represented as the expansion

$$\eta_E(x) = \int \frac{d^4p}{(2\pi)^2} \alpha_E^\lambda(p) w^\lambda(p) e^{-ipx}. \quad (6.287)$$

where λ is summed over all four solutions to the eigenvalue problem.

The Euclidean field (6.287) is spanned by the same set of functions appearing in the Minkowski field expansion (6.255), and the relationship between the coefficients in the two expansions is similar to that of (6.281):

$$\alpha_E^\lambda(p) = i\sqrt{\frac{m}{\epsilon_p}} \left[\beta_s(\mathbf{p}, i\omega) w^{\lambda\dagger}(p) u_s(p) + \alpha_s^*(-\mathbf{p}, -i\omega) w^{\lambda\dagger}(p) v_s(-p)\right]. \quad (6.288)$$

The measure for the Euclidean version of the path integral is written

$$\mathcal{D}\eta = N \prod_{p,\lambda} d\alpha_E^\lambda(p)\, d\alpha_E^{\lambda*}(p). \quad (6.289)$$

As a result, the measure developed in the previous subsection and the Euclidean measure discussed here are equivalent.

There is a subtlety to the spinor action that makes it necessary to transform the Euclidean measure. In the action both η and $\bar{\eta}$ appear, and it is useful, for reasons that will become clear in Chapters 8 and 9, to expand $\bar{\eta}$ in terms of $w^\dagger$ rather than $\bar{w}$. This is accomplished by writing

$$\bar{\eta}_E(x) = \int \frac{d^4p}{(2\pi)^2} \alpha_E^{\lambda*}(p) w^{\lambda\dagger}(p)\gamma^0 e^{ipx} = \int \frac{d^4p}{(2\pi)^2} \beta_E^\lambda(p) w^{\lambda\dagger}(p) e^{ipx}. \quad (6.290)$$

The relation of α^* to β is given by projecting $\bar{w}$ onto $w^\dagger$ using the completeness of the w eigenspinors. It follows that

$$\beta_E^\lambda(p) = \alpha_E^{\lambda'*}(p) \bar{w}^{\lambda'}(p) w^\lambda(p). \quad (6.291)$$

Field Theory

> **Exercise 6.51**: Calculate the Jacobian of the transformation given by (6.291) and show that it is nonsingular.

The transformed measure is given by

$$\mathcal{D}\eta = N \prod_{p,\lambda} d\alpha_E^\lambda(p)\, d\beta_E^\lambda(p) \,. \tag{6.292}$$

One of the advantages to the Euclidean approach is that it can be extended to allow the presence of a Hermitian potential. The potential may arise from some direct or indirect nonlinear coupling to other fields, such as the gauge fields to be discussed in the next chapter. This allows transcending the free field decomposition of the fields and measure and incorporates nonperturbative effects into the path integral. The Euclidean formulation of the measure allows the time dependence of the potential to be subsumed into the Hermitian eigenvalue equation, which is assured of possessing real eigenvalues and an associated complete set of eigenfunctions. Once evaluated in the Euclidean region, the path integral can be continued back to Minkowski space.

In the scalar field case this potential is denoted $V(x)$, and is such that under a Lorentz transformation $V'(x') = V(x)$. The eigenvalue problem then becomes

$$\left[\Box_E + m^2 + V(x)\right]\psi_n(x) = \lambda_n \psi_n(x) \,, \tag{6.293}$$

and may include bound state solutions, depending upon the form for the potential V. The Euclidean measure is then defined as an integration over the eigenmodes that satisfy (6.293), and such a measure may be drastically different from the free field measure encountered in this section. In particular, if there are negative eigenmodes the theory is pathological since the integration over such a mode is not defined. Vanishing eigenvalues in the Euclidean region, so-called zero modes, serve as the ultimate source of anomalies in a theory where Dirac fields are coupled to gauge fields. This will be seen in detail in Sec. 9.5.

6.8.4 Configuration Measure

In Sec. 6.1 the measure for the path integral was given by a configuration space representation derived from the continuum limit of a many-body problem. Such a measure was realized by making space-time into a lattice, and treating the value of the fields at each node of the lattice as an independent variable. Such a measure is used at Euclidean times in numerical

Sec. 6.8 The Path Integral for Field Theory

approaches to evaluating the path integral on computers. The phenomenological successes in deriving masses and other properties of particles by this method represent strong support for both the path integral and the configuration space measure.

The configuration space measure does not directly address the question of the class of functions over which the path integral is actually summed. However, in Sec. 3.3 a similar situation for a quantum mechanical system was treated. There, a configuration space measure consisting of variables associated with the lattice of the time interval was transformed into a measure over the modes of the associated Fourier series. Such a transformation induced a complicated Jacobian into the measure, but this Jacobian was independent of the time interval being "latticized".

It is in this spirit that the field variables at each space-time point, either in the Minkowski or Euclidean region, are treated as independent variables. This measure possesses a Fourier transform in the limit that the partition of space-time becomes arbitrarily fine, and the coefficients of this Fourier transform coincide with the measure (6.240), developed by partitioning the evolution operator using a projection operator built up from *free field states*. It is critical to stress the fact that this configuration space measure can be shown to be formally equivalent to the Fourier component definition only for such a derivation of the path integral; there is no proof that a configuration space measure takes the same form for the case that interactions dictate a mode expansion that differs from (6.240), nor are there counterexamples that expose calamities associated with the use of a configuration space form for the measure.

It should be possible to change between the two forms of the measure. The change of variables from the configuration space form to the momentum space form must be accompanied by a complicated Jacobian, whose form is currently unkown, but one that is assumed to be regulated, i.e., kept finite, by some limiting procedure. This Jacobian can be absorbed into an overall factor that will later be cancelled through normalization of the transition element. This *assumes* that the transition element is to be normalized, so that an overall constant factor is not relevant to physical processes.

Nevertheless, both types of measure occur in applications of path integrals, and configuration space measure is very intuitive since the action appearing in the path integral is always written in configuration space. The field variables at each point can be viewed as a generalization of the p and q variables of quantum mechanics, and this allows many of the results obtained for quantum mechanical path integrals to be carried over into the field theory versions. It is therefore important to present a consistent set of rules for configuration space manipulations of path integrals.

The basic approach is similar to that of momentum space versions, so that integrals are rewritten as Riemann sums over the space-time lattice. The elements of measure appearing in the integral of the action, say d^4x in the case of a four-dimensional theory, are treated as small, but nonvanishing. In order to obtain complete generality this measure element will be written dx, regardless of the dimensionality. Dirac deltas will be assumed to carry the same dimensionality as space-time, unless specifically indicated to be otherwise. This procedure allows integrals over space-time to be broken into a sum, so that

$$\int dx\, f(x) \rightarrow \sum_j dx_j\, f(x_j)\,, \qquad (6.294)$$

where x_j is a point in the space-time lattice. Doing so creates the matrix equivalent of functionals, a method already discussed in Sec. 1.4 for the case of functions of a single variable.

As an example, for the case where $\psi(x)$ is a real-valued scalar function, the commonly occurring Gaussian path integral is understood to be evaluated by the replacement

$$\int \mathcal{D}\psi \exp\left[-\int dx\, dy\, \psi(x) D(x,y) \psi(y)\right] =$$
$$\int \prod_j d\psi(x_j) \exp\left[-\sum_{j,k} dx_j\, dx_k\, \psi(x_j) D(x_j, x_k) \psi(x_k)\right]\,, \qquad (6.295)$$

thereby reducing it to the familiar matrix form (1.104). If $D(x,y)$ is a *Hermitian* matrix (6.295) can be evaluated. The result is

$$\int \prod_j d\psi(x_j) \exp\left[\sum_{j,k} dx_j\, dx_k\, \psi(x_j) D(x_j, x_k) \psi(x_k)\right]$$
$$= N[\det D]^{-\frac{1}{2}}\,, \qquad (6.296)$$

where N refers to the collection of products of dx and constants such as 2π that are generated by performing the infinite sequence of Gaussian integrals. This collection of constants will almost always be irrelevant to the calculation of a physical quantity since it is usually cancelled by normalizing the path integral. Again, this *assumes* that the path integral represents the vacuum transition element, or some other normalized quantity.

Sec. 6.8 The Path Integral for Field Theory   225

Result (6.296) is valid only for the case that D is Hermitian and possesses positive-definite eigenvalues. For this reason configuration space measure is well defined only in the Euclidean region. To expose why this is so, the determinant of a differential operator must be examined, and this is a special case of the determinant of a continuum function. This determinant is understood in the same sense as it was in (1.81), so that

$$\det D = \exp(\operatorname{Tr} \ln D) = \exp\left[\sum_{n=1}^{\infty} \frac{(-1)^n}{n} \operatorname{Tr}(D-\delta)^n\right]. \tag{6.297}$$

The trace operation is understood as an integration, so that, for example,

$$\operatorname{Tr} D(x,y) = \int dx\, D(x,x), \tag{6.298}$$

while a power of a function of two arguments is understood as a sequence of integrals,

$$D^n(x,y) = \int dz_1 \cdots dz_{n-1}\, D(x,z_1) D(z_1,z_2) \cdots D(z_{n-1},y). \tag{6.299}$$

For the case that $D(x,y) = L_x \delta(x-y)$, where L_x is some self-adjoint linear operator, the equivalent form of (1.86) gives

$$\det L = \prod_j \lambda^{(j)}, \tag{6.300}$$

where the product runs over all the eigenvalues $\lambda^{(j)}$ of L_x that are consistent with the boundary conditions and the underlying space-time manifold associated with the integral appearing in the exponential. This shows that the configuration space form results in

$$\int \mathcal{D}\psi \exp\left[-\int dx\, dy\, \psi(x) L_x \delta(x-y) \psi(y)\right] = \left[\prod_j \lambda^{(j)}\right]^{-1/2}, \tag{6.301}$$

and this is defined only if the eigenvalues of L_x are real and positive-definite. Recalling the discussion of the Euclidean form of $\Box$ that showed its eigenvalues are positive-definite, it is apparent that Euclidean continuation is a necessity to make the configuration space path integral measure well defined. Because of the intimate relationship between the eigenvalues and eigenfunctions of a differential operator and the manifold over which it is defined, it is not surprising that path integrals are imbued with many of the topological properties of the space-time manifold of the quantum theory.

An exception to this is the case of integration over complex fields. Using the definition of the Dirac delta gives

$$\int \mathcal{D}\phi \mathcal{D}\phi^* \exp\left\{i\int dx\, dy\, \phi^*(x)D(x,y)\phi(y)\right\} =$$

$$\int \prod_j d\phi(x_j)\, d\phi^*(x_j)\, \exp\left\{i\sum_{j,k} dx_j\, dx_k\, \phi^*(x_j)D(x_j,x_k)\phi(x_k)\right\} =$$

$$N_1 \int \prod_j d\phi(x_j)\, \delta\left(\sum_k D(x_j,x_k)\phi(x_k)\right) = N_1 N_2 (\det D)^{-1}, \quad (6.302)$$

where N_1 and N_2 are collections of products of the measure elements and π. Result (6.302) can be used to demonstrate the result

$$\det\left|\frac{\delta G(x)}{\delta \psi(y)}\right| \prod_x \delta\left(G(x)\right) = \prod_x \delta\left(\psi_c(x)\right), \quad (6.303)$$

where $\psi_c(x)$ is the solution to $G(x) = 0$, assumed to be unique. This form is critical to generalizing the results of Sec. 4.6 on constraints to the field theory case, which will be done in the next chapter.

> **Exercise 6.52**: Verify (6.303) using result (6.302).

Similar results follow for Grassmann variables and are left as an exercise.

> **Exercise 6.53**: If $\eta(x)$ is a Grassmann valued function, show that, up to constant factors,
>
> $$\int \mathcal{D}\eta\, \exp\left\{\int dx\, dy\, \eta(x)D(x,y)\eta(y)\right\} \sim (\det D)^{\frac{1}{2}}, \quad (6.304)$$
>
> and that, for a complex Grassmann function,
>
> $$\int \mathcal{D}\eta \mathcal{D}\eta^*\, \exp\left\{\int dx\, dy\, \eta^*(x)D(x,y)\eta(y)\right\} \sim \det D. \quad (6.305)$$

A subtlety occurs when the momentum space bispinor measure derived in this section is expressed in a continuum form. The result (6.292) combined with the expansion (6.290) shows that the Fourier components of the measure are derived from η and $\bar\eta$, and *not* $\eta^\dagger$. Therefore, the continuum measure for spinor fields is written

$$\mathcal{D}\eta = \prod_x d\bar\eta_a(x)\, d\eta_a(x)\,, \qquad (6.306)$$

which gives eight independent components as before.

It is also possible to include commutators in the continuum measure approach by exploiting their relationship to the classical Poisson brackets. However, for continuum measure, and hence Euclidean times, the Poisson bracket is understood as an integral over all of Euclidean space. For a field theory the Poisson bracket of A and B, assumed to be functionals of the fields ϕ_α and their canonical momenta π_α, is defined for four-dimensional Euclidean space as

$$\{A,B\}_{\phi,\pi} = \int d^4x \left[\frac{\delta A}{\delta \phi_\alpha(x)}\frac{\delta B}{\delta \pi_\alpha(x)} - \frac{\delta B}{\delta \phi_\alpha(x)}\frac{\delta A}{\delta \pi_\alpha(x)}\right]. \qquad (6.307)$$

Equal-time commutators can be analyzed using the c-number functions appearing in the path integral by simply constraining the times of the operators to coincide. Of course, for a general Poisson bracket this need not be the case.

However, the use of Poisson brackets dictates that the form for symmetry generators must be altered. From the discussion of Noether's theorem in Sec. 6.5, the generator of a symmetry in operator formulations of field theory is written as an integral over the spatial volume only, typically taking the form

$$U(t) = \exp\left\{i\int d^3x\, \lambda(\mathbf{x},t) G(\mathbf{x},t)\right\}, \qquad (6.308)$$

where λ are the transformation parameters and G is some collection of field operators determined from Noether's theorem. The commutators that represent the unitary symmetry transformation are performed at equal times using equal-time commutation or anticommutation relations. In the Poisson bracket formulation the generator is written as an integral over all of Euclidean space. The definition of the Poisson bracket gives

$$\{A(\mathbf{x},t), Q\} =$$
$$\int d^4y\, d^4z\, \lambda(z) \left[\frac{\delta A(x)}{\delta \psi_\alpha(y)}\frac{\delta G(z)}{\delta \pi_\alpha(y)} - \frac{\delta A(x)}{\delta \pi_\alpha(y)}\frac{\delta G(z)}{\delta \psi_\alpha(y)}\right]. \qquad (6.309)$$

If G and A contain no integrations over time, then the rules of functional differentiation automatically align the times of the two forms A and λG through the occurrence of Dirac deltas, so that

$$\{A(x), Q\} = A'(x)\,, \qquad (6.310)$$

where $A'(x)$ is the form of $A(x)$ after an infinitesimal transformation. The Poisson bracket is highly useful since it places space and time on a truly equal footing in the functional formulation.

The main goal of this section has been to obtain a set of guidelines for quantizing a classical field theory system using the path integral formalism. The results can now be summarized. The starting point is the classical action, which is considered a functional of some set of generalized coordinates ψ_α. From the action the form of the canonical momentum, π_α, is obtained, and it is used to construct the Hamiltonian $H[\psi_\alpha, \pi_\alpha]$ in terms of a Hamiltonian density $\mathcal{H}(\psi_\alpha, \pi_\alpha)$. If the system of equations that determines $\dot\psi_\alpha$ in terms of ψ_α and π_α is underdetermined, or if there are constraints on the system, then it is necessary to implement the techniques for constrained path integrals derived in Sec. 4.6. Deferring the discussion of this problem to the next chapter, the quantum transition element between two classical configurations ψ_- and ψ_+ is given by the path integral

$$Z = \int_{\psi_-}^{\psi_+} \mathcal{D}\pi \mathcal{D}\psi \exp\left\{ i \int dx\, \mathcal{L}(\psi, \pi) \right\}, \qquad (6.311)$$

where the action appearing in the path integral is given by

$$\mathcal{L}(\psi, \pi) = \tfrac{1}{2}\pi_\alpha \dot\psi_\alpha - \tfrac{1}{2}\dot\pi_\alpha \psi_\alpha - \mathcal{H}(\psi, \pi). \qquad (6.312)$$

The measure is understood to be an integration over all possible intermediate configurations of the functions ψ and π. There is a variable of integration for every canonical degree of freedom appearing in the path integral. Barring subtleties induced by Fermi fields and possible constraints, this recipe is applicable in almost all systems.

Unfortunately, the validity of the field theoretic path integral procedure for quantization is even murkier than its quantum mechanical counterpart. It can be argued that the derivation of this section relied solely upon the existence of a well-defined perturbative solution to the field theory in the form of the unitary transformation U that generates the interaction picture, and that therefore the path integral representation of a theory will exist whenever a consistent perturbative representation can be defined. However, when coupled with the negative aspects of perturbation theory enunciated by Haag's theorem [29], this does little to strengthen one's confidence in the "greater validity" of the path integral. Certainly, many aspects of modern field theory, ranging from the role of classical solutions of the equations of motion to the quantization of nonabelian gauge fields, find their most compact and elegant expression in terms of a path integral. It is also clear that compactness and elegance will never constitute a proof of the path integral's validity. Indeed, there exist aspects of field theory problems that are as obscure in the path integral formalism as they are in the operator and state formalism. In particular, the identification of the spectrum of most interacting nonlinear theories continues to defy analytical approaches originating in either school. If any development serves to strengthen a belief in the "greater validity" of the path integral, it is the application of numerical techniques to extract dynamical information from the path integral. The best that can be said is that the general validity of the path integral remains an open question.

References

[1] A. Einstein, Ann. Physik **17**, 891 (1905).

[2] E. Schrödinger, Ann. Physik **81**, 109 (1926); W. Gordon, Z. Physik **40**, 117 (1926); O. Klein, Z. Physik **41**, 407 (1927).

[3] P.A.M. Dirac, Proc. Roy. Soc. (London) **A117**, 610 (1928); Proc. Roy. Soc. (London) **A118**, 351 (1928).

[4] W. Heisenberg and W. Pauli, Z. Physik **56**, 1 (1929).

[5] C. Itzykson and J.-B. Zuber, *Quantum Field Theory*, McGraw-Hill, New York, 1980.

[6] L.D. Faddeev and A.A. Slavnov, *Gauge Fields: Introduction to Quantum Theory*, Benjamin/Cummings, Reading, Massachusetts, 1980.

[7] T.D. Lee, *Particle Physics and Introduction to Field Theory*, Harwood, Chur, Switzerland, 1981.

[8] L.H. Ryder, *Quantum Field Theory*, Cambridge University Press, Cambridge, 1985.

[9] S. Pokorski, *Gauge Field Theories*, Cambridge University Press, Cambridge, 1987.

[10] P. Ramond, *Field Theory: A Modern Primer, Second Edition*, Addison-Wesley, Redwood City, California, 1989.

[11] H. Goldstein, *Classical Mechanics, Second Edition*, Wiley, New York, 1983.

[12] M. Hamermesh, *Group Theory and Its Application to Physical Problems*, Addison-Wesley, Reading, Mass., 1962.

[13] B. Wybourne, *Classical Groups for Physicists*, Wiley, New York, 1974.

[14] The classification of the representations of the Poincaré group was completed by Wigner. See E.P. Wigner, Ann. Math. **40**, 1 (1939);

Nuovo Cimento **3**, 517 (1956).

[15] An excellent review of variational techniques and Noether's theorem [17] in field theory is found in E.L. Hill, Rev. Mod. Phys. **23**, 253 (1951).

[16] H. Weyl, Z. Physik **56**, 330 (1929).

[17] E. Noether, Nachr. Ges. Wiss. Göttingen, 171 (1918); see also [15].

[18] The subject of general relativity lies outside the scope of this book. The reader is recommended to R. Wald, *General Relativity*, University of Chicago Press, Chicago, 1984.

[19] V. Fock, Z. Physik **75**, 622 (1932).

[20] W. Pauli, Phys. Rev. **58**, 716 (1940).

[21] R.F. Streater and A.S. Wightman, *PCT, Spin and Statistics, and All That*, Benjamin, New York, 1964.

[22] The S-matrix formulation originated with W. Heisenberg, Z. Physik **120**, 513 (1943).

[23] H. Lehmann, K. Symanzik, and W. Zimmermann, Nuovo Cimento **1**, 205 (1955).

[24] S. Tomonaga, Prog. Theor. Phys. **1**, 27 (1946).

[25] J. Schwinger, Phys. Rev. **74**, 1439 (1948).

[26] R.P. Feynman, Phys. Rev. **76**, 769 (1949).

[27] F.J. Dyson, Phys. Rev. **75**, 486 (1949).

[28] G.C. Wick, Phys. Rev. **80**, 268 (1950).

[29] R. Haag, Dan. Mat. Fys. Medd. **29**, 12 (1955).

[30] R.J. Glauber, Phys. Rev. **130**, 2529 (1963).

[31] J.R. Klauder, Ann. Phys. **11**, 123 (1960); J.R. Klauder, Phys. Rev. **D19**, 2349 (1979).

[32] The approach to constructing the field theory path integral presented here follows that of A.A. Slavnov, Teoret. i Mat. Fiz. **22**, 177 (1975); see also the monograph [6].

[33] Other authors have employed coherent states to examine aspects of constructing the path integral; see, for example, C.L. Hammer, J.C. Schrauner, and B. DeFacio, Phys. Rev. **D18**, 373 (1978); M.S. Swansong, Phys. Rev. **D24**, 2132 (1981); R.E. Pugh, Phys. Rev. **D33**, 1027 (1986); R.E. Pugh, Phys. Rev. **D33**, 1033 (1986).

Chapter 7

Gauge Field Theory

The triumphs of modern particle physics, in particular the standard model, have as their wellspring the concept of the gauge field. In this chapter gauge field theory will be developed at the classical and quantum levels for both the abelian and nonabelian case. Such a topic is vast, and the presentation here will not be exhaustive by any stretch of the imagination. A particular course will be chosen to arrive at the desired destination, that of a path integral representation of the gauge field S-matrix that is unitary, and unfortunately many interesting aspects of the problem will be ignored for lack of space.

In Sec. 7.1 the problem is motivated by considering the canonical quantization of the free electromagnetic field, which is the archetype for all gauge theories. The dual aspects of gauge invariance and ghosts are discussed. In Sec. 7.2 the path integral formulation of quantum electrodynamics is developed as an example of a constrained path integral. The technique is the generalization to field theory of the quantum mechanical method for implementing a gauge condition developed in Sec. 4.6. The first example of factorizing gauge volumes is presented. In Sec. 7.3 the properties of Lie algebras are briefly discussed. In Sec. 7.4 the nonabelian extension of the electromagnetic field, the Yang–Mills field, is introduced, and the technique of minimal coupling is developed for nonabelian gauge-matter interactions. In Sec. 7.5 the technique of constrained path integrals is again applied to the path integral representation of the Yang–Mills field. The result is the Fadeev–Popov method for consistently quantizing the theory for a particular choice of gauge fixing term or gauge condition. The chapter closes in Sec. 7.6 with an introduction to the concept of the gauge field as a one-form and a discussion of some of the topological properties of gauge fields.

7.1 The Maxwell Field

In this section classical electrodynamics is first reviewed and then cast into the form of an action functional approach. The first step in quantization is to construct a minimally coupled theory that possesses abelian gauge invariance. Aspects of gauge invariance and gauge fixing are discussed at the classical level and the ramifications of these ideas for the quantized version of electrodynamics are presented. The ghost structure present in the Fock space of the manifestly covariant version of quantum electrodynamics is discussed.

7.1.1 The Classical Action

The classical theory of electrodynamics has been in its final form for over a century [1]. The theory was first formulated in terms of the electric field **E** and the magnetic field **B**, and these obey Maxwell's equations,

$$\nabla \cdot \mathbf{B} = 0, \quad \nabla \times \mathbf{B} = \mathbf{J} + \frac{\partial \mathbf{E}}{\partial t},$$
$$\nabla \cdot \mathbf{E} = \rho, \quad \nabla \times \mathbf{E} = -\frac{\partial \mathbf{B}}{\partial t}, \qquad (7.1)$$

where ρ is the charge density and **J** is the spatial current density of electric charges, assumed to obey the conservation law

$$\frac{\partial \rho}{\partial t} + \nabla \cdot \mathbf{J} = 0. \qquad (7.2)$$

The two equations not involving the electric charges are automatically satisfied by assuming that **B** and **E** are obtained from a scalar potential ϕ and a spatial vector potential **A** according to

$$\mathbf{E} = -\nabla \phi - \frac{\partial \mathbf{A}}{\partial t}, \quad \mathbf{B} = \nabla \times \mathbf{A}. \qquad (7.3)$$

The remaining two equations may be cast into relativistic form by defining the contravariant four-vector A, whose components are given by $A^\mu = (\phi, \mathbf{A})$, so that $A_\mu = g_{\mu\nu} A^\nu = (\phi, -\mathbf{A})$. The four-vector electric current J is defined as $J^\mu = (\rho, \mathbf{J})$, so that $J_\mu = (\rho, -\mathbf{J})$. The conservation of electric charge, (7.2), is written $\partial_\mu J^\mu = 0$. The final forms for Maxwell's equations are

$$\Box A_i = J_i + \nabla_i (\partial_\mu A^\mu), \qquad (7.4)$$
$$\nabla^2 A_0 = -J_0 - \nabla \cdot \dot{\mathbf{A}}. \qquad (7.5)$$

Equation (7.5) is Gauss's law rewritten in terms of the potential. In manifestly covariant notation Gauss's law can be written

$$\Box A_0 = J_0 + \partial_0(\partial_\mu A^\mu) \,, \tag{7.6}$$

so that all of the previous equations may be expressed as one equation,

$$\Box A_\mu = J_\mu + \partial_\mu(\partial_\nu A^\nu) \,. \tag{7.7}$$

For the purposes of quantization the equation (7.7) must be derived from an action principle. In order to do this the antisymmetric tensor F is defined as

$$F_{\mu\nu} = \partial_\mu A_\nu - \partial_\nu A_\mu \,. \tag{7.8}$$

From (7.3) it follows that F is related to the three-dimensional **E** and **B** field strengths according to

$$E_i = F_{0i} \,, \quad B_i = -\tfrac{1}{2}\varepsilon_{ijk}F_{jk} \,. \tag{7.9}$$

The Lagrangian density that reproduces Maxwell's equations is given by

$$\mathcal{L} = -\tfrac{1}{4}F_{\mu\nu}F^{\mu\nu} - J_\mu A^\mu \,. \tag{7.10}$$

Exercise 7.1: Verify that the action (7.10) reproduces relations (7.4) and (7.5).

It now remains to construct a form for J_μ that is conserved. Noether's theorem dictates that every conserved current must be associated with some symmetry of the action. The symmetry relevant to electrodynamics is identified by noting that the definitions of the **E** and **B** fields in terms of A_μ are such that they are unchanged by a *gauge transformation of the second kind* for A_μ [2], defined by (4.63). In relativistic notation a gauge transformation takes the form

$$A_\mu(x) \rightarrow A_\mu(x) - \partial_\mu \Lambda(x) \,, \tag{7.11}$$

where Λ is a function of x satisfying $[\partial_\mu, \partial_\nu]\Lambda = 0$. The definition of $F_{\mu\nu}$ guarantees that it is also invariant under (7.11). It is important to note that a term of the form $m^2 A_\mu A^\mu$ would spoil the gauge invariance of the action for the vector potential.

In order that gauge transformations leave Maxwell's equations invariant it must also be true that the electric current is invariant under a gauge transformation. In (4.66) it was shown that the wave functions of charged particles must also be transformed simultaneously under a gauge transformation according to

$$\psi(x) \to e^{ie\Lambda(x)}\psi(x) \, . \tag{7.12}$$

Since the field is the generalization of the wave function, it follows that the action involving both the electromagnetic field and the charged or matter field must be *gauge invariant* under the simultaneous *gauge transformations* (7.11) and (7.12). The Dirac action of Sec. 6.3 may be combined with the action (7.10) to obtain a gauge invariant form. It is not difficult to see that a gauge invariant action is given by

$$\mathcal{L} = \bar{\psi}(i\slashed{\partial} - m - e\slashed{A})\psi - \tfrac{1}{4}F_{\mu\nu}F^{\mu\nu} \, , \tag{7.13}$$

where $\slashed{A} = \gamma^\mu A_\mu$.

Exercise 7.2: Use Noether's theorem to show that the conserved current associated with invariance of (7.13) under gauge transformations is

$$J^\mu = e\bar{\psi}\gamma^\mu\psi \, . \tag{7.14}$$

The action (7.13) contains the free Dirac action as well as the term $F_{\mu\nu}F^{\mu\nu}$ that is quadratic in the A_μ field. It also contains the nonquadratic term $A^\mu J_\mu$, which describes a coupling between the Dirac field and the electromagnetic vector potential, rather than the electric or magnetic field. Because it is derived from the constraint of gauge invariance, it is usually referred to as *minimal coupling*. Because of its association with gauge transformations A^μ is called a *gauge field*. Since the current is conserved the coupling term is itself gauge invariant when integrated over space-time. Under a gauge transformation the integrated interaction changes to

$$\int d^4x \, A_\mu J^\mu \to \int d^4x \, A_\mu J^\mu - \int d^4x \, \partial_\mu(\Lambda J^\mu) \, , \tag{7.15}$$

and the total divergence vanishes for a class of gauge functions that drops off to zero at the boundaries of space-time.

In order to understand what the electric current represents, J_0 can be evaluated in terms of the spinor free field expansions derived in Sec. 6.5 to obtain

$$Q = \int d^3x \, J_0 = e \int d^3p \, \left[b_s^\dagger(\mathbf{p})b_s(\mathbf{p}) - d_s^\dagger(\mathbf{p})d_s(\mathbf{p})\right] \, . \tag{7.16}$$

Since the number operator appears in Q it follows that the Fock states are eigenstates of Q, and that b states have charge $+e$, while the d states have charge $-e$. It is also important to note that the electric charges of multiparticle Fock states are simply the additive sum of the individual charges.

It is convenient to introduce the gauge covariant derivative, defined by

$$D_\mu = \partial_\mu + ieA_\mu , \tag{7.17}$$

and it is easy to see that its action on a matter field undergoing a gauge transformation of the form (7.12) is given by

$$\gamma^\mu D_\mu \psi \to e^{ie\Lambda} \gamma^\mu D_\mu \psi , \tag{7.18}$$

so that a theory may be minimally coupled to the electromagnetic field by the simple replacement $\partial_\mu \to D_\mu$ in the matter field action.

> **Exercise 7.3**: Use property (7.18) to obtain the form of minimal coupling for the complex scalar field of (6.67) and find the conserved current [3].

7.1.2 Gauge Invariance and Gauge Fixing

By construction the classical equations of motion for the gauge field are gauge invariant, and so are the forms for **E** and **B**. In order to solve Maxwell's equations at the classical level, some consistent set of initial and boundary conditions for **E** and **B** must be chosen. However, once these boundary conditions are chosen, the corresponding initial and boundary conditions for A_μ are not unique since each member of the family of functions given by $A_\mu - \partial_\mu \Lambda$ corresponds to the same **E** and **B**.

The source of this ambiguity lies in a mismatch between the numbers of degrees of freedom available to A_μ to match boundary conditions and those available to the original **E** and **B**. Initially A_μ possesses four degrees of freedom and satisfies one equation of motion, (7.7), leaving three final degrees of freedom. On the other hand, the original **E** and **B** initially possess six degrees of freedom and satisfy the four independent Maxwell equations, thereby reducing the final degrees of freedom to two. The freedom to gauge transform A_μ corresponds to the fact that one of the four degrees of freedom that A_μ represents is extraneous and unphysical. At the classical level the freedom to perform a gauge transformation allows the removal of one of the degrees of freedom by a judicious choice for the function Λ. However, removing a quantum degree of freedom requires more care.

It is instructive to consider this aspect of the electromagnetic field in the noninteracting case where $J^\mu = 0$. The **E** and **B** fields may be decomposed in a gauge invariant manner into their *longitudinal* and *transverse* pieces, denoted respectively by $\mathbf{E}^L$ and $\mathbf{E}^T$ for the electric field, in such a way that $\mathbf{E} = \mathbf{E}^L + \mathbf{E}^T$. These pieces satisfy $\nabla \cdot \mathbf{E}^T = 0$ and $\nabla \times \mathbf{E}^L = 0$, and represent three degrees of freedom, two transverse and one longitudinal, for each of the fields. A representation of these pieces, for the case of an infinite spatial volume or free space, is given by the integral expressions

$$\mathbf{E}^T(\mathbf{x},t) = \mathbf{E}(\mathbf{x},t) - \nabla \int d^3y\, G(\mathbf{x}-\mathbf{y})\, \nabla \cdot \mathbf{E}(\mathbf{y},t)\,, \quad (7.19)$$

$$\mathbf{E}^L(\mathbf{x},t) = \nabla \int d^3y\, G(\mathbf{x}-\mathbf{y})\, \nabla \cdot \mathbf{E}(\mathbf{y},t)\,, \quad (7.20)$$

where G is the familiar Coulomb Green's function,

$$G(\mathbf{x}-\mathbf{y}) = -\int \frac{d^3k}{(2\pi)^3} \frac{1}{k^2} \exp i\mathbf{k}\cdot(\mathbf{x}-\mathbf{y}) = -\frac{1}{4\pi|\mathbf{x}-\mathbf{y}|}\,, \quad (7.21)$$

which satisfies

$$\nabla^2 G(\mathbf{x}-\mathbf{y}) = \delta^3(\mathbf{x}-\mathbf{y})\,. \quad (7.22)$$

It is convenient to introduce a streamlined notation for such integral equations, since they will become common throughout what follows. Result (7.22) represents the inversion of the operator ∇^2 in the functional sense of Sec. 1.6, and so (7.19) and expressions like it may be written

$$\mathbf{E}^T = \mathbf{E} - \nabla \left(\frac{1}{\nabla^2}\right) \nabla \cdot \mathbf{E} \quad (7.23)$$

with no loss of generality. It is important to remember that there are implicit integrations present in such a notation.

Having made this decomposition into longitudinal and transverse parts, two of Maxwell's equations are satisfied by the choice $\mathbf{E}^L = \mathbf{B}^L = 0$, while the remaining two Maxwell equations relate the four transverse pieces of the **E** and **B** fields. There are then two remaining degrees of freedom in these fields that can be chosen to satisfy the boundary conditions in free space, and it will be seen momentarily that these correspond to the two helicity projections available to the photon in the quantized theory.

The upshot of this analysis is that the formulation of Maxwell's equations in terms of the gauge field A_μ requires a subsidiary condition, some other equation, to remove one more degree of freedom so that a match between the two versions of electromagnetism is possible. There are many so-called *gauge conditions* [4], often referred to as *gauges*, that are employed, and each has its advantages. The Lorentz gauge condition, $\partial_\mu A^\mu = 0$, has the advantage of manifest covariance. The radiation or Coulomb gauge condition, $\nabla \cdot \mathbf{A} = 0$, is not covariant, but is manifestly transverse. The temporal gauge condition, $A_0 = 0$, is very easy to implement.

Gauge conditions are *constraints*, as in Sec. 4.6, and therefore the quantization of the theory and the identification of the physical subspace requires more care than that of the scalar and spinor fields. These conditions must be implemented so that a physical quantity, at both the classical and quantum levels, is independent of the choice of gauge. This is because the field in one gauge may be transformed into the field in another gauge by implementing a gauge transformation of the appropriate form. For example, a field in the Lorentz gauge, $A_\mu^{(L)}$, may be converted into a field in the Coulomb gauge, $A_\mu^{(C)}$, by choosing the gauge function Λ to be given by

$$\Lambda(\mathbf{x}, t) = \int_{-\infty}^{t} d\tau \, A_0^{(L)}(\mathbf{x}, \tau) \,, \qquad (7.24)$$

so that $\partial_0 \Lambda = A_0^{(L)}$. Under this gauge transformation $A_0^{(L)} \to 0$, so that the Lorentz condition becomes $\partial^\mu A_\mu^{(L)} = \nabla \cdot (A^{(L)} - \nabla \Lambda) = 0$, the Coulomb condition. Therefore, the gauge transformation of (7.24) transforms the Lorentz gauge field to the Coulomb gauge field, $A_j^{(C)} = A_j^{(L)} - \partial_j \Lambda$.

This raises a second point: once a gauge condition has been chosen, there may be a restricted class of gauge transformations that preserves the gauge condition. For example, the temporal gauge condition is invariant under gauge transformations that satisfy $\partial_0 \Lambda = 0$. The Lorentz gauge condition is invariant under transformations that satisfy $\Box \Lambda = 0$, while the Coulomb gauge is invariant under transformations that satisfy $\nabla^2 \Lambda = 0$. For an infinite spatial volume there are no solutions to the latter equation, $\nabla^2 \Lambda = 0$, that are normalizable. As a result, the Coulomb gauge possesses no residual gauge invariance when the theory is considered for an infinite spatial volume. Residual gauge invariance presages the existence of ghost or zero norm states in the gauge field sector of the quantized theory, but the discussion of this must wait until the end of this section.

The third point of importance emerges when the canonical momenta are identified. The action (7.10) gives

$$\frac{\partial \mathcal{L}}{\partial(\partial_0 A_\mu)} = F^{\mu 0} \equiv \Pi^\mu \,, \qquad (7.25)$$

so that $\Pi^0 = 0$. The absence of a momentum canonically conjugate to A_0

requires implementing a gauge constraint. The general method for doing this was developed for the quantum mechanical path integral in Sec. 4.6. The standard approach is to add a *gauge fixing* term to the Lagrangian in order to create a momentum canonically conjugate to A_0. The theory is quantized by restricting it to the physical subspace where this momentum vanishes in the weak sense. It is almost always the case that enforcing the constraint $\Pi_0 = 0$ ensures that the gauge fixing term, from which it was derived, also vanishes, thereby regaining the original form of the Lagrangian density. The gauge fixing term therefore plays the role of the canonical transformation to a new momentum that was discussed in Sec. 4.6.

However, once Π_0 does not vanish, additional problems appear. The quantized theory must obey the standard ETCR, given by

$$[A_\mu(\mathbf{x},t), \Pi^\nu(\mathbf{y},t)] = i\delta^\nu{}_\mu \delta^3(\mathbf{x}-\mathbf{y}) \,, \tag{7.26}$$

and this yields

$$[A_\mu(\mathbf{x},t), \Pi_\nu(\mathbf{y},t)] = ig_{\mu\nu}\delta^3(\mathbf{x}-\mathbf{y}) \,. \tag{7.27}$$

Result (7.27) shows that the commutation relation of A_0 with its canonical momentum will have the opposite sign of the other three components, and it will be seen momentarily that this results in the presence of negative norm states or ghosts in the Fock space. A Fock space with negative norm states is said to have an *indefinite metric*. This creates new difficulties, that of maintaining unitarity of the S-matrix and positivity of the Hamiltonian. These problems require some care in implementing the quantization conditions, especially in the nonabelian extensions of electrodynamics.

An example of this procedure is given by using the Lorentz gauge condition and changing the action to read

$$\mathcal{L} = \bar{\psi}(i\slashed{\partial} - m - e\slashed{A})\psi - \tfrac{1}{4}F_{\mu\nu}F^{\mu\nu} - \tfrac{1}{2}\kappa(\partial_\mu A^\mu)^2 \,, \tag{7.28}$$

where κ is an arbitrary real number. For the case that the Lorentz gauge condition $\partial_\mu A^\mu = 0$ is satisfied, the action (7.28) coincides with the original action (7.13). However, the addition of this term has created

$$\Pi_0 = -\kappa \partial_\mu A^\mu \,, \tag{7.29}$$

so that the Lorentz condition and Π_0 coincide. However, a nonzero value for Π_0 now allows the standard canonical quantization procedure to be followed. The gauge condition must be understood as a constraint on the states of the theory, and the physical subspace of the theory is defined as the set of states for which the expectation value of Π_0, whatever form it takes, vanishes. Satisfying this constraint in the canonical quantization approach will be discussed in the next subsection.

The method of adding terms to the Lagrangian, which are then set to zero in the physical subspace, is referred to as gauge fixing since adding terms to the action forces a specific choice of gauge condition to regain the original form of the action. For example, the Coulomb gauge condition is forced by the addition of the term

$$\mathcal{L}_{gf} = \dot{A}_0 \nabla \cdot \mathbf{A} \tag{7.30}$$

to the Lagrangian. Such a term gives $\Pi_0 = \nabla \cdot \mathbf{A}$, and the gauge condition is therefore given by $\Pi_0 = \nabla \cdot \mathbf{A} = 0$. It does not matter that the gauge fixing term is not Lorentz invariant. As long as it is zero in the physical subspace of the theory, Lorentz invariance of the action is regained.

In the presence of the specific gauge fixing term of (7.28) the equations of motion (7.7) are changed to

$$\Box A_\mu = J_\mu + (1-\kappa)\partial_\mu(\partial_\nu A^\nu) , \tag{7.31}$$

where, for the moment, the constraint $\Pi_0 = -\kappa \partial_\mu A^\mu = 0$ has not been implemented. Taking the divergence of (7.31) and using the conservation of the current gives

$$\Box \partial_\mu A^\mu = \frac{1}{\kappa} \partial_\mu J^\mu = 0 . \tag{7.32}$$

Thus, even in the presence of coupling to the electric current J_μ, the quantity $\partial_\mu A^\mu$ behaves as a free scalar quantity, totally decoupled from the theory, when the conservation of the electric current is recalled. This has the great advantage that, once $\partial_\mu A^\mu$ is fixed to be zero everywhere at some initial time, it remains zero at all later times, thereby simplifying the implementation of the constraint. The particular choice of $\kappa = 1$ is known as the Feynman gauge, and it will be used throughout the remainder of this section for the covariant theory.

7.1.3 Quantization of the Free Gauge Field

In this subsection the matter-free Feynman gauge field, denoted a_μ, will be treated by canonical quantization [5], in order that contact can be made to the particle content of the theory when the path integral is treated. Along the way some aspects of the Coulomb gauge quantization of the free field will be discussed. The coupled field, A_μ, will be discussed briefly in terms of the canonical formalism in the next section, but its full quantization will be obtained by the path integral.

240   Gauge Field Theory

The starting point is the Lagrangian, given by

$$\mathcal{L} = -\tfrac{1}{4}F_{\mu\nu}F^{\mu\nu} - \tfrac{1}{2}(\partial_\mu a^\mu)^2 \ . \tag{7.33}$$

The momenta canonically conjugate to the field a^μ, denoted π_μ, are identified by (7.25). The explicit forms are given by

$$\pi_j = -\pi^j = -\dot{a}_j + \partial_j a_o \ , \tag{7.34}$$
$$\pi^o = \pi_o = -\partial_\mu a^\mu = -\partial_j a^j - \dot{a}_o = \partial_j a_j - \dot{a}_o \ , \tag{7.35}$$

and the fields and their canonical momenta are assumed to satisfy the ETCR,

$$[a_\mu(\mathbf{x},t), \pi_\nu(\mathbf{y},t)] = ig_{\mu\nu}\delta^3(\mathbf{x}-\mathbf{y}) \ . \tag{7.36}$$

Using the definitions of the momenta the Feynman gauge Hamiltonian is found to be

$$H = \int d^3x \ [\pi^\mu \dot{a}_\mu - \mathcal{L}]$$
$$= \int d^3x \ [-\tfrac{1}{2}\pi_o^2 + \pi_o \partial_j a_j + \tfrac{1}{2}\pi_j \pi_j - \pi_j \partial_j a_o + \tfrac{1}{4}F_{ij}F_{ij}] \ , \tag{7.37}$$

where only covariant components have been used to express H.

Using (7.37), it follows that the primary constraint, $\pi_o = -\partial_\mu a^\mu = 0$, gives rise to the secondary constraint

$$\dot{\pi}_o = i[H, \pi_o] = -\partial_j \pi_j = 0 \ . \tag{7.38}$$

This secondary constraint is precisely the original form of Gauss's law (7.5) with $J_o = 0$. The relationship of the primary and secondary constraint in the Feynman gauge can also be demonstrated by using the definition (7.34) to find

$$\partial_j \pi_j = \nabla^2 a_o - \partial_j \dot{a}_j = \partial_\mu \dot{a}^\mu - \Box a_o = -\dot{\pi}_o \ , \tag{7.39}$$

where the equation of motion $\Box a_o = 0$ has been anticipated.

Both these constraints will be realized weakly by constructing a physical subspace where they both hold, and for that reason it is important to note that they are first-class constraints in the Feynman gauge since their equal-time commutator vanishes in the physical subspace. This is seen by applying the Dirac prescription and using the form of the Feynman gauge Hamiltonian, which gives

$$\dot{a}_o = i[H, a_o] = -\pi_o + \partial_j a_j \ . \tag{7.40}$$

Sec. 7.1 The Maxwell Field 241

In the physical subspace, where $\pi_0 = 0$, this reduces to $\dot{a}_0 = \partial_j a_j$. Therefore, the commutator of the constraints becomes

$$[\pi_0, \nabla \cdot \pi] = [\partial_j a_j, \nabla \cdot \pi] - [\dot{a}_0, \nabla \cdot \pi] = 0 . \tag{7.41}$$

Such a result is not obtained for the Coulomb gauge, where $\pi_0 = \nabla \cdot a$. The Coulomb gauge Hamiltonian is constructed using the gauge fixing term (7.30), so that

$$H_C = \int d^3x \left[\tfrac{1}{2} \pi_j \pi_j - \pi_j \partial_j a_0 + \tfrac{1}{4} F_{ij} F_{ij} \right] , \tag{7.42}$$

As in the Lorentz gauge, the secondary constraint is Gauss's law,

$$\dot{\pi}_0 = i[H_C, \pi_0] = -\partial_j \pi_j . \tag{7.43}$$

Now, however, these constraints are second class, since their equal-time commutator gives

$$i[\nabla \cdot a(\mathbf{x},t), \nabla \cdot \pi(\mathbf{y},t)] = \nabla^2 \delta^3(\mathbf{x}-\mathbf{y}) . \tag{7.44}$$

Result (7.44) shows that π_0 and π_j are not independent degrees of freedom in the Coulomb gauge. As a result, it is necessary to replace the standard commutators with Dirac commutators in the Coulomb gauge in order to guarantee the independence of the two constraints. This will be shown to cause the fields a_j to become effectively transverse.

The free, uncoupled Feynman gauge field a_μ is quantized *in the physical subspace* where $\partial_\mu a^\mu$ and its derivatives vanish, so that it satisfies the equation of motion

$$\Box a_\mu(x) = 0 . \tag{7.45}$$

In order to satisfy the canonical commutation relations the Fock space operators $a_\mu(\mathbf{k})$ are introduced, and these satisfy the algebra

$$[a_\mu(\mathbf{k}), a_\nu^\dagger(\mathbf{p})] = -g_{\mu\nu} \delta^3(\mathbf{k}-\mathbf{p}) . \tag{7.46}$$

The field a_μ is written as a wave packet,

$$a_\mu(x) = \int \frac{d^3k}{(2\pi)^{3/2}} \frac{1}{\sqrt{2\omega_k}} \left[a_\mu(\mathbf{k}) e^{-ikx} + a_\mu^\dagger(\mathbf{k}) e^{ikx} \right] , \tag{7.47}$$

where $\omega_k = +|\mathbf{k}|$. The equation of motion for the a_μ is satisfied by setting $k_0 = \omega_k$, so that $k^2 = \omega_k^2 - \mathbf{k}^2 = 0$. The canonical momenta are written

242   Gauge Field Theory

as Fourier expansions by using these wave packets in the defining relations (7.34). This gives

$$\pi_j(x) = i \int \frac{d^3k}{(2\pi)^{3/2}} \frac{1}{\sqrt{2\omega_k}} \left[\left(\omega_k a_j(\mathbf{k}) - k_j a_0(\mathbf{k}) \right) e^{-ikx} \right.$$
$$\left. - \left(\omega_k a_j^\dagger(\mathbf{k}) - k_j a_0^\dagger(\mathbf{k}) \right) e^{ikx} \right] , \qquad (7.48)$$

$$\pi_0(x) = i \int \frac{d^3k}{(2\pi)^{3/2}} \frac{1}{\sqrt{2\omega_k}} \left[k^\mu a_\mu^\dagger(\mathbf{k}) e^{ikx} - k^\mu a_\mu(\mathbf{k}) e^{-ikx} \right] . \qquad (7.49)$$

Exercise 7.4: Verify that the algebra (7.46) guarantees that the canonical commutation relations (7.36) are satisfied.

In order to satisfy both constraints it is convenient to introduce the *polarization vectors*, $\varepsilon_\mu^{(\lambda)}(k)$. The $\lambda = 1, 2$ polarization vectors are *transverse* to the spatial part of k, denoted $\mathbf{k}$. The polarization vector $\varepsilon^{(3)}(k)$ is parallel to $\mathbf{k}$, and so it is referred to as the *longitudinal* polarization, while $\varepsilon^{(0)}(k)$ points in the *timelike* direction of k. The polarization vectors form a very convenient basis in momentum space. These definitions give

$$k^\mu \varepsilon_\mu^{(\lambda)}(k) = \begin{cases} 0 & \lambda = 1, 2 \\ \omega_k & \lambda = 0, 3 \end{cases} . \qquad (7.50)$$

The polarization vectors are complete and orthonormal, so that

$$g^{\mu\nu} \varepsilon_\mu^{(\lambda)}(k) \varepsilon_\nu^{(\lambda')}(k) = g^{\lambda'\lambda} , \qquad (7.51)$$

$$g_{\lambda'\lambda} \varepsilon_\mu^{(\lambda)}(k) \varepsilon_\nu^{(\lambda')}(k) = g_{\mu\nu} . \qquad (7.52)$$

The timelike polarization can be expressed as $\varepsilon^{(0)}(k) = (1, 0)$, while the longitudinal polarization vector is represented by $\varepsilon^{(3)}(k) = (0, \mathbf{k}/\omega_k)$.

The polarization vectors are used to give the Fock space operators of (7.46) the expansions

$$a_\mu(\mathbf{k}) = \varepsilon_\mu^{(\lambda)}(k) \alpha_\lambda(\mathbf{k}) , \qquad (7.53)$$

where, in order to maintain the algebra of (7.46), the new operators α_λ must satisfy the algebra

$$[\alpha_\lambda(\mathbf{k}), \alpha_{\lambda'}^\dagger(\mathbf{p})] = -g_{\lambda\lambda'} \delta^3(\mathbf{k} - \mathbf{p}) . \qquad (7.54)$$

In terms of the new Fock space operators the two constraints become, in the physical subspace,

$$\partial_j \pi_j(x) = -\int \frac{d^3k}{(2\pi)^{3/2}} \sqrt{\omega_k^3} \left[\alpha_g(\mathbf{k}) e^{-ikx} + \alpha_g^\dagger(\mathbf{k}) e^{ikx} \right] , \quad (7.55)$$

$$\pi_o(x) = i \int \frac{d^3k}{(2\pi)^{3/2}} \sqrt{\omega_k} \left[\alpha_g(\mathbf{k}) e^{-ikx} - \alpha_g^\dagger(\mathbf{k}) e^{ikx} \right] , \quad (7.56)$$

where

$$\alpha_g(\mathbf{k}) = \frac{1}{\sqrt{2}} [\alpha_o(\mathbf{k}) - \alpha_3(\mathbf{k})] . \quad (7.57)$$

The two forms (7.55) and (7.56) clearly satisfy (7.39).

The Fock space for the theory is constructed by the usual assumption that there is a vacuum state $|0\rangle$ that is annihilated by the α_μ operators. However, this Fock space possesses an indefinite metric, and this is easily seen from the commutation relations of the α_o operator. The state defined by

$$|\mathbf{k}, \lambda = 0\rangle = \alpha_o^\dagger(\mathbf{k})|0\rangle , \quad (7.58)$$

has the property

$$\langle \mathbf{k}, \lambda = 0 | \mathbf{p}, \lambda = 0 \rangle = -\delta^3(\mathbf{k} - \mathbf{p}) . \quad (7.59)$$

The presence of negative norm states threatens any probabilistic interpretation of the theory and, in its coupled form, transition elements. The solution is to use only the subspace that satisfies the constraints.

The first step in identifying this subspace is to note that the α_g operator, which is the constituent of both constraints (7.55) and (7.56), has the property

$$[\alpha_g(\mathbf{k}), \alpha_g^\dagger(\mathbf{p})] = [\alpha_o(\mathbf{k}), \alpha_o^\dagger(\mathbf{p})] + [\alpha_3(\mathbf{k}), \alpha_3^\dagger(\mathbf{p})] = 0 , \quad (7.60)$$

so that the state $\alpha_g^\dagger(\mathbf{k})|0\rangle$ has *zero* norm. For that reason, states created by $\alpha_g^\dagger$ are referred to as *ghosts*. Because the ghost annihilation and creation operators commute, and because the two constraints are composed solely of ghost operators, it follows that the two Lorentz gauge constraints commute in the physical subspace. This is consistent with result (7.41). For the same reasons, a state composed of any number of transverse excitations, created by $\alpha_1^\dagger$ and $\alpha_2^\dagger$, and any number of ghosts will automatically satisfy both of the constraints. This set of states breaks into two orthogonal pieces: the states containing solely transverse excitations and therefore possessing positive-definite norm, and all the remaining states, containing one or more ghosts and therefore possessing zero norm. The restriction to this subspace, so that $\partial_\mu a^\mu$ vanishes, is identical to the Gupta–Bleuler condition [6] for identifying the physical subspace states $|\psi\rangle_p$, which is usually expressed

$$\partial_\mu a^{\mu(+)}|\psi\rangle_P = 0 , \qquad (7.61)$$

where $\partial_\mu a^{\mu(+)}$ is the annihilation operator piece of $\partial_\mu a^\mu$. Key to the success of the Gupta–Bleuler condition is result (7.32), which allows a time-independent separation of $\partial_\mu A^\mu$ into positive and negative frequency modes even in the presence of interactions with the conserved electric current. The physical subspace contains no negative norm states, but there remain the zero norm states, and in a coupled theory these can still be problematic. This will be discussed in the next section.

It is also important to determine the positive-definiteness of the Hamiltonian since it is unclear how the negative norm states contribute to the energy. This is readily done by employing the expansions for the fields and momenta in the defining form (7.37). It is left as an exercise to show that

$$H = \int d^3k\, \omega_k \left[\sum_{\lambda=1}^{2} \alpha_\lambda^\dagger(\mathbf{k})\alpha_\lambda(\mathbf{k}) - \alpha_b^\dagger(\mathbf{k})\alpha_g(\mathbf{k}) - \alpha_g^\dagger(\mathbf{k})\alpha_b(\mathbf{k}) \right] , \qquad (7.62)$$

where

$$\alpha_b(\mathbf{k}) = \frac{1}{\sqrt{2}}[\alpha_0(\mathbf{k}) + \alpha_3(\mathbf{k})] . \qquad (7.63)$$

Exercise 7.5: Verify (7.62).

The operator (7.63) is also a ghost operator since

$$[\alpha_b^\dagger(\mathbf{k}), \alpha_b(\mathbf{p})] = 0 . \qquad (7.64)$$

However, the two types of ghost *do not* commute. It is easy to see that

$$[\alpha_b(\mathbf{k}), \alpha_g^\dagger(\mathbf{p})] = -\delta^3(\mathbf{k} - \mathbf{p}) . \qquad (7.65)$$

The ghost states associated with α_b are banned from the physical subspace. It therefore follows from the definition of the physical subspace and (7.60) that the potentially threatening negative energy terms in (7.62) vanish from matrix elements of the Hamiltonian in the physical subspace. In effect, the physical subspace is constrained to have equal numbers of negative energy timelike and positive energy longitudinal excitations, and the net energy of these two polarizations is therefore zero.

Sec. 7.1 The Maxwell Field

In the Coulomb gauge it is necessary to calculate using Dirac commutators. This requires adapting the techniques of Sec. 4.6 to the case of field theory, but this is straightforward using the continuum methods presented in Sec. 6.8. In such an approach the constraint equations are implemented at each space-time point, just as the fields are considered quantum mechanical variables at each space-time point. This allows the notation of Sec. 4.6 to be taken over with only a few modifications. The sum over constraints must be replaced with integrations, and the inversion of matrices is understood to be in the continuum sense. The constraints at each point are denoted $\theta_1 = \pi_0$ and $\theta_2 = \nabla \cdot \pi$, and the commutators (7.44) of these second-class constraints give the t matrix:

$$t = \begin{pmatrix} 0 & -i\nabla^2 \delta^3(\mathbf{x} - \mathbf{y}) \\ i\nabla^2 \delta^3(\mathbf{x} - \mathbf{y}) & 0 \end{pmatrix}. \tag{7.66}$$

Using the streamlined notation of (7.23), the inverse of this matrix is written

$$c = \left(\frac{1}{\nabla^2}\right) \begin{pmatrix} 0 & -i \\ i & 0 \end{pmatrix}, \tag{7.67}$$

where $1/\nabla^2$ is understood to be the Coulomb Green's function, $G(\mathbf{x} - \mathbf{y})$, of (7.21). The matrix c is the inverse of t in the continuum sense, so that

$$(t \cdot c)_{ik}(\mathbf{x}, \mathbf{z}) = \int d^3y \, t_{ij}(\mathbf{x} - \mathbf{y}) \, c_{jk}(\mathbf{y} - \mathbf{z}) = \delta_{ik} \delta^3(\mathbf{x} - \mathbf{z}). \tag{7.68}$$

Therefore, the equal-time Dirac commutators of B and C are given by

$$\begin{aligned}[] [B, C]_D &= [B, C] + i[B, \nabla \cdot \pi]\left(\frac{1}{\nabla^2}\right)[\partial_j a^j, C] \\ &\quad - i[B, \partial_j a^j]\left(\frac{1}{\nabla^2}\right)[\nabla \cdot \pi, C]. \end{aligned} \tag{7.69}$$

The integrations implicit in (7.69) are best exposed by an example. The equal-time Dirac commutator for the spatial part of the Coulomb gauge field and its canonical momentum are given by

$$\begin{aligned}[] [a_j(\mathbf{x}, t), \pi_k(\mathbf{y}, t)]_D &= [a_j(\mathbf{x}, t), \pi_k(\mathbf{y}, t)] \\ &\quad + i \int d^3z \, d^3z' \, [a_j(\mathbf{x}, t), \nabla \cdot \pi(\mathbf{z}, t)] G(\mathbf{z} - \mathbf{z}')[\nabla \cdot a(\mathbf{z}', t), \pi_k(\mathbf{y}, t)] \\ &\quad - i \int d^3z \, d^3z' \, [a_j(\mathbf{x}, t), \nabla \cdot a(\mathbf{z}, t)] G(\mathbf{z} - \mathbf{z}')[\nabla \cdot \pi(\mathbf{z}', t), \pi_k(\mathbf{y}, t)] \\ &= -i\delta_{jk}\delta^3(\mathbf{x} - \mathbf{y}) + i\partial_j \partial_k G(\mathbf{x} - \mathbf{y}) \\ &\equiv -i\delta^{\text{Tr}}_{jk}(\mathbf{x} - \mathbf{y}). \end{aligned} \tag{7.70}$$

The latter function, δ^{Tr}_{jk}, is the so-called *transverse delta*, and clearly obeys

$$\partial_j \, \delta^{\text{Tr}}_{jk}(\mathbf{x} - \mathbf{y}) = \delta_{ij}\partial_j\delta^3(\mathbf{x}-\mathbf{y}) - \partial_i\nabla^2 G(\mathbf{x}-\mathbf{y}) = 0 \,. \tag{7.71}$$

The effect of the Dirac commutator is to force the spatial part of the field and its canonical momentum to be transverse. The Dirac commutator for the Coulomb gauge can be replaced with the standard commutator by using Lorentz gauge expansions with the longitudinal and timelike parts of all field operators suppressed. The revised Coulomb gauge field operator is therefore written

$$a_j(x) = \int \frac{d^3k}{(2\pi)^{3/2}} \frac{1}{\sqrt{2\omega_k}} \sum_{\lambda=1}^{2} \varepsilon_j^{(\lambda)}(k) \left[\alpha_\lambda(\mathbf{k})e^{-ikx} + \alpha_\lambda^\dagger(\mathbf{k})e^{ikx} \right] \,. \tag{7.72}$$

This field is explicitly transverse, satisfying $\nabla \cdot a = 0$. This allows the Dirac commutator to be replaced by the standard commutator. The canonical momentum must also be altered in the Coulomb gauge, and this is left as an exercise.

Exercise 7.6: Determine the form for the Coulomb gauge momenta, π_j, and express the Coulomb gauge Hamiltonian in terms of Fock space operators.

The final step in the Coulomb gauge is to note that the equation of motion for a_o is

$$\nabla^2 a_o = 0 \,, \tag{7.73}$$

and this possesses no normalizable solutions in free space. As a result, there is no expansion for a_o, and it must be set to zero.

Exercise 7.7: Calculate the Dirac commutation relations necessary to enforce the constraint $a_o = 0$.

Thus, in the Lorentz gauge the allowed spectrum of states consists of transverse excitations and zero norm ghosts. The ghost states possess zero energy and are not physical. In the Coulomb gauge only transverse excitations are present. From this point forward only transverse excitations are considered to be physical in any gauge.

Exercise 7.8: Using Noether's theorem, show that the components of the spin of the electromagnetic field are given by

$$S_i = \int d^3x \, \epsilon_{ijk} \pi_j(x) a_k(x) \,. \tag{7.74}$$

Exercise 7.9: Show that the helicity, i.e., the projection of the spin along the direction of motion, is conserved since its commutator with the free Hamiltonian vanishes, and that its eigenstates, the transverse excitations, have the eigenvalues ± 1.

The results of these exercises show that the electromagnetic field, modelled by the gauge theory, is associated with a massless particle, of course referred to as a *photon*, and that the photon has only two helicities available. For the massless free theory there can be no longitudinal component for the field, so that there is no zero helicity state for the photon. This is because it is impossible to use a Lorentz transformation to boost an observer into the photon's rest frame. If such a boost were possible, then **p** would vanish, and the projection of the spin along **p** would also vanish. As in the classical theory there are therefore only two degrees of freedom available to the photon, and the implementation of the gauge condition and the subsequent constraints has led directly to the representation of these two degrees of freedom as the two helicities of the photon. However, it should be remembered that for a *massive* gauge field this argument fails, and the longitudinal component must be present since it is possible to boost into the rest frame of a massive particle. The construction of a consistent *interacting* massive gauge field theory is a nontrivial matter, and its discussion will be deferred to Chapter 9.

It is natural to wonder why canonical quantization in the Feynman gauge is accompanied by the appearance of ghost states. The answer lies in the theory's residual gauge invariance. The gauge condition, $\partial_\mu a^\mu = 0$, as well as the rest of the action, is invariant under gauge transformations satisfying $\Box \Lambda = 0$. Using Noether's theorem the generator of gauge transformations, simply the conserved charge associated with the residual gauge invariance, is found for the general case to be

$$Q_\Lambda = \int d^3x \left[\partial_j \Lambda \, \pi_j + \dot\Lambda \, \pi_0 \right] \,. \tag{7.75}$$

An integration by parts yields

$$Q_\Lambda = \int d^3x \left[\dot\Lambda\, \pi_0 - \Lambda\, \nabla\cdot\pi\right] . \tag{7.76}$$

The constraints of the theory, $\pi_0 = 0$ and Gauss's law, $\nabla\cdot\pi = 0$, have appeared as the coefficients of the Λ. While this has been derived within the context of the Feynman gauge, form (7.76) will result in *any* gauge where there is a residual gauge invariance. In the physical subspace of the theory both π_0 and Gauss's law $\nabla\cdot\pi$ must vanish. Using the forms for the canonical momenta in the Feynman gauge it is straightforward to show that Q_Λ becomes

$$Q_\Lambda = \int d^3x \left[\Lambda\, \partial_\mu \dot a^\mu - \dot\Lambda\, \partial_\mu a^\mu\right] , \tag{7.77}$$

a result satisfying $\dot Q_\Lambda = 0$.

Since Q_Λ generates the gauge transformation, and all matrix elements of the physically relevant fields of the theory must be gauge invariant, it follows that the matrix elements of Q_Λ and its powers must vanish, for otherwise the theory would fail to be gauge invariant. If the local operators that make up Q_Λ are to possess Fourier expansions that involve both creation and annihilation operators, and if powers of Q_Λ are to vanish in the physical subspace, then it follows that these annihilation and creation operators must commute. This implies that the local operators that constitute Q_Λ must be associated with ghosts.

It is a hallmark of gauge theories that the original form of Gauss's law, i.e., the form derived prior to any kind of gauge fixing and subsidiary condition, appears in the generator of gauge transformations. Identifying the physical subspace through the implementation of this form of Gauss's law is therefore intimately connected to any residual gauge invariance associated with a particular choice of gauge condition. Enforcing the residual gauge invariance is accomplished by choosing a Fock space decomposition that causes the gauge generator to consist solely of ghosts. Therefore, the gauge condition will be associated with zero norm excitations if there is a residual gauge invariance. With certain modifications these observations form the basis of BRST invariance [7] as a means of enforcing unitarity in gauge theory, and this is discussed briefly in Sec. 8.2. However, the path integral representation of gauge theories presents a very efficient vehicle to discuss almost all aspects of gauge theory, and so the BRST approach will not be explored in detail.

7.2 QED as a Path Integral

In this section the coupled version of electromagnetism [8], referred to as quantum electrodynamics (QED), will be analyzed by using a path integral. It should be apparent from the previous section that canonical quantization can be complicated for gauge theories. In fact, mild controversies still persist over the proper implementation of some gauges in the canonical treatment of QED [4]. More confusing is the fact that some derivations of perturbation theory for QED, to be discussed in the next chapter, result in Feynman rules that are *apparently* gauge dependent. This stems from the fact that QED is a very pliable theory, and the calculation of *physical* transition elements in QED is immune to ignoring the gauge condition. The ultimate source of this immunity is that the abelian gauge field is coupled to a conserved current, so that all transition elements conserve electric charge. This indifference to the gauge condition will be seen shortly.

A first guess at a spectrum for the coupled theory would be the tensor product of the physically allowed gauge states with the charged spinor or scalar states to which the gauge field is coupled. The course of action in the canonical framework is to derive an interaction picture representation of the evolution operator as in Sec. 6.7. The evolution operator then represents the effects of the coupling on the time development of the system. The ghost states in the physical subspace, present in the manifestly covariant approach discussed in the previous section, are an immediate concern in such an approach. These states pose a threat to unitarity in the following way. The inner product of a ghost in state can be written

$$0 = {}_{\text{in}}\langle G | G \rangle_{\text{in}} = \sum_n |{}_{\text{in}}\langle G | n \rangle_{\text{out}}|^2 = \sum_n |{}_{\text{in}}\langle G | S | n \rangle_{\text{in}}|^2 , \qquad (7.78)$$

where the sum is over the allowed physical states. If the S-operator does not commute with the ghost operators, or if its commutator does not vanish in the physical subspace, then the right-hand side of (7.78) will not vanish, in contradiction to the left-hand side. If the right-hand side does not vanish, then using the completeness of the in states immediately yields

$${}_{\text{in}}\langle G | G \rangle_{\text{in}} \neq {}_{\text{in}}\langle G | SS^\dagger | G \rangle_{\text{in}} , \qquad (7.79)$$

demonstrating that the zero norm states threaten unitarity of the S-matrix unless the commutator of S with the ghost operators vanishes in the physical subspace.

At first blush S is given by the evolution operator, defined by (6.214) and derived from all terms cubic and higher in the coupled action, so that

$$S = T\left\{ \exp\left[-i \int d^4x \, j_\mu(x) \, a^\mu(x) \right] \right\} . \qquad (7.80)$$

Here j^μ is the current associated with the charged field, and for the bispinor case would be given by (7.14) written in terms of free Dirac fields. The proof of unitarity reduces to showing that the ghost-S commutator satisfies

$$\left[\partial_\nu a^\nu(x), \int d^4y\, j_\mu(y) a^\mu(y)\right] = 0\,, \tag{7.81}$$

when evaluated between in and out states in the physical subspace.

Exercise 7.10: Using Lorentz gauge expansions, show that (7.81) vanishes if $\partial_\mu j^\mu(x) = 0$ and the electric charges of the in and out state are identical.

Therefore, conservation of electric charge allows the naive S-matrix to be used without jeopardizing unitarity. Such is not always the case, particularly in Yang–Mills gauge theories. There are methods for treating the constraints consistently in manifestly covariant formulations of the coupled theory using the canonical operator formalism, but these will not be presented here.

A possible approach to constructing the QED path integral would be a repetition of the techniques of Sec. 6.8, where the evolution operator was partitioned by using complete sets of intermediate coherent states. This is tedious and has the additional task of first constructing the evolution operator that is consistent with the choice of gauge in the coupled problem. Such derivations are found in the literature [9]. However, from the general properties of the gauge field action discussed in the previous section, it follows that such a derivation must result in a constrained path integral of the form discussed in Sec. 4.6 [10, 11]. Because the constraints are most conveniently expressed in configuration space, configuration measure will be used. The form of the QED path integral is determined by combining the configuration measure discussed in Sec. 6.8.4, as well the rules for gauge constraints derived in the quantum mechanical context in Sec. 4.6. The result for the vacuum transition element is

$$_+\langle 0|0\rangle_- = \int_0^0 \mathcal{D}\pi\, \mathcal{D}A\, \mathcal{D}\eta\, \delta(\chi)\, \delta(G)\, \det\{\chi, G\} \exp\left\{i \int dx\, \mathcal{L}_\chi(\eta, A, \pi)\right\}\,. \tag{7.82}$$

The explanation of this form is straightforward. The measure is understood to be the configuration space form discussed in Sec. 6.8.4, since this is easiest to use with the configuration space constraints. The measure over

π and A contains all four modes of each, while the spinor measure is identical to the configuration form discussed in Sec. 6.8.4. Because the gauge constraint $\pi_0 = 0$ exists at every space-time point, the constraint measure of Sec. 4.6 must be generalized to the field theory case. In Sec. 4.6 it was argued that a canonical transformation or a gauge fixing term may be used to redefine π_0 into χ, and that χ may appear in the transformed action, denoted $\mathcal{L}_\chi$. The technique of adding a gauge fixing term was discussed in detail in the previous section for the Lorentz gauge. An example of a canonical transformation for π_0 is the temporal gauge, where the canonical transformation is generated by adding the following total divergence to the action,

$$\int d^4x \, \tfrac{1}{2}\kappa \frac{\partial}{\partial t}(A_o{}^2) = \int d^4x \, \kappa A_o \dot{A}_o \, . \tag{7.83}$$

This immediately gives $\pi_0 = \kappa A_o = 0$, which is the temporal gauge condition.

As a result, the action $\mathcal{L}_\chi$ is a combination of the spinor and the gauge field forms constructed using the recipe of Sec. 6.8.4, consistent with the goal of deriving a vacuum transition element, and is written

$$\mathcal{L}_\chi = \mathcal{L}_f(\bar{\eta},\eta) + \pi^\mu \dot{A}_\mu - J^\mu A_\mu - \mathcal{H}_\chi(\pi, A) \, . \tag{7.84}$$

$\mathcal{L}_f$ is the free spinor action, familiar from Sec. 6.8.2,

$$\mathcal{L}_f = \bar{\eta}(i\slashed{\partial} - m)\eta \, , \tag{7.85}$$

while $\mathcal{H}_\chi$ is the Hamiltonian density for the free gauge field. It is constructed from the canonically transformed or gauge fixed classical action, and therefore depends upon the gauge condition χ.

The presence of the primary constraint, $\chi = 0$, creates the secondary constraint, $G = \dot{\chi} = 0$. From Sec. 4.6 the constraint G is given by

$$G = \dot{\chi} = -\frac{\delta \mathcal{H}_\chi}{\delta A_o} \, , \tag{7.86}$$

so that G is revealed as Gauss's law for the choice of gauge associated with the condition $\chi = 0$. The vanishing of π_0, or its canonical transform χ, and G at every space-time point is accomplished by defining the Dirac deltas of the gauge condition χ and the secondary constraint G as products over all space-time, since this is the continuum limit of the constrained variables of integration. The determinant of the Poisson bracket of χ and G, dictated by inverting Gauss's law, was generalized to the continuum case in (6.307).

It is important to recall from the analysis of Sec. 4.6 that the additional factors in the measure, the Dirac deltas and the determinant, arise in the following way. The initial gauge constraint is $\pi_0 = 0$, and this can be implemented in the absence of gauge fixing or canonical transformations. It gives the secondary constraint of Gauss's law, $G = \dot{\pi}_0 = 0$. This is solved for A_0, the variable canonically conjugate to π_0, and applied as another constraint. If A_0^c is the solution to this secondary constraint, the measure factors due to the constraints are then given by

$$\delta(\pi_0)\,\delta(G)\,\det\frac{\delta G}{\delta A_0} = \delta(\pi_0)\delta(A_0 - A_0^c)\,, \qquad (7.87)$$

and the action appearing in the path integral has the Hamiltonian density $\mathcal{H}$, calculated from the action without a gauge fixing term. If a gauge fixing term or canonical transformation is invoked, then $\pi_0 \to \chi$, while the determinant becomes [10]

$$\det\frac{\delta G}{\delta A_0} = \det\{\pi_0, G\} \to \det\{\chi, G\}\,. \qquad (7.88)$$

The Hamiltonian density appearing in the path integral action is now determined from the original action with the gauge fixing term present, and the path integral takes the form given by (7.82). The form of the path integral defined by either method must yield identical results.

A subtle but important problem occurs if the secondary constraint G does not possess a *unique* solution for A_0. Such a secondary constraint renders the relationship (7.87) ambiguous, since it is unclear which of the solutions is being implemented as the constraint. This will be discussed again when Yang–Mills fields are quantized.

It is recalled from the discussion in Sec. 6.8.4 that configuration measure for an interacting field theory requires the Wick rotation in order that integrations and determinants are well defined. For a gauge theory there are more vectors than simply x^μ, so that the appropriate Wick rotation to Euclidean space is given by the prescription

$$t \to -it,\quad A_0 \to iA_0,\quad \pi_0 \to i\pi_0,\quad J_0 \to iJ_0\,. \qquad (7.89)$$

The prescription of (7.89) is chosen so that the canonical commutation relations in Euclidean space are given by

$$[A_\mu, \pi_\nu] = -i\delta_{\mu\nu}\,. \qquad (7.90)$$

The determinant of the Poisson bracket appearing in (7.82) is understood only in the Euclidean formulation. In this chapter all integrations and evaluations of determinants are performed by first Wick rotating the theory

Sec. 7.2 QED as a Path Integral

to Euclidean space and, after the evaluation, continuing the result back to Minkowski space to obtain the final result.

The first step in deriving the Feynman rules for QED or any theory is to perform the momentum integrations appearing in the path integral. The remaining steps will be taken in the next chapter. This integration will be done in the Feynman gauge, where $\chi = -\partial_\mu A^\mu$, along with the appropriate measure factors in (7.87). The configuration space form for the Hamiltonian in the Feynman gauge is given by

$$H = H_f + H_\chi + H_I \,, \tag{7.91}$$

where H_f is the Hamiltonian for the matter field and has no reference to A_μ, and

$$H_I = \int d^3x \, J^\mu A_\mu \,, \tag{7.92}$$

$$H_\chi = \int d^3x \left[-\tfrac{1}{2}\pi_0^2 + \pi_0 \partial_j A_j + \tfrac{1}{2}\pi_j \pi_j - \pi_j \partial_j A_0 + \tfrac{1}{4} F_{ij} F_{ij} \right] \,. \tag{7.93}$$

The secondary constraint is obtained as

$$G = \dot{\pi}_0 = -\frac{\delta H}{\delta A_0} = -\partial_j \pi_j - J_0 = 0 \,. \tag{7.94}$$

and, using the identification $\pi_j = \dot{A}^j + \partial_j A_0$, the secondary constraint is immediately seen to be the original form of the coupled Gauss's law,

$$G = \partial_j \dot{A}_j - \nabla^2 A_0 - J_0 = 0 \,. \tag{7.95}$$

This can be uniquely inverted to give

$$A_0^c = \left(\frac{1}{\nabla^2}\right)\left(\partial_j \dot{A}_j - J_0\right) \,, \tag{7.96}$$

showing that the replacement (7.87) is valid for QED.

In the Feynman gauge it is necessary to calculate the determinant that appears in the measure. This is done by noting that, using (7.95), G can be written in the Feynman gauge as

$$G = -\partial_j \pi_j - J_0 = -\partial_\mu \dot{A}^\mu + \Box A_0 - J_0 = \Box A_0 - J_0 \,. \tag{7.97}$$

In Euclidean space the determinant appearing in the measure is therefore given by

$$\Delta_\chi \equiv \det\left[\frac{\delta G(x)}{\delta A_0(y)}\right] = \det \Box_E \,. \tag{7.98}$$

The absolute necessity of the Wick rotation is now apparent. If $\Box$ is not continued to Euclidean space, it possesses a multiplicity of zero eigenvalues, rendering the determinant zero. The continuation to Euclidean space allows the constrained measure of the path integral to remain well defined. Since the determinant has no reference to any of the fields, for the case of QED it will be cancelled by normalization and there is no harm in ignoring it entirely.

It is also possible to obtain the same determinant from the definition of the Poisson bracket. In what follows the evaluation is understood to be in the Euclidean region. The definition of the determinant gives

$$\Delta_\chi = \det\{\chi, G\} = \det \int d^4 x \left[\frac{\delta \chi}{\delta A_\mu} \frac{\delta G}{\delta \pi^\mu} - \frac{\delta G}{\delta A_\mu} \frac{\delta \chi}{\delta \pi^\mu} \right] . \quad (7.99)$$

Using $\chi = -\partial^\mu A_\mu$ gives

$$\frac{\delta \chi(y)}{\delta A_\mu(x)} = -\partial_y^\mu \delta^4(y - x) , \quad (7.100)$$

while the form $G = \partial_k \pi^k - J_o$ gives

$$\frac{\delta G(z)}{\delta \pi^j(x)} = \partial_j \delta^4(z - x) . \quad (7.101)$$

Extracting the π_o dependence of G requires using the expressions for the π^j in G to obtain the form (7.97) prior to setting $\partial_\mu A^\mu = 0$,

$$G = -\partial_\mu \dot{A}^\mu + \Box A_o - J_o = \dot{\pi}_o + \Box A_o - J_o , \quad (7.102)$$

or recalling that, by definition, the secondary constraint was determined by setting $G = \dot{\pi}_o$ in the physical subspace. As a result,

$$\frac{\delta G(z)}{\delta \pi_o(x)} = \partial_o \delta^4(z - x) , \quad (7.103)$$

so that placing these results back into (7.99) gives the previously obtained result,

$$\Delta_\chi = \det \Box_E . \quad (7.104)$$

In Euclidean space the π_o momentum integration is a standard Gaussian. Isolating only the terms involving π_o and using the prescriptions (7.89), the relevant integration, *modulo* normalization factors, is given in

Euclidean space by

$$\int \mathcal{D}\pi_0 \exp\left\{i\int d^4x\left[\tfrac{1}{2}\pi_0{}^2 + (\dot{A}_0 - \partial_j A_j)\pi_0\right]\right\} \xrightarrow{E}$$
$$\int \mathcal{D}\pi_0 \exp\left\{-\int d^4x\left[\tfrac{1}{2}\pi_0{}^2 + i(\dot{A}_0 + \partial_j A_j)\pi_0\right]\right\} =$$
$$\exp\left\{-\int d^4x\, \tfrac{1}{2}(\dot{A}_0 + \partial_j A_j)^2\right\} \xrightarrow{M}$$
$$\exp\left\{-i\int d^4x\, \tfrac{1}{2}(\partial_\mu A^\mu)^2\right\} . \tag{7.105}$$

where $\xrightarrow{E,M}$ will henceforth represent a continuation to Euclidean/Minkowski space. The result shows that the π_0 integration has reproduced the gauge fixing term that was present in the original classical action.

The resulting path integral is now given by

$$+\langle 0|0\rangle_- = \int_0^0 \mathcal{D}\vec{\pi}\mathcal{D}A\mathcal{D}\eta\,\Delta_\chi\,\delta(\partial_\mu A^\mu)\delta(G)\exp\left\{i\int d^4x\left[\mathcal{L}(\eta)\right.\right.$$
$$\left.\left. -\tfrac{1}{2}\pi_j\pi_j - \pi_j\dot{A}_j + \pi_j\partial_j A_0 - \tfrac{1}{4}F_{jk}F^{jk}\right.\right.$$
$$\left.\left. - J_\mu A^\mu - \tfrac{1}{2}(\partial_\mu A^\mu)^2\right]\right\}, \tag{7.106}$$

where $\mathcal{D}\vec{\pi}$ indicates the absence of the π_0 measure.

The secondary constraint $\delta(G)$ can now be eliminated from the measure by absorbing it into the action. The first step is to rewrite the product of Dirac deltas as a product of integral representations of the Dirac delta:

$$\prod_x \delta\left(G(x)\right) = \int \mathcal{D}\Lambda \exp\left\{i\int d^4x\, \Lambda(x)G(x)\right\}. \tag{7.107}$$

Using the definition of G allows (7.107) to be rewritten

$$\int \mathcal{D}\Lambda \exp\left\{i\int d^4x\, \Lambda(x)G(x)\right\}$$
$$= \int \mathcal{D}\Lambda \exp\left\{i\int d^4x\, (\pi_j\partial_j\Lambda - J_0\Lambda)\right\}. \tag{7.108}$$

Because the constraint has been exponentiated, the space-time integral appearing in (7.108) can be added to the action of the path integral (7.106), while $\mathcal{D}\Lambda$ becomes part of the measure in the path integral. The argument of the space-time integral in (7.108) exactly matches the form of two of the terms where A_0 appears in the action of the path integral (7.106), with Λ playing the role of A_0. The change of variables $A_0 \to A_0 - \Lambda$ therefore causes both terms appearing in the exponential (7.108) to disappear while inducing no change in the measure of the path integral. However, this same change of variables affects the gauge fixing term and the gauge condition, since

$$\partial_\mu A^\mu \to \partial_\mu A^\mu - \dot{\Lambda} \,. \qquad (7.109)$$

The π_j integrations are also Gaussian and after evaluating these the path integral becomes

$$_+\langle 0|0\rangle_- = \int_0^0 \mathcal{D}A\,\mathcal{D}\eta\,\mathcal{D}\Lambda\,\Delta_\chi\,\delta(\partial_\mu A^\mu - \dot{\Lambda}) \exp\left\{i\int d^4x \left[\mathcal{L}(\eta)\right.\right.$$
$$\left.\left. - \tfrac{1}{4}F_{\mu\nu}F^{\mu\nu} - J_\mu A^\mu - \tfrac{1}{2}(\partial_\mu A^\mu - \dot{\Lambda})^2\right]\right\} \,. \qquad (7.110)$$

Exercise 7.11: Verify (7.110).

Except for the modified gauge fixing term and gauge condition, the remainder of the action and measure appearing in (7.110) is gauge invariant. This gauge invariance allows the appearance of Λ to be removed from the action and gauge condition. The variables undergo the gauge transformation $\eta \to \exp(ie\lambda)\eta$ and $A_\mu \to A_\mu - \partial_\mu \lambda$, where λ is chosen to solve

$$\Box\lambda = -\dot{\Lambda} \,. \qquad (7.111)$$

The resulting path integral becomes

$$_+\langle 0|0\rangle_- = \int \mathcal{D}\Lambda \int_0^0 \mathcal{D}A\,\mathcal{D}\eta\,\Delta_\chi\,\delta(\partial_\mu A^\mu) \exp\left\{i\int d^4x \left[\mathcal{L}(\eta)\right.\right.$$
$$\left.\left. - \tfrac{1}{4}F_{\mu\nu}F^{\mu\nu} - J_\mu A^\mu - \tfrac{1}{2}(\partial_\mu A^\mu)^2\right]\right\} \,. \qquad (7.112)$$

Result (7.112) is a path integral with the coupled form of the action (7.13), complete with the gauge fixing term and the gauge condition enforced in the measure in the form of the Dirac delta over all space-time points.

However, two new factors have emerged. The first is the overall factor $\int \mathcal{D}\Lambda$. The meaning of this factor can be understood when its origin is examined. It arises from enforcing the secondary Gauss's law constraint by expressing it as an exponential form of the Dirac delta. Because Gauss's law generates gauge transformations in the physical subspace, this form of the constraint is equivalent to integrating over all possible gauge configurations, and $\int \mathcal{D}\Lambda$ is therefore the "gauge volume" of the theory. In effect, the process of implementing the constraints has factored this gauge volume from the path integral, and at first glance this may seem strange. However, recalling the ambiguity in the classical form of electrodynamics that gauge invariance represents, it follows that this factorization is precisely what is needed to make the quantized theory *unique*. Once the overall factor of the gauge volume is divided out, the measure of (7.112) is over gauge equivalent configurations.

The second factor, Δ_χ, is less easily understood, but its presence is dictated by the need to maintain a unitary theory. Its origin is in the inversion of the Dirac delta that ensures $\dot{\pi}_0 = 0$. Such a secondary constraint is paramount to decoupling the ghosts of the theory from the S-matrix, of which the path integral is a representation. In the case of QED it seems to take a rather unexciting form, but it can be rewritten in a more illuminating way by introducing two Grassmann variables $\bar{c}(x)$ and $c(x)$. Using the results of Sec. 6.8.4, the determinant can be written in Euclidean space as an integration over these Grassmann variables,

$$\Delta_\chi = \det \Box_E = \int \mathcal{D}\bar{c}\,\mathcal{D}c \, \exp\left\{\int d^4x\, \bar{c}\, \Box_E\, c\right\}$$
$$\xrightarrow{M} \int \mathcal{D}\bar{c}\,\mathcal{D}c \, \exp\left\{-i \int d^4x\, \partial_\mu \bar{c}\, \partial^\mu c\right\}. \quad (7.113)$$

The argument of the exponential appearing in (7.113),

$$\mathcal{L}_c = -\partial_\mu \bar{c}\, \partial^\mu c, \quad (7.114)$$

can be added to the action, while $\mathcal{D}\bar{c}\,\mathcal{D}c$ is added to the measure of the path integral. The argument of the exponential serves to define a Minkowski space quantum field theory of anticommuting scalars, and, for one last time, the canonical quantization formalism will be invoked to understand these new fields. If the action (7.114) is to be real, then $\mathcal{L}_c^*$ must be $\mathcal{L}_c$. The properties of Grassmann variables derived in Sec. 5.1 show that under complex conjugation the action (7.114) becomes

$$\mathcal{L}_c^* = -\partial^\mu c^*\, \partial_\mu \bar{c}^* = \partial_\mu \bar{c}^*\, \partial^\mu c^*. \quad (7.115)$$

In order for the action to be real, one of the Grassmann variables must be pure imaginary while the other is pure real. The Grassmann variable $\bar{c}$ will be chosen to have the property $\bar{c}^* = -\bar{c}$, while $c^* = c$. According to the action both fields satisfy the equation of motion

$$\Box c = \Box \bar{c} = 0. \quad (7.116)$$

Therefore, they can be given the decomposition

$$c(x) = \int \frac{d^3k}{(2\pi)^{3/2}} \frac{1}{\sqrt{2\omega_k}} \left[c(\mathbf{k})e^{-ikx} + c^\dagger(\mathbf{k})e^{ikx} \right] ,$$

$$\bar{c}(x) = \int \frac{d^3k}{(2\pi)^{3/2}} \frac{1}{\sqrt{2\omega_k}} \left[\bar{c}(\mathbf{k})e^{-ikx} - \bar{c}^\dagger(\mathbf{k})e^{ikx} \right] . \quad (7.117)$$

The forms for $c(x)$ and $\bar{c}(x)$ as well as the quantization relations for their Fourier components are dictated by three facts. First, $c(x)$ must be a real Grassmann variable, while $\bar{c}(x)$ must be pure imaginary. Second, $\partial_o c$ is the momentum canonically conjugate to $\bar{c}$, while $-\partial_o \bar{c}$ is the momentum canonically conjugate to c. Third, these fields must be quantized with equal-time anticommutation relations since they are Grassmann variables in the path integral. The ETAR are satisfied by the following anticommutation relations, along with their Hermitian conjugates,

$$\{c(\mathbf{k}), c(\mathbf{p})\} = \{c(\mathbf{k}), \bar{c}(\mathbf{p})\} = \{c^\dagger(\mathbf{k}), c(\mathbf{p})\} = 0 ,$$
$$\{\bar{c}^\dagger(\mathbf{k}), c(\mathbf{p})\} = \{c^\dagger(\mathbf{k}), \bar{c}(\mathbf{p})\} = \delta^3(\mathbf{k} - \mathbf{p}) . \quad (7.118)$$

Exercise 7.12: Verify that the equal-time anticommutation relations $-\{c(\mathbf{x}, t), \partial_o \bar{c}(\mathbf{y}, t)\} = \{\bar{c}(\mathbf{x}, t), \partial_o c(\mathbf{y}, t)\} = i\delta^3(\mathbf{x} - \mathbf{y})$ are satisfied.

The algebra dictated by the canonical anticommutation relations (7.118) has the result that the states $c^\dagger|0\rangle$ and $\bar{c}^\dagger|0\rangle$ are both ghosts, since

$$\langle 0 | c(\mathbf{k}) c^\dagger(\mathbf{p}) | 0 \rangle = -\langle 0 | c^\dagger(\mathbf{p}) c(\mathbf{k}) | 0 \rangle = 0 , \quad (7.119)$$

where the standard creation and annihilation roles have been assumed. Using anticommutation relations for a scalar field is a breakdown of the spin and statistics theorem. In three spatial dimensions this can occur only in a Fock space with indefinite metric [12], and this has forced the Grassmann fields to be ghosts.

The Faddeev–Popov method [13] is an equivalent technique for quantizing gauge theories based on the results obtained so far, but a method that is easier to apply in most cases. The basic problem in quantizing any theory with a gauge degree of freedom is to apply a gauge condition to remove the ambiguity in the equations of motion. In the case of gauge field theories the simultaneous problem of $\pi_0 = 0$ is typically solved by mapping π_0 into the gauge condition. In the case of QED, the consistent application of the gauge condition leads to a path integral with the expected form for the action, but with the gauge volume factored and the gauge condition appearing in the measure. In addition, the determinant Δ_χ is present.

Faddeev and Popov demonstrated that these results could be obtained in the following way. The starting point is the path integral with the momenta integrated and the gauge fixing term absent,

$$_+\langle 0|0\rangle_- = \int_0^0 \mathcal{D}A\,\mathcal{D}\eta \, \exp\left\{i\int d^4x\, \mathcal{L}(\eta, A)\right\} . \tag{7.120}$$

Such a path integral is invariant under the full set of gauge transformations, and so the argument goes, overcounts the number of field configurations. The gauge volume must be factored out, but in a manner consistent with the gauge condition $\chi = 0$. In order to accomplish this the following factor of unity,

$$\Delta_\chi \int \mathcal{D}\Lambda\, \delta(\chi^\Lambda) = 1 , \tag{7.121}$$

is inserted into the measure, where the measure in (7.121) is understood in the continuum sense,

$$\mathcal{D}\Lambda = \prod_x d\Lambda(x) . \tag{7.122}$$

Relation (7.121) simply serves to define the quantity Δ_χ. The notation χ^Λ represents the form of the gauge condition χ after a gauge transformation parameterized by the function Λ. Obviously, there is a Λ for every gauge degree of freedom, and it is assumed there is a gauge condition for every gauge degree of freedom. It is also assumed that there is an inverse gauge transformation that undoes the gauge transformation Λ. Because the measure and the action are invariant under this inverse gauge transformation it is possible to transform χ^Λ back to χ for all values of Λ, and this allows the path integral to be written

$$\begin{aligned}_+\langle 0|0\rangle_- &= \int_0^0 \mathcal{D}A\,\mathcal{D}\eta\, \Delta_\chi\, \mathcal{D}\Lambda\, \delta(\chi^\Lambda)\, \exp\left\{i\int d^4x\, \mathcal{L}(\eta, A)\right\} \\ &= \int \mathcal{D}\Lambda \int_0^0 \mathcal{D}A\,\mathcal{D}\eta\, \Delta_\chi\, \delta(\chi)\, \exp\left\{i\int d^4x\, \mathcal{L}(\eta, A)\right\} . \end{aligned} \tag{7.123}$$

A critical point is that Δ_χ is invariant under the inverse gauge transformations, and this follows since the right-hand side of

$$\Delta_\chi^{-1} = \int \mathcal{D}\Lambda\, \delta(\chi^\Lambda) \tag{7.124}$$

is invariant under gauge transformations because the integral runs over all gauge configurations. This is manifestly clear in QED, where a gauge transformation on χ may be compensated for by a simple translation of each variable of integration in (7.124). The generalization of a measure over the gauge space, known as Haar measure, to the case of Yang-Mills fields requires some discussion, but this must await further developments.

The factor of the gauge volume has appeared, and now all that remains to calculate is Δ_χ. This is accomplished by the change of variables from Λ to χ^Λ, so that

$$\Delta_\chi^{-1} = \int \mathcal{D}\Lambda\, \delta(\chi^\Lambda) = \int \mathcal{D}\chi^\Lambda\, \det\left[\frac{\delta\Lambda}{\delta\chi^\Lambda}\right] \delta(\chi^\Lambda)\,. \qquad (7.125)$$

where the determinant is simply the Jacobian of the change of variables. By virtue of the Dirac delta this integration is quite easy, yielding

$$\Delta_\chi = \det\left[\frac{\delta\chi^\Lambda}{\delta\Lambda}\right]\bigg|_{\chi^\Lambda=0}. \qquad (7.126)$$

In the case of the Lorentz condition, in Euclidean space it follows that

$$\chi^\Lambda = -\partial_\mu A^\mu + \Box_E \Lambda \;\Rightarrow\; \Delta_\chi = \det \Box_E\,, \qquad (7.127)$$

which is the result obtained earlier. The factor Δ_χ is known as the Faddeev-Popov determinant, and the related ghost variables introduced in (7.113), $\bar{c}$ and c, are known as Faddeev-Popov ghosts. After integrating the momenta the final form for the path integral in an arbitrary Lorentz gauge is given by

$$_+\langle 0|0\rangle_- = \int \mathcal{D}\Lambda \int_0^0 \mathcal{D}\bar{c}\,\mathcal{D}c\,\mathcal{D}A\,\mathcal{D}\eta\, \delta(\partial_\mu A^\mu)\, \exp\left\{i\int d^4x\, (\mathcal{L}\right.$$
$$\left. - \tfrac{1}{2}\kappa\, (\partial_\mu A^\mu)^2 - \partial_\mu \bar{c}\, \partial^\mu c)\right\}\,, \qquad (7.128)$$

where terms involving $\partial_\mu A^\mu$ can be added freely to the action due to the presence of the Dirac delta in the path integral measure. This is the same result obtained by applying Gauss's law as a secondary constraint.

It would be remiss not to evaluate the path integral using the implementation of the constraint measure given by (7.87) and compare the result to the form of the path integral already obtained. It is straightforward to integrate the momenta from the path integral, in the absence of the gauge fixing term, to obtain

$$_+\langle 0|0\rangle_- = \int \mathcal{D}\eta\,\mathcal{D}A\, \delta(A_0 - A_0^c)\, \exp\left\{i\int d^4x\, \mathcal{L}(\eta,A)\right\}\,, \qquad (7.129)$$

Sec. 7.2 QED as a Path Integral

where A_0^c is given by (7.96). The A_0 variable can now be trivially integrated out of the path integral. This variable occurs in two terms in the action, $-\frac{1}{2}F_{0i}F^{0i}$ and $-J_0 A_0$. Upon insertion of A_0^c, the former term becomes

$$-\tfrac{1}{2}F_{0i}F^{0i} = \tfrac{1}{2}\dot{A}_i^T \dot{A}_i^T - \tfrac{1}{2}J_0\left(\frac{1}{\nabla^2}\right)J_0 \ . \qquad (7.130)$$

where A_i^T is the transverse part of A_i, obeying $\partial_i A_i^T = 0$. Throughout this and the following integration by parts will be valid since these terms are in the action. The second term becomes

$$-J_0 A_0 = -J_0\left(\frac{1}{\nabla^2}\right)\partial_j \dot{A}_j + J_0\left(\frac{1}{\nabla^2}\right)J_0 \ . \qquad (7.131)$$

The first term on the right-hand side of (7.131) cancels against the longitudinal part of the term $-J^i A_i$, which can be written

$$-J^i A_i = -J^i A_i^T + \partial_j J^j \left(\frac{1}{\nabla^2}\right)\partial_i A_i \ . \qquad (7.132)$$

Combining the two terms involving longitudinal parts of the vector field yields

$$\partial_j J^j \left(\frac{1}{\nabla^2}\right)\partial_j A_j - J_0\left(\frac{1}{\nabla^2}\right)\partial_j \dot{A}_j = \partial_\mu J^\mu \left(\frac{1}{\nabla^2}\right)\partial_j A_j \ . \qquad (7.133)$$

These terms can be discarded since it can be shown that

$$\int_0^0 \mathcal{D}\eta\, \mathcal{D}A\, \partial_\mu J^\mu(y) \exp\left\{i \int d^4x\, \mathcal{L}(\eta, A)\right\} = 0 \ . \qquad (7.134)$$

Exercise 7.13: The technique for verifying that Noether's theorem holds at the quantized level will be detailed in Chapter 8. In anticipation of this, adapt the techniques of Sec. 4.3 to verify (7.134).

The remainder of the gauge field action is transverse, since

$$F_{ij} = \partial_i A_j - \partial_j A_i = \partial_i A_j^T - \partial_j A_i^T \ . \qquad (7.135)$$

As a result, the path integral becomes

$$_+\langle 0|0\rangle_- =$$
$$\int_0^0 \mathcal{D}\eta\, \mathcal{D}\mathbf{A}\, \exp\left\{i \int d^4x \left[\mathcal{L}(\eta, \mathbf{A}^T) - \tfrac{1}{2}J_0\left(\frac{1}{\nabla^2}\right)J_0\right]\right\} \ . \qquad (7.136)$$

262 Gauge Field Theory

Only transverse parts of **A** appear in the action, while A_0 has disappeared to be replaced by the Coulomb interaction of J_0, the charge density. However, the measure contains an integration over all three spatial components of **A**. Therefore, the measure breaks into an integration over the two transverse parts of **A**, which are physically meaningful since they occur in the action, and an integration over the longitudinal part. The latter factorizes out of the path integral since there is no occurrence of the longitudinal degree of freedom in the action, and this factorization reproduces the gauge volume that was removed by other means in the covariant approach. The final result (7.136) is precisely the Coulomb gauge path integral for the coupled theory. Because all fields appearing in the action are transverse, any presence of the gauge constraint $\delta(\nabla \cdot A)$ would be completely redundant.

The two forms of the path integral, (7.128) and (7.136), are entirely equivalent, although at first glance the covariant version appears to have twice as many components for the gauge field and two additional degrees of fermionic freedom in the form of the Faddeev–Popov ghosts. The reason for this has already been discussed. The ghosts in the gauge sector of covariant versions of QED are not dangerous because their sole coupling is to a conserved current. Therefore, their presence can be tolerated and their contributions to any physical process, i.e., one where there are equal numbers of longitudinal and timelike photons in either the in or out state, will cancel. This is reflected in the fact that the Faddeev–Popov ghosts in QED are free, totally decoupled from the theory. The presence of Faddeev–Popov ghosts is precisely to cancel any terms that threaten unitarity, and in a covariant gauge this means countering any terms generated by the negative and zero norm states available to the theory. Since this problem does not really occur in QED the Faddeev–Popov ghosts have no real function. Although the determinant Δ_χ found by the Faddeev–Popov method is fairly uninteresting for QED, this will not be the case for other theories. It is possible to view the Faddeev–Popov ghosts as negative degrees of freedom since their function is to cancel the ghosts present in the gauge sector. Using this reasoning shows that the two forms of the path integral, the Lorentz gauge and Coulomb gauge versions, have equal numbers of degrees of freedom.

At a more subtle level of inquiry it is worth pondering why the Faddeev–Popov determinant obtained by transforming the gauge condition reproduces the measure found by applying the secondary constraint of Gauss's law. It will be proved that the two methods are equivalent for a fairly unrestrictive set of circumstances; however, to make the discussion more interesting requires the generalization of gauge transformations to the nonabelian case. The proof of equivalence is given in Sec. 7.5.

7.3 Lie Algebras

In the next section the QED gauge field analysis of the previous section will be applied to the Yang–Mills gauge field. The key step in constructing the Yang–Mills field is the generalization of the concept of a gauge transformation to include *nonabelian* cases. Rudimentary aspects of nonabelian groups were presented in Sec. 6.2, where the Lorentz group was introduced, but now a more general discussion of the types of groups and their associated algebras must be presented [14, 15]. Limitations of space prevent an adequate treatment, and it is hoped that the presentation will at least clarify notation.

Most groups may be understood in terms of classes of square matrices. For example, the set of $n \times n$ matrices that are unitary and possess determinant one forms a group known as the *special unitary group*, $SU(n)$, while the set of $n \times n$ orthogonal matrices with determinant one forms the *special orthogonal group*, $SO(n)$. Relaxing the restriction on the value of the determinant to nonzero values defines the groups $U(n)$ and $O(n)$.

> **Exercise 7.14**: Show that the members of $SO(n)$ with real entries are characterized by $\frac{1}{2}n(n-1)$ free parameters, while $SU(n)$ has $n^2 - 1$ free parameters.

The number of parameters needed to characterize a group element is known as the *order* of the group. In both of the previous cases the parameters could vary continuously without removing the matrix from the group. For this reason these groups are referred to as *continuous groups*. Other continuous groups of importance in physics are the *symplectic groups*, $Sp(2n, C)$, and the *exceptional* groups, of which there are only five. The symplectic group is a $2n \times 2n$ complex matrix that acts on the $2n$ dimensional representation space $x = (x_1, \ldots, x_n, x'_1, \ldots, x'_n)$ in such a way that it preserves the antisymmetric product of two vectors, $x_i y'_i - x'_i y_i$. The symplectic group $Sp(2n, C)$ has $2n(2n+1)$ free parameters. The exceptional groups are not readily defined in terms of a class of matrices. The five exceptional groups are denoted G_2, F_4, E_6, E_7, and E_8. These groups have 14, 52, 78, 133, and 248 parameters, respectively.

The space of allowed values for the parameters of a group defines a manifold. If this manifold is compact, then the group itself is said to be *compact*. The Lorentz group discussed in Sec. 6.2 is not compact, while the group $SU(n)$ is compact. Although not necessary, it is standard practice to associate the group element with all parameters set to zero with the identity element of the group. Therefore, a member $A(\alpha)$ of some group, where $\alpha = (\alpha_1, \ldots, \alpha_p)$ are the p parameters of the group, can be expanded in the vicinity of the identity,

$$A(\alpha) \approx A(0) + i\alpha_k \left.\frac{\partial A}{\partial \alpha_k}\right|_{\alpha=0} \equiv I + i\alpha_a T^a . \qquad (7.137)$$

The p matrices T^a are known as the *generators* of the group, and their multiplication table defines the *local* properties of the group. The number of generators must match the order of the group. The presence of i in (7.137) is for later convenience.

If $A(\alpha)$ and $A(\beta)$ are two elements both close to the identity, then by the group closure property there must be two other elements of the group such that $A(\alpha)A(\beta) = A(\gamma)$, where $\gamma = \gamma(\alpha, \beta)$, and $A(\beta)A(\alpha) = A(\gamma')$, where $\gamma' = \gamma(\beta, \alpha)$. These two products should also be close to the identity by the assumption of continuity. By the definition of the identity it follows that $\gamma(0,0) = 0$. The function γ can therefore be expanded in a Taylor series in the vicinity of the identity to obtain

$$\gamma_c(\alpha, \beta) \approx \alpha_a \left.\frac{\partial \gamma_c}{\partial \alpha_a}\right|_{\alpha,\beta=0} + \beta_a \left.\frac{\partial \gamma_c}{\partial \beta_a}\right|_{\alpha,\beta=0} + \alpha_a \beta_b \left.\frac{\partial^2 \gamma_c}{\partial \alpha_a \partial \beta_b}\right|_{\alpha,\beta=0} . \qquad (7.138)$$

To first order in α it is true that $A^{-1}(\alpha) = A(-\alpha)$, and this shows that

$$\gamma_c(\alpha, -\alpha) \approx \alpha_a \left.\left(\frac{\partial \gamma_c(\alpha, \beta)}{\partial \alpha_a} - \frac{\partial \gamma_c(\alpha, \beta)}{\partial \beta_a}\right)\right|_{\alpha,\beta=0} = 0 , \qquad (7.139)$$

and this can be true only if

$$\left.\frac{\partial \gamma_c(\alpha, \beta)}{\partial \alpha_a}\right|_{\alpha,\beta=0} = \left.\frac{\partial \gamma_c(\alpha, \beta)}{\partial \beta_a}\right|_{\alpha,\beta=0} . \qquad (7.140)$$

If the commutator of $A(\gamma)$ and $A(\gamma')$ is evaluated using the infinitesimal forms (7.137), then

$$[A(\alpha), A(\beta)] = A(\gamma) - A(\gamma') \Rightarrow$$
$$\alpha_a \beta_b [T^a, T^b] = -i(\gamma_c - \gamma'_c)T^c = -i[\gamma_c(\alpha, \beta) - \gamma_c(\beta, \alpha)]T^c . \qquad (7.141)$$

Using the expansion for γ and the identity (7.140) it follows that, for the infinitesimal case, it is possible to write

$$\gamma_c(\alpha, \beta) - \gamma_c(\beta, \alpha) = \left.\left(\frac{\partial^2 \gamma_c(\alpha, \beta)}{\partial \alpha_a \partial \beta_b} - \frac{\partial^2 \gamma_c(\alpha, \beta)}{\partial \alpha_b \partial \beta_a}\right)\right|_{\alpha,\beta=0} \alpha_a \beta_b$$
$$\equiv -f^c_{ab} \alpha_a \beta_b , \qquad (7.142)$$

Sec. 7.3 Lie Algebras   265

so that the f^c_{ab} are a set of constants that characterize the group composition property. Placing this back into (7.141) immediately yields the definition of a *Lie algebra*:

$$[T^a, T^b] = i f^c_{ab} T^c \ . \tag{7.143}$$

The f^c_{ab} are known as the *structure constants* of the algebra and are antisymmetric, $f^c_{ab} = -f^c_{ba}$. Since $\text{Tr}\,[T^a, T^b] = 0$ by cyclicity of the trace, all the generators must be traceless.

> **Exercise 7.15**: Show that the generators of the group SU(2) are σ_j, the Pauli matrices, and that the structure constants are $f^i_{jk} = 2\varepsilon_{ijk}$.

While the idea of a Lie algebra has its origin in the continuous groups defined by matrices, it should be viewed as an independent concept. The generators of a Lie algebra may be thought of as a set of operators acting on a representation space. Once a specific form for the representation space is chosen the generators may be cast into matrix form by using the inner product present in the representation space, and this set of matrix generators can be used to reproduce the algebra of one of the classical groups. This is analogous to the representation of quantum mechanical operators in matrix form. For example, the three angular momentum operators familiar from quantum mechanics form a Lie algebra and have many different representations depending upon the representation space they are rotating or *transforming*. For that reason a set of matrices satisfying the Lie algebra is referred to as a *representation* of that algebra. It comes as no surprise that different matrix groups may share a common algebra. The most familiar example of this in physics is the identity between the algebra of SO(3), representing the orbital angular momentum, and the algebra of SU(2), representing the spin angular momentum of the electron. This common algebra means that the behavior of the groups SU(2) and SO(3) are identical *local* to the identity, and that differences between the two must be *global* properties, visible only when finite elements are considered.

It follows by substituting (7.143) into the identity

$$\big[[T^a, T^b], T^c\big] + \big[[T^b, T^c], T^a\big] + \big[[T^c, T^a], T^b\big] = 0 \tag{7.144}$$

that the structure constants must satisfy the *Jacobi identity*:

$$f^a_{bc} f^d_{ea} + f^a_{ce} f^d_{ba} + f^a_{eb} f^d_{ca} = 0 \ . \tag{7.145}$$

The Jacobi identity guarantees that every Lie algebra has at least one representation. By defining the set of matrices T^a whose elements are given by $T^a_{(bc)} = -if^c_{ab}$, the identity (7.145) reproduces the algebra (7.143). This is called the *adjoint representation*.

The generators allow the finite elements, ones where the parameters are not close to zero, to be built up from repeated application of infinitesimal elements. If Λ_a is the finite parameter, then it may be reached by N applications of the infinitesimal element associated with Λ_a/N. In the limit that $N \to \infty$, this yields the finite element U

$$U \equiv \lim_{N \to \infty} (I + \frac{i}{N}\Lambda_a T^a)^N = \exp i\Lambda_a T^a . \qquad (7.146)$$

If the matrices representing the T^a are Hermitian, then (7.146) will be a unitary matrix. The formal resemblance to the unitary transformations familiar from quantum theory is not accidental since groups and their algebras are associated with continuous sets of transformations on the representation space.

Among the set of all generators of an algebra there will be a *maximal* subset whose members commute among themselves, so that the subset forms an abelian subalgebra. It may be that there is only one member of this set, as in the case of the angular momentum operators associated with the algebra SO(3) or the spin operators associated with SU(2). This maximal subset of generators is called the *Cartan subalgebra*. Because all members of this subalgebra commute, a representation space may be chosen so that every state in the representation space is an eigenstate of these generators. This means that the states of the representation space are characterized by the eigenvalues of the members of the Cartan subalgebra. If the representation space is a quantum mechanical Hilbert space, then the eigenvalues of the Cartan subalgebra become quantum numbers associated with the states, and the Cartan subalgebra may be added to the set of observables. The number of generators in the Cartan subalgebra is referred to as the *rank* of the Lie algebra, and it follows that each state in the representation space is characterized by a set of eigenvalues whose number matches the rank, but whose range is determined by the specific properties of the Lie algebra.

The members of the Cartan subalgebra are denoted H^a. If the rank of the algebra is r, then the abstract representation space may be written $|\mu_1, \ldots, \mu_r\rangle$, where μ_a is one of the eigenvalues of H^a. However, it will usually be the case that not all possible combinations of the eigenvalues appear in the allowed states of the representation space. An allowed eigenvalue combination is known as a *weight*. To see this structure, it is necessary to consider an algebra that is *not* rank one. For example, SU(3) is a rank 2 algebra, and in its smallest representation consists of $3^2 - 1 = 8$

3×3 matrices. An explicit representation of the Cartan subalgebra is given by

$$H_1 = \begin{pmatrix} 1 & 0 & 0 \\ 0 & -1 & 0 \\ 0 & 0 & 0 \end{pmatrix}, \quad H_2 = \frac{1}{\sqrt{3}} \begin{pmatrix} 1 & 0 & 0 \\ 0 & 1 & 0 \\ 0 & 0 & -2 \end{pmatrix}, \quad (7.147)$$

while the remaining matrices are given, in the Gell-Mann representation, by

$$\lambda_1 = \begin{pmatrix} 0 & 1 & 0 \\ 1 & 0 & 0 \\ 0 & 0 & 0 \end{pmatrix} \quad \lambda_2 = \begin{pmatrix} 0 & -i & 0 \\ i & 0 & 0 \\ 0 & 0 & 0 \end{pmatrix} \quad \lambda_3 = \begin{pmatrix} 0 & 0 & 1 \\ 0 & 0 & 0 \\ 1 & 0 & 0 \end{pmatrix},$$

$$\lambda_4 = \begin{pmatrix} 0 & 0 & -i \\ 0 & 0 & 0 \\ i & 0 & 0 \end{pmatrix} \quad \lambda_5 = \begin{pmatrix} 0 & 0 & 0 \\ 0 & 0 & 1 \\ 0 & 1 & 0 \end{pmatrix} \quad \lambda_6 = \begin{pmatrix} 0 & 0 & 0 \\ 0 & 0 & -i \\ 0 & i & 0 \end{pmatrix}. \quad (7.148)$$

This particular representation is characterized by $\mathrm{Tr}\, T^a T^b = 2\delta_{ab}$. The allowed weights are read off from the diagonals of the Cartan subgroup as

$$\left(1, \frac{1}{\sqrt{3}}\right), \quad \left(-1, \frac{1}{\sqrt{3}}\right), \quad \left(0, -\frac{2}{\sqrt{3}}\right). \quad (7.149)$$

Since the SU(3) algebra closes, it must be that the other six generators "raise" and "lower" the weights through the allowed set of (7.149). Therefore, it must be possible to extend to SU(3) the same treatment that is accorded to angular momentum in quantum mechanics, where one generator, typically J_z, is diagonalized, and the remaining two operators are written $J_+ = J_x + iJ_y$ and $J_- = J_x - iJ_y$. The latter two operators raise and lower the eigenvalue of J_z, which is the one member of the Cartan subalgebra of SO(3) or SU(2). The allowed weights of these two algebras are simply all possible eigenvalues of J_z, and the weight space is one dimensional. In the case of SU(3) the weight space is two dimensional, and the remaining generators supply the difference between the allowed weights. These differences are known as *roots* and are found by treating the weights as vectors and finding all nonzero differences between them. For the case of SU(3) there are three weights, and therefore there are six roots, given by

$$\left(1, -\sqrt{3}\right), \quad \left(1, \sqrt{3}\right), \quad (2, 0), \quad (7.150)$$

along with their three negative counterparts. By construction there will always be an equal number of positive and negative roots. By discarding all roots whose first nonzero entry is a negative number, one is left with the *positive roots*, and for SU(3) these are given by (7.150). The final step is to exclude all positive roots that can be written as the sum, as opposed to the difference, of two other positive roots. In the case of SU(3) this removes the third root of (7.150). The remaining roots,

$$\left(1, \sqrt{3}\right), \quad \left(1, -\sqrt{3}\right), \tag{7.151}$$

are referred to as the *simple roots* of SU(3).

Such a treatment is possible for all Lie algebras of importance in physics. In the Cartan basis for a Lie algebra the generators not belonging to the Cartan subalgebra are denoted E_α, where α is a positive root with components α_a. For each generator E_α there is a generator, $E_{-\alpha}$, corresponding to the negative root. The commutation relations between H^a and $E_{\pm\alpha}$ are the multidimensional extensions of the relationship between J_z and $J_\pm$,

$$[H^a, E_\alpha] = \alpha_a E_\alpha . \tag{7.152}$$

The E_α satisfy

$$[E_\alpha, E_\beta] = F(\alpha, \beta) E_{\alpha+\beta}, \quad \alpha \neq -\beta, \tag{7.153}$$

where $F(\alpha, \beta)$ is some number depending upon the two roots, while

$$[E_\alpha, E_{-\alpha}] = \alpha_a H^a . \tag{7.154}$$

Once the allowed weights of the group are known from the Cartan subalgebra, the identification of all the roots is straightforward. From the set of all roots it is always possible to find the the subset of the α that corresponds to the simple roots.

It is the study of the root structure, and in particular the simple roots, that allows the complete classification of Lie algebras, and theorems abound. It is possible to show that the number of simple roots for a Lie algebra matches the rank r of the algebra. Furthermore, it is possible to classify the simple roots by the angle between them in the r-dimensional weight space. Given two simple roots, α and β, this angle is given by

$$\theta = \arccos\left(\frac{\alpha \cdot \beta}{|\alpha||\beta|}\right) . \tag{7.155}$$

For the case of SU(3) there are only the two two-dimensional simple roots of (7.151), and the angle between them is given by $\theta = \arccos(-\frac{1}{2}) = 2\pi/3$. It can be shown that the angle between the simple roots of a Lie algebra must be one of the set $(\pi/2, 2\pi/3, 3\pi/4, 5\pi/6)$, and that no other angles are possible. The graphical expression of the relative lengths and

Sec. 7.3 Lie Algebras 269

the angles between the simple roots is called the *Dynkin diagram* of the algebra. The study of all possible Dynkin diagrams, or equivalently root structures, allows the identification of all possible Lie algebras. It was by the study of root structure that the existence of the exceptional groups was first demonstrated. Unfortunately, further discussion of the root structure of Lie algebras or a detailed discussion of Dynkin diagrams is beyond the scope of this book.

Given a representation of a Lie algebra in the Cartan basis, it is always possible to transform by a unitary transformation to a basis where the generators satisfy

$$\text{Tr}\,(T^a T^b) = \tfrac{1}{2}\delta_{ab}\,, \tag{7.156}$$

if the algebra is *semisimple*. There are a number of ways to determine if an algebra is semisimple. The simplest way is to examine the *metric tensor* of the algebra. The metric tensor g_{ab} is found from the structure constants of the algebra by

$$g_{ab} = g_{ba} = f^c_{ad} f^d_{bc}\,. \tag{7.157}$$

The algebra is semisimple if and only if $\det g \neq 0$. In effect, if the algebra is semisimple it means that there is no generator that commutes with all other generators, although proving this requires some effort. It can be shown that $SU(n)$, $SO(n)$, $Sp(2n, C)$, and the exceptional groups are all semisimple.

Restricting attention to compact semisimple algebras has many advantages. For example, the structure constants of a compact semisimple algebra must obey

$$f^a_{bc} = \tfrac{1}{2}\text{Tr}\,(f^d_{bc} T^d T^a) = -\tfrac{1}{2}i\text{Tr}\,([T^b, T^c]T^a)\,. \tag{7.158}$$

From the invariance of the trace under cylic permutations, it follows that the structure constants of a semisimple algebra satisfy $f^a_{bc} = f^c_{ab} = f^b_{ca}$, which shows that the structure constants are completely antisymmetric in all indices. Furthermore, it can be shown that all irreducible representations of compact semisimple algebras possess Hermitian generators. This latter property is easily seen in the case of the adjoint representation by using the cyclicity of the structure constants. There, the matrix representing the generator is given by $T^a_{(bc)} = -if^c_{ab} = -if^a_{bc}$, which is a Hermitian matrix by virtue of the antisymmetry of f^a_{bc}.

Once the metric tensor is known, the *Casimir operator*, defined by $C = g_{ab}T^a T^b$, can be found.

Exercise 7.16: Prove that the Casimir operator commutes will all elements of the Lie algebra.

270   Gauge Field Theory

The proofs of these statements, as well as many other important properties of Lie algebras and their representations, may be found in any of the standard texts on the subject [14, 15].

7.4 Classical Yang–Mills Fields

In this section the generalization of the QED gauge field is presented. The basic idea is to generalize the gauge transformation of QED to the non-abelian case. The gauge transformation will be associated with some Lie algebra, and the representation space will be a set of matter fields, commonly referred to as a *multiplet*.

The starting point is the physical observation that many of the particles appearing in nature share similar properties. The most obvious cases are particles that share similar mass and spin. The structure of the Dirac field and its quantization has already necessitated placing a spinor particle and its antiparticle, which has the same mass and spin but opposite electric charge, into the same multiplet. Long before the standard model appeared, such similar particles were often placed into multiplets whose components are labelled by some new index, usually referred to as an *isospin* index, which reflects the existence of an *internal* symmetry in the system. Interactions between the multiplets, in the form of terms nonlinear in the fields, can be constructed to be invariant under *global* isospin transformations. In what follows the spinor and Lorentz indices will be suppressed for notational simplicity. If ψ^k is a field in the multiplet, then an isospin transformation takes the form $\psi^j \to U_{jk}\psi^k$, where $\mathbf{U}$ is an arbitrary member of some group whose representation space is the multiplet. Typical early applications of this idea placed the neutron and proton into an SU(2) multiplet [16], and the lower mass hadrons into an SU(3) multiplet [17]. It was known that these must be, at best, approximate symmetries because of the mass differences between the particles. However, experimental evidence showed that the couplings of the members of the multiplet to other particles are very nearly identical. Once the symmetry was postulated, Noether's theorem then allowed the calculation of generalized charges associated with the approximate symmetry, and the quasi-conservation of these charges and their associated quantum numbers was used to make predictions about particle dynamics.

The transformation $\mathbf{U}$ considered so far is global in the sense that it takes the same form regardless of the space-time argument of the fields. Yang and Mills [18] took the step of constructing actions that are invariant for the case that the parameters appearing in the form of **U** were space-time dependent functions. This is the nonabelian version of the gauge transformation familiar from QED. There, the gauge transformation takes the form

$$\psi(x) \to e^{ie\Lambda(x)}\psi(x) \,. \tag{7.159}$$

Such a transformation is abelian in the sense that the order of two gauge transformations is irrelevant to the final phase of the transformed field. In the terminology of Lie algebras, the gauge transformations at each space-time point form a representation of the abelian group U(1), the set of all unitary transformations with one parameter. The group U(1) is simply the set of all possible phases.

The gauge transformation of QED is the global U(1) symmetry, which guarantees a conserved charge, elevated to a *local* U(1) symmetry by allowing the single parameter of U(1) to become space-time dependent. It may be remembered from the discussion of the complex scalar field that a theory may be globally U(1) or phase invariant without the presence of the gauge field. However, in order for a theory to be locally U(1) invariant it is necessary to introduce an additional field, the abelian gauge field of QED. Furthermore, imposing this local symmetry severely restricts the possible forms of the coupling between the matter and gauge field that can be present. In the same way, Yang–Mills gauge fields must be present when a nonabelian symmetry is elevated from a global to a local symmetry, and the interactions consistent with this local symmetry are quite restricted in form.

For the remainder of this book only compact semisimple algebras will be considered. This has the advantage that there is a basis satisfying (7.156). It can also be shown that any irreducible representation of the generators takes the form of Hermitian matrices with finite dimension. This means that the finite transformations of (7.146) are all *unitary* matrices. If **U** is an element of some irreducible representation of the compact semisimple Lie algebra G, then **U** takes the form $\mathbf{U} = \exp ig\Lambda^a T^a$ in terms of the parameters Λ^a and the generators T^a. The real number g is a coupling constant, the nonabelian counterpart of the electric charge.

For the sake of simplicity, a specific case will be considered. A multiplet of massless spinor fields, the elements of which are denoted ψ^j, can be used to construct an action invariant under global transformations. Obviously, the dimension of the multiplet must match the dimension of the representation. The transformation $\psi \to \mathbf{U}\psi \Rightarrow \bar\psi \to \bar\psi \mathbf{U}^\dagger$. For the semisimple algebras $\mathbf{U}^\dagger \mathbf{U} = \mathbf{I}$, so that the action $i\bar\psi^j \slashed\partial \psi^j$ is both Lorentz invariant and invariant under transformations satisfying $\partial_\mu \mathbf{U} = 0$. The obvious step in elevating this to a local symmetry is to allow the group parameters to become space-time dependent, so that $\partial_\mu \Lambda^a \neq 0 \Rightarrow \partial_\mu \mathbf{U} \neq 0$. The action as it stands is no longer invariant, and under this local transformation generates an additional term

$$i\bar{\psi}^j \not{\partial} \psi^j \to i\bar{\psi}^j \not{\partial} \psi^j + i\bar{\psi}^j \gamma^\mu U_{jk}^{-1} \partial_\mu U_{kl} \psi^l . \tag{7.160}$$

It is important to remember that the γ matrices act solely upon the spinor indices, which are suppressed, while $\mathbf{U}$ acts upon the isospin index, which will usually be suppressed in the notation from this point forward as well.

The nonabelian result (7.160) is quite different from the abelian case of U(1). Because of the nonabelian nature of the algebra it is not possible to write $\partial_\mu \mathbf{U} = i\partial_\mu \Lambda^a T^a \mathbf{U}$. Motivated by the abelian case, the obvious step in regaining invariance of the action is to introduce the space-time dependent matrices $A_\mu(x)$ with the same dimension as $\mathbf{U}$. The combination

$$\mathcal{L} = i\bar{\psi}\gamma^\mu (\partial_\mu + igA_\mu)\psi \tag{7.161}$$

is once again invariant if the A_μ matrix is simultaneously transformed according to

$$A'_\mu = \mathbf{U} A_\mu \mathbf{U}^{-1} + \frac{i}{g}(\partial_\mu \mathbf{U})\mathbf{U}^{-1} . \tag{7.162}$$

Exercise 7.17: Verify (7.162).

For the case that $\mathbf{U}$ is a member of U(1), and therefore takes the form $\mathbf{U} = \exp ig\Lambda$, the familiar gauge field transformation law of QED emerges from (7.162). When $\mathbf{U}$ is infinitesimal, $\mathbf{U} \approx 1 + ig\Lambda^a T^a$, (7.162) becomes

$$A'_\mu \approx A_\mu + ig\Lambda^a [T^a, A_\mu] - \partial_\mu \Lambda^a T^a + O(\Lambda^2) . \tag{7.163}$$

In the limit that $g \to 0$ the commutator in (7.163) is suppressed, so that this limit corresponds to a set of copies of the abelian case, equal in number to the order of the group. Examining the third term in (7.163), which resembles the standard abelian gauge transformation, shows that in the abelian limit A_μ must have the expansion

$$A_\mu = T^a A^a_\mu , \tag{7.164}$$

in order to complete the abelian analogy. The gauge field A^a_μ is then defined as

$$A^a_\mu = 2\operatorname{Tr}(T^a A_\mu) . \tag{7.165}$$

Therefore, the number of gauge fields matches the order of the group. By using the Lie algebra definition (7.143) in (7.163) the infinitesimal form of the gauge transformation reduces to

$$A_\mu^a{}' T^a = A_\mu^a T^a - \partial_\mu \Lambda^a T^a - g f_{ab}^c \Lambda^a A_\mu^b T^c , \qquad (7.166)$$

so that the field itself has the infinitesimal transformation property

$$A_\mu^a{}' = A_\mu^a - \partial_\mu \Lambda^a - g f_{bc}^a \Lambda^b A_\mu^c . \qquad (7.167)$$

Even at lowest order this transformation mixes the gauge fields and is therefore nonlinear.

As in the case of U(1) QED, it is necessary to construct an action for the gauge fields. This is facilitated by noting that a gauge covariant derivative D_μ can be defined in the nonabelian case in much the same way as it was for QED,

$$D_\mu = \partial_\mu + ig A_\mu . \qquad (7.168)$$

The covariant derivative has the property that

$$D_\mu' \psi' = D_\mu' \mathbf{U} \psi = \left(\partial_\mu + ig A_\mu' \right) \mathbf{U} \psi = \mathbf{U} D_\mu \psi . \qquad (7.169)$$

Another way of expressing (7.169) is to write

$$D_\mu' = \mathbf{U} D_\mu \mathbf{U}^{-1} . \qquad (7.170)$$

While (7.170) has been derived for the special case of the covariant derivative, any object that transforms in this manner is said to be *gauge covariant*. Using such objects it is very simple to construct gauge invariant structures by noting that the trace of any product of gauge covariant objects is gauge invariant,

$$\text{Tr}\,(A \cdots B) \to \text{Tr}\,(\mathbf{U} A \mathbf{U}^{-1} \cdots \mathbf{U} B \mathbf{U}^{-1}) = \text{Tr}\,(A \cdots B) . \qquad (7.171)$$

This property allows the action to be constructed from the covariant derivative in the following way. One begins by noting that in QED the $F_{\mu\nu}$ tensor can be written

$$\frac{1}{ig}[D_\mu, D_\nu] = \partial_\mu A_\nu - \partial_\nu A_\mu . \qquad (7.172)$$

Since the nonabelian gauge covariant derivative reduces to this form in the abelian limit, it is natural to form the same object for the nonabelian case. Since D_μ is gauge covariant its commutator is also guaranteed to be gauge covariant. The result is the nonabelian definition of the matrix $F_{\mu\nu}$:

$$F_{\mu\nu} = \frac{1}{ig}[D_\mu, D_\nu] = \partial_\mu A_\nu - \partial_\nu A_\mu + ig[A_\mu, A_\nu] . \qquad (7.173)$$

Exercise 7.18: Show that (7.173) is explicitly gauge covariant under the gauge transformation (7.162).

Result (7.171) shows that the quantity

$$\mathcal{L} = -\tfrac{1}{2}\text{Tr}\, F_{\mu\nu} F^{\mu\nu} \qquad (7.174)$$

is both Lorentz invariant and gauge invariant. It can be rewritten by performing the trace using the condition (7.156). The result is

$$\mathcal{L} = -\tfrac{1}{4} F^a_{\mu\nu} F^{a\mu\nu} , \qquad (7.175)$$

where the antisymmetric tensor $F^a_{\mu\nu}$ is given by

$$F^a_{\mu\nu} = \partial_\mu A^a_\nu - \partial_\nu A^a_\mu - g f^a_{bc} A^b_\mu A^c_\nu . \qquad (7.176)$$

This shows that the action (7.174) reduces to the abelian case in the limit $g \to 0$.

Exercise 7.19: Verify (7.175) and (7.176).

The demand for nonabelian gauge invariance has led to a minimal form for the gauge field action that is not quadratic, and therefore the gauge action itself already describes an interacting and nontrivial theory. The Euler–Lagrange equations of motion in the absence of matter coupling can be written

$$\partial_\mu F^{a\mu\nu} - g f^a_{bc} A^b_\mu F^{c\mu\nu} = 0 . \qquad (7.177)$$

It is convenient to use a notation that encapsulates the Lie algebra structure by an analogy to SO(3). If the Lie algebra were either SU(2) or SO(3), the structure constants would be ε_{abc}, and the second term in (7.177) would resemble a vector product in group space. It is easy to think of the second term as the generalization of this vector product to more complicated systems. As a first step the covariant derivative is written in terms of the group indices,

$$D^{ab}_\mu = \delta^{ab} \partial_\mu - g f^a_{cb} A^c_\mu . \qquad (7.178)$$

Using the analogy to the vector product, this covariant derivative will be written with the group indices suppressed as

$$D_\mu = \partial_\mu - g A_\mu \times , \qquad (7.179)$$

so that (7.177) becomes

$$\partial_\mu F^{\mu\nu} - gA_\mu \times F^{\mu\nu} = D_\mu F^{\mu\nu} = 0 , \qquad (7.180)$$

allowing a more economical notation. The infinitesimal form of the gauge transformation of (7.167) can now be written

$$A_\mu \to A_\mu - D_\mu \Lambda . \qquad (7.181)$$

The matter coupling present in (7.161) may be written in terms of the gauge field components (7.165). Explicitly displaying the indices on the spinor and gauge fields, this coupling takes the form

$$g\bar{\psi}\gamma^\mu A_\mu \psi = g\bar{\psi}^j T^a_{(jk)} \gamma^\mu \psi^k A^a_\mu \equiv j^{a\mu} A^a_\mu . \qquad (7.182)$$

The coupled equations of motion for the gauge field are given by

$$D_\mu F^{\mu\nu} = j^\nu \equiv j^{a\nu} T^a \qquad (7.183)$$

in the index suppressed form. The equations of motion for the spinor fields can be used to show that the matter currents j^a_μ are *not* conserved. However, it is possible to find the conserved current associated with gauge invariance by using the antisymmetry of the $F^a_{\mu\nu}$ and the equation of motion for the matter coupled theory. It follows that the current J^a_μ, defined as

$$J^{a\mu} \equiv \partial_\nu F^{a\nu\mu} = gf^a_{bc} A^b_\nu F^{c\nu\mu} + j^{a\mu} , \qquad (7.184)$$

satisfies $\partial_\mu J^{a\mu} = \partial_\mu \partial_\nu F^{a\mu\nu} = 0$ by virtue of the antisymmetry of $F^a_{\mu\nu}$.

Exercise 7.20: Using the infinitesimal form for gauge transformation, show that Noether's theorem gives J^a_μ as the conserved current associated with the nonabelian gauge invariance.

The matter current $j_\mu = j^a_\mu T^a$ transforms in a gauge covariant manner. This can be seen by transforming (7.183), which gives

$$j'_\nu = D'_\mu F^\mu{}_\nu{}' = \mathbf{U} D_\mu F^\mu{}_\nu \mathbf{U}^{-1} = \mathbf{U} j_\nu \mathbf{U}^{-1} . \qquad (7.185)$$

While not conserved, the matter current does satisfy the gauge covariant extension of the conservation law,

$$D_\mu j^\mu = \partial_\mu j^\mu - gA_\mu \times j^\mu = 0 . \qquad (7.186)$$

276   Gauge Field Theory

Under an infinitesimal gauge transformation j^a_μ transforms according to

$$j^a_\mu \rightarrow j^a_\mu + gf^a_{bc} j^b_\mu \Lambda^c . \tag{7.187}$$

> **Exercise 7.21**: Verify (7.186) and (7.187).

Unlike the matter field, the conserved current $J_\mu = J^a_\mu T^a$ is not gauge covariant. This dichotomy appears again when the gauge fields are attached to an external source current. Either the coupling to the source is gauge invariant while the source current is not conserved, or the source current is conserved while the coupling to the source is not gauge invariant. This is discussed again in Chapter 8.

It is not difficult to understand why the nonabelian matter currents are not conserved while the QED matter current is. The abelian gauge field carries no electric charge, and therefore all charge resides in the matter sector of the theory. This is not so in the nonabelian case, where the gauge fields, through self-coupling, carry the nonabelian charge just as the matter current does. Therefore, it is a combination of matter and gauge field configurations that represents the conserved flow of nonabelian charge.

> **Exercise 7.22**: Using a multiplet of complex scalars as the matter fields, determine the form of a gauge invariant minimal coupling to the Yang–Mills fields.

7.5 Quantized Yang–Mills Fields

The quantization of the Yang–Mills field is best accomplished within the framework of the path integral and serves as an example of a theory where canonical quantization techniques are difficult to use as an *a priori* approach. The starting point is the classical action for a Yang–Mills field coupled to a multiplet of spinor fields:

$$\mathcal{L} = -\tfrac{1}{4} F^a_{\mu\nu} F^{a\mu\nu} + \mathcal{L}(\bar\psi, \psi) - j^a_\mu A^{a\mu} , \tag{7.188}$$

where $\mathcal{L}(\psi)$ is the free spinor action for each member of the multiplet. The techniques developed in Sec. 7.2 for the abelian path integral will be extended to the nonabelian case. Again, continuum measure will be

Sec. 7.5 Quantized Yang–Mills Fields 277

used, and integrations and determinants are understood to be performed or evaluated in Euclidean space.

The problems begin when the canonical momenta are determined. It follows from the gauge field action that

$$\pi^a_\mu = \frac{\partial \mathcal{L}}{\partial(\partial_0 A^{a\mu})} = F^a_{\mu 0} = \partial_\mu A^a_0 - \partial_0 A^a_\mu - g f^a_{bc} A^b_\mu A^c_0 , \qquad (7.189)$$

so that once again $\pi^a_0 = 0$, and the implementation of a gauge condition and its secondary constraint is necessary. As before, the theory will be quantized in the Feynman gauge. For the nonabelian case the action is altered by adding the term $-\text{Tr}\,(\partial_\mu A^\mu)^2$, which reduces to $-\frac{1}{2}\partial_\mu A^{a\mu}\partial_\nu A^{a\nu}$. The canonical momentum is then given by $\pi^a_0 = -\partial_\mu A^{a\mu}$. The nonabelian Feynman gauge Hamiltonian density is then given by

$$\begin{aligned}\mathcal{H}_\chi &= -\tfrac{1}{2}\pi^a_0\pi^a_0 + \tfrac{1}{2}\pi^a_j\pi^a_j + \pi^a_0\partial_j A^a_j - \pi^a_j\partial_j A^a_0 + g f^a_{bc}\,\pi^a_j A^b_j A^c_0 \\ &\quad + \tfrac{1}{4}F^a_{jk}F^a_{jk} + j^a_\mu A^{a\mu} + \mathcal{H}(\psi^j,\bar\psi^j) ,\end{aligned} \qquad (7.190)$$

where $\mathcal{H}(\psi^j)$ is the standard free spinor Hamiltonian density. The action appearing in the Yang–Mills path integral prior to integrating the canonical momenta is given by

$$\mathcal{L}_\chi = \pi^{a\mu}\dot A^a_\mu - H_\chi + \mathcal{L}(\psi^j) . \qquad (7.191)$$

Assuming that an integration by parts is possible in the Hamiltonian, the secondary constraint yields the nonabelian analog of Gauss's law,

$$\dot\pi^a_0 = G^a = -\frac{\partial \mathcal{H}}{\partial A^a_0} = -\partial_j \pi^a_j + g f^a_{bc} A^b_j \pi^c_j - j^a_0 = 0 , \qquad (7.192)$$

or, in the isospin suppressed notation,

$$G = -\partial_j \pi_j + g A_j \times \pi_j + j_0 = -D_j \pi_j - j_0 = 0 . \qquad (7.193)$$

Using the definitions of the canonical momenta the secondary constraint becomes

$$G = \partial_j \dot A_j - \nabla^2 A_0 + g\partial_j(A_j \times A_0) + g A_j \times (\partial_j A_0 - \partial_0 A_j) - j_0 = 0 . \quad (7.194)$$

The nonlinear secondary constraint (7.194) is a tremendous deviation from the abelian case of QED. There, the secondary constraint was linear in the gauge field, and once the charge distribution was specified it was easily solved to obtain a unique form for the constrained value of A_0. The nonabelian case is dramatically different, and a *unique* solution does not

278   Gauge Field Theory

exist. At the classical level, solutions to (7.194) can be found even for the case $j_o = 0$, and this creates a difficulty in the theory that transcends the method of quantization. In the path integral formalism this difficulty manifests itself in the breakdown of the equivalence of the constraint measures,

$$\delta(\pi_o) \det\left(\frac{\delta G}{\delta A_o}\right) \delta(G) = \delta(\pi_o)\delta(A_o - A_o^c), \qquad (7.195)$$

since the inversion of $\delta(G)$ is ambiguous in the presence of nonunique solutions to the secondary constraint $G = 0$. This difficulty is referred to as the *Gribov ambiguity* [19] and will be discussed in more detail in the next section when some topological aspects of the classical structure of the theory are presented. For now, this problem will be ignored, and the general technique for quantization will be applied. Since the classical gauge field configurations that are problematic will be shown to be "large," it is usually argued that ignoring the Gribov ambiguity is acceptable in the "small" field or perturbative region [11].

It is now time to consider the equivalence between the Faddeev–Popov method of quantizing gauge field theories and the method derived by implementing Gauss's law as a constraint on the measure of the path integral. Previous analysis has shown that for a general set of gauge constraints, denoted χ^a, the secondary constraint of Gauss's law is implemented by employing the constraint measure

$$\mathcal{D}^c \pi \, \mathcal{D}^c A = \mathcal{D}\pi \, \mathcal{D}A \, \delta(\chi) \, \delta(G) \, \det\{\chi, G\}, \qquad (7.196)$$

where the Dirac deltas are understood as a product over all isospin indices as well as space-time points. In order for the two approaches to be equivalent, the determinants generated by either method must be identical. In addition, the secondary constraint on the path integral measure must be shown to be irrelevant since it is not present in the Faddeev–Popov method. The key step is to note that the secondary constraints, G^a, are Gauss's law. Noether's theorem shows that these G^a are the crucial element of the generators of the nonabelian gauge transformations of the theory. The generators are given explicitly by

$$\begin{aligned} U_\Lambda &= \exp\left\{i \int d^3x \, \left[\pi_j^a(\partial_j \Lambda^a + g f_{bc}^a A_j^b \Lambda^c)\right.\right. \\ &\quad \left.\left. - \pi_o^a(\dot{\Lambda}^a - g f_{bc}^a A_o^b \Lambda^c)\right]\right\}. \end{aligned} \qquad (7.197)$$

Integrating by parts and dropping the surface terms gives

$$U_\Lambda = \exp\left\{i \int d^3x \, \left[\Lambda^a G^a - \pi_o^a(\dot{\Lambda}^a - g f_{bc}^a A_o^b \Lambda^c)\right]\right\}. \qquad (7.198)$$

Once again, the primary constraint, π_0^a, and the secondary constraint, G^a, have appeared in the unitary operator that generates the gauge transformation. Modulo the gauge-fixing term, the action is constructed to be invariant under the gauge transformations associated with the unitary transformation U_Λ. However, it is assumed that the choice of gauge condition, $\delta(\chi)$, forces the gauge-fixing term to zero, whatever it is. This must be true if the final results are to reproduce the original theory. As a result, any potentially noninvariant part of the action is effectively zero. Of course, the gauge condition $\delta(\chi)$ is not gauge invariant.

On the face of it, demanding that the primary and secondary constraints are satisfied in the physical subspace appears to force the theory to be gauge invariant, since U_Λ is the unit operator when the constraints are satisfied. This argument breaks down if the surface terms that were dropped in going from (7.197) to (7.198) do not vanish. If it is true that configurations available to the gauge fields are nontrivial at the boundary of the manifold, then it is no longer possible to argue that the form (7.197) for U_Λ is the unit operator. Regardless of the surface terms, form (7.197) still generates the gauge transformations, and it must therefore leave physical transition amplitudes invariant. This can still be true if the action of the form (7.197) for U_Λ generates a phase on the states of the theory, since all transition elements are invariant under simultaneous phase changes of the Fock space and its dual space. Such a situation will be discussed in detail in the next section.

The Poisson bracket appearing in the measure (7.196), created by inverting the Gauss's law constraint, can now be understood in the following way. No matter what form the gauge condition χ_a may take, the fact that the G^a generate the gauge transformations gives the result of the Poisson bracket,

$$\chi_b^\Lambda(y) = \int d^4x\, \Lambda^a(x)\{\chi_b(y), G^a(x)\}\,, \tag{7.199}$$

where χ_b^Λ is the form of χ_b after it has undergone an *infinitesimal* gauge transformation, regardless of how "large" Λ is. At first glance the absence of π_0 from (7.199) appears to render the statement true only for the spatial part of χ_b, i.e., only those terms involving A_j, since terms proportional to π_0^a are present in (7.198) while they are not present in (7.199). However, the full gauge transformation appears for the following reasons. First, the Poisson bracket will be evaluated in the physical subspace, and this allows an arbitrary function of π_0 to be added to G^a, so that the gauge covariant replacement $G^a \to G^a - gf_{bc}^a A_0^b \pi_0^c$ can be made. Second, by definition $G^a(x) = \dot{\pi}_0^a(x)$, and since a space-time integral is being used, (7.199) can be integrated by parts to obtain the equivalent form,

$$\int d^4x\, \Lambda^a(x)\{\chi_b(y), G^a(x)\} =$$
$$-\int d^4x\, \Lambda^a(x)\{\chi_b(y), g f^a_{dc} A^d_0(x)\pi^c_0(x)\} - \int d^4x\, \dot\Lambda^a(x)\{\chi_b(y), \pi^a_0(x)\} =$$
$$-\int d^4x\, D_0^{ac}\Lambda^c(x)\{\chi_b(y), \pi^a_0(x)\}\,. \qquad (7.200)$$

Result (7.199) therefore gives the correct transformation property for the A_0 component as well.

It is now trivial to note that (7.199) gives

$$\left.\frac{\delta \chi_b^\Lambda}{\delta \Lambda^a}\right|_{\Lambda=0} = \{\chi_b, G^a\} = \frac{\delta G^a}{\delta A^b_0}\,, \qquad (7.201)$$

where the right-hand side of (7.201) follows from the definition of χ_b as the canonically transformed or gauge fixed form of π^b_0. The A^b_0 appearing in (7.201) is the variable canonically conjugate to χ_b.

The next step in the proof of equivalence between the two approaches to quantization is to demonstrate that the left-hand side of (7.201) is precisely the argument of the Faddeev–Popov determinant. This can be shown by direct comparison of the results from the two approaches. For Yang–Mills theory in the Feynman gauge the Faddeev–Popov determinant is given in Euclidean space by

$$\Delta_\chi = \det\left(\frac{\delta \chi^\Lambda}{\delta \Lambda}\right)$$
$$= \det\left[\frac{\delta}{\delta \Lambda^b(y)}\left(-\partial_\mu A^{a\mu}(x) + \Box_E \Lambda^a(x) - g f^a_{de} A^{d\mu}(x)\partial_\mu \Lambda^e(x)\right)\right]$$
$$= \det\left[\delta_{ab}\Box_E \delta^4(x-y) - g f^a_{cb} A^{c\mu}(x)\partial_\mu \delta^4(x-y)\right]$$
$$= \det(\partial_\mu D^\mu)\,. \qquad (7.202)$$

This result can be obtained from the Poisson bracket formulation by using the replacement $G^a = \dot\pi^a_0 - g f^a_{bc} A^b_0 \pi^c_0 = D_0 \pi_0$ in the physical subspace. It can also be obtained from the form (7.194) as well.

Exercise 7.23: Show that (7.202) is identical to

$$\Delta_\chi = \det\left|\frac{\delta G^b}{\delta A^a_0}\right| = \det\{-\partial_\mu A^{a\mu}, G^b\}\,, \qquad (7.203)$$

when Gauss's law G^b is evaluated in the Feynman gauge.

Thus, the determinant generated by inverting the Gauss's law constraint is exactly equivalent to the Faddeev–Popov determinant Δ_χ.

To complete the demonstration of equivalence, the Gauss's law constraints $\delta(G)$ are absorbed into the action by writing

$$\delta(G) = \int \mathcal{D}\Lambda \, \exp\left\{ i \int d^4x \, \Lambda^a G^a \right\} . \qquad (7.204)$$

The measure takes the form of an integration over the parameter space of the Lie algebra associated with the gauge transformations,

$$\mathcal{D}\Lambda = \prod_{x,a} d\Lambda^a(x) , \qquad (7.205)$$

and this is the nonabelian form for the volume of gauge space. There is an extremely important subtlety in this statement. The integral form of the Dirac delta requires that the range of the integration be infinite. However, the nonabelian groups under consideration are such that their parameters are restricted to a *compact* manifold, and the corresponding gauge volume must be finite. The measure for such a compact group manifold is therefore quite different than (7.205).

The measure for the manifold of the group is known as *Haar measure* and will be denoted $d\mu(g)$, where g denotes some member of the group. If the group element g is specified by the set of parameters Λ^a, then $d\mu(g)$ is an integration over the Λ^a with an appropriate weighting function of these parameters. The weighting function is chosen to make the Haar measure invariant under the group algebra, so that if g' is another member of the group, then $d\mu(g'g) = d\mu(g)$. The volume of the group manifold is therefore invariant under reparameterizations consistent with the group algebra. This property is essential to a demonstration that the nonabelian Faddeev–Popov determinant is gauge invariant. The Haar measure for a specific group, SU(2), will be discussed in detail in the next section. However, it should be obvious that it will be more complicated than the form (7.205), simply because of the possible compact structure of the group manifold. This jeopardizes the technique of implementing the Gauss's law constraint by absorbing the exponential of (7.204) into the action through a redefinition of A_0. On the face of it, it seems that either this method of implementing Gauss's law or the use of Haar measure must be abandoned in the case that the gauge group is compact. The resolution of this dilemma is given in the next section, where topological aspects of the gauge field and the gauge group are discussed. There, an SU(2) invariant field theory

Gauge Field Theory

is analyzed, and a new parameter, θ, characterizing the ground state, is present when Gauss's law is enforced. The ultimate source of θ is the compactness of the SU(2) group manifold, and this compactness creates nontrivial field configurations at the boundary of the manifold.

For the moment, this problem will be ignored, and it will be assumed that it is possible to write

$$\delta(G) = \int \mathcal{D}\mu(g) \exp\left\{i \int d^4x\, \Lambda^a G^a\right\}, \tag{7.206}$$

where $\mathcal{D}\mu(g)$ is the product of the Haar measure for all space-time points. The choice of the Haar measure guarantees that the number of integration variables matches the number of Gauss's law constraints. The invariance of Δ_χ under gauge transformations can be demonstrated by using the invariance of the Haar measure and the Faddeev–Popov procedure. The Faddeev–Popov form for the determinant is given by

$$\Delta_\chi^{-1} = \int \mathcal{D}\mu(g)\, \delta(\chi^\Lambda), \tag{7.207}$$

where $\delta(\chi)$ stands for the product of Dirac deltas for all space-time points and for all gauge constraints. Under a gauge transformation this becomes

$$\Delta_{\chi^\alpha}^{-1} = \int \mathcal{D}\mu(g)\, \delta(\chi^{(\alpha,\Lambda)}), \tag{7.208}$$

where (α, Λ) is the group composition of α and Λ. Of course, this defines a new set of parameters $\Lambda' = (\alpha, \Lambda)$, and this defines a new element of the group, g'. The invariance of the Haar measure allows the change

$$\Delta_{\chi^\alpha}^{-1} = \int \mathcal{D}\mu(g')\, \delta(\chi^{\Lambda'}) = \Delta_\chi^{-1}. \tag{7.209}$$

As a result, the Faddeev–Popov determinant is gauge invariant.

Once again, the Faddeev–Popov determinant is evaluated by making a change of variables in its defining integral. The change of integration variables from the group parameters to the constraints is possible because the two sets are equal in number. The change of variables is accompanied by the Jacobian, and this is assumed to take the form

$$d\mu(g)\, \delta(\chi^g) = d\chi^\Lambda \det\left|\frac{\delta\Lambda}{\delta\chi^\Lambda}\right| \delta(\chi^\Lambda). \tag{7.210}$$

This must be checked by using the actual form of the Haar measure. However, assuming that it is correct, the Faddeev–Popov determinant is given

Sec. 7.5 Quantized Yang–Mills Fields

by the nonabelian extension of the QED result:

$$\Delta_\chi = \det \left| \frac{\delta \chi^\Lambda}{\delta \Lambda} \right| . \tag{7.211}$$

This result coincides with (7.201).

As in the QED case, the argument of the integral appearing in the exponential of (7.204), $\Lambda^a G^a$, can be added to the action, while the Haar measure at each point is added to the measure of the path integral. Comparing the form for G^a, given by (7.193), with the form of the action, obtained by inserting (7.190) into (7.191), shows that the redefinition of the variable of integration $A_0^a \to A_0^a - \Lambda^a$ removes all terms of the form $\Lambda^a G^a$, just as it did in the QED case. However, translating A_0^a by Λ^a changes the Dirac delta of the gauge condition,

$$\delta(\chi^a) \to \delta(\chi^a + \dot{\Lambda}^a) , \tag{7.212}$$

as well as the $\pi_0^a \dot{A}_0^a$ term in the action,

$$\pi_0^a \dot{A}_0^a \to \pi_0^a \dot{A}_0^a - \pi_0^a \dot{\Lambda}^a . \tag{7.213}$$

The determinant Δ_χ is invariant under this translation since the determinant is invariant under gauge transformations, and it will be seen momentarily that the presence of Λ may be removed by a gauge transformation. At this point all the canonical momenta are integrated, and the resulting action in the exponential of the path integral is manifestly gauge invariant except for the translated gauge fixing term that results from integrating π_0.

The action appearing in the path integral is the original action with the gauge fixing term (7.188), but the gauge fixing term is in the translated form,

$$\mathcal{L}_\chi = \mathcal{L} - \tfrac{1}{2}(\partial_\mu A^{a\mu} - \dot{\Lambda}^a)^2 . \tag{7.214}$$

For the case of n gauge fields, the gauge field measure appearing in the path integral is now given by

$$\mathcal{D}A = \prod_x \prod_{a=1}^{n} \prod_{\mu=0}^{3} dA_\mu^a(x) . \tag{7.215}$$

Exercise 7.24: Verify that the gauge field measure (7.215) and the spinor measure are both manifestly gauge invariant.

The $\dot\Lambda$ appearing in the gauge condition and the gauge fixing term may be removed by a gauge transformation on the fields, so that

$$\partial_\mu A^{a\mu} - \dot\Lambda^a \to \partial_\mu A^{a\mu} - \partial^\mu D_\mu^{ab}\Gamma^b - \dot\Lambda^a = \partial_\mu A^{a\mu}, \qquad (7.216)$$

where Γ is chosen to satisfy $\partial_\mu D^\mu \Gamma = -\dot\Lambda$. The determinant Δ_χ is invariant under this gauge transformation. There is a conceptual ambiguity present in (7.216) in that the infinitesimal form for the gauge transformation is being used to remove $\dot\Lambda$, which can be arbitrarily large. It will be assumed without proof that finite gauge transformations U may be used as well, although the differential equation to be solved for U to remove a large value for $\dot\Lambda$ is quite formidable. The final result is the path integral

$$_+\langle 0|0\rangle_- = \int \mathcal{D}\mu(g) \int_0^0 \mathcal{D}A\mathcal{D}\psi\, \delta(\chi)\, \Delta_\chi\, \exp\left\{i\int d^4x\, \mathcal{L}_\chi\right\}, \qquad (7.217)$$

where $\mathcal{L}_\chi$ is given precisely by the gauge fixed action (7.188). Once again, the gauge volume of the theory has been factorized, leaving behind the gauge condition and the Faddeev–Popov determinant Δ_χ.

As in the QED case, the Faddeev–Popov determinant can be absorbed into the action by using Faddeev–Popov ghosts, $\bar c^a$ and c^a, identical in number to the gauge fields, to express the determinant as an exponential. The result is obtained by continuing the Euclidean result into Minkowski space:

$$\begin{aligned}\Delta_\chi &= \int \mathcal{D}\bar c \mathcal{D}c\, \exp\left\{\int d^4x\, \bar c^a(\delta_{ab}\Box_E - gf_{ab}^c A^{c\mu}\partial_\mu)c^b\right\} \\ &\stackrel{M}{\to} \int \mathcal{D}\bar c\mathcal{D}c\, \exp\left\{-i\int d^4x\, \partial^\mu \bar c^a\, D_\mu^{ab} c^b\right\},\end{aligned} \qquad (7.218)$$

where the gauge condition has been used to drop a term generated by the integration by parts. For a nonabelian gauge theory in the Feynman gauge the ghost action is therefore given by

$$\mathcal{L}_g = -\partial_\mu \bar c\, D^\mu c = -\partial_\mu \bar c^a\, \partial^\mu c^a + gf_{bc}^a\, \partial_\mu \bar c^a A^{b\mu} c^c. \qquad (7.219)$$

Exercise 7.25: Show that there are no Faddeev–Popov ghosts in the temporal gauge, $A_o^a = 0$.

In the nonabelian gauge theory in the Feynman gauge the ghosts are no longer free, possessing nontrivial interactions with the gauge fields. This occurs because the zero norm states associated with the gauge fields in the Feynman gauge are no longer coupled to a conserved current, and thus no longer commute with the naive S-matrix. The Faddeev–Popov determinant arises precisely to force $\tilde{\pi}_0$ to vanish, thereby assuring unitarity. In the nonabelian case this requires a mechanism to cancel contributions to the vacuum transition element made by the zero norm states that are implicit in the terms nonlinear in the gauge fields. Although this does not constitute a proof that the theory, given by

$$_+\langle 0|0\rangle_- = \int_0^0 \mathcal{D}\mu(g)\,\mathcal{D}A\,\mathcal{D}\psi\,\mathcal{D}\bar{c}\,\mathcal{D}c\,\delta(\chi)$$

$$\times \exp\left\{i\int d^4x\,[\mathcal{L}_\chi(A,\psi) + \mathcal{L}_g(\bar{c},c,A)]\right\}, \quad (7.220)$$

is unitary, it does give a starting point for analysis. Perturbative calculations and arguments based on BRST invariance [7], both defined in Chapter 8, verify that the theory of (7.220) is indeed unitary. It is an interesting historical sidelight that the first formulation of Faddeev–Popov ghosts was presented by Feynman [20], who introduced anticommuting scalars *ad hoc* into the action in order to solve unitarity problems in quantized gravity.

The vacuum transition element described by (7.220) can be adapted to a specific gauge theory known as *quantum chromodynamics*, or QCD [21]. This theory uses SU(3) as the gauge group, so that there are eight gauge fields, referred to as *gluons*. The fundamental representation of SU(3) is used, so that the matter fields, representing spinor particles called *quarks*, have an internal symmetry index that ranges over three values, known as the *color* index [22]. The quarks also possess a flavor index, but it is assumed that each flavor of quark interacts with the gluon fields in an identical manner.

Quarks are spin $\frac{1}{2}$ particles, and any s-wave bound state of two quarks or antiquarks would therefore have either spin 0 or spin 1, while a bound state of three quarks and antiquarks could have spin $\frac{1}{2}$ or $\frac{3}{2}$. It is the intent of QCD to create a theory where all experimentally observed *hadrons*, strongly interacting particles, can be understood as bound states of two or three quarks and antiquarks. The individual quarks, for reasons beyond the scope of this text [23], must have electric charge of $\frac{2}{3}e$ and $-\frac{1}{3}e$. There has never been a direct experimental observation of this very distinctive unit of electric charge. For this reason it is widely believed that quarks must exist only in bound states with other quarks, a property known as *confinement* [24], and that the behavior of the gluonic fields, induced by the nonlinear terms in the action, is therefore quite different from that of the photon field. Applications of analytic techniques to the QCD path integral to demonstrate this behavior have produced little tangible success. With the advent of high-speed computers and sophisticated algorithms, numerical analysis of the Euclidean QCD path integral, restricted to a lattice, has given strong evidence that the theory reproduces the experimentally observed spectrum of hadronic matter [25]. Nevertheless, there is a place for perturbation theory in QCD, and this will be discussed in Chapter 8.

7.6 Topological Aspects of Gauge Fields

In this section a brief discussion of the relation of gauge fields to the topology of manifolds is given. It has long been known that mathematicians, working on the problem of understanding the relationship between local and global structures of manifolds, have employed constructions identical to the the Yang–Mills field. As a result, the Yang–Mills field has a rich geometric interpretation that adds considerable insight into the meaning of gauge theories.

The starting point is the theory of *forms* used in differential geometry [26], and these are related to the *wedge product*. For the sake of visualization, these will be presented in the context of three-dimensional space, although it does not need to be the simple case of $\mathbf{R}^3$. If a path in this space is considered, an infinitesimal oriented element of the path can be written as some combination of the dx^i at each point. If an oriented two-dimensional surface is considered, its elements can be written as dS^i. The dx^i can be used to form oriented surfaces through the wedge product, written $dx^i \wedge dx^j$. The wedge product is antisymmetric in its arguments, so that $dx^i \wedge dx^j = -dx^j \wedge dx^i$. In this particular case, the relation between dS^i and the wedge product is given by $dS^i = \frac{1}{2}\varepsilon^{ijk} dx^j \wedge dx^k$ or $dx^i \wedge dx^j = \varepsilon^{ijk} dS^k$. Of course, the next higher-order product gives the wedge product $dx^i \wedge dx^j \wedge dx^k$, and this is the antisymmetric product of an oriented surface with an oriented line element. It follows that this is given by $dx^i \wedge dx^j \wedge dx^k = \varepsilon^{ijk} d^3x$, an oriented volume element. For three dimensions all higher order products vanish due to the antisymmetry of the wedge product.

Continuing to restrict attention to three dimensions, forms may be defined in the following way. A *zero-form* φ is a differentiable function $\varphi(x)$ defined everywhere on the manifold. A *one-form* A is understood as an oriented line-element for a vector field, $A = A_i(x)dx^i$. A *two-form* F can be visualized as an element of flux through an oriented surface, $F = F_{ij}(x) dx^i \wedge dx^k$. A *three-form* B is a volume density, $B = B_{ijk}(x) dx^i \wedge dx^j \wedge dx^k$. For three dimensions the allowed types of forms are now exhausted. The quantities φ, A_i, F_{ij}, and B_{ijk} are the *components*

Sec. 7.6 Topological Aspects of Gauge Fields 287

of the forms and are assumed to be differentiable functions on the manifold. It follows from the antisymmetry of the wedge products appearing in the definitions of the forms that the components of the forms must be completely antisymmetric in their indices. For the case considered here, the definition of the forms can be made invariant under spatial rotations by allowing the components to transform according to the inverse rotation of the dx^i. The definition of a form therefore transcends the particular coordinate system chosen to represent it.

The *exterior derivative* d is defined to create a $(n+1)$-form from an n-form. If T is an n-form, then it can be written $T = T_{j \cdots k}(x)\, dx^j \wedge \cdots \wedge dx^k$, where there are n elements in the wedge product and n subscripts. The exterior derivative of T is defined as

$$dT = \frac{\partial T_{j \cdots k}}{\partial x^i}\, dx^i \wedge dx^j \wedge \cdots \wedge dx^k \,. \tag{7.221}$$

Due to the antisymmetry of the wedge product only the antisymmetric part of the derivative survives. The components of the $(n+1)$-form derived by this method are therefore antisymmetric. An important result is obtained immediately, and that is

$$d^2 T = 0 \,, \tag{7.222}$$

for all forms.

Exercise 7.26: Prove (7.222).

Examples of forms abound in gauge theory. In the three dimensional case electric and magnetic fields can be related through the exterior derivative. The scalar potential φ provides an excellent example of a zero-form. Its exterior derivative is the one-form

$$d\varphi = \frac{\partial \varphi}{\partial x^i}\, dx^i = -\left(E_i + \frac{\partial A_i}{\partial t} \right) dx^i \,, \tag{7.223}$$

where the relation of the electric field **E** to the scalar and vector potential has been used. The second exterior derivative yields

$$d^2 \varphi = -\frac{\partial (E_i + \dot{A}_i)}{\partial x^j}\, dx^j \wedge dx^i = \left(\nabla \times \mathbf{E} + \frac{\partial \mathbf{B}}{\partial t} \right) \cdot d\mathbf{S} = 0 \,, \tag{7.224}$$

where one of Maxwell's equations, Faraday's law, has been used along with the relation of oriented surfaces to the wedge product. Likewise, the vector potential can be used to define the one-form $A = A_i dx^i$. Its exterior derivative gives

$$dA = \frac{\partial A_i}{\partial x^j} dx^j \wedge dx^i = \tfrac{1}{2}\varepsilon^{jik}\frac{\partial A_i}{\partial x^j} dS^k = -\tfrac{1}{2}\mathbf{B}\cdot d\mathbf{S} , \qquad (7.225)$$

so that

$$d^2 A = -\nabla\cdot\mathbf{B}\, d^3 x = 0 . \qquad (7.226)$$

Forms enter into the generalization of Stokes's theorem. If $\mathcal{M}$ is an n-dimensional manifold and A is an $(n-1)$-form, then this theorem states

$$\int_{\mathcal{M}} dA = \int_{\partial\mathcal{M}} A , \qquad (7.227)$$

where $\partial\mathcal{M}$ is the boundary of the manifold $\mathcal{M}$. The topological structure of a manifold is partially characterized by the nature of its boundaries. For example, a disk and a cylinder are two-dimensional surfaces that possess boundaries, while the sphere and the torus are two-dimensional surfaces that do not.

The reader may recall that the proof of Stokes's theorem in elementary calculus involves breaking the surface into a set of interlocked triangles. The generalization of this triangulation to more complex surfaces and volumes is known as *homology* theory. Clearly, the way a generalized surface may be triangulated depends upon its dimension p, its boundaries, and its overall topological shape. The set of generalized p-dimensional triangles required to cover a manifold, along with the lower-dimensional sides, faces, and vertices of these triangles, is called a *simplicial complex*. It is possible to classify the members of the simplicial complex needed to cover the manifold, along with their lower-dimensional members, into *homology classes*. In effect, this classification can occur because some of the simplicial complex members have no boundary since they form part of the manifold's boundary, while other members have no boundary because they are already the boundary of a member of the simplicial complex. Stokes's theorem points out a deep connection between forms and the manifolds over which they are defined, and therefore to the homology classes of the manifold. For this reason the relationship is known as *de Rham cohomology theory*. This relationship occurs because the boundary operator is very similar to the exterior derivative, since the boundary of a boundary vanishes, just as d^2 vanishes.

While a full explanation of this relation is beyond the scope of this book [27], a cursory sketch will be given. Given a manifold $\mathcal{M}$, there is a set of p-forms $Z^p(\mathcal{M})$ whose members ω satisfy $d\omega = 0$. These forms are said to be *closed*. Some of these forms are closed because they can be written $\omega = d\eta$, where η is a $(p-1)$-form. The closed forms that can be written $\omega = d\eta$ are said to be *exact*, and the set of all exact forms is denoted $B^p(\mathcal{M})$. The cohomology classes of all closed p-forms are defined by grouping together into an equivalence class all closed forms whose difference is an exact form. In other words, two forms, ω and ω', are equivalent, $\omega \sim \omega'$, i.e., placed in the same equivalence class, if $\omega - \omega' = d\eta$. The general classes of the closed p-forms are denoted as the quotient space $H^p(\mathcal{M}) = Z^p(\mathcal{M})/B^p(\mathcal{M})$. Stokes's theorem shows that the structure of these classes is *dual* to that of the homology classes of the p-dimensional simplicial complexes of the manifold. The homology class structure therefore reveals the cohomology structure of the closed forms on the manifold.

This has direct implications for gauge theory, and they are best demonstrated by an example. Restricting attention again to three dimensions, it is obvious that a one-form A can be associated with the vector potential **A**. Two one-forms, A and A', are equivalent if the difference between them is the exterior derivative of a zero-form, $d\Lambda = \partial_i \Lambda \, dx^i$. This means that all vector potentials differing only by a gauge transformation are placed into the same equivalence class. Once the vector potential one-form is known, the magnetic field is obtained according to dA. Therefore, for the case that $\mathbf{B} = 0$ everywhere on the manifold, the vector potentials fall into the general topological classes given by the first cohomology group $H^1(\mathcal{M})$ of the manifold.

The magnetic field furnishes a natural two-form, $B = F_{jk} \, dx^j \wedge dx^k$. In general, the magnetic field is derived from the one-form vector potential, which is assumed to be everywhere differentiable. If this latter restriction is relaxed, it is possible to find singular functions that yield $dB \neq 0$ for some set of points, and this corresponds to the presence of a magnetic monopole. For the moment, the exact form of the vector potential that results in the monopole will not be important. Instead, if the points where $dB \neq 0$ are deleted from the manifold, then, over the remainder of the manifold, it is true that $dB = 0$. For a pointlike monopole of charge g at the origin, equivalent to a magnetic field that satisfies

$$\nabla \cdot \mathbf{B} = 4\pi g \, \delta^3(\mathbf{x}) \,, \tag{7.228}$$

this amounts to puncturing $\mathbf{R}^3$ at the origin, i.e., deleting the origin $\mathbf{x} = 0$. It can be shown that the remaining manifold is topologically equivalent to the sphere, denoted S_2. Therefore, the equivalence classes of two-forms B that satisfy (7.228) are the same as those of the closed two-forms on S_2 and are thus given by the second cohomology classes $H^2(S_2)$. $H^2(S_2)$ has a one-to-one correspondence with the integers, and this is the first sign that

290   Gauge Field Theory

a magnetic monopole is accompanied by a quantization condition. It will be seen shortly that these integers correspond to the Dirac quantization condition.

Another place where boundaries play a significant role is in integration by parts in the action. So far, all surface terms have been dropped under the assumption that the field configurations drop off at infinity rapidly enough to be suppressed. However, if magnetic monopoles are present this is not true. One of the important integrations by parts in the gauge sector of QED involved the term $\pi_j \partial_j A_0$. Integrating by parts for the purpose of absorbing the Gauss's law constraint involved dropping a surface term of the form

$$\int dt \int d^3x \, \nabla \cdot (\pi A_0) = \int dt \oint d\mathbf{S} \cdot \pi A_0 \,, \tag{7.229}$$

under the assumption that the surface lay at infinity. It is also clear that the classical canonical momentum π_j is precisely the magnetic field B_j. For the case where there is a single static electric charge and a single static magnetic charge present, the classical field configuration is given by

$$A_0 = \frac{e}{r}, \quad \pi = \mathbf{B} = \frac{g}{r^2} \hat{r} \,, \tag{7.230}$$

and the integral of (7.229) becomes

$$\int dt \oint d\mathbf{S} \cdot \mathbf{B} A_0 = \frac{4\pi e g}{R} \int dt \,, \tag{7.231}$$

where R is the radius of the surface. In the limit that $R \to \infty$ while $\int dt \to R$, the two infinite terms cancel, and the result is

$$\int dt \oint d\mathbf{S} \cdot \mathbf{B} A_0 = 4\pi e g \,. \tag{7.232}$$

The restriction $\int dt = R$ is equivalent to defining the boundary of the manifold to be a three-sphere. The definition of the three-sphere and the explanation of why it is the boundary of the manifold will be given momentarily. However, even without the exact definitions, the gauge condition allows (7.229) to be written in a manifestly covariant way as

$$\int_\mathcal{M} d^4x \, \partial_\mu (\pi^\mu A_0) = \int_{\partial \mathcal{M}} dS^\mu \, \pi_\mu A_0 \,, \tag{7.233}$$

where the integration is over the surface bounding the original volume. This result places the space and time components on an equal footing when the boundary limit is taken, showing that the restriction $\int dt = R$ is part of this process.

Sec. 7.6 Topological Aspects of Gauge Fields

Since the role of classical solutions will be discussed in Chapter 9, it is not clear at this point what such a term has to do with the path integral. It can be shown that the presence of both static electric and magnetic charges in the in and out states will generate such terms in the path integral, but there they are exponentiated. This should be evident from the discussion of the limits on the path integral in its general form made in Sec. 6.8. The integration by parts generates the overall phase factor

$$\exp(iS) \to \exp(i4\pi eg) \exp(iS') \ . \tag{7.234}$$

This phase factor can be rendered harmless if $eg = \frac{1}{2}n$, where n is an arbitrary integer, since for that condition the phase is fixed to be unity. This is the Dirac quantization condition [28], and the integers appearing in it are the ones predicted by the cohomology argument.

It is logical to consider the possibility that there are other field configurations, perhaps classical solutions to the equations of motion, for which surface terms are not negligible. To examine this possibility the concept of forms is extended to Euclidean space. Euclidean space is, at least locally, equivalent to $\mathbf{R}^4$. This allows the introduction of the one-form abelian gauge field $A = A_\mu dx^\mu$, from which the two-form $F = \frac{1}{2}F_{\mu\nu} dx^\mu \wedge dx^\nu$ can be derived by exterior differentiation, $F = dA$. Clearly, the $F_{\mu\nu}$ so obtained coincides with the $F_{\mu\nu}$ appearing in the Euclidean QED action. It follows that $dF = d^2 A = 0$, and this reproduces the two Maxwell equations that have no reference to the electric charge as purely *geometric* aspects of forms. Equations of motion that are the result of purely geometric identities occur in gravitation as well, and there, as well as here, they are known as *Bianchi identities* [29]. In the case of Yang–Mills fields $F_{\mu\nu}$ is not derived by exterior differentiation. Instead, using the matrix form of the field, it can be shown that the one-form matrix $A = A_\mu dx^\mu = A_\mu^a T^a dx^\mu$ is related to the two-form $F = \frac{1}{2}F_{\mu\nu} dx^\mu \wedge dx^\nu$ by

$$F = dA + ig\, A \wedge A \ . \tag{7.235}$$

The relation (7.235) can be viewed as the gauge covariant version of the exterior derivative.

However, the QED action is *not* $F \wedge F$. Using the four-dimensional definition

$$dx^\mu \wedge dx^\nu \wedge dx^\rho \wedge dx^\sigma = \varepsilon^{\mu\nu\rho\sigma} d^4x \ , \tag{7.236}$$

shows that

$$F \wedge F = \varepsilon^{\mu\nu\rho\sigma} F_{\mu\nu} F_{\rho\sigma} d^4x \ . \tag{7.237}$$

The tensor $^*F^{\mu\nu} = \frac{1}{2}\varepsilon^{\mu\nu\rho\sigma} F_{\rho\sigma}$ is referred to as the *dual* of $F_{\mu\nu}$. For the case of QED the dual tensor is obtained by simply interchanging **E** and **B** in the definition of F. The Euclidean identity, $\varepsilon_{\mu\nu\rho\sigma}\varepsilon_{\rho\sigma\alpha\beta} = 2[\delta_{\mu\alpha}\delta_{\nu\beta} - \delta_{\mu\beta}\delta_{\nu\alpha}]$, can be used to show that $**F = F$.

> **Exercise 7.27**: Show that the QED action is proportional to $*F \wedge F$.

It is natural to consider the possibility of adding a term proportional to $F \wedge F$ to the action. It follows from the definition of $*F$ that such a term is a *pseudoscalar* since $\varepsilon^{\mu\nu\rho\sigma}$ changes sign under an improper Lorentz transformation, and therefore its contribution will violate parity invariance. The first step in analyzing such a term's possible contribution to the action is to show that it can be written as a total divergence. This will be done for the general nonabelian case. The proof begins by noting that the quartic term must vanish since the properties of the trace give

$$\varepsilon^{\mu\nu\rho\sigma}\text{Tr}\,(A_\mu A_\nu A_\rho A_\sigma) = \varepsilon^{\mu\nu\rho\sigma}\text{Tr}\,(A_\sigma A_\mu A_\nu A_\rho)$$
$$= \varepsilon^{\nu\rho\sigma\mu}\text{Tr}\,(A_\mu A_\nu A_\rho A_\sigma) = -\varepsilon^{\mu\nu\rho\sigma}\text{Tr}\,(A_\mu A_\nu A_\rho A_\sigma)\,. \quad (7.238)$$

The same cyclicity of the trace gives the identity

$$\varepsilon^{\mu\nu\rho\sigma}\,\text{Tr}\,(A_\mu A_\nu \partial_\rho A_\sigma) = \tfrac{1}{3}\partial_\rho\,[\varepsilon^{\mu\nu\rho\sigma}\text{Tr}\,(A_\mu A_\nu A_\sigma)]\,. \quad (7.239)$$

Finally, it follows from $\varepsilon^{\mu\nu\rho\sigma}\partial_\nu\partial_\rho A_\sigma = 0$ that

$$\varepsilon^{\mu\nu\rho\sigma}\text{Tr}\,(\partial_\mu A_\nu \partial_\rho A_\sigma) = \partial_\rho[\varepsilon^{\mu\nu\rho\sigma}\text{Tr}\,(\partial_\mu A_\nu A_\sigma)]\,. \quad (7.240)$$

Using the definition (7.173) in combination with these results gives the result

$$\tfrac{1}{4}\varepsilon^{\mu\nu\rho\sigma}\text{Tr}\,(F_{\mu\nu}F_{\rho\sigma}) = \partial_\rho\left\{\varepsilon^{\rho\sigma\mu\nu}\text{Tr}\,[A_\sigma\partial_\mu A_\nu + \tfrac{2}{3}ig\,A_\sigma A_\mu A_\nu]\right\}\,. \quad (7.241)$$

The right-hand side of (7.241) gives the components of a three-form K, and this three-form satisfies

$$\tfrac{1}{2}\text{Tr}\,(F_{\mu\nu}\,{}^*F^{\mu\nu}) = dK\,, \quad (7.242)$$

where

$$K = \text{Tr}\,(A_\sigma\partial_\mu A_\nu + \tfrac{2}{3}ig A_\sigma A_\mu A_\nu)\,dx^\sigma \wedge dx^\mu \wedge dx^\nu\,. \quad (7.243)$$

K is known as the *Chern–Simons form* [27], and in three-dimensional manifolds it can serve as an action for the gauge sector, resulting in a theory with many exotic properties [30]. From Stokes's theorem it follows that

$$\int_M \tfrac{1}{2}\text{Tr}\,(F\,{}^*F) = \int_{\partial M} K\,, \quad (7.244)$$

Sec. 7.6 Topological Aspects of Gauge Fields

showing that this term is sensitive only to the behavior of the field configurations at the boundary of the manifold. In terms of the exterior derivative and matrix forms the Chern–Simons form can be written

$$K = \text{Tr}\left(AdA + \tfrac{2}{3}igA \wedge A \wedge A\right). \tag{7.245}$$

At first glance it would appear that the term $F \wedge F$ can be discarded from the action since its entire contribution lies at the boundary of the manifold, and up to this point it has been *pro forma* to discard such surface terms as giving zero contribution. However, in nonabelian theories greater care must be taken.

Since the evaluation of gauge theory path integrals and determinants is performed in Euclidean space, the starting manifold $\mathcal{M}$ is $\mathbf{R}^4$. The boundary of $\mathbf{R}^4$ is the three-sphere, S_3. This is the logical continuation of the progression that the boundary of $\mathbf{R}^2$ is the circle S_1 and the boundary of $\mathbf{R}^3$ is the sphere S_2. The three-sphere is defined by *identifying*, i.e., making the same, all the points at infinity in $\mathbf{R}^3$. This is the logical continuation of the definitions of S_1, identifying the points at infinity in $\mathbf{R}^1$, and S_2, identifying the points at infinity in $\mathbf{R}^2$. The Chern–Simons form is therefore to be evaluated on S_3, and a point on S_3 is characterized by three angles, again, in analogy with the progression $S_1 \to S_2 \to S_3$. The generalization of an integration over the angles parameterizing S_n is given by

$$d\Omega_n = d\phi \sin\theta_1 \, d\theta_1 \sin^2\theta_2 \, d\theta_2 \cdots \sin^{n-1}\theta_{n-1} \, d\theta_{n-1}, \tag{7.246}$$

where the limits on ϕ are 0 to 2π, while all other angles range from 0 to π.

> **Exercise 7.28**: If the radius of S_3 is R, show that the volume of S_3 is given by $2\pi^2 R^3$.

The Chern–Simons form first appeared in differential topology [27], and it has a deep relationship to the topological structure of the gauge group as well as to the manifold. To expose this by example, it is possible to consider very simple solutions to the gauge field equations of motion in the absence of matter coupling. Such solutions are referred to as "pure gauge" configurations, and in matrix notation take the form

$$A_\mu = \frac{1}{ig}(\partial_\mu \mathbf{U})\mathbf{U}^{-1}, \tag{7.247}$$

where $\mathbf{U}$ is an arbitrary member of the gauge group of the theory.

294   Gauge Field Theory

> **Exercise 7.29**: Verify that (7.247) is a solution to the equations of motion for A_μ in the absence of a gauge fixing term.

Solutions to the equations of motion in the Euclidean region are generically referred to as *instantons*. The name originates from the fact that many of the gauge field configurations are localized in both space and time in the Euclidean region. However, the term *instanton* is commonly used to designate any Euclidean solution to the equations of motion in any theory.

> **Exercise 7.30**: Use (7.247) in (7.245) to show that the pure gauge form gives
> $$K = -\tfrac{1}{3}ig \int_{S_3} \text{Tr}\,(A \wedge A \wedge A)\,. \tag{7.248}$$

For the case of the pure gauge configuration, form (7.248) reduces to

$$K = \frac{1}{3g^2} \int_{S_3} dS_\mu\, \varepsilon^{\mu\nu\rho\sigma} \text{Tr}\left[(\partial_\nu \mathbf{U})\mathbf{U}^{-1}(\partial_\rho \mathbf{U})\mathbf{U}^{-1}(\partial_\sigma \mathbf{U})\mathbf{U}^{-1}\right]\,. \tag{7.249}$$

To make the example more specific, the gauge group will be chosen to be SU(2). In the Euclidean region, it is easy to construct forms for $\mathbf{U}$ in terms of the coordinates that exhibit interesting results for (7.249) [31]. The form

$$\mathbf{U}_n = r^{-n}(x_0 \mathbf{1} + ix_j \sigma_j)^n\,, \quad r^2 = x_0^2 + x_1^2 + x_2^2 + x_3^2\,, \tag{7.250}$$

where n is an arbitrary integer, automatically satisfies $\mathbf{U}_n^\dagger \mathbf{U}_n = 1$ and $\det \mathbf{U}_n = 1$, and therefore is a member of the SU(2) gauge group.

> **Exercise 7.31**: Calculate the form of A_μ^a associated with the pure gauge solution (7.247) and show that it satisfies the Euclidean form of the Lorentz condition.

In the $r \to \infty$ limit, corresponding to the boundary of Euclidean space, the value of the Chern–Simons form for $\mathbf{U}_n$ is given by inserting (7.250) into (7.249). The evaluation is simplified by noting that the Euclidean surface element can be written $dS^\mu = d\Omega\, r^2\, x^\mu$, where the total solid angle is given by $\int d\Omega = 2\pi^2$. In the $r \to \infty$ limit, the integral (7.249) becomes

$$K_n = \frac{1}{3g^2} \int dS^\mu\, K_\mu = \frac{1}{3g^2} \int dS_\mu\, \frac{12n x^\mu}{r^4}\,, \tag{7.251}$$

Sec. 7.6 Topological Aspects of Gauge Fields 295

resulting in

$$K_n = \frac{8\pi^2}{g^2} n \ . \tag{7.252}$$

Exercise 7.32: Verify (7.251) and (7.252).

For this case of the pure gauge solutions considered here, the relation of K to $F \wedge F$ shows that

$$\frac{g^2}{16\pi^2} \int_{\mathcal{M}} \text{Tr}\left(F_{\mu\nu} \, {}^*F^{\mu\nu}\right) = n \ . \tag{7.253}$$

Because this integral results in an integer, it is said to represent the topological property of the solution known as the *Pontryagin index*. This index could have been predicted by an extension of the homotopy arguments first used in Sec. 3.4. The parameters of SU(2) are the three Euler angles, ξ_1, ξ_2, and ξ_3, and therefore the manifold of parameters available to SU(2) is identical to the manifold S_3. From a topological point of view SU(2) is therefore equivalent to S_3, SU(2) $\sim S_3$. For a pure gauge configuration in the local SU(2) group manifold, it follows that the Pontryagin index, via the Chern–Simons form, corresponds to a mapping of the S_3 parameters of SU(2) onto the S_3 manifold that forms the boundary of Euclidean space and the domain of the integration for the Chern–Simons form. This mapping is therefore characterized by the homotopy group $\pi_3(S_3)$, and this is in a one-to-one correspondence with the integers. These integers represent the winding number of the mapping, i.e., how many times S_3 is wrapped around itself by the pure gauge solution being evaluated. These winding numbers are the same integers obtained from the Pontryagin index. The pure gauge configurations of a nonabelian field theory therefore can have large r behavior that is quite different than that of QED.

The form of the Haar measure for SU(2) will now be found [32], further illuminating the results obtained so far, and exposing new properties of the nonabelian ground state. The Haar measure must take the form

$$d\mu(g) = W(\xi_1, \xi_2, \xi_3) \, d\xi_1 \, d\xi_2 \, d\xi_3 \ , \tag{7.254}$$

where the function W is chosen to make the Haar measure invariant under group reparameterizations. This is understood in the following way. If $\tilde{g}$ is any member of SU(2), then the product $\tilde{g}g$ gives g', another member of SU(2). If the set of parameters (ξ_1, ξ_2, ξ_3) characterize the group element g

while the set of parameters (ξ_1', ξ_2', ξ_3') characterize the group element g', then the function W must be chosen so that

$$W(\xi_1', \xi_2', \xi_3') \, d\xi_1' \, d\xi_2' \, d\xi_3' = W(\xi_1, \xi_2, \xi_3) \, d\xi_1 \, d\xi_2 \, d\xi_3 \ . \tag{7.255}$$

The solution for SU(2) is to choose

$$W(\xi_1, \xi_2, \xi_3) = \varepsilon_{ijk} \mathrm{Tr} \left[g^{-1} \frac{\partial g}{\partial \xi_i} g^{-1} \frac{\partial g}{\partial \xi_j} g^{-1} \frac{\partial g}{\partial \xi_k} \right] \ . \tag{7.256}$$

To demonstrate that this form for W is precisely what is needed, the action of the group is calculated. If $\tilde{g} g = g'$, then $g = \tilde{g}^{-1} g'$ and $g^{-1} = (g')^{-1} \tilde{g}$. Substituting these into (7.256) and using the chain rule for derivatives gives

$$W(\xi_1, \xi_2, \xi_3) = \varepsilon_{ijk} \frac{\partial \xi_l'}{\partial \xi_i} \frac{\partial \xi_m'}{\partial \xi_j} \frac{\partial \xi_p'}{\partial \xi_k} \mathrm{Tr} \left[g'^{-1} \frac{\partial g'}{\partial \xi_l'} g'^{-1} \frac{\partial g'}{\partial \xi_m'} g'^{-1} \frac{\partial g'}{\partial \xi_p'} \right] \ . \tag{7.257}$$

The inverse of the Jacobian for the change of variables has appeared, since

$$\varepsilon_{ijk} \frac{\partial \xi_l'}{\partial \xi_i} \frac{\partial \xi_m'}{\partial \xi_j} \frac{\partial \xi_p'}{\partial \xi_k} = \varepsilon_{lmp} \det \left| \frac{\partial \xi'}{\partial \xi} \right| \ , \tag{7.258}$$

and this precisely cancels the Jacobian that arises from the remainder of the measure,

$$d\xi_1 \, d\xi_2 \, d\xi_3 = \det \left| \frac{\partial \xi}{\partial \xi'} \right| d\xi_1' \, d\xi_2' \, d\xi_3' \ . \tag{7.259}$$

The nature of the Haar measure in the vicinity of the identity can now be established to validate the change of variables used to evaluate the Faddeev–Popov determinant.

Exercise 7.33: Using an infinitesimal form for g in the definition of W appearing in the SU(2) Haar measure, show that $W \approx 1$ in the vicinity of the identity.

The Haar measure for SU(2) can be shown to coincide with the result (7.249). This is accomplished by setting $\mu = 0$ and using the fact that $\varepsilon_{o\nu\rho\sigma} \to \varepsilon_{ijk}$. The volume element becomes $dS^o = R^3 d\Omega$, where Ω is the solid angle of S_3. The final step is to rewrite the derivatives in terms of the dimensionless Euler angles, $x_i = R \xi_i$. The integral of the Chern–Simons form over S_3 is now seen to coincide with the SU(2) group volume. The Chern–Simons form, when integrated over S_3, can be thought of as the topological charge Q_T of the state or gauge field configuration. For the pure gauge configurations considered in this section, the charge Q_T is quantized. It is not difficult to show that $[H, Q_T] = 0$, so that the Hamiltonian and Q_T share the same eigenstates.

Despite the fact that $F \wedge F$ is a total divergence, it is now clear that the properties of the simplest solutions to an SU(2) gauge theory prevent disregarding its possible contribution to the action. In fact, it will be demonstrated momentarily that the existence of the winding number n associated with SU(2) pure gauge configurations breaks the measure of the path integral into disjoint topological sectors, just as it did for the simple quantum mechanical model analyzed in Sec. 3.4, resulting in a periodic structure for the nonabelian vacuum. In addition, it is necessary to resolve the difficulty in enforcing the Gauss's law constraint induced by the fact that the SU(2) group manifold is compact. It is now quite clear for SU(2) that the statement (7.206) does not hold.

To demonstrate the existence of this periodic structure the implications of gauge invariance for the vacuum will be studied. Since enforcing Gauss's law is equivalent to enforcing gauge invariance, the two problems may be resolved simultaneously. Because the topologically nontrivial solutions are pure gauge, it is possible to use the generator of gauge transformations (7.197), denoted $\mathbf{G}_\Lambda$, to translate the gauge field by the pure gauge solution, denoted $\mathbf{U}_n$. Using operator notation this is written

$$(\mathbf{G}_{\Lambda_n})^{-1} A_\mu \mathbf{G}_{\Lambda_n} = (\mathbf{U}_n)^{-1} A_\mu \mathbf{U}_n + \frac{1}{ig} \partial_\mu (\mathbf{U}_n)^{-1} \mathbf{U}_n , \qquad (7.260)$$

where Λ_n is the gauge function associated with the winding number n pure gauge solution $A_\mu = (ig)^{-1} \partial_\mu (\mathbf{U}_n)^{-1} \mathbf{U}_n$. The normalized state $|0\rangle$ denotes the winding number zero vacuum of the theory, so that $\langle 0|Q_T|0\rangle = 0$, while the zero point energy of the Hamiltonian H is chosen so that $\langle 0|H|0\rangle = 0$. Using (7.260) gives

$$\langle 0|(\mathbf{G}_{\Lambda_n})^{-1} Q_T \mathbf{G}_{\Lambda_n}|0\rangle = n , \qquad (7.261)$$

revealing that Q_T is not gauge invariant when the gauge function corresponds to a pure gauge configuration with a nontrivial winding number. This result can be demonstrated directly from the definition of Q_T in terms of the integral of the Chern–Simons density. However, this result also reveals that the state $\mathbf{G}_{\Lambda_n}|0\rangle \equiv |n\rangle$ obeys $\langle n|Q_T|n\rangle = n$. Therefore, the state $|0\rangle$ is itself not gauge invariant for any nonabelian theory possessing topologically nontrivial pure gauge solutions. The source of this problem is the breakdown of the Gauss's law constraint. Such a failure could have been predicted from the compact nature of the group manifold, and it is precisely this compact nature that led to the possibility of topologically nontrivial pure gauge solutions with asymptotic behavior that precludes dropping surface terms generated by integrating the gauge transformation operator G_Λ by parts. Therefore, any theory possessing pure gauge solutions with nontrivial winding numbers must be altered in order that Gauss's law can hold while simultaneously allowing the theory to be gauge invariant.

The solution is found by examining the nature of the vacuum. The states $|n\rangle$ are characterized by the winding number n. However, because the generator of gauge transformations commutes with the Hamiltonian, it follows that

$$\langle n|H|n\rangle = \langle 0|(G_{\Lambda_n})^{-1}HG_{\Lambda_n}|0\rangle = \langle 0|H|0\rangle = 0, \qquad (7.262)$$

revealing an infinite set of states, all of which are energetically degenerate with the topologically trivial vacuum state. Because they create the topologically nontrivial sectors of the theory from the topologically trivial sector, the G_{Λ_n} are referred to as *large*-gauge transformations.

Next, a topological operator $\hat{N}$ is defined such that $\hat{N}|n\rangle = n|n\rangle$. The large-gauge transformations act as raising and lowering operators for $\hat{N}$, since

$$\hat{N}|n\rangle = n|n\rangle \;\Rightarrow\; [\hat{N}, G_{\Lambda_n}]|0\rangle = nG_{\Lambda_n}|0\rangle$$
$$\Rightarrow\; [\hat{N}, G_{\Lambda_n}] = nG_{\Lambda_n} . \qquad (7.263)$$

This result can be used to show that the large-gauge transformations are topologically additive, in the sense that

$$\hat{N}\, G_{\Lambda_m} G_{\Lambda_n}|0\rangle = (n+m)G_{\Lambda_m} G_{\Lambda_n}|0\rangle , \qquad (7.264)$$

so that the large-gauge transformations act as topological translation operators,

$$G_{\Lambda_m}|n\rangle = |n+m\rangle . \qquad (7.265)$$

Next, the operator $\hat{\theta}$ is introduced. It is canonically conjugate to $\hat{N}$, so that it satisfies $[\hat{\theta}, \hat{N}] = i$. The eigenstates of $\hat{\theta}$ are denoted $|\theta\rangle$. The commutation relation between $\hat{\theta}$ and $\hat{N}$ is identical to the quantum mechanical commutator of an angle and an angular momentum, as in the example of Sec. 3.4, so that the state $|\theta\rangle$ has the expansion in terms of $|n\rangle$ states given by

$$|\theta\rangle = \sum_n |n\rangle\langle n|\theta\rangle = \sum_n e^{in\theta}|n\rangle = \sum_n e^{in\theta}G_{\Lambda_n}|0\rangle . \qquad (7.266)$$

This state has two important properties: first, it is energetically degenerate with the topological vacuums, since the fact that H commutes with the

Sec. 7.6 Topological Aspects of Gauge Fields   299

large-gauge transformations gives $\langle \theta|H|\theta\rangle = 0$ regardless of the value of θ; second, it is an eigenstate of the large-gauge transformations,

$$\mathbf{G}_{\Lambda_m}|\theta\rangle = \sum_n e^{in\theta}\mathbf{G}_{\Lambda_m}|n\rangle = \sum_n e^{in\theta}|n+m\rangle = e^{-im\theta}|\theta\rangle. \quad (7.267)$$

The latter property is the key step toward satisfying Gauss's law. Because the state $|\theta\rangle$ develops only a phase under the action of the large-gauge transformations, it possesses the property that the expectation value of a gauge invariant operator A in the state $|\theta\rangle$ is truly gauge invariant under a large-gauge transformation,

$$\langle \theta|(\mathbf{G}_{\Lambda_n})^{-1}A\mathbf{G}_{\Lambda_n}|\theta\rangle = \langle \theta|A|\theta\rangle. \quad (7.268)$$

The final step is to construct a state energetically degenerate with the topological vacuums that is gauge invariant up to a phase. This is done by using the state $|\theta\rangle$ and a hybrid of the Faddeev–Popov method. The *theta vacuum* [33] is defined as

$$|0,\theta\rangle = \int d\mu(g_o)\,\mathbf{G}_{\Lambda_0}|\theta\rangle, \quad (7.269)$$

where the Haar measure and the gauge parameters Λ_0 are in the topologically trivial sector, i.e., winding number zero. To see that it is truly gauge invariant, it is necessary to calculate the action of an arbitrary gauge transformation. An arbitrary gauge transformation can be written as the product of two transformations,

$$\mathbf{G}_\Lambda = \mathbf{G}_{\Lambda_0}\mathbf{G}_{\Lambda_m}, \quad (7.270)$$

where $\mathbf{G}_\Lambda$ has been assumed to be in the mth topological sector of the theory. Applying this gauge transformation to the theta vacuum gives

$$\mathbf{G}_\Lambda|0,\theta\rangle = e^{-im\theta}\int d\mu(g_o')\,\mathbf{G}_{\Lambda_0}\mathbf{G}_{\Lambda_0'}|\theta\rangle. \quad (7.271)$$

Using the fact that
$$\mathbf{G}_{\Lambda_0}\mathbf{G}_{\Lambda_0'} = \mathbf{G}_{\Lambda_0''}, \quad (7.272)$$

and the invariance of the Haar measure,

$$d\mu(g_o') = d\mu(g_o''), \quad (7.273)$$

immediately establishes the gauge invariance of the theta vacuum, up to the harmless phase factor $\exp(-im\theta)$. However, the gauge invariant ground state of the nonabelian theory is now seen to possess the new parameter θ, resulting in periodicity of the vacuum under the translation $\theta \to \theta + 2\pi$.

The form of the theta vacuum reveals the correct method for enforcing the constraint of Gauss's law or gauge invariance in the case of a compact group. The state $|0, \theta\rangle$ can be written

$$|0,\theta\rangle = \sum_n \int d\mu(g_o)\, \mathbf{G}_{\Lambda_0} e^{in\theta} |n\rangle$$

$$= \sum_n \int d\mu(g_o)\, e^{in\theta} \mathbf{G}_{\Lambda_0} \mathbf{G}_{\Lambda_n} |0\rangle . \qquad (7.274)$$

Using the fact that the generator of gauge transformations can be expressed in terms of the functional of Gauss's law,

$$\mathbf{G}_{\Lambda_n} = \exp\left\{ i \int d^4x\, \Lambda_n^a G^a \right\}, \qquad (7.275)$$

it follows that the gauge invariant measure appearing in (7.274) can be read from the form

$$|0,\theta\rangle = \sum_n \int d\mu(g_o)\, e^{in\theta} \exp\left\{ i \int d^4x\, (\Lambda_0^a + \Lambda_n^a) G^a \right\} |0\rangle . \qquad (7.276)$$

Gauss's law will be satisfied, up to a phase, by replacing the Dirac delta in the measure of the path integral with

$$\delta(G) \to \sum_n \int \mathcal{D}\mu(g_o)\, e^{in\theta} \exp\left\{ i \int d^4x\, (\Lambda_0^a + \Lambda_n^a) G^a \right\}, \qquad (7.277)$$

where θ is an arbitrary angle. The exponential of Gauss's law can now be absorbed into the action and removed by redefining A_0, leaving the correct compact group volume factorized. The price paid for this is the remaining sum over all topological sectors, as well as the term $\exp(in\theta)$. The correct gauge field measure appearing in the path integral is given by

$$\mathcal{D}A^c = \sum_n \mathcal{D}\mu(g_o)\, \mathcal{D}A\, \delta(\chi)\, \Delta_\chi\, e^{in\theta} . \qquad (7.278)$$

The constraint measure is now periodic, just as the θ vacuum is.

It is standard procedure to suppress the sum over n by assuming it is implicit in the gauge field measure. The factor $\exp(in\theta)$ can be absorbed into the action by recalling that the winding number n is obtained from the gauge field configurations in terms of the Pontryagin index, so that

$$e^{in\theta} = \exp\left\{ \frac{i\theta g^2}{16\pi^2} \int d^4x\, \mathrm{Tr}\, (F_{\mu\nu}\, {}^*F_{\mu\nu}) \right\}, \qquad (7.279)$$

Sec. 7.6 Topological Aspects of Gauge Fields   301

when the gauge field A_μ is in the nth topological sector.

As a result, the action for a nonabelian field theory invariant under a compact group necessarily contains the theta term of (7.279), where θ is an arbitrary angle. Because this term is a pseudoscalar, it induces violation of CP invariance with measurable predictions [34]. This is known as the *strong CP* problem, since it is present in the QCD sector of the standard model. Experimentally determined bounds require that the value of θ be very small. However, setting θ to zero is an *ad hoc* solution to this problem. It was for this reason that the Peccei–Quinn mechanism [35] was devised to constrain θ to vanish in a more natural way. Unfortunately, further discussion of this problem lies outside the scope of this book.

It is possible to return to the Gribov ambiguity [19] and understand it in terms of the presence of topological sectors in the theory. The Gribov ambiguity occurs if the process of inverting the Gauss's law constraint in a gauge theory is not unique. It is straightforward to characterize the circumstances under which this may happen by the following simple argument. If λ is an $n \times n$ matrix, then a set of n linear constraints may be written

$$\lambda_{jk} x_k - a_j = 0 \ . \tag{7.280}$$

This set of constraints may be uniquely inverted if λ has an inverse, so that the constrained coordinates are given by

$$x_j^c = (\lambda^{-1})_{kj} a_j \ . \tag{7.281}$$

However, if λ does not possess an inverse this process is not possible, and for that case there may be more than one set of solutions to the constraint equations or even families of solutions. The existence of a zero eigenvalue for λ is a sufficient condition for this calamity, since for that case $\det \lambda = 0$, thereby preventing the existence of an inverse.

The generalization of this argument to the gauge field case is straightforward. The role of the matrix is played by the differential operator appearing in the Faddeev–Popov determinant, since this is the object that must be inverted to solve the Gauss's law constraint. It follows that the failure of uniqueness occurs if this differential operator possesses any zero eigenvalues, since then its determinant vanishes. Here is further evidence of the necessity of the Wick rotation, since without the continuation into the Euclidean region, the presence of $\Box$ in the differential operator that defines the Faddeev–Popov determinant will generate zero eigenvalues even for QED. The Wick rotation places the Coulomb gauge and the Lorentz gauge on an even footing in the path integral, removing the artificial distinctions induced by Minkowski space. The existence of such zero-modes for the Faddeev–Popov determinant is easiest to demonstrate in the

Coulomb gauge, where the differential operator to be examined is given by $\delta_{ab}\nabla^2 + g f^a_{cb} A^c_j \partial_j$, while the gauge field must be transverse, $\partial_j A^a_j = 0$, due to the constraint. These can be viewed as the time-independent versions of the corresponding Euclidean Lorentz gauge condition, which also suffers from the Gribov ambiguity.

The starting point is to note that the differential operator may be factorized, due to the transversality of A_j, into $\partial_j(\delta_{ab}\partial_j + g f^a_{cb} A^c_j)$. The existence of a zero-mode is assured if there exists a function h^b that satisfies

$$(\delta_{ab}\partial_j + g f^a_{cb} A^c_j) h^b = 0 \ . \tag{7.282}$$

It is easy to see that a pure gauge configuration for both h^b and A^c_j satisfies (7.282). This is best demonstrated by rewriting (7.282) as a matrix equation in the adjoint representation,

$$(\partial_j + ig A_j) h = 0 \ , \tag{7.283}$$

where $A_j = A^a_j T^a$ and $h = h^a T^a$, with $T^a_{(bc)} = -i f^c_{ab}$. Equation (7.283) has the solution

$$h = \mathbf{U} \ , \quad A_j = \frac{i}{g}(\partial_j \mathbf{U}) \mathbf{U}^{-1} \ . \tag{7.284}$$

The individual forms for A^a_j and h^b can then be found from projecting the gauge components by taking the trace against the group generators.

Proving the existence of a zero-mode reduces to showing that a pure gauge configuration can be found that gives a transverse potential A^a_j. For the abelian case a pure gauge form is given by $\partial_j \Lambda$, so that the abelian problem is equivalent to finding a nonzero Λ such that $\nabla^2 \Lambda = 0$. It was pointed out in Sec. 7.1 that it is impossible to find a normalizable solution to this equation, and therefore the Gribov ambiguity is not present in QED. However, for the case that the gauge group is SU(2) such a pure gauge potential can be found [36]. It is a radially symmetric function of r derived from

$$\mathbf{U} = \mathbf{1} \cos\tfrac{1}{2}g\omega + i\frac{x_j \sigma_j}{r} \sin\tfrac{1}{2}g\omega \ , \tag{7.285}$$

where ω is a function of r to be determined. In what follows the coupling constant g is set to unity for convenience. The demand that the associated gauge field be transverse gives, after some manipulation, the following equation for ω,

$$\frac{d^2\omega}{dt^2} + \frac{d\omega}{dt} - \sin 2\omega = 0 \ , \quad t = \ln r \ . \tag{7.286}$$

Sec. 7.6 Topological Aspects of Gauge Fields

> **Exercise 7.34**: Verify that the demand for transversality gives equation (7.286).

The boundary conditions on ω are determined by the fact that $\mathbf{U}$ must possess an inverse at all values of r. The point of danger is the limit $r = 0$. The restriction that $\omega(r = 0) = \omega(t = -\infty) = 2\pi n$, where n is an integer, guarantees that $\mathbf{U}(r=0) = \pm 1$. Equation (7.286), along with the boundary conditions just derived, is the Euler–Lagrange equation of motion for a classical damped simple pendulum at the surface of the earth beginning at its position of unstable equilibrium. The equation is therefore guaranteed to possess solutions, and depending on the initial velocity chosen, the solution can describe only one of two behaviors. Either the pendulum stays at its position of unstable equilibrium, or it eventually reaches its position of stable equilibrium $\omega = \pm \pi$ as $t \to \infty$ after executing n full rotations. The gauge field configurations associated with this $\mathbf{U}$ are given by

$$A_j = -i \exp(-in\omega) \frac{1}{r} x_k \sigma_k \partial_j \exp\left(i \frac{n\omega}{r} x_l \sigma_l \right), \qquad (7.287)$$

where the n correspond to the various boundary conditions. Inserting these into the Pontryagin index gives

$$\frac{1}{16\pi^2} \int_{\mathcal{M}} \operatorname{Tr} F^* F = \tfrac{1}{2} n, \qquad (7.288)$$

demonstrating, at least for this case, that the Gribov ambiguity is tied to the topological nature of the nonabelian field.

The final topic to be briefly discussed is the geometric nature of the covariant derivative. The equation of a spinor field ψ minimally coupled to an abelian gauge field A_μ is given by

$$\gamma^\mu (\partial_\mu + ieA_\mu)\psi = 0. \qquad (7.289)$$

If the form of the field ψ that solves this equation for a given A_μ is known at some point x_p, denoted ψ_p, then it can be found at another point x by "dragging" the function ψ_p along a path P connecting the two points. This is accomplished by writing

$$\psi(x) = \exp\left\{-ie \int_{x_p}^{x} dy^\mu A_\mu(y)\right\} \psi_p(x), \qquad (7.290)$$

where dy^μ is a parameterization of the path chosen between x_p and x. In effect, the form for A_μ serves as a way to connect various points, and for that reason, A_μ is known in differential geometry as a *connection* [27].

Result (7.290) manifests itself in the Aharanov–Bohm effect [37]. If ψ is the *wave function* of a charged particle, then the expression equivalent to (7.290) is given by

$$\psi(\mathbf{x}) = \exp\left\{-ie \int_{\mathbf{x}_p}^{\mathbf{x}} dy^j \, A_j(\mathbf{y})\right\} \psi_p(\mathbf{x}) \,, \qquad (7.291)$$

where the path is entirely spatial. By analogy with the path integral the quantum mechanical wave function at another point is found by summing over all possible paths between the two points. Limiting this sum to two equally probable paths gives the following result. The wave function becomes

$$\psi(x) = \frac{1}{\sqrt{2}} \left(\psi_1(x) + \psi_2(x)\right) \,, \qquad (7.292)$$

and using (7.291), the associated probability density at the point x is given by

$$|\psi(x)|^2 = \tfrac{1}{2}|\psi_p(x)|^2 \left(1 + \sin e \oint_P dy^j \, A_j(y)\right) \,, \qquad (7.293)$$

where the closed path P passes through x_p and x.

Exercise 7.35: Verify (7.293).

From Stokes's theorem the integral appearing in (7.293) reduces to

$$\oint_P dy^j \, A_j = \int_{S(P)} d\mathbf{S} \cdot \nabla \times \mathbf{A} = \int_{S(P)} d\mathbf{S} \cdot \mathbf{B} = \Phi_{S(P)} \,, \qquad (7.294)$$

where $\Phi_{S(P)}$ is the magnetic flux through the surface defined by the closed path.

This gives the remarkable result that the quantum mechanical interference pattern, determined by those points where the probability density vanishes, is influenced by the presence of any magnetic field, regardless of how *localized* the magnetic field is. Even if the trajectory of a wave packet lies entirely in the region where **B** vanishes, resulting in zero Lorentz force on the charge, there is still a measurable effect in the form of an interference pattern. This is because the wave packet is moving in the region where **A** is nonzero, and, in contradiction to classical expectations, the vector potential can create physically measurable effects. This constitutes the observed Aharanov–Bohm effect.

Sec. 7.6 Topological Aspects of Gauge Fields

Using a similar argument it is possible to rederive the Dirac quantization condition for magnetic monopoles. The phase induced on the wave-function of a charge dragged in a closed loop around the monopole is given by

$$\psi \to \exp\left\{ie \oint_P dx^j A_j\right\}\psi = e^{i\lambda}\psi . \tag{7.295}$$

Once again, using Stokes's theorem the phase factor can be written as

$$i\lambda \equiv ie \oint_P dx^j A_j = ie \int_{S(P)} d\mathbf{S} \cdot \mathbf{B} . \tag{7.296}$$

The effect on the wave function must be independent of choice of surface, $S(P)$, to evaluate the integral, since only the path is relevant. The change in the wave function must be the same for *any* spatial surface bounded by the path P. The difference in phase between any two surfaces, S and S', bounded by P is given by

$$i(\lambda - \lambda') = ie \int_{S(P)} d\mathbf{S} \cdot \mathbf{B} - ie \int_{S'(P)} d\mathbf{S} \cdot \mathbf{B} . \tag{7.297}$$

Because the surfaces S and S' share a common perimeter, they define a closed surface. Since they are arbitrary, they can be chosen to bound an arbitrary volume $\mathcal{V}$ in the interior of the original path P. The difference in the phase can now be written

$$i(\lambda - \lambda') = ie \oint_{S(\mathcal{V})} d\mathbf{S} \cdot \mathbf{B} = ie \int d^3x \, \nabla \cdot \mathbf{B} , \tag{7.298}$$

where the divergence theorem has been used. A magnetic monopole solution is characterized by a static magnetic field,

$$\mathbf{B} = \frac{g}{r^2}\hat{r} \Rightarrow \nabla \cdot \mathbf{B} = 4\pi g \, \delta^3(\mathbf{x}) , \tag{7.299}$$

and this immediately gives

$$i(\lambda - \lambda') = i4\pi eg . \tag{7.300}$$

The difference in phase that results from using different surfaces in Stokes's theorem is rendered irrelevant if e and g satisfy

$$4\pi eg = 2\pi n \Rightarrow eg = \tfrac{1}{2}n , \tag{7.301}$$

since then $\exp i(\lambda - \lambda') = 1$. This is Dirac's quantization condition again.

The quantity appearing in this analysis may be written in a covariant form,

$$W = \exp\left\{ie \oint_P dx^\mu A_\mu(x)\right\}, \qquad (7.302)$$

and is referred to as a *Wilson loop*. The Wilson loop may be generalized to the nonabelian case of the gauge group G by the definition

$$W_G = \exp\left\{ig \oint_P dx^\mu A_\mu\right\}, \qquad (7.303)$$

where $A_\mu = A_\mu^a T^a$. Since W_G constitutes a mapping of a closed loop into the manifold of the gauge group, it is subject to analysis by homotopy arguments, in particular it must be characterized by $\pi_1(G)$. Further analysis lies outside the scope of this book.

> **Exercise 7.36**: Verify that W_G is gauge covariant, so that under a gauge transformation of A_μ it transforms according to $W_G' = \mathbf{U} W_G \mathbf{U}^{-1}$.

Because the covariant derivative characterizes the way in which objects, such as wave functions and fields, are affected by being dragged along paths, it can be used to define *curvature*. The idea is that the change in an object resulting from being dragged around a closed loop, called *holonomy*, is the result of curvature, and the connection characterizes this curvature in a computable way. This concept has its origin in the differential geometric approach to parallel transport on curved manifolds. The systematic development of these ideas led Riemann to formulate the concept of the metric tensor, and this gives the natural language of general relativity. While the explication of general relativity is also outside the scope of this book, it should be mentioned that the curvature associated with a connection is computed from the covariant derivative D_μ and is given by

$$F_{\mu\nu} = [D_\mu, D_\nu]. \qquad (7.304)$$

A moment's reflection shows that this gives the holonomy associated with dragging an object around an infinitesimal closed loop. It also shows that the two-form F, introduced as a gauge covariant extension of the **E** and **B** fields in physics, is the curvature two-form of differential geometry [27].

References

[1] See, for example, J.D. Jackson, *Classical Electrodynamics, Second Edition*, Wiley, New York, 1975.

[2] Gauge invariance as an important symmetry was first pointed out by E. Noether, Nachr. Ges. Wiss. Göttingen, 235 (1918). The term *gauge invariance* (sometimes translated from the German as "measure invariance") was coined by Weyl in his work attempting to unite gravitation and electromagnetism at the classical level using the framework of Riemannian geometry; see H. Weyl, Ann. Physik **59**, 101 (1919); H. Weyl, Z. Physik **56**, 330 (1929).

[3] The first use of the gauge covariant derivative for charged scalar particles is found in V. Fock, Z. Physik **39**, 226 (1926); the modern formulation is found in P.T. Matthews, Phys. Rev. **80**, 292 (1950).

[4] An encyclopedic review of gauge conditions is found in G. Leibbrandt, Rev. Mod. Phys. **59**, 1067 (1987).

[5] J.M. Jauch and F. Rohrlich, *The Theory of Photons and Electrons*, Springer-Verlag, Berlin, 1985.

[6] S.N. Gupta, Proc. Phys. Soc. London **A63**, 681 (1950); K. Bleuler, Helv. Phys. Acta **23**, 567 (1950).

[7] C. Becchi, A. Rouet, and R. Stora, Ann. Phys. **98**, 287 (1976); I.V. Tyupin, Lebedev preprint, FIAN No. 39 (1975), unpublished.

[8] The original articles by the early modern masters who formulated covariant perturbation theory for QED are collected in J. Schwinger, *Quantum Electrodynamics*, Dover, New York, 1958; see also R.P. Feynman, *Quantum Electrodynamics*, Benjamin, New York, 1961.

[9] M.S. Swanson, Phys. Rev. **D24**, 2132 (1981); R.E. Pugh, Phys. Rev. **D33**, 1033 (1986).

[10] L.D. Faddeev, Teoret. i Mat. Fiz. **1**, 3 (1969).

[11] K. Sundermeyer, *Constrained Dynamics*, Springer-Verlag, Berlin, 1983.

[12] R.F. Streater and A.S. Wightman, *PCT, Spin and Statistics, and All That*, Benjamin, New York, 1964.

[13] L.D. Faddeev and V.N. Popov, Phys. Lett. **25B**, 29 (1967).

[14] B. Wybourne, *Classical Groups for Physicists*, Wiley, New York, 1974.

[15] H. Georgi, *Lie Algebras in Particle Physics*, Benjamin/Cummings, Reading, Mass., 1982.

[16] A review of early nuclear isospin models can be found in J.D. Bjorken and S.D. Drell, *Relativistic Quantum Mechanics*, McGraw-Hill, New York, 1965.

[17] See, for example, *The Eight-Fold Way*, ed. M. Gell-Mann and Y.

Ne'eman, Benjamin, New York, 1964.

[18] The first discussion of nonabelian gauge symmetry was by O. Klein in *New Theories in Physics, Proceedings of the Conference Organized by the International Union of Physics and the Polish Intellectual Cooperation Committee*, Warsaw, May 30–June 3, 1939, International Institute of Intellectual Co-operation (Scientific Collection), Paris, 1939. The modern formulation by Yang and Mills is found in C.N. Yang and R.L. Mills, Phys. Rev. **96**, 379 (1954).

[19] V.N. Gribov, Nucl. Phys. **B139**, 1 (1978).

[20] R.P. Feynman, Acta Physica Polonica **24**, 697 (1963).

[21] J. Pati and A. Salam, Phys. Rev. **D8**, 1240 (1973); H. Fritzsch, M. Gell-Mann, and H. Leutwyller, Phys. Lett. **B47**, 365 (1973); S. Weinberg, Phys. Rev. Lett. **31**, 494 (1973).

[22] O. Greenberg, Phys. Rev. Lett. **13**, 598 (1964); M.Y. Han and Y. Nambu, Phys. Rev. **139B**, 1006 (1965).

[23] See, for example, L.B. Okun, *Leptons and Quarks*, North-Holland, Amsterdam, 1982; F. Halzen and A.D. Martin, *Quarks and Leptons*, Wiley, New York, 1984; O. Nachtmann, *Elementary Particle Physics: Concepts and Phenomena*, Springer-Verlag, New York, 1985.

[24] K. Wilson, Phys. Rev. **D3**, 1818 (1971).

[25] See, for example, *Lattice Gauge Theory—A Challenge in Large Scale Computing*, ed. B. Bunk, K.H. Mütter, and K. Schilling, Plenum, New York, 1986.

[26] An excellent introduction to differential forms can be found in L.H. Ryder, *Quantum Field Theory*, Cambridge University Press, Cambridge, 1985.

[27] See, for example, T. Eguchi, P.B. Gilkey, and A.J. Hanson, Phys. Rep. **66**, 213 (1980); C. Nash and S. Sen, *Topology and Geometry for Physicists*, Academic Press, New York, 1983; M. Nakahara, *Geometry, Topology, and Physics*, Adam Hilger, Bristol, 1990.

[28] P.A.M. Dirac, Proc. Roy. Soc. (London) **A133**, 60 (1931).

[29] See, for example, R. Wald, *General Relativity*, University of Chicago Press, Chicago, 1984.

[30] For a treatment of gauge field theory using the Chern–Simons term as an action see E. Witten, Commun. Math. Phys. **121**, 351 (1989).

[31] R. Jackiw and C. Rebbi, Phys. Rev. Lett. **37**, 172 (1976).

[32] An outstanding monograph on many topological aspects of quantum

field theory and gauge groups is R. Rajaraman, *Solitons and Instantons: An Introduction to Solitons and Instantons in Quantum Field Theory*, Elsevier, New York, 1987.

[33] G. 't Hooft, Phys. Rev. **D14**, 3432 (1976); C. Callan, R. Dashen, and D. Gross, Phys. Lett. **63B**, 334 (1976).

[34] See, for example, R.N. Mohapatra, *Unification and Supersymmetry*, Springer-Verlag, Berlin, 1986.

[35] R.D. Peccei and H. Quinn, Phys. Rev. Lett. **38**, 1440 (1977).

[36] The discussion presented here follows the one found in P. Ramond, *Field Theory: A Modern Primer, Second Edition*, Addison-Wesley, Redwood City, California, 1989.

[37] Y. Aharanov and D. Bohm, Phys. Rev. **115**, 485 (1959).

Chapter 8

Perturbation Theory

In this chapter the path integral will be used to give a perturbative representation to the time-ordered products that constitute the dynamics of the S-matrix. Developing the general path integral technique for doing this is the chief goal of the chapter. The result is a recipe for computing the form of contributions from arbitrary orders of perturbation theory known as the *Feynman rules* of the theory. This leads to a useful and intuitive graphical technique for representing these terms. Unfortunately, these terms are plagued with divergent and poorly defined expressions, and the perturbation expansion must be *renormalized* in order to extract any physical meaning.

It has been mentioned in previous chapters that a perturbative analysis of any nonlinear field theory is fraught with the potential for misleading and meaningless results [1]. Those theories that may be analyzed accurately by perturbative techniques include QED, where the techniques for renormalization were first developed. Some processes associated with QCD are amenable to perturbative analysis as well, although care must be taken not to extrapolate the predictions to regions outside a limited domain of validity. Unfortunately, the limited space and scope of this chapter precludes presenting the rules for the application of perturbation theory to the evaluation of physically measurable quantities such as lifetimes and cross sections, or even to deal in a systematic way with the perturbation expansions for gauge field theories. The reader whose interests lie in this direction is urged to consult the many excellent texts [2] that feature perturbative calculations of scattering, decay, and self-energies in models such as QED, QCD, and the standard model. The goal of the presentation in this chapter is solely to develop the necessary tools for evaluating certain aspects of the path integral.

In Sec. 8.1 the generating functionals for quantum field theory are defined from the path integral by appending source terms for each field. Various properties of these generating functionals are derived. In Sec. 8.2 BRST invariance of gauge theories is discussed and the Ward–Takahashi identities for the generating functionals are derived. These identities are critical to the renormalization of gauge theories. In Sec. 8.3 the method for deriving Feynman rules from the path integral is presented and used to derive the perturbative representation of several theories. In Sec. 8.4 the general methods of renormalization are sketched and the concepts of bare parameters and fields are introduced. A necessary condition for renormalizability is derived by the technique of *power counting*. In addition, the rudiments of the renormalization group are discussed.

8.1 Generating Functionals

The time-ordered products of fields may be derived from the path integral by attaching a source term. This is precisely the field theoretic analog of the generating functional approach for quantum mechanical systems discussed in Sec. 3.2. The techniques presented here derive from the original work of Schwinger, Symanzik, and Jona-Lasinio [3, 4, 5, 6].

As a simple example, the path integral for the vacuum persistence functional of a scalar field theory in the presence of a scalar source J is given by the path integral

$$Z[J] = \langle 0|0\rangle_J = \int_0^0 \mathcal{D}\phi \, \exp\left\{i \int d^4x \left[\mathcal{L}(\phi) + J\phi\right]\right\}, \quad (8.1)$$

where the form of the path integral with the momentum integrated has been used. The steps used to derived the path integral can be examined to show that functional derivatives of $Z[J]$ give the vacuum expectation value of time-ordered products of the theory,

$$i^n \langle 0|T\{\phi(x_1)\cdots\phi(x_n)\}|0\rangle_J = \frac{\delta^n Z[J]}{\delta J(x_1)\cdots\delta J(x_n)}$$

$$= i^n \int_0^0 \mathcal{D}\phi \, \phi(x_1)\cdots\phi(x_n) \exp\left\{i \int d^4x \left[\mathcal{L}(\phi) + J\phi\right]\right\}. \quad (8.2)$$

Contact with the original theory is achieved when the functional derivatives are evaluated at $J = 0$.

From Sec. 6.7 it is evident that the physically important quantity is the normalized expectation value, obtained by dividing (8.2) by the vacuum

Sec. 8.1 Generating Functionals

transition element $Z[J]$. For that reason it is convenient to introduce the generating functional $W[J] = -i \ln Z[J]$, the field theoretic counterpart of the quantum mechanical generating functional introduced in Sec. 3.2, and begin to develop a graphical representation. The n-point function $G_c(x_1, \ldots, x_n)$ is defined as

$$G_c(x_1, \ldots, x_n) = \frac{\delta^n W[J]}{\delta J(x_1) \cdots \delta J(x_n)}, \tag{8.3}$$

and is represented graphically in Fig. 8.1 by a blob in configuration space with n legs attached to the n points that constitute its arguments. The blob represents the effects of the nonlinear dynamics, which in turn depend upon the specific form of the theory.

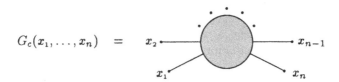

Fig. 8.1. The graphical representation of the n-point function.

It is useful to examine the relation of the G_c to the functional derivatives of $Z[J]$. The first derivative gives

$$G_c(x) = \frac{\delta W[J]}{\delta J(x)} = -i \frac{1}{Z[J]} \frac{\delta Z[J]}{\delta J(x)} = \frac{\langle 0|\phi(x)|0\rangle_J}{\langle 0|0\rangle_J} \equiv \phi_0(x), \tag{8.4}$$

giving the normalized vacuum expectation value of the field ϕ. If this is not zero in the limit $J = 0$, then the relation of the vacuum expectation value to the S-matrix shows that a single particle state can be excited from the vacuum. This represents an *unstable* vacuum, and in general serves as a mechanism for the spontaneous breakdown of symmetry for the theory. This will be discussed in detail in Chapter 9, since it requires further developments.

The second derivative yields

$$G_c(x_1, x_2) = -i \frac{1}{Z[J]} \frac{\delta^2 Z[J]}{\delta J(x_1) \delta J(x_2)} + i \frac{1}{(Z[J])^2} \frac{\delta Z[J]}{\delta J(x_1)} \frac{\delta Z[J]}{\delta J(x_2)}$$

$$= i \frac{\langle 0|T\{\phi(x_1)\phi(x_2)\}|0\rangle_J}{\langle 0|0\rangle_J} - i \phi_0(x_1) \phi_0(x_2). \tag{8.5}$$

Perturbation Theory

This gives the normalized time-ordered product, the object that enters into the S-matrix, with the vacuum expectation values of the fields subtracted. Therefore, $G_c(x_1, x_2)$ represents the *connected* part of the two-point function. This is a general feature of all higher-order n-point functions. Because of their relation to the S-matrix the n-point functions are sometimes called *generalized Green's functions*.

> **Exercise 8.1**: To expose this structure further evaluate the three-point function in terms of its relationship to the normalized time-ordered product of three fields and the lower-order two- and one-point functions.

The functional $W[J]$ is therefore the generator for the *connected* pieces of the n-point function that enters the S-matrix. From a graphical standpoint it is an important analytical aid to break the connected n-point function into *propagators* and *vertices*. As in Sec. 4.5, this is accomplished by introducing the *effective action*, $\Gamma[\phi_0]$, through a functional Legendre transformation,

$$\Gamma[\phi_0] = W[J] - \int d^4y\, J(y)\, \phi_0(y) \,. \tag{8.6}$$

It follows from its definition that

$$\frac{\delta \Gamma[\phi_0]}{\delta \phi_0(x)} = \int d^4y\, \frac{\delta W[J]}{\delta J(y)} \frac{\delta J(y)}{\delta \phi_0(x)} - \int d^4y\, \frac{\delta J(y)}{\delta \phi_0(x)} \phi_0(y) - J(x) \,, \tag{8.7}$$

where the functional chain rule has been used. Using the definition of $\phi_0(x)$, given by (8.4), reduces (8.7) to

$$\frac{\delta \Gamma[\phi_0]}{\delta \phi_0(x)} = -J(x) \,. \tag{8.8}$$

Relation (8.8) gives an instrument to probe the vacuum structure of the theory, and it will be evaluated in later sections for several model field theories.

The relation of the higher-order functional derivatives of Γ to the connected n-point functions can now be demonstrated. The identity obtained from the chain rule, when combined with (8.5) and (8.8), gives

$$\begin{aligned}
\delta^4(x_1 - x_2) &= \frac{\delta J(x_1)}{\delta J(x_2)} = \int d^4y\, \frac{\delta J(x_1)}{\delta \phi_0(y)} \frac{\delta \phi_0(y)}{\delta J(x_2)} \\
&= -\int d^4y\, \frac{\delta^2 \Gamma[\phi_0]}{\delta \phi_0(x_1)\, \delta \phi_0(y)}\, G_c(y, x_2) \,,
\end{aligned} \tag{8.9}$$

and this shows that the second derivative of the effective action is the functional inverse of the connected two-point function. The connected two-point function has a graphical representation as a line between its arguments. For that reason $G_c(x_1, x_2)$ will be referred to as the propagator from this point on, since it allows a visual interpretation as the motion of a particle between the two points. This is reinforced by its relation to the S-matrix element that describes the overlap of single-particle in and out states.

The nth derivative of Γ with respect to ϕ_0 is denoted $A(x_1, \ldots, x_n)$. Taking the functional derivative of both sides of (8.9) with respect to $J(x_3)$ and using the chain rule gives

$$\begin{aligned}
0 = & \int d^4y_1\, d^4y_2\, A(x_1, y_1, y_2)\, G_c(y_2, x_2)\, G_c(y_1, x_1) \\
& + \int d^4y_1 A(x_1, y_1)\, G_c(y_1, x_2, x_3)\,.
\end{aligned} \qquad (8.10)$$

Relation (8.9) allows (8.10) to be inverted, yielding

$$G_c(x_1, x_2, x_3) = \int d^4y_1\, d^4y_2\, d^4y_3\, G_c(y_1, x_1)\, G_c(y_2, x_2)\, G_c(y_3, x_3)\, A(y_1, y_2, y_3)\,. \qquad (8.11)$$

Result (8.11) has the graphical representation shown in Fig. 8.2.

Fig. 8.2. The decomposition of the three-point function.

The functional derivatives of Γ yield the *vertices* of the theory, and their visual representations summarize the effects of nonlinear dynamics in processes connecting the external lines. The propagator represents the generalized probability amplitude of a particle's transition between the two points that form its argument. The vertices occurring in a given theory cannot be decomposed any further, so that there is no occurrence of lower-order vertices connected by a single propagator occurring within a given

316   Perturbation Theory

vertex. For this reason it must be that the vertices cannot be broken, i.e., rendered into disjoint pieces, by cutting one line representing a propagator. Therefore, the vertices are referred to as *one-particle-irreducible*.

> **Exercise 8.2**: Show that the four-point function decomposes according to Fig. 8.3.

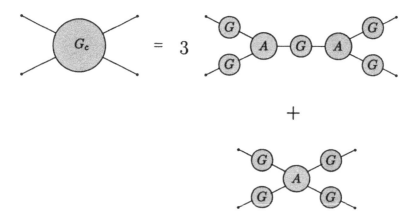

Fig. 8.3. The decomposition of the four-point function.

Although these results have been derived for the case of a single scalar field, it is straightforward to generalize them to the case of spinor and vector fields. In the case of spinor fields it is necessary to introduce Grassmann valued sources K and $\bar{K}$ that transform as do the spinors, while for the case of the vector field the source J_μ must transform as a vector.

For the case of a single scalar field the effective action may be represented as a functional power series around the value $\phi_0 = 0$ with the form

$$\Gamma[\phi_0] = \sum_{n=0}^{\infty} \frac{1}{n!} \int d^4x_1 \cdots d^4x_n \, \phi_0(x_1) \cdots \phi_0(x_n) \, A_0(x_1, \ldots, x_n) , \quad (8.12)$$

where

$$A_0(x_1, \ldots, x_n) = \left. \frac{\delta^n \Gamma[\phi_0]}{\delta \phi_0(x_1) \cdots \delta \phi_0(x_n)} \right|_{\phi_0 = 0} . \quad (8.13)$$

Sec. 8.1 Generating Functionals

It is an important observation that the vertices of the theory evaluated at nonzero ϕ_0 can be obtained from the $\phi_0 = 0$ vertices of the theory. That this is so is seen from the power series representation (8.12). For example, the source term at nonzero ϕ_0 can be written

$$A_{\phi_0}(x) = \sum_{n=0}^{\infty} \frac{1}{n!} \int d^4x_1 \cdots d^4x_n \, \phi_0(x_1) \cdots \phi_0(x_n) \, A_0(x, x_1, \ldots, x_n) \, . \quad (8.14)$$

This means that a theory can be renormalized for the case $\phi_0 = 0$, and the vertices of the $\phi_0 \neq 0$ theory will also be well behaved, *modulo* possible problems with convergence of the power series.

As in the quantum mechanical case the effective action may be expanded around a constant value for ϕ_0. The vertices are usually translationally invariant functions, although this need not always be the case. In particular, if the ground state of the theory is not translationally invariant, then neither are the vertices. However, if they are, then it is necessary to factor a space-time volume out of the effective action when it is expanded around a constant form for ϕ_0. This is easily seen at lowest order, noting that translational invariance means that $A(x)$ is a constant. Therefore, the *effective potential* $V_{\text{eff}}(\phi_0)$ is defined as

$$\Gamma[\phi_0] \approx -V_{\text{eff}}(\phi_0) \int d^4x \, . \quad (8.15)$$

The effective potential is a function of ϕ_0, but it inherits many characteristics of its functional predecessor, the effective action.

The relation of the effective potential to the vertices of the theory can be seen by examining the power series representation of the effective action and employing the general representation of a translationally invariant function,

$$A(x_1, \ldots, x_n) = \int \frac{d^4p_1}{(2\pi)^4} \cdots \frac{d^4p_n}{(2\pi)^4} (2\pi)^4 \delta^4(p_1 + \cdots + p_n)$$

$$\times \exp\left\{i \sum_{j=1}^{n} p_j x_j\right\} \tilde{A}(p_1, \ldots, p_n) \, , \quad (8.16)$$

where $\tilde{A}$ is the Fourier transform of the function A into momentum space. The presence of the Dirac delta in (8.16) assures that this function is invariant under the translation of all its arguments by a constant vector, $A(x_1, \ldots, x_n) \to A(x_1 + a, \ldots, x_n + a)$. Such an invariance ensures that the theory conserves four-momentum in all processes, and this must be true if the ground state or vacuum expectation value of the field is to be translationally invariant. Using (8.16) in (8.12) and treating ϕ_0 as a constant gives

$$\Gamma[\phi_0] \approx \sum_{n=0}^{\infty} \frac{1}{n!} (\phi_0)^n \tilde{A}(0,\ldots,0) (2\pi)^4 \delta^4(0) . \qquad (8.17)$$

Using the definition of the effective potential, (8.15), and the identity,

$$\delta^4(0) = \frac{1}{(2\pi)^4} \int d^4x , \qquad (8.18)$$

gives the effective potential in terms of the zero momentum value of the vertices,

$$V_{\text{eff}}(\phi_0) = -\sum_{n=0}^{\infty} \frac{1}{n!} (\phi_0)^n \tilde{A}(0,\ldots,0) . \qquad (8.19)$$

The method for constructing the effective potential is now clear. The vertices of the theory are found and their Fourier transform, evaluated at zero momentum, is inserted into the expansion (8.19). In the limit that the source term J is set to zero, giving the original theory, this function must satisfy (8.8), so that

$$\frac{\partial V_{\text{eff}}(\phi_0)}{\partial \phi_0} = 0 . \qquad (8.20)$$

Form (8.20) then gives the first criterion for a translationally invariant nonzero vacuum expectation value for ϕ: it must be an extremum of the effective potential. The zero momentum vertices for the nonzero ϕ_0 case are then obtained from (8.19) by differentiation and evaluation at the values found from the condition (8.20).

A second restriction is found by examining the special case is the second derivative. The translationally invariant propagator has the Fourier representation

$$G_c(x,y) = G_c(x-y) = \int \frac{d^4p}{(2\pi)^4} \frac{d^4k}{(2\pi)^4} \tilde{G}_c(p,k) \delta^4(p+k) e^{i(px+ky)}$$

$$\equiv \int \frac{d^4p}{(2\pi)^4} \tilde{G}(p) e^{ip(x-y)} . \qquad (8.21)$$

Recalling that $A(x,y)$ is the inverse of the propagator, this shows that

$$\tilde{A}(0,0) = \tilde{G}^{-1}(p=0) . \qquad (8.22)$$

Furthermore, it shows that that the second derivative of V_{eff} is

$$\frac{\partial^2 V_{\text{eff}}(\phi_0)}{\partial \phi_0^2} = -\tilde{G}^{-1}(p=0) . \qquad (8.23)$$

It will be seen that the right-hand side of (8.23) corresponds to the mass of the theory, and for a stable theory this must be positive. This completes the analogy to a potential, since it shows that a stable vacuum must be at least a relative minimum of the effective potential. The higher-order derivatives can be used to renormalize the effective potential, but a full explanation must await further developments.

It is possible to generalize these results to include the presence of composite sources of the form $K(x,y)\phi(x)\phi(y)$. While the identities obeyed by the generating functionals corresponding to the effective action and the effective potential become more complicated for such a case, they eventually lead to a set of self-consistent nonlinear integral equations involving not only the vacuum expectation of the field but additionally the propagator for the fields [7].

8.2 Ward–Takahashi Identities

The action appearing in the path integral may possess symmetries, and these have ramifications for the Green's functions and vertices of the theory. This was demonstrated in Sec. 4.3 for the quantum mechanical case, where it was seen that the presence of a symmetry is accompanied by a set of identities for the time-ordered products and expectation values. The extensions of these ideas to the field theory context are known variously as the Ward–Takahashi and Slavnov–Taylor identities. The latter are usually associated with BRST symmetry [8] in gauge theories. These identities are best expressed in terms of the generating functionals defined in the previous section.

As in Sec. 6.4, where Noether's theorem was discussed, a symmetry is some variation of the fields and their arguments that leaves the action invariant. For the purposes of this section this variation will be restricted to those that change the form of the field. In the infinitesimal case this can be written $\Psi_\alpha(x) \to \Psi'_\alpha = \Psi_\alpha(x) + \delta\Psi_\alpha(x)$, while the case of a finite transformation often takes the form of a unitary matrix $\Psi_\alpha(x) \to U_{\alpha\beta}\Psi_\beta(x)$. A counterexample is found in the Lorentz transformation properties of the Weyl spinor discussed in Sec. 6.2, involving a unimodular, but not unitary, matrix. However, it is usually the case that an infinitesimal symmetry transformation can be written as an expansion of the matrix about the identity $U_{\alpha\beta} \approx \delta_{\alpha\beta} + \delta U_{\alpha\beta}$.

From the discussion of the previous section it is apparent that the generating functionals for the time-ordered products require the presence of source terms that are linear in the fields. A transformation *solely of the fields* may leave the original source-free action and the measure of the path integral invariant, but in most cases it will not leave the source terms invariant. For example, the term $J\phi$ is not invariant under the transformation $\phi \to e^{i\alpha}\phi$ unless the source J is also transformed simultaneously according to $J \to e^{-i\alpha}\phi$. That this is not always necessary can be seen from the case of QED, where a photon source term takes the form of the integral of $J^\mu A_\mu$. The invariance of this source term under gauge transformations will be discussed momentarily.

For a theory whose path integral representation is invariant under the action of some global Lie group, the generalization of this argument allows it to have a source attached in an invariant manner. This is done by enforcing a transformation property for the source that cancels the transformation of the fields. For the moment consideration will be limited to a multiplet of scalar fields, ϕ_j. If the term $J_j\phi_j$ is attached to the action, which is otherwise invariant under the transformation $\phi'_j = U_{jk}\phi_k$, then the source must transform according to $J'_j = (\mathbf{U}^{-1})_{jk}J_k$ in order to maintain overall invariance. The inverse transformation has the infinitesimal form $(\mathbf{U}^{-1})_{jk} \approx \delta_{jk} - \delta U_{jk}$. Reversing the argument, it follows that the generating functional for the connected n-point functions is invariant under this transformation on the source since it can be compensated by a change in the field variables of integration. The measure is assumed to be invariant under the transformation, and this will occur in most cases for the reason that the Jacobian at each time slice of the measure yields

$$\prod_j d\phi_j \to \det \mathbf{U} \prod_j d\phi'_j \,, \qquad (8.24)$$

and $\det \mathbf{U} = \pm 1$ from the assumption that $\mathbf{U}$ is unitary. It follows that

$$W[J'] = W[J] \quad \Rightarrow \quad W[J - \delta U J] - W[J] = 0 \,. \qquad (8.25)$$

Expanding this in a functional power series gives a generator for identities of the Green's functions,

$$\int d^4x \, \delta U_{jk} J_k(x) \frac{\delta W[J]}{\delta J_j(x)} = 0 \,. \qquad (8.26)$$

For the case that $\mathbf{U}$ is a member of a Lie group the infinitesimal transformation takes the form $\delta U_{jk} = i\Lambda^a T^a_{jk}$. Since the infinitesimal parameters Λ^a may be chosen arbitrarily, the number of symmetry generators matches the order of the group,

$$\int d^4x \, T^a_{jk} J_k(x) \frac{\delta W[J]}{\delta J_j(x)} = 0 \,, \quad \forall a \,. \qquad (8.27)$$

Sec. 8.2 Ward–Takahashi Identities

Functional differentiation of (8.27) yields a set of relations between the various n-point functions, and these are *exact*.

A more important set of identities that relate the vertices and inverse propagators can be obtained from the effective action. The Lorentz gauge action for QED is invariant under gauge transformations that satisfy $\Box \Lambda = 0$. In order to construct a generating functional for QED it is necessary to attach the source terms $\bar{K}\psi + \bar{\psi}K + J^\mu A_\mu$, where K and $\bar{K}$ are Grassmann valued sources. In what follows the spinor indices on the fields and sources are suppressed. Under a gauge transformation the spinor sources, K and $\bar{K}$, must also be transformed, $K \to \exp(-i\Lambda)K$, so that the spinor source terms do not destroy gauge invariance. In order for the gauge source term to maintain gauge invariance it is necessary to assume that the external source J^μ is conserved, $\partial_\mu J^\mu = 0$. The result of a gauge transformation on A_μ can then be integrated by parts and discarded,

$$\int d^4x\, J^\mu \partial_\mu \Lambda = \int d^4x\, \partial_\mu (J^\mu \Lambda) = 0 , \qquad (8.28)$$

under the assumption that the gauge functions are suitably well behaved at infinity.

This invariance of the action with sources allows the effective action to be invariant under the transformation as well. The effective action is constructed from $W[J^\mu, K, \bar{K}]$, which in turn is found from the Lorentz gauge path integral for the vacuum persistence functional. For Lorentz gauge QED the effective action is therefore given by

$$\Gamma[a_\mu, \eta, \bar{\eta}] = W[J^\mu, K, \bar{K}] - \int d^4x\, (\bar{K}\eta + \bar{\eta}K + J^\mu a_\mu) , \qquad (8.29)$$

where

$$a_\mu(x) = \frac{\delta W}{\delta J^\mu(x)}, \quad \eta(x) = \frac{\delta W}{\delta \bar{K}(x)}, \quad \bar{\eta}(x) = -\frac{\delta W}{\delta K(x)} . \qquad (8.30)$$

The minus sign appearing in the last expression of (8.30) stems from the Grassmann nature of both K and η. It follows from the definition (8.29) that a_μ, η, and $\bar{\eta}$ may undergo a gauge transformation identical to their field counterparts, and that this transformation leaves the effective action invariant because of the transformation properties of the source terms. This invariance immediately yields

$$\Gamma[a_\mu - \partial_\mu \Lambda, e^{ie\Lambda}\eta, e^{-ie\Lambda}\bar{\eta}] = \Gamma[a_\mu, \eta, \bar{\eta}] . \qquad (8.31)$$

By expanding this expression in a functional power series for an infinitesimal gauge transformation and performing an integration by parts, the following identity is obtained,

$$\int d^4x\, \Lambda(x) \left[ie\frac{\delta\Gamma}{\delta\eta(x)}\eta(x) - ie\frac{\delta\Gamma}{\delta\bar{\eta}(x)}\bar{\eta}(x) + \partial_\mu\frac{\delta\Gamma}{\delta a_\mu(x)} \right] = 0. \quad (8.32)$$

Because $\Lambda(x)$ is an arbitrary local function, the expression in brackets in (8.32) must vanish everywhere. The local gauge symmetry has therefore created a local identity for the generating functional.

The true vacuum of the theory is obtained in the $J^\mu = K = \bar{K} = 0$ limit, and for QED this is assumed to correspond to the stable vacuum characterized by $\eta = \bar{\eta} = a_\mu = 0$. Therefore, applying the functional derivatives $\delta/\delta\eta(y)$ and $\delta/\delta\bar{\eta}(z)$ to the local expression in (8.32) and evaluating the result in the true vacuum yields the identity

$$-ie\frac{\delta^2\Gamma}{\delta\bar{\eta}(z)\,\delta\eta(x)}\delta^4(x-y) + ie\frac{\delta^2\Gamma}{\delta\bar{\eta}(x)\,\delta\eta(y)}\delta^4(x-z)$$
$$+ \partial_\mu\frac{\delta^3\Gamma}{\delta\bar{\eta}(z)\,\delta\eta(y)\,\delta a_\mu(x)} = 0. \quad (8.33)$$

The derivation of the signs appearing in (8.33) rests upon the anticommutativity of the Grassmann derivatives.

The identity (8.33) relates the spinor propagator to the gauge–spinor vertex. To see this, the translationally invariant spinor propagator, with spinor indices exhibited, is defined as

$$S_{ab}(x-y) = \frac{\langle 0|T\{\bar{\psi}_a(x)\psi_b(y)\}|0\rangle}{\langle 0|0\rangle} = \frac{\delta^2 W}{\delta K_a(x)\,\delta\bar{K}_b(y)}, \quad (8.34)$$

and possesses the Fourier transform

$$S_{ab}(x-y) = \int \frac{d^4p}{(2\pi)^4}\, \tilde{S}_{ab}(p)\, e^{ip(x-y)}. \quad (8.35)$$

It is straightforward to repeat the steps of (8.9) to show that

$$\int d^4y\, S_{ab}(x-y)\frac{\delta^2\Gamma}{\delta\bar{\eta}_b(y)\,\delta\eta_c(z)} = \delta_{ac}\delta^4(x-z), \quad (8.36)$$

so that the second derivative of Γ is the inverse spinor propagator.

Exercise 8.3: Show that the Fourier transform of (8.33) yields

$$\tilde{S}_{ab}^{-1}(p+q) - \tilde{S}_{ab}^{-1}(p) = q_\mu \tilde{A}_{ab}^\mu(p,q,-(p+q)), \quad (8.37)$$

where $\tilde{A}_{ab}^\mu$ is the Fourier transform of the gauge-spinor vertex.

Result (8.37) is an example of a Ward–Takahashi identity for QED [9]. Its generalization to nonabelian theories is critical to the renormalization of those theories. This will be discussed briefly in the next section.

There is another invariance of the QED Feynman gauge path integral that involves the Faddeev–Popov ghosts. It is recalled that, after gauge fixing, the Lagrangian of the theory takes the form $\mathcal{L} = \mathcal{L}_{GI} - \frac{1}{2}(\partial_\mu A^\mu)^2 - \partial_\mu \bar{c}\, \partial^\mu c$, where $\mathcal{L}_{GI}$ is the gauge invariant part of the action, involving the spinor and gauge fields and their coupling. This action is invariant under the local BRST transformation, given by

$$\psi \to e^{iec}\psi, \quad A_\mu \to A_\mu - \partial_\mu c, \quad \bar{c} \to \bar{c} + \partial_\mu A^\mu. \tag{8.38}$$

Exercise 8.4: Use the Grassmann nature of c to verify that (8.38) is an exact invariance of the gauge-fixed Feynman gauge action and the path integral measure, where the gauge invariant spinor action is given by $\mathcal{L}_{GI} = i\bar{\psi}\gamma^\mu D_\mu \psi$.

The transformation (8.38) mixes the c-number and a-number variables appearing in the action, and therefore has many of the properties of the supersymmetry transformations discussed in Sec. 5.6. The charge associated with this invariance is calculated using Noether's theorem. The result in the operator formalism is the generator of the Grassmann-valued BRST transformation,

$$Q = \int d^3x \left(c\, \partial_\mu \dot{A}^\mu - \dot{c}\, \partial_\mu A^\mu \right), \tag{8.39}$$

where the A^μ appearing in (8.39) solves the Lorentz gauge equations of motion. The path integral generalization of this generator is the Euclidean form

$$Q = \int d^4x \left(c\, \partial_j \pi_j - \pi_{\bar{c}}\, \pi_0 \right), \tag{8.40}$$

where $\pi_{\bar{c}}$ is the momentum conjugate to $\bar{c}$. Because the two constraints of QED have appeared in Q it follows that both the expectation value of Q and its Poisson bracket with field variables will vanish. Because of the Grassmann nature of c it follows that Q anticommutes with itself, so that $Q^2 = 0$. This property is similar to the vanishing of the exterior derivative, $d^2 = 0$, discussed in Sec. 7.6, and has been used to establish a cohomology structure for the states of a gauge theory [10]. Because of the appearance of the constraints in Q, and because Q generates the gaugelike BRST symmetry transformation, Q can be used to select the

324   Perturbation Theory

physical spectrum of a gauge theory. The physical subspace must satisfy

$$Q|\text{physical}\rangle = 0 . \tag{8.41}$$

Some operator approaches to constructing the physical spectrum make (8.41) an *a priori* postulate [11]. The generating functional approach to identities, used to derive the Ward–Takahashi identity, can also take advantage of the BRST transformation invariance to derive identities involving the Faddeev–Popov ghosts. These are more useful for nonabelian gauge theories, and so their derivation will be deferred to later in this section.

Another set of important identities can be derived from the translational invariance of the measure in the path integral. If any subset of fields appearing in the measure is translated by c-number or a-number functions, chosen to match the type of field being translated, the resulting path integral is unchanged. This is the field theoretic counterpart of the technique used in Sec. 4.3 to derive Ehrenfest's theorem for quantum mechanics. For example, the gauge fields appearing in the Lorentz gauge path integral may undergo the gauge transformation $A_\mu \to A_\mu - \partial_\mu \Lambda$, where Λ is an arbitrary solution to $\Box \Lambda = 0$. The resulting change in the action in the path integral is

$$S \to S + \int d^4x\, \Lambda\, \partial_\mu J^\mu , \tag{8.42}$$

where J^μ is the spinor current, not to be confused with the external source. Since the path integral is invariant under this change, it follows that the vacuum expectation value of $\partial_\mu J^\mu$ must vanish, since

$$\left. \frac{\delta}{\delta \Lambda(x)}{}_+\langle 0|0\rangle_- \right|_{\Lambda=0} = {}_+\langle 0|\partial_\mu J^\mu(x)|0\rangle_- \equiv \langle \partial_\mu J^\mu \rangle = 0 . \tag{8.43}$$

This generalizes to the case of time-ordered products that include $\partial_\mu J^\mu$.

Exercise 8.5: Show that the expectation value of the Euler–Lagrange equation vanishes.

The BRST transformation can be extended to the action appearing in the path integral form for a nonabelian gauge theory after the Faddeev–Popov determinant has been written in terms of ghosts. For the Feynman gauge with minimally coupled matter fields ψ_j the corresponding BRST transformations are

$$\delta A_\mu^a = -\partial_\mu c^a + g f_{bc}^a A_\mu^b c^c = -D_\mu^{ab} c^b , \quad \delta \psi_j = ig\, c^a\, T_{jk}^a\, \psi_k ,$$
$$\delta \bar{c}^a = \partial^\mu A_\mu^a , \quad \delta c^a = \tfrac{1}{2} g f_{bc}^a c^b c^c . \tag{8.44}$$

> **Exercise 8.6**: Verify that the BRST transformations of (8.44) leave the Lorentz gauge action invariant and have a Jacobian of unity.

It is now possible to construct the generating functional for the Feynman gauge version of the nonabelian gauge theory by attaching external sources for all the fields. This means adding to the total action the terms

$$\int d^4x \left(J^a_\mu A^{a\mu} + \bar{K}_j \psi_j + \bar{\psi}_j K_j + S^a c^a + \bar{S}^a \bar{c}^a \right) , \tag{8.45}$$

where $S, \bar{S}, K,$ and $\bar{K}$ are Grassmann-valued sources. In order that the source terms remain invariant under the BRST transformations it is necessary that the sources also be transformed, and their transformations are given by

$$\delta \bar{S}^a = 0 , \quad \delta S^a = \tfrac{1}{2} g f^a_{bc} S^b c^c ,$$
$$\delta \bar{K}_j = -ig \bar{K}_k T^a_{kj} c^a , \quad \delta K_j = ig c^a T^a_{jk} K_k , \tag{8.46}$$

It is also necessary that the J^μ satisfy the covariant conservation law

$$\partial_\mu J^{a\mu} - g f^a_{bc} A^b_\mu J^{c\mu} = D^{ab}_\mu J^{b\mu} = 0 . \tag{8.47}$$

The absence of a transformation property for $\bar{S}$ is due to the Lorentz gauge condition $\delta(\partial_\mu A^{a\mu})$ present in the measure.

> **Exercise 8.7**: Verify that the transformations (8.44) and (8.46), coupled with the condition (8.47), leave the source terms invariant.

This allows the nonabelian analog of the effective action to be defined, from which a generator for local Ward–Takahashi identities can be derived.

> **Exercise 8.8**: Derive the nonabelian analog of (8.32) using the BRST invariance of the action and source terms.

8.3 Deriving the Feynman Rules

A perturbative representation of the the general graphical forms defined in Sec. 8.1 can now be derived from the path integral. The basic idea is

326   Perturbation Theory

straightforward, and is the field theoretic extension of the idea presented in Sec. 3.2 for the quantum mechanical path integral. All terms higher than quadratic in the fields are written as functional derivatives with respect to the sources, and these serve to define the "bare" vertices of the theory. The remaining path integral, now solely quadratic and linear in the fields and sources, can be integrated exactly in the Gaussian approximation. In so doing, a set of "bare" propagators are derived for each field in the theory. The path integral is then represented by a power series in the bare vertices of the theory.

For the moment, consideration will be limited to a very simple theory of one scalar field ϕ. The starting point of this perturbative procedure is the separation of the action into three general pieces,

$$\mathcal{L} = \mathcal{L}_0 + \mathcal{L}_I + J\phi \ . \tag{8.48}$$

The term $\mathcal{L}_0$ represents the "basis" Lagrangian and is usually chosen to define a quadratic or free field theory. For the scalar field such a case occurs when $\mathcal{L}_0$ is given by

$$\mathcal{L}_0 = \tfrac{1}{2}\partial_\mu \phi \, \partial^\mu \phi - \tfrac{1}{2}m^2\phi^2 \ . \tag{8.49}$$

The term $\mathcal{L}_I$ contains all the terms cubic and higher in the fields, and for the sake of making a specific example, it will be assumed to take the form

$$\mathcal{L}_I = -\frac{1}{3!}g\phi^3 \ . \tag{8.50}$$

where g is a constant with the units of (length)$^{-1}$. The theory defined by this interaction is not consistent, and the reason for this is easy to see. Because the interaction is cubic, the energy of the theory is not bounded from below, and so no stable ground state in the fully interacting theory can exist. For that reason, this theory should be viewed only as a heuristic device for demonstrating the derivation of the perturbation series. It should also serve as a warning that a perturbative representation of a theory is almost always possible, but in some cases is entirely meaningless.

The generating functional for the vacuum expectation values can be written

$$Z[J] = \int_0^\circ \bar{\mathcal{D}}\phi \, \exp\left\{i \int d^4x \left(\mathcal{L}_0 + \mathcal{L}_I + J\phi\right)\right\} =$$
$$\exp\left\{i \int d^4x \, \mathcal{L}_I \left(\frac{\delta}{i\delta J(x)}\right)\right\} \int_0^\circ \bar{\mathcal{D}}\phi \, \exp\left\{i \int d^4x \left(\mathcal{L}_0 + J\phi\right)\right\} \ , \tag{8.51}$$

where $\delta/i\delta J(x)$ has been substituted as the argument of $\mathcal{L}_I$. Expanding the exponential containing the functional derivatives in a power series and

Sec. 8.3 Deriving the Feynman Rules

applying the result to the truncated path integral results in the original path integral. Therefore, the two forms for the path integral appearing in (8.51) are, at least formally, identical.

The basis path integral, denoted

$$Z_0[J] = \int_0^0 \bar{\mathcal{D}}\phi \, \exp\left\{ i \int d^4x \, (\mathcal{L}_0 + J\phi) \right\}, \tag{8.52}$$

will now be evaluated. This will first be done using the momentum space version of the Minkowski space theory derived in Sec. 6.8.1. The Euclidean space version will be evaluated later. The limits on the path integral allow an integration by parts, so that the action becomes

$$S_0 = -\int d^4x \, \left[\tfrac{1}{2}\phi(\Box + m^2)\phi - J\phi \right]. \tag{8.53}$$

The field ϕ and the source J are expanded in terms of the orthonormal eigenmodes according to the Fourier expansions

$$\phi(x) = \int \frac{d^4p}{(2\pi)^2} \, \tilde{\phi}(p) \, e^{ipx}, \quad J(x) = \int \frac{d^4p}{(2\pi)^2} \, \tilde{J}(p) \, e^{ipx}. \tag{8.54}$$

Denoting $\lambda_p = p^2 - m^2$ and substituting the expansions (8.54) into (8.53) reduces the action to

$$S_0 = \int d^4p \, \left[\tfrac{1}{2}\lambda_p \, \tilde{\phi}(-p) \, \tilde{\phi}(p) + \tilde{J}(-p) \, \tilde{\phi}(p) \right]. \tag{8.55}$$

From Sec. 6.8.3 the measure of the path integral is, up to an overall normalization factor,

$$\bar{\mathcal{D}}\phi = \prod_p d\tilde{\phi}(p). \tag{8.56}$$

The integration is performed by splitting the momentum space into two parts, one with $\omega > 0$, the other with $\omega < 0$. The measure can then be written

$$\bar{\mathcal{D}}\phi = \prod_{p(\omega>0)} d\tilde{\phi}(p) \, d\tilde{\phi}(-p), \tag{8.57}$$

while the action becomes

$$S_0 = \int_{p(\omega>0)} d^4p \, \left[\lambda_p \, \tilde{\phi}(p)\tilde{\phi}(-p) + \tilde{J}(-p)\tilde{\phi}(p) + \tilde{J}(p)\tilde{\phi}(-p) \right]. \tag{8.58}$$

Breaking the integral over p into a Riemann sum and suppressing the factors of d^4p, the integrations over $d\tilde{\phi}(p)$ yield a product of Dirac deltas of the form $\delta(\lambda_p \tilde{\phi}(-p) + \tilde{J}(-p))$. Using the properties of the Dirac delta the integrals over $d\tilde{\phi}(-p)$ are then performed, with the result that the entire path integral reduces to

$$Z_o[J] = N \left[\prod_{p(\omega>0)} \frac{1}{\lambda_p} \right] \exp\left\{ -i \int_{p(\omega>0)} d^4p\, \tilde{J}(-p) \frac{1}{\lambda_p} \tilde{J}(p) \right\}. \qquad (8.59)$$

where N is the overall normalization factors, independent of J.

There is a subtlety in this result stemming from the fact that there is a large domain of p for which $\lambda_p = 0$. In fact, for any $\omega = \pm\sqrt{m^2 + \mathbf{p}^2}$, referred to as *on mass shell*, it follows that $\lambda_p = 0$, and the formal integrations performed to evaluate the path integral break down. This can be remedied by performing the integrations in the Euclidean region, where $\lambda_p^E > m^2$, or it can be remedied by adding a small imaginary term to λ_p, so that $\lambda_p = p^2 - m^2 + i\epsilon$. This prevents λ_p from vanishing and allows the path integration to remain well defined. It is understood that the limit $\epsilon \to 0$ is to be taken at the end of all calculations. The latter technique will be employed here and will be referred to as *regulating* the inversion.

The product and integration in (8.59) can be extended to the entirety of momentum space, giving the form

$$Z_o[J] = N \left[\prod_p \frac{1}{\lambda_p} \right]^{\frac{1}{2}} \exp\left\{ \tfrac{1}{2} i \int d^4x\, d^4y\, J(x) \Delta(x-y) J(y) \right\} \qquad (8.60)$$

where the function Δ is given by

$$\Delta(x-y) = -\int \frac{d^4p}{(2\pi)^4} \frac{1}{\lambda_p} e^{ip(x-y)} = -\int \frac{d^4p}{(2\pi)^4} \frac{e^{ip(x-y)}}{p^2 - m^2 + i\epsilon}. \qquad (8.61)$$

In the limit $\epsilon \to 0$ Δ satisfies

$$(\Box_x + m^2)\Delta(x-y) = \delta^4(x-y), \qquad (8.62)$$

so that $\Delta(x-y)$ is the Green's function for the Klein–Gordon operator.

Equation (8.60) is the momentum space form of the configuration space result (6.301). Using the compact continuum notation of Chapter 7, the Green's function can be written

$$\Delta = \frac{1}{\Box + m^2 - i\epsilon}, \qquad (8.63)$$

while the overall factor can be written

$$\left[\prod_p \frac{1}{\lambda_p} \right]^{\frac{1}{2}} = \{\det(\Box + m^2 - i\epsilon)\}^{-\frac{1}{2}}. \qquad (8.64)$$

Sec. 8.3 Deriving the Feynman Rules   329

This is a useful result, since it allows a generalization of the path integral to other quadratic systems. In effect, the quadratic path integral serves to invert the differential operator present in the action.

It is straightforward to perform the integration over ω appearing in (8.61) using the Cauchy residue theorem and the definition of the step function (2.18) to obtain

$$\Delta(x_1 - x_2) = i \int \frac{d^3p}{(2\pi)^3 2\epsilon_p} \left[\theta(t_1 - t_2) e^{-ip(x_1-x_2)} + \theta(t_2 - t_1) e^{ip(x_1-x_2)} \right], \tag{8.65}$$

where p_0 is now "on shell," satisfying $p_0 = \epsilon_p = +\sqrt{m^2 + \mathbf{p}^2}$.

Exercise 8.9: Verify (8.65).

It is now possible to show that the Δ appearing in (8.65) is the vacuum expectation value of the time-ordered product of free scalar fields,

$$\Delta(x - y) = i \langle 0 | T\{\phi(x) \phi(y)\} | 0 \rangle . \tag{8.66}$$

This can be used to demonstrate that the path integral is reproducing the Dyson–Wick expansion for the time-ordered product of fields discussed at the end of Sec. 6.7. It also follows from (8.60) that for the free theory, $\mathcal{L}_I = 0$, (8.61) is the propagator defined in Sec. 8.1. Therefore, (8.61) is represented graphically by a line with no blob, since it is the lowest-order approximation to the propagator if interactions are present. Result (8.66) is true only for the particular method of regulation used in (8.61). Other methods of regulation lead to other methods of ordering the fields appearing in the perturbative expansion of the S-matrix. The time-ordered product of (8.66) is known as the *causal* Green's function despite the fact that it does not vanish for spacelike separations of its arguments. A discussion of the other types of ordering, such as *retarded* and *advanced*, as well as the space-time behavior of the different types of Green's functions lies outside the scope of this book [12].

The Euclidean space version of the path integral will now be evaluated to establish the relation of the two forms. After an integration by parts, the Euclidean action can be written

$$S_E = i \int d^4x \left[\tfrac{1}{2} \phi (\Box_E + m^2) \phi - J\phi \right] . \tag{8.67}$$

The real Euclidean field ϕ and the real source J possess the expansions

$$\phi(x) = \int \frac{d^4p}{(2\pi)^2} \phi_p e^{ipx}, \quad J(x) = \int \frac{d^4p}{(2\pi)^2} J_p e^{ipx} . \tag{8.68}$$

Perturbation Theory

In order that $\phi^* = \phi$, it is necessary that $\phi_p = \phi_{-p}$, with a similar relation holding for the source J_p. Inserting (8.68) into (8.67) therefore gives the result

$$S_E = i \int d^4p \left(\tfrac{1}{2} \lambda_p^E \phi_p^2 + J_{-p}\,\phi_p \right) , \qquad (8.69)$$

where $\lambda_p^E = p_E^2 + m^2$. The measure of the path integral is given by

$$\mathcal{D}\phi = \prod_p d\phi_p . \qquad (8.70)$$

The path integral is now a Gaussian form, and is integrated to obtain

$$\begin{aligned}Z[J] &= N \prod_p \left[\frac{1}{\lambda_p}\right]^{\frac{1}{2}} \exp\left\{ \tfrac{1}{2} \int d^4p\, J_p \frac{1}{\lambda_p^E} J_{-p} \right\} \\ &= \exp\left\{ \tfrac{1}{2} \int d^4x\, d^4y\, J(x) \Delta_E(x-y) J(y) \right\} , \end{aligned} \qquad (8.71)$$

where the Euclidean Green's function is given by

$$\Delta_E(x-y) = \int \frac{d^4p}{(2\pi)^4} \frac{e^{ip(x-y)}}{\lambda_p^E} . \qquad (8.72)$$

The Euclidean Green's function requires no regulation since the denominator possesses no poles for the $m \neq 0$ case. The result may be continued back to Minkowski space by the usual replacement $\omega \to -i\omega$ and $t \to it$. The result is the *unregulated* Minkowski space propagator derived before.

The Feynman rules of the simple scalar field theory can now be derived, but before doing so a step backward will be taken. Planck's constant will be reinserted into the path integral, so that

$$Z[J] = \int_0^0 \tilde{\mathcal{D}}\phi \, \exp\left\{ \frac{i}{\hbar} \int d^4x\, (\mathcal{L} + J\phi) \right\} . \qquad (8.73)$$

It is also necessary to insert $\hbar$ into the measure of the path integral as well, and this is equivalent to multiplying the scalar measure by $1/\sqrt{\hbar}$ at each value of p. It is not difficult to retrace the steps of this section to show that the the perturbation series is given by

$$\begin{aligned}Z[J] &= N \exp\left\{ \frac{i}{\hbar} \int d^4x\, \mathcal{L}_I\left(\frac{\hbar \delta}{i\,\delta J(x)}\right) \right\} \\ &\quad \times \exp\left\{ \frac{i}{2\hbar} \int d^4x\, d^4y\, J(x)\, \Delta(x-y)\, J(y) \right\} . \end{aligned} \qquad (8.74)$$

Sec. 8.3 Deriving the Feynman Rules

The overall factor N, representing measure factors and the determinant of $\Box + m^2 - i\epsilon$, can be ignored. This is because it has no reference to J and will therefore be cancelled by the normalization procedure.

The time-ordered products are obtained by functionally differentiating (8.74). If there are n fields in the time-ordered product, then the path integral must be differentiated n times initially. Following that, the formal expansion of the first exponential and its application to the second exponential gives a power series in the coupling constant g appearing in (8.50). In the limit that $J = 0$, this gives a perturbative representation of the time-ordered product of the original theory.

This expansion can be given a graphical representation known as the *Feynman rules* [13]. The rules are derived by noting that the act of functionally differentiating creates "half" of a bare propagator Δ, and that Δ is represented graphically as a bare line joining its two arguments. In the limit that $J = 0$, this half line must connect to another half line to complete the propagator. Each term in the power series expansion of the first exponential, for this special case, starts three half lines from the same spacetime point, and this represents the "bare vertex" of the theory. Therefore, the rth term in its expansion must have r points with three lines emanating from them. If there are r bare vertices, this corresponds to the rth term in the expansion of the exponential, and so it must be accompanied with an overall factor of $(-ig\hbar^2/3!)^r/r!$. Since $W[J]$ generates the connected n-point functions, all external points, representing the arguments of the n-point function, must connect to one of the half lines emanating from a vertex. The n-point function is therefore the sum of all Feynman graphs with n external lines or "legs" that can be drawn consistent with the Feynman rules of the theory. The Feynman rules for any field theory consist of the bare vertices and the propagators. For the ϕ^3 theory these are summarized in Fig. 8.4.

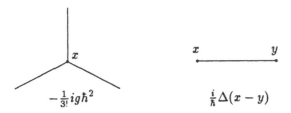

Fig. 8.4. The Feynman rules for the ϕ^3 theory.

Perturbation Theory

Each propagator line is accompanied by an overall factor of $i/\hbar$. Each graph associated with an n-point function also receives an overall factor of $(\hbar/i)^n$ due to the initial differentiations. The final overall numerical factor associated with a connected graph is an outgrowth of *combinatorics*, since there are many ways the lines leaving vertices may be connected to other vertices and external lines. The combinatoric factors associated with a particular graph are generated by the rules of differentiation and the overall powers of fields appearing in the bare vertex.

As a specific example, the $O(g^2)$ correction to the full propagator, or two-point function, is given in the ϕ^3 theory by the graph in Fig. 8.5.

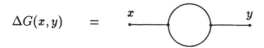

Fig. 8.5. The lowest-order correction to the propagator in ϕ^3 theory.

Since each vertex is accompanied by an integration over all space-time, and because the propagators possess a Fourier transform (8.61), the graph may be viewed in momentum space as well. Eventually the lines attached to the external points will be put "on mass shell," since the n-point function corresponds to a reduced S-matrix element, and this involves an integration against a plane wave. However, for now the lines can be viewed as carrying an arbitrary four-momentum. Because each vertex is integrated over all space-time the four-momentum is conserved at each vertex. It is left as an exercise to show that the graph of Fig. 8.5 has the Fourier transform given by

$$\Delta G(p) = \hbar \frac{1}{p^2 - m^2 + i\epsilon} \Sigma(p) \frac{1}{p^2 - m^2 + i\epsilon}, \qquad (8.75)$$

where the *self-energy* is defined as

$$\Sigma(p) = \tfrac{1}{2} i g^2 \hbar \int \frac{d^4k}{(2\pi)^4} \frac{1}{[k^2 - m^2 + i\epsilon][(p+k)^2 - m^2 + i\epsilon]}. \qquad (8.76)$$

Exercise 8.10: Verify (8.75) and (8.76).

The large k behavior of (8.76) shows four powers of k in the numerator and four powers of k from the measure. The result is an integral that

Sec. 8.3 Deriving the Feynman Rules

is logarithmically divergent, since in the limit that $k \to \infty$ the integrand behaves like dk/k. This is an example of the technique known as *power counting*. The evaluation and renormalization of this divergent example will be presented in the next section. However, it is important to note that the self-energy diagram alters the pole structure of the propagator from that of a free theory. This is seen from the fact that there is an infinite sequence of graphs that contain the self-energy diagram as a subgraph. These are depicted in Fig. 8.6.

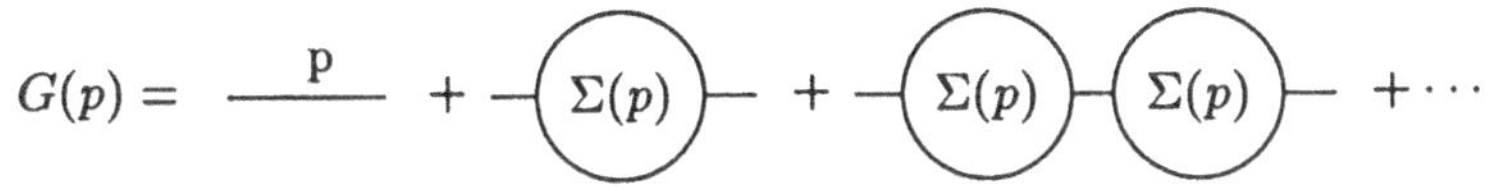

$$G(p) = \quad \frac{\quad p \quad}{\quad} \quad + \quad -\!\!\bigcirc\!\!\Sigma(p)\!\!\bigcirc\!\!- \quad + \quad -\!\!\bigcirc\!\!\Sigma(p)\!\!\bigcirc\!\!-\!\!\bigcirc\!\!\Sigma(p)\!\!\bigcirc\!\!- \quad + \cdots$$

Fig. 8.6. The self-energy correction to the poles of the propagator.

It is not difficult to show that the graphs depicted give the geometric series

$$G(p) = \frac{1}{p^2 - m^2 + i\epsilon} \sum_{n=0}^{\infty} \left[\frac{\Sigma(p)}{p^2 - m^2 + i\epsilon}\right]^n = \frac{1}{p^2 - \Sigma(p) - m^2 + i\epsilon}.$$
(8.77)

Such a general form for the self-energy corrections to the propagator is true in all field theories, and it shows that quantum or nonlinear effects can drastically alter the mass spectrum and even the stability of a theory.

The powers of $\hbar$ associated with a specific graph representing an N-point function can be computed in the following way. The graph consists of a set of N bare propagators, attached to the N external points that define the N-point function, I internal bare propagators not attached to the external points, and V bare vertices. Each propagator contributes a power of $\hbar$, while each vertex contributes an inverse power of $\hbar$. Therefore, the graph has the power of $\hbar$ given by $N + I - V$. The minimum value of $\hbar$ occurs for the case that I is zero. This can only occur if $V = 1$, so that the minimum power of $\hbar$ is $N - 1$. However, the definition of the generating functional $W[J]$ shows that $N - 1$ powers of $\hbar$ are cancelled by the powers of $\hbar$ resulting from the functional derivatives. Therefore, the net power of $\hbar$ associated with a graph is $I - V + 1$.

Of all the graphs, those containing L distinct closed loops, corresponding to an independent momentum integration in the integral representation of the graph, play a special role. Each distinct closed loop occurring in a graph corresponds to a factor of $\hbar$ in the overall power $I - V + 1$. This can be seen by splitting an internal propagator line making up part of the loop, thus removing the loop from the graph. The act of doing this creates two propagators from the one original propagator, and these attach to two new external points. In effect, the graph goes from L to $L-1$ and from I to $I-1$. The overall effect of losing the loop is the change to a graph with the power of $\hbar$ given by $(I-1) - V + 1 = I - V$, which is a graph with one less power of $\hbar$. Thus, each loop must correspond to a power of $\hbar$. A graph with zero loops corresponds to zero powers of $\hbar$, since the case where $I = 0$ and $V = 1$ corresponds to $L = 0$. From this it is inferred that the number of loops corresponds to the total power of $\hbar$ of the graph, and therefore $L = I - V + 1$.

This result demonstrates that the quantum effects are contained entirely in the graphs where loops occur. The other graphs consist of lines emerging from vertices and eventually joining to the external lines. These graphs are, for obvious visual reasons, referred to as *tree graphs*. The contribution of the tree graphs to a process does not vanish in the limit $\hbar \to 0$, and therefore corresponds to the *classical* limit of the process being calculated. An alternate and useful method of organizing the perturbation expansion is to group graphs by the number of loops present. This gives the power of $\hbar$ at which the graph contributes, and is known as the *loop expansion*. Of course, the results derived so far are valid only for the simple scalar field theory, but they may be generalized to other theories.

The Feynman rules for other theories may be derived in a similar manner, and the procedure for doing this, with few exceptions [14], begins by identifying the form of the quadratic path integral with a source attached. The evaluation of this path integral defines the propagators necessary for the perturbative expansion of the theory. This will now be done for the typical forms for the quadratic parts of the spinor, gauge, and ghost field actions. In order to simplify this derivation continuum methods will be employed. Either Euclidean or Minkowski space techniques can be used for the noninteracting theory, but for the evaluation of the spinor propagator Euclidean methods will be employed. This means the replacement $t \to -it$ to reach the Euclidean region. Minkowski space is regained by the replacement $t \to it$ at the end of the calculation.

The spinor propagator is defined by an integration over Grassmann variables of the form

$$Z[K, \bar{K}] = \int_0^0 \mathcal{D}\bar\psi \mathcal{D}\psi \, \exp\left\{ \frac{1}{\hbar} \int d^4x \left(\bar\psi i D_E \psi + \bar K \psi + \bar\psi K \right) \right\}, \quad (8.78)$$

where D_E is the manifestly Hermitian Euclidean Dirac operator, given by (6.282),

$$D_E = (i\gamma_0 \partial_0 + \gamma \cdot \nabla + im). \quad (8.79)$$

Sec. 8.3 Deriving the Feynman Rules

Performing the integral using the form (6.305) yields

$$Z[K, \bar{K}] = N \det(iD_E) \exp\left\{-\frac{1}{\hbar} \int d^4x\, d^4y\, \bar{K}(x)(iD_E)^{-1}(x,y)K(y)\right\}, \quad (8.80)$$

where N is the collection of normalization factors in the measure.

The inverse operator can be continued back to Minkowski space, but this requires care on two points. The first point is that the Euclidean differential operator being inverted has been written

$$iD_E(x, y) = (-\gamma_0 \partial_0 + i\gamma \cdot \nabla - m)\delta^4_E(x-y), \quad (8.81)$$

where there is a factor in the Dirac delta defined for *Euclidean time*. When the differential operator is continued to Minkowski time a factor of i appears,

$$\delta_E(t - t') = -i\,\delta(t - t'). \quad (8.82)$$

The second difficulty in performing this continuation is the fact that the Minkowski space Dirac differential operator may possess eigenvalues that are zero, and this renders the inversion poorly defined. Recalling the solution to this problem for the scalar field, the differential operator is modified to read $iD_E \to iD_E + i\epsilon$, where the limit $\epsilon \to 0$ is understood. The final result of continuing the path integral result back to Minkowski space is

$$Z[K, \bar{K}] = N \det(iD + i\epsilon)$$
$$\times \exp\left\{-\frac{i}{\hbar} \int d^4x\, d^4y\, \bar{K}(x)(iD + i\epsilon)^{-1}(x,y)K(y)\right\}, \quad (8.83)$$

where $iD = i\slashed{\partial} - m$ is the standard Dirac operator.

Inverting this operator is straightforward. For the boundary conditions implicitly present in the path integral, this inverse, denoted $S(x - y)$, must satisfy

$$(iD(x) + i\epsilon)S(x - y) = \delta^4(x - y), \quad (8.84)$$

where the Dirac delta is given by the standard Fourier transform. The inverse is thus given by

$$\begin{aligned} S(x - y) &= \int \frac{d^4p}{(2\pi)^4} \frac{e^{-ip(x-y)}}{\slashed{p} - m + i\epsilon} \\ &= \int \frac{d^4p}{(2\pi)^4} \frac{\slashed{p} + m}{p^2 - m^2 + i\epsilon} e^{-ip(x-y)}. \end{aligned} \quad (8.85)$$

The original form (8.78) shows that

$$\hbar^2 \frac{1}{Z} \frac{\delta^2 Z}{\delta \bar{K}(x)\,\delta K(y)}\bigg|_{K=0} = \frac{\langle 0|T\{\psi(x)\bar{\psi}(y)\}|0\rangle}{\langle 0|0\rangle}. \tag{8.86}$$

Using the form (8.83) shows that

$$\hbar^2 \frac{1}{Z} \frac{\delta^2 Z}{\delta \bar{K}(x)\,\delta K(y)}\bigg|_{K=0} = i\hbar\, S(x-y), \tag{8.87}$$

where, again, the anticommutativity of the Grassmann sources plays a critical role.

> **Exercise 8.11:** Using the free field decomposition of the spinor fields found in Sec. 6.5, along with the definition of time-ordering appropriate for anticommuting fields, show that the canonical formalism yields
>
> $$\langle 0|T\{\psi(x)\bar{\psi}(y)\}|0\rangle = i\hbar\, S(x-y). \tag{8.88}$$

In order to define the gauge field propagator a choice of gauge must be made. In the remainder of this section the Lorentz gauge will be used, and the derivation of the gauge field propagator in the Coulomb gauge will be left as an exercise. The Lorentz gauge propagator is defined by the Minkowski space path integral

$$Z[J_\mu] = \int_0^\circ \mathcal{D}A\, \delta(\partial_\mu A^\mu)\, \exp\left\{-\frac{i}{\hbar}\int d^4x\left(\tfrac{1}{2}A^\mu \Box A_\mu - J_\mu A^\mu\right)\right\}. \tag{8.89}$$

The gauge condition may be absorbed into the action by first writing the Dirac delta as an exponential and then continuing the path integral to Euclidean space. Writing the action in terms of the covariant Euclidean components gives

$$Z_E[J_\mu] = \int_0^\circ \mathcal{D}A\,\mathcal{D}\gamma\, \exp\left\{-\frac{1}{\hbar}\int d^4x\left(\tfrac{1}{2}A_\mu \Box_E A_\mu + (J_\mu^E + \kappa \partial_\mu \gamma)A_\mu\right)\right\}, \tag{8.90}$$

where an integration by parts has been performed on the γ Lagrange multipliers used to exponentiate the constraint. Because of the possibilty of scaling the argument of the Dirac delta by an arbitrary factor, the value of the real number κ appearing in (8.90) is also arbitrary. Such an ambiguity is useful, since the final results obtained for a physical process or a physically measurable quantity must be independent of κ, or any gauge

Sec. 8.3 Deriving the Feynman Rules

condition for that matter, and including it in the definition of the gauge field propagator gives a handy way to check the validity of a calculation [15]. For simplicity, κ will be set to unity in what follows. Use also has been made of the fact that $A_\mu^E A_E^\mu = -A_\mu A_\mu$. The source has absorbed the factor of i that characterized the Euclidean A_E^0 to become the Euclidean source, where $J_0^E = -iJ_0$.

The gauge field may now be integrated to obtain

$$Z_E[J_\mu] = \left[\det\left(\frac{1}{\Box_E}\right)\right]^{\frac{1}{2}} \int_0^0 \mathcal{D}\gamma$$

$$\times \exp\left\{-\frac{1}{2\hbar}\int d^4x\, d^4y\, (J_\mu^E - \partial_\mu\gamma)_x \left(\frac{1}{\Box_E}\right)_{x,y} (J_\mu^E - \partial_\mu\gamma)_y\right\}, \quad (8.91)$$

where the Euclidean inverse is given by

$$\frac{1}{\Box_E} = \int \frac{d^4p}{(2\pi)^4}\, \frac{e^{ip(x-y)}}{p_E^2}. \quad (8.92)$$

The Euclidean form p_E^2 becomes $-p^2$ when continued back to Minkowski space. The terms that with no reference to γ may be continued back to Minkowski space to give

$$\exp\left\{-\frac{1}{2\hbar}\int d^4x\, d^4y \left[J_j(x)\left(\frac{1}{\Box_E}\right)_{x,y} J_j(y) - J_0(x)\left(\frac{1}{\Box_E}\right)_{x,y} J_0(y)\right]\right\}$$

$$\xrightarrow{M} \exp\left\{-\frac{i}{2\hbar}\int d^4x\, d^4y \left[J_\mu(x)\left(\frac{1}{\Box - i\epsilon}\right)_{x,y} J^\mu(y)\right]\right\}. \quad (8.93)$$

where the inverse Minkowski operator has been regulated in the standard way.

The γ dependent terms in (8.91) may now be treated separately and integrated. The first step is to make an integration by parts in the exponential and to use the fact that

$$\overset{\rightarrow}{\partial}_\mu^x \left(\frac{1}{\Box_E}\right)_{x,y} \overset{\leftarrow}{\partial}_\mu^y = \delta^4(x-y). \quad (8.94)$$

This follows from the representation (8.92). The γ dependent terms in the path integral then take the form

$$\int_0^0 \mathcal{D}\gamma \exp\left\{-\frac{1}{\hbar}\int d^4x\, d^4y \left[\tfrac{1}{2}\gamma(x)\delta^4(x-y)\gamma(y)\right.\right.$$

$$\left.\left. -\gamma(x)\left(\partial_j^x\left(\frac{1}{\Box_E}\right)_{(x,y)} J_j(y) + \partial_o\left(\frac{1}{\Box_E}\right)_{(x,y)} iJ_0(y)\right)\right]\right\}. \quad (8.95)$$

Perturbation Theory

This is another Gaussian integral that, when integrated and continued back to Minkowski space, gives

$$N \exp\left\{\frac{i}{2\hbar} \int d^4x\, d^4y\, d^4z\, J^\mu(x) \partial^x_\mu \left(\frac{1}{\Box}\right)_{(x,z)} \left(\frac{1}{\Box}\right)_{(z,y)} \overleftarrow{\partial}^y_\nu J^\nu(y)\right\}. \quad (8.96)$$

Exercise 8.12: Verify result (8.96).

As it stands, result (8.96) is poorly defined. In momentum space it is possible to consolidate the integral over z to obtain

$$\int d^4z\, \partial^x_\mu \left(\frac{1}{\Box}\right)_{(x,z)} \left(\frac{1}{\Box}\right)_{(z,y)} \overleftarrow{\partial}^y_\nu = \int \frac{d^4p}{(2\pi)^4} \frac{p_\mu p_\nu}{p^4 + i\epsilon} e^{ip(x-y)}, \quad (8.97)$$

where the integral has been regulated by the factor of $i\epsilon$.

Putting (8.96) together with (8.93) in (8.91) gives the final result

$$Z[J_\mu] = \left[\det\left(\frac{1}{\Box - i\epsilon}\right)\right]^{\frac{1}{2}}$$
$$\times \exp\left\{\frac{i}{2\hbar} \int d^4x\, d^4y\, J^\mu(x) \Delta_{\mu\nu}(x-y) J^\nu(y)\right\}, \quad (8.98)$$

where

$$\Delta_{\mu\nu}(x-y) = -\int \frac{d^4p}{(2\pi)^4} \left[\frac{g_{\mu\nu}}{p^2 + i\epsilon} - \frac{p_\mu p_\nu}{p^4 + i\epsilon}\right] e^{ip(x-y)}. \quad (8.99)$$

It is to be noted that $\Delta_{\mu\nu}$ satisfies $\partial_\mu \Delta^{\mu\nu} = 0$. However, it is also true that $\Box \Delta_{\mu\nu}(x-y) \neq g_{\mu\nu} \delta^4(x-y)$.

This propagator gives a way of verifying that the ghosts in a manifestly covariant gauge have been decoupled from the S-matrix. The LSZ reduction formulas, presented for the scalar and spinor case in Sec. 6.6, can be derived for the gauge field case as well. The physical spectrum of the in and out states consists of transverse excitations and the single variety of allowed ghosts, which are associated with the Fourier transform of $\partial_\mu A^\mu$. A transverse excitation can be reduced from the in state to give

$$_{\text{out}}\langle A | \mathbf{p}, \gamma, B \rangle_{\text{in}} =$$
$$\frac{i}{\sqrt{Z}} \int d^4x\, \frac{1}{\sqrt{16\pi^3 \omega_k}} \epsilon^\lambda_\mu(p) e^{ipx} \Box_x \,_{\text{out}}\langle A | A^\mu(x) | B \rangle_{\text{in}}, \quad (8.100)$$

Sec. 8.3 Deriving the Feynman Rules

where ε^λ is the polarization vector introduced in Sec. 7.1, so that transverse excitations correspond to $\lambda = 1, 2$. A ghost may be reduced from the in state by using its relation to the Fourier transform of $\partial_\mu A^\mu$. This gives

$$\text{out}\langle A | \mathbf{p}, B \rangle_{\text{in}} =$$
$$\frac{i}{\sqrt{Z}} \int d^4x \, \frac{1}{\sqrt{16\pi^3 \omega_k{}^3}} e^{ipx} \Box_x \, \text{out}\langle A | \partial_\mu A^\mu(x) | B \rangle_{\text{in}} \, . \quad (8.101)$$

The latter equation vanishes by virtue of the fact that the QED gauge condition satisfies $\Box \partial_\mu A^\mu = 0$, or by treating the propagator for the ghost operator $\partial_\mu A^\mu$ as the divergence of the propagator for A^μ. Because the A^μ propagator is manifestly transverse, $\partial_\mu \Delta^{\mu\nu} = 0$, it follows that an external line corresponding to a ghost will vanish.

The Faddeev–Popov ghost propagator can be derived by similar methods. The generating functional is given by

$$Z[S, \bar{S}] = \int_0^0 \mathcal{D}c \, \mathcal{D}\bar{c} \, \exp\left\{ i \int d^4x \, \left(-\partial_\mu \bar{c} \, \partial^\mu c + S\bar{c} + \bar{S}c \right) \right\} . \quad (8.102)$$

It is straightforward to use the rules of Grassmann integration and the methods of this section to evaluate (8.102). The result is

$$Z[S, \bar{S}] = N \det\left(\Box - i\epsilon \right) \exp\left\{ \frac{i}{\hbar} \int d^4x \, d^4y \, \bar{S}(x) \Delta_g(x - y) S(y) \right\} , \quad (8.103)$$

where the ghost propagator Δ_g has the regulated Minkowski space representation

$$\Delta_g(x - y) = \int \frac{d^4p}{(2\pi)^4} \, \frac{e^{ip(x-y)}}{p^2 + i\epsilon} \, . \quad (8.104)$$

Exercise 8.13: Verify (8.103) and (8.104).

Exercise 8.14: Derive the Feynman rules for a nonabelian gauge theory in the Feynman gauge.

It is also possible to use the path integral representation of the generating functional to derive a nonlinear integral equation for the propagator of the theory. For simplicity, attention will be restricted to the case of a single scalar field theory. After adding a source term $J\phi$ to the action, the action

Perturbation Theory

is separated into two pieces, $\mathcal{L}_o + \mathcal{L}_I$. The path integral is then integrated to give the functional

$$Z[J] = N \exp\left\{i \int d^4z\, \mathcal{L}_I\left(\frac{\delta}{i\delta J(z)}\right)\right\}$$
$$\times \exp\left\{\tfrac{1}{2}i \int d^4x\, d^4y\, J(x)\Delta(x-y)J(y)\right\}, \qquad (8.105)$$

where Δ is the inverse of the differential operator appearing in $\mathcal{L}_o$, assumed to be $\Box + m^2$ for simplicity. The second functional derivative of (8.105) evaluated at $J = 0$ gives

$$-i\frac{1}{Z[J]}\frac{\delta^2 Z[J]}{\delta J(x)\,\delta J(y)}\bigg|_{J=0} = \Delta(x-y)$$
$$+ \int d^4z\, \Delta(x-z)\frac{1}{Z[J]}\mathcal{L}'_I\left(\frac{\delta}{i\delta J(z)}\right)\frac{\delta Z[J]}{\delta J(y)}\bigg|_{J=0}, \qquad (8.106)$$

where

$$\mathcal{L}'_I(\phi) = \frac{\partial \mathcal{L}_I}{\partial \phi}. \qquad (8.107)$$

The derivation of (8.106) relies upon the commutator

$$\left[\exp\left\{i \int d^4z\, \mathcal{L}_I\left(\frac{\delta}{i\delta J(z)}\right)\right\}, J(y)\right]$$
$$= \mathcal{L}'_I\left(\frac{\delta}{i\delta J(y)}\right)\exp\left\{i \int d^4z\, \mathcal{L}_I\left(\frac{\delta}{i\delta J(z)}\right)\right\}. \qquad (8.108)$$

The left-hand side of (8.106) is the propagator $G(x,y)$. Using this fact allows the definition of the self-energy $\Sigma(x,y)$ to be made through the identification

$$\frac{1}{Z[J]}\mathcal{L}'_I\left(\frac{\delta}{i\delta J(z)}\right)\frac{\delta Z[J]}{\delta J(y)}\bigg|_{J=0} = -\int d^4x\, \Sigma(z,x)\, G(x,y). \qquad (8.109)$$

Using the property of the Δ function that

$$(\Box + m^2)\Delta(x-y) = \delta^4(x-y), \qquad (8.110)$$

as well as the definition (8.109), reduces (8.106) to the integral equation

$$(\Box + m^2)G(x,y) = \delta^4(x-y) - \int d^4z\, \Sigma(x,z) G(z,y). \qquad (8.111)$$

Result (8.111) is known as the Dyson–Schwinger equation [16]. It may be inverted by assuming that $G(x,y)$ and $\Sigma(x,y)$ possess translationally invariant Fourier transforms, $\tilde{G}(p)$ and $\tilde{\Sigma}(p)$. Under those conditions (8.111) reduces to

$$\tilde{G}^{-1}(p) = -p^2 + m^2 + \tilde{\Sigma}(p)\,, \qquad (8.112)$$

which is the form already encountered in (8.77).

Exercise 8.15: Verify (8.112).

8.4 Renormalization

In this section a very brief and incomplete description of the process of renormalization [17] will be given. The reader wishing more details is referred to the many excellent texts on this subject [18]. The need for renormalization stems from the fact that the integrals that occur in the perturbation series are divergent. Renormalization is a consistent method for extracting finite and measurable results from these divergent integrals.

There are two steps to renormalization. The first step is to render finite or *regularize* the divergent expressions in a manner consistent with relativistic invariance and possible symmetries present in the theory. The most widely used method to do this is *dimensional regularization*. The second step is to remove the terms that are infinite in the absence of regularization by introducing *counterterms* and wave function renormalization factors. The demand that the result be unique leads to a set of criteria for determining which theories are renormalizable.

It is through renormalization that the quantum effects manifest themselves, and so renormalizability may be viewed as a criterion for the quantum mechanical consistency of a theory, at least in the absence of a nonperturbative solution. These quantum effects represent the deviation of the quantum mechanical system from its classical counterpart. One of these effects is to make the strength of interactions, defined through an *effective* coupling constant, momentum dependent [19]. Such a result might be expected since the quantum effects arise from the nonlinearity of the system.

However, it is this aspect of the renormalized perturbation expansion that allows the theory to make contact with physically measurable parameters. The physically measured values of transition elements, via cross sections and lifetimes, are correlated to the value of vertices and inverse propagators calculated using the perturbation series. These experimentally measured values give a definition of the coupling constants and masses appearing in the perturbative representation of these same vertices. As an example, the classical method of determining the value of the fundamental unit of electric charge e is to measure the force between *static* electrons at fixed separations. However, in quantum field theory it is necessary to define e by calculating a cross section for scattering or radiation and matching it to the measured value at some value of the momenta of the participating particles. This is referred to as an *effective coupling constant*. Values of the effective coupling constant may be defined at different values of momenta by different experiments, but the theory should predict the *scaling* between all possible values once all arbitrary parameters in the theory have been fixed.

As a simple example, the scalar theory considered in the previous section, defined by the interaction $g\phi^3/3!$, will be examined, for the moment ignoring the theory's ultimate instability. There are two arbitrary parameters in the theory, g and m^2, corresponding to a coupling and a mass, and these are initially unknown. The theory is used to calculate perturbation expansions for vertices and propagators to some order in g. If this theory were being proposed as a model for a physical situation, it would be necessary to match the theory's predictions to the observations of the physical system. This requires fixing the values of m and g used in the perturbation expansion. By analogy with the bare propagator, the pole of the full two-point function's Fourier transform should occur at the observed mass of the particle, denoted m_{ph}. This is equivalent to the statement

$$G^{-1}(p^2 = m_{\text{ph}}^2) = 0 \,, \tag{8.113}$$

where, of course, m_{ph}^2 must be finite and, for a stable particle, real. Likewise, a second physical measurement of the particle's behavior could be used to determine g. A possible measurement might be related to a physical process involving the one-particle-irreducible vertex with three lines at zero momentum, so that

$$\tilde{A}(0,0,0) = g_{\text{ph}} \,. \tag{8.114}$$

Once these two values are determined, the theory must be able to predict without ambiguity the values of all other n-point functions at all momenta. If this is not the case, and other parameters are needed, the theory is *nonrenormalizable*. The stipulations of (8.113) and (8.114) form a possible set of *renormalization conditions* for the ϕ^3 theory.

If the expressions appearing in the perturbation series representation of the vertices and n-point functions were not divergent, this procedure would

Sec. 8.4 Renormalization

be straightforward. For a general case one would calculate the n-point functions' values in terms of a well-behaved and convergent power series in the r parameters appearing in the action. The r parameters would be fixed by matching them to a set of r renormalization conditions, defined in terms of appropriate n-point functions evaluated at some set of momenta. In principle this would then uniquely fix the r parameters appearing in the expansions. Such a situation would correspond to a *finite renormalization* of the theory.

However, the fact that the perturbative expressions for the n-point functions are divergent in most theories creates an additional problem and ambiguity. To examine this aspect of the problem the result (8.76) for the ϕ^3 theory is recalled. To $O(g^2)$ in the initial coupling constant g, the inverse propagator is shifted by the self-energy to become

$$G^{-1}(p^2) = p^2 - m^2 - \Sigma(p^2, m^2, g^2) , \qquad (8.115)$$

where Σ is a divergent function of the *initial* parameters m and g.

Because Σ is divergent, any analysis of its properties requires that it be *regulated* by using some formal procedure to render it finite temporarily. This aspect of renormalization is present even at the classical level in electrodynamics, where the calculation of the energy present in the electrostatic field of a pair of charges requires that a cutoff be placed on the lower limit of the spatial integral used to calculate it [20]. One method of regularization is to place a *cutoff*, i.e., an upper limit, on the momentum space integrations that define Σ. If this renders the integrals finite, then the divergence is said to be *ultraviolet*. It is possible the integrals also diverge in the lower limit on the momenta, and if this is the case, the theory is said to possess *infrared* divergences. The latter are associated with theories where some or all of the particles are massless, and although the path integral is an excellent tool to attack the infrared divergences [14], they will be only briefly discussed later in this section and the next chapter. Of course, at the end of the calculation the cutoff should be *lifted*, i.e., allowed to become infinite, but in the interim the manipulations of the formally divergent expressions are well defined. A disadvantage of placing a cutoff on integrals is that it violates Poincaré invariance as well as other symmetries that are used to determine the finite part of a graph. For that reason, dimensional regularization is the preferred method in most calculations, and an example will be given later in this section.

Returning to the ϕ^3 theory, the regulated self-energy can be written

$$\Sigma(p^2, m^2, g^2) = \Sigma_d(p^2, m^2, g^2) + \Sigma_f(p^2, m^2, g^2) , \qquad (8.116)$$

where all the divergent contributions to the self-energy, Σ_d, have been uniquely identified and rendered finite temporarily by the regularization procedure being employed. The term Σ_f corresponds to the remainder of the self-energy that is finite in the absence of regularization. For the case of a divergent self-energy, the renormalization procedure consists of introducing a mass *counterterm*, $-\delta m^2 \phi^2$, into the set of *interactions*. This counterterm introduces a new bare vertex into the theory, depicted in Fig. 8.7. In what follows the $O(g^2)$ approximation to Σ, given by (8.76), will renormalized.

$$G_\delta(p) = \frac{p}{} + \frac{\delta m^2}{\times} + \frac{\delta m^2\ \delta m^2}{\times\ \ \times} + \cdots$$

Fig. 8.7. The mass counterterm corrections to the propagator.

It is easy to show that this vertex generates a geometric power series for the perturbative representation of the two-point function in the same manner as the original self-energy graphs, so that the full two-point function becomes

$$G^{-1}(p^2) = p^2 - m^2 - \Sigma_d(p^2, m^2, g^2) - \Sigma_f(p^2, m^2, g^2) - \delta m^2 . \quad (8.117)$$

Now, the demand that the propagator possesses a pole at $p^2 = m_{\text{ph}}^2$ gives

$$m_{\text{ph}}^2 - m^2 - \Sigma_d(m_{\text{ph}}^2, m^2, g^2) - \Sigma_f(m_{\text{ph}}^2, m^2, g^2) - \delta m^2 = 0 . \quad (8.118)$$

Since this relation must be sensible in the limit that the regularization procedure is lifted, the divergent part of the self-energy must be removed by the mass counterterm. This can be accomplished by choosing, to $O(g^2)$,

$$\delta m^2 = -\Sigma_d(m_{\text{ph}}^2, m^2, g^2) + g^2 C , \quad (8.119)$$

where C is a finite but *arbitrary* number. Relation (8.118) becomes

$$m_{\text{ph}}^2 - m^2 - \Sigma_f(m_{\text{ph}}^2, m^2, g^2) - g^2 C = 0 , \quad (8.120)$$

and this is assumed to be a well-defined equation for m^2 in terms of m_{ph}, g, and C. It is always possible to force the identification $m = m_{\text{ph}}$ by choosing C to satisfy

$$g^2 C = -\Sigma_f(m_{\text{ph}}^2, m_{\text{ph}}^2, g^2) , \quad (8.121)$$

but this is *not* necessary. The choice of $C = 0$ is referred to as *minimal subtraction*, since for that case the counterterm δm^2 solely removes the divergent part of the self-energy. The method of self-energy subtraction is quite similar to the necessity at the classical level of subtracting the electrostatic energy associated with both members of a pair of charges in order to obtain a finite answer for the additional energy induced by the pair [20].

It is customary to refer to $m^2 + \delta m^2 = m_0^2$ as the *bare mass*. The bare mass must be independent of the value of C for reasons that will become evident. For the simple example considered here it is true that, to $O(g^2)$, the bare mass is independent of the choice of C. This can be seen by combining (8.119) and (8.120) to find that

$$m_0^2 = m_{\text{ph}}^2 - \Sigma_d(m_{\text{ph}}^2, m^2, g^2) - \Sigma_f(m_{\text{ph}}^2, m^2, g^2) \,. \tag{8.122}$$

From the fact that Σ is itself proportional to g^2, the part of m^2 proportional to $g^2 C$ creates an $O(g^4)$ correction in the bare mass. The renormalization procedure is therefore consistent so far only to $O(g^2)$.

The inverse propagator has now become

$$\begin{aligned}G^{-1}(p^2) &= p^2 - m^2 - g^2 C - \Sigma_f(p^2, m^2, g^2) \\ &\quad - \left[\Sigma_d(p^2, m^2, g^2) - \Sigma_d(m_{\text{ph}}^2, m^2, g^2)\right] \,,\end{aligned} \tag{8.123}$$

and, unfortunately, the problems with divergences may not have been cured solely by a mass renormalization. It is possible that the quantity

$$\Sigma_d(p^2, m^2, g^2) - \Sigma_d(m_{\text{ph}}^2, m^2, g^2) \tag{8.124}$$

is still divergent for $p^2 \neq m_{\text{ph}}^2$. If this is the case, then any integration over p is doomed to give nonsense. However, it is possible to remove this divergence for special cases. Since Σ_d is regularized, its power series expansion in terms of p^2 around the physical value m_{ph}^2 is well defined. The lowest-order term in this expansion, independent of p^2, vanishes by construction, so that

$$\Sigma_d(p^2, m^2, g^2) - \Sigma_d(m_{\text{ph}}^2, m^2, g^2) = (p^2 - m_{\text{ph}}^2) \left.\frac{\partial \Sigma_d}{\partial p^2}\right|_{p^2 = m_{\text{ph}}^2}$$
$$+ \Sigma_R(p^2, m_{\text{ph}}^2, m^2, g^2) \,, \tag{8.125}$$

where the terms of order higher than p^2 represented by Σ_R are assumed to be *finite* when the regularization is lifted. The propagator has now become

$$G^{-1}(p^2) = p^2 - m^2 - g^2 C - (1-Z)p^2 - \Sigma_R - \Sigma_f \,, \tag{8.126}$$

where Z is defined as

$$1 - Z = \left.\frac{\partial \Sigma_d}{\partial p^2}\right|_{p^2 = m_{\text{ph}}^2} , \tag{8.127}$$

and is *apparently* divergent in the event that the regularization is lifted.

This divergence may be removed by adding the following counterterms to the Lagrangian, in addition to the mass counterterm:

$$\mathcal{L}_c = -\tfrac{1}{2}(m^2 + \delta m^2)(Z-1)\phi^2 + \tfrac{1}{2}(Z-1)\partial_\mu \phi\, \partial^\mu \phi \,. \tag{8.128}$$

Again, these generate a geometric series for the two-point function that may be summed exactly. The second term generates a correction to p^2, and the final result for the propagator is

$$G^{-1}(p^2) = p^2 - m^2 - g^2 C - \Sigma_R - \Sigma_f \,. \tag{8.129}$$

Exercise 8.16: Verify (8.129).

Form (8.129) is renormalized by noting that, at this level of approximation, both Σ_R and Σ_f are proportional to g^2, and this constant may be redefined. The coupling constant is scaled according to $g^2 \to g^2 Z$, where g remains finite, so that $\Sigma_R \to Z\Sigma_R$ and $\Sigma_f \to Z\Sigma_f$. The final form for the propagator becomes

$$G^{-1} = Z(p^2 - m^2 - g^2 C - \Sigma_R - \Sigma_f) \,. \tag{8.130}$$

The constant Z is the wave-function renormalization constant first introduced in Sec. 6.6. One need only recall that all fields are scaled by $\sqrt{Z}$ in the perturbation expansion, so that each propagator is scaled by an overall factor of Z. Therefore, the two factors of Z, one from the numerator, the other from the denominator, cancel. Because this scaling leaves the residue at the pole $p^2 = m_{\rm ph}^2$ unity for the propagator, it is known as wave-function renormalization. The observant or knowledgeable reader will note that Z appears on both sides of the defining equation (8.127) after the coupling constant is scaled by Z. This allows a solution for Z in terms of the divergent quantity proportional to p^2. This solution gives

$$1 - Z = Z \left.\frac{\partial \Sigma_d}{\partial p^2}\right|_{p^2 = m_{\rm ph}^2} \;\Rightarrow\; Z = \left(1 + \frac{\partial \Sigma_d}{\partial p^2}\right)^{-1} \,. \tag{8.131}$$

If the derivative of the self-energy is divergent, then this gives the well-known result that $Z \to 0$ as the regularization is lifted. It should be noted that the scaling of g does not alter the equation (8.120) for m^2.

The counterterms introduced so far bear a striking resemblance to the terms already in the action. This is because the counterterms must come

Sec. 8.4 Renormalization

from the original Lagrangian by suitable scalings of the fields and redefinitions of the constants. The original action was given by

$$\mathcal{L} = \tfrac{1}{2}\partial_\mu \phi\, \partial^\mu \phi - \tfrac{1}{2}m^2\phi^2 - \frac{1}{3!}g\phi^3 \;. \tag{8.132}$$

In the renormalization procedure the fields and constants present in (8.132) are understood as the bare field, ϕ_B, the bare coupling, g_B, and the bare mass, m_B^2. These are related to the *renormalized* fields, ϕ, and parameters, m^2 and g, by the relations

$$\phi_B = \sqrt{Z}\phi, \quad g_B = g + \delta g, \quad m_B^2 = m^2 + \delta m^2 \;. \tag{8.133}$$

Substituting (8.133) into (8.132) gives the basis action

$$\mathcal{L}_B = \tfrac{1}{2}\partial_\mu \phi\, \partial^\mu \phi - \tfrac{1}{2}m^2\phi^2 \;, \tag{8.134}$$

as well as the interaction vertices

$$\begin{aligned}\mathcal{L}_I &= \tfrac{1}{2}(Z-1)\partial_\mu\phi\,\partial^\mu\phi - \tfrac{1}{2}\delta m^2 \phi^2 \\ &\quad - \tfrac{1}{2}(m^2 + \delta m^2)(Z-1)\phi^2 - \frac{1}{3!}(g+\delta g)Z^{3/2}\phi^3 \;.\end{aligned} \tag{8.135}$$

All the terms necessary to renormalize the simple model up to $O(g^2)$ are present in (8.135). In particular, the scaling of the coupling constant is given by the choice

$$\delta g = \left(\frac{1-Z}{Z}\right)g \;. \tag{8.136}$$

The proof that this set of counterterms is adequate to renormalize the theory to all orders of perturbation theory is beyond the scope of this book [18]. However, the arbitrary term $g^2 C$ present in the renormalization of the self-energy can now be seen to correspond to the arbitrariness in how the bare mass is split into m^2 and δm^2. It is always possible to shift these terms by the same finite quantity, since it is necessary only that δm^2 removes the divergent terms. This one-parameter set of different renormalization conditions is known as the *renormalization group*. If the bare mass is not independent of the way it is split into m^2 and δm^2, then the theory is nonrenormalizable.

Such a procedure is the general method for renormalizing a theory, and if the set of all counterterms generated by these definitions is inadequate to remove all the divergences, i.e., if yet more counterterms are necessary, then the theory is perturbatively nonrenormalizable. A basic criterion for the renormalizability of a theory can be established by a *power counting* argument. This argument begins by noting that the perturbative corrections to vertices and propagators must match, to all orders in perturbation theory, the dimensions of the vertex or propagator being corrected. For example, all self-energy corrections to a scalar propagator must have the dimension (length)$^{-2}$, the same as (mass)^2. For a general vertex, Γ, the dimension or power of length associated to the vertex is denoted $D(\Gamma)$. The next step is to note that an arbitrary perturbative correction always takes the form of a product of coupling constants multiplied by a product of integrals over momentum space functions. By a power counting argument, this set of integrals will be ultraviolet divergent if its overall dimension, denoted $\delta(\Gamma)$, is less than or equal to zero. This is because, if $\delta(\Gamma)$ is less than or equal to zero, it indicates that the power of momentum appearing in the numerator is greater than or equal to the power of momentum appearing in the denominator. Such a result means that the integrals are at least logarithmically divergent when the cutoffs are lifted. If the dimension associated with the coupling constants appearing in Γ is denoted $C(\Gamma)$, it follows that $D(\Gamma) = C(\Gamma) + \delta(\Gamma)$, from which it follows that $\delta(\Gamma) = D(\Gamma) - C(\Gamma)$.

Since $D(\Gamma)$ is fixed, the overall dimensional behavior of the momentum integrals is determined by the dimensions of the coupling constants. If there exist coupling constants with positive dimension, then $C(\Gamma)$ can be made arbitrarily large and positive by considering higher and higher order graphs in perturbation theory. Therefore, the presence of coupling constants with positive dimension indicates that the perturbation expansion will eventually generate integrations over momenta that have an arbitrarily high degree of divergence in the ultraviolet region. For the case of the corrections to the propagator considered so far, this would eventually lead to a self-energy whose power series expansion would possess divergent coefficients beyond the p^2 term. The renormalization of this situation would require the introduction of an infinite set of new counterterms with derivatives of the fields and with coefficients that must be chosen order by order. Therefore, barring a set of "miraculous cancellations," the presence of positive dimension couplings will render a theory nonrenormalizable in the sense that it would require an infinite number of renormalization conditions to specify these new terms uniquely, thereby destroying the theory's predictive power. This does not constitute a proof that theories without positive dimension couplings are renormalizable, but in general it can be shown that the absence of such couplings is a sufficient condition to allow all divergences to be removed by mass, wave-function, and coupling constant renormalizations.

For example, in four dimensions the scalar field has the units (length)$^{-1}$. For a term with n powers of the scalar field, the coupling constant must have the units (length)$^{(n-4)}$. As a result, for $n > 4$ the coupling constant has positive dimension and the theory is nonrenormalizable.

> **Exercise 8.17**: Demonstrate that arbitrary powers of the scalar field may be present in a renormalizable scalar field theory in two space-time dimensions.

Another subtlety occurs in nonabelian gauge theories, and that is the fact that in the Lorentz gauge the same coupling constant appears in at least three vertices, and four if matter coupling is present. Considering only the gauge part of the theory, it can be shown that the basis action

$$\mathcal{L}_0 = \tfrac{1}{4}(\partial_\mu A^a_\nu - \partial_\nu A^a_\mu)(\partial^\mu A^{a\nu} - \partial^\nu A^{a\mu})$$
$$- \tfrac{1}{2}(\partial_\mu A^{a\mu})^2 - \partial_\mu \bar{c}^a \, \partial^\mu c^a , \qquad (8.137)$$

along with the vertices

$$\mathcal{L}_I = \tfrac{1}{4}(Z_3 - 1)(\partial_\mu A^a_\nu - \partial_\nu A^a_\mu)(\partial^\mu A^{a\nu} - \partial^\nu A^{a\mu})$$
$$- \tfrac{1}{2} Z_1 \, g f^a_{bc}(\partial_\mu A^a_\nu - \partial_\nu A^a_\mu) A^{b\mu} A^{c\nu}$$
$$+ \tfrac{1}{4} Z_4 \, g^2 f^a_{bc} f^a_{de} A^b_\mu A^c_\nu A^{d\mu} A^{e\nu}$$
$$- (\tilde{Z}_3 - 1)\, \partial_\mu \bar{c}^a \, \partial^\mu c^a + \tilde{Z}_1 \, g f^a_{bc} A^{a\mu} \partial_\mu \bar{c}^b \, c^c . \qquad (8.138)$$

defines a renormalizable theory. This is accomplished by calculating perturbative corrections, regularizing them, and defining the renormalization factors $Z_1, Z_3, Z_4, \tilde{Z}_3,$ and $\tilde{Z}_1$ to render the corrections finite. Demonstrating such calculations lies outside of the scope of this book [2, 18]. However, it is to be noted that these vertices can be obtained from the original bare Lagrangian by the rescalings

$$A^{a\mu}_B = \sqrt{Z_3}\, A^{a\mu}, \quad \bar{c}^a_B = \sqrt{\tilde{Z}_3}\, \bar{c}^a ,$$
$$c^a_B = \sqrt{\tilde{Z}_3}\, c^a , \quad g_B = Z_1 Z_3^{-3/2}\, g , \qquad (8.139)$$

along with the result that the term $\tfrac{1}{2}(\partial_\mu A^{a\mu})^2$ is not renormalized since it is zero. In order for this to be true it is necessary that the following relations hold:

$$\frac{Z_4}{Z_1} = \frac{Z_1}{Z_3} = \frac{\tilde{Z}_1}{\tilde{Z}_3} . \qquad (8.140)$$

The relations of (8.140) comprise the Slavnov–Taylor identities [21] and can be demonstrated to be a consequence of the Ward–Takahashi identities derived from the BRST invariance of the path integral. Nonabelian

350 Perturbation Theory

gauge theories are therefore made finite by the *multiplicative renormalization* prescription defined by (8.139) [22]. Similar results can be obtained for the matter-coupled nonabelian gauge theories.

The dimensional regularization procedure [23, 24] can now be demonstrated for the logarithmically divergent self-energy of the ϕ^3 theory [18], derived in the previous section,

$$\Sigma(p) = \tfrac{1}{2} ig^2 \hbar \int \frac{d^4k}{(2\pi)^4} \frac{1}{[k^2 - m^2 + i\epsilon][(p+k)^2 - m^2 + i\epsilon]} . \qquad (8.141)$$

The idea of dimensional regularization is to formally extend the divergent integrals into an arbitrary Euclidean space-time with noninteger dimension n. The divergences are then shown to manifest themselves as poles at the physically meaningful dimension, $n = 4$. This procedure breaks down when the important properties of a quantity being calculated are manifestly dependent upon the dimension of space-time. For example, the definition of the matrix $\gamma_5 = \gamma_0 \gamma_1 \gamma_2 \gamma_3$ does not allow an easy generalization to an arbitrary fractional dimension. Nevertheless, in most cases of interest the dimensional regularization technique is applicable.

To demonstrate this technique, the self-energy propagators are continued to the Euclidean region, so that $p^2 \to -p_E^2$ is negative for all values, and written as exponentials by using the form

$$\frac{1}{(m^2 - p^2 - i\epsilon)} = \int_0^\infty da \; \exp\{-a(m^2 - p^2 - i\epsilon)\} . \qquad (8.142)$$

After exchanging the order of integrals, in itself an act of analytic continuation, the self-energy becomes

$$\Sigma(p) = -\frac{\hbar g^2 \mu^{(4-n)}}{2(2\pi)^n} \int_0^\infty da \int_0^\infty db \int d^n k \; \exp\{-(a+b)m^2 + bp^2 + 2bp \cdot k + (a+b)k^2\} . \qquad (8.143)$$

The new constant μ has the units of mass, and its appearance is necessary because the overall coupling constant must have the units of $(\text{length})^{(n/2-3)}$ in the space-time dimension n, so that the coupling constant squared has the units $(\text{length})^{(n-6)}$.

Exercise 8.18: Show that the coupling constant has the units $(\text{length})^{(n/2-3)}$ in dimension n.

Sec. 8.4 Renormalization

By introducing μ the coupling constant g appearing in (8.143) is allowed to retain its four dimensional units, (length)$^{-1}$ and therefore its $n = 4$ value. However, μ is completely arbitrary. At first glance this step may seem pointless since the exponent of μ will vanish for $n = 4$. However, it will be seen that the presence of poles at $n = 4$ can make the choice of μ relevant to the scaling behavior of the integral, in particular for a massless theory.

The integrals appearing in (8.143) may be redefined by translating each k^μ by $p^\mu b/(a+b)$, valid due to the limits on the k integrals. The new variables $z = a + b$ and $x = a/z$ are introduced, and the integral becomes

$$\Sigma = -\frac{\hbar g^2 \mu^{(4-n)}}{2(2\pi)^n} \int_0^1 dx \int_0^\infty dz\, z \int d^n k\, \exp\{-z[m^2 - p^2 x(1-x)] + zk^2\}. \tag{8.144}$$

The Euclidean integrals over k may now be performed using the result

$$\int d^n k\, \exp(zk^2) = \left(\frac{2\pi}{z}\right)^{n/2}, \tag{8.145}$$

so that the self-energy becomes

$$\Sigma = -\frac{\hbar g^2 \mu^{(4-n)}}{2(4\pi)^{n/2}} \int_0^1 dx \int_0^\infty dz\, z^{(1-n/2)} \exp\{-z[m^2 - p^2 x(1-x)]\}. \tag{8.146}$$

The integral over z may be performed by noting its relation to the Gamma function of *arbitrary* argument, defined by

$$\Gamma(\alpha) = \int_0^\infty dx\, e^{-x} x^{\alpha-1}. \tag{8.147}$$

The result is

$$\Sigma = -\frac{\hbar g^2 \mu^{(4-n)}}{2(4\pi)^{n/2}} \Gamma\left(2 - \frac{n}{2}\right) \int_0^1 dx\, [m^2 - p^2 x(1-x)]^{(n/2-2)} \tag{8.148}$$

The divergence of the original integral has appeared in the form of a pole of the Gamma function. The presence of this pole is readily seen from the fact that the Gamma function satisfies the relation $\Gamma(n+1) = n\Gamma(n)$, along with the result that $\Gamma(1) = 1$. In the vicinity of $n = 0$, it can be shown that the Gamma function has the form

$$\Gamma(n) = \frac{1}{n} - \gamma_E + O(n), \tag{8.149}$$

where $\gamma_E = 0.577\ldots$, and is known as Euler's constant.

Perturbation Theory

The self-energy will be renormalized by performing a minimal subtraction. It is evident that the divergent part of the self-energy is associated with the pole of the Gamma function, and this divergent part is removed by defining the mass counterterm:

$$\delta m^2 = \frac{\hbar g^2}{32\pi^2} \frac{1}{(2 - \frac{n}{2})} . \qquad (8.150)$$

However, care must be taken in the limit $n \to 4$. This is seen from the fact that, to the lowest order in $(n-4)$,

$$[m^2 - p^2 x(1-x)]^{(n/2-2)} \to 1 + \left(\frac{n}{2} - 2\right) \ln [m^2 - p^2 x(1-x)]$$

$$\Gamma\left(2 - \frac{n}{2}\right) \to \frac{1}{2 - \frac{n}{2}} - \gamma_E$$

$$\mu^{(4-n)} \to 1 - \left(\frac{n}{2} - 2\right) \ln \mu^2 . \qquad (8.151)$$

Because of the pole in the Gamma function, the limit of the products of these expressions yields nonvanishing crossterms and is *not* the product of the individual limits. The final form for the minimally subtracted self-energy is

$$\Sigma = \frac{\hbar g^2}{32\pi^2} \int_0^1 dx \left\{ \ln \left[\frac{m^2 - p^2 x(1-x)}{\mu^2}\right] + \gamma_E \right\} . \qquad (8.152)$$

Exercise 8.19: Verify (8.152).

Remarkably, the arbitrary constant μ has entered into the final form of the mass-renormalized self-energy. Of course, a physical quantity, such as a cross section or lifetime, must be independent of the value of μ since the value chosen for it is entirely arbitrary. This leads directly to a differential equation determined by the following argument. If σ is a physically observable quantity that depends on the choice of parameters μ, g, and m, then the fact that it must be independent of μ is expressed by demanding

$$\mu \frac{d\sigma}{d\mu} = 0 . \qquad (8.153)$$

This yields the differential equation

$$\left(\mu \frac{\partial}{\partial \mu} + \beta(g,m) \frac{\partial}{\partial g} - \gamma_m(g,m) \frac{\partial}{\partial m^2}\right) \sigma = 0 , \qquad (8.154)$$

where the chain rule relates the coefficients according to

$$\beta(g,m) = \mu\frac{dg(\mu)}{d\mu}, \quad \gamma_m(g,m) = -\mu\frac{dm^2}{d\mu}. \tag{8.155}$$

Results (8.154) and (8.155) constitute the simplest form for the *renormalization group* or *Callan-Symanzik* equations [25]. These equations represent the fact that the values of g and m possess an arbitrariness in their finite parts. The freedom to choose any value for μ forces m and g to be related through the choice of μ. The method of applying the renormalization group is straightforward. For the case of ϕ^3 theory the bare coupling g_B and bare mass m_B^2 are determined by renormalizing the theory. Their derivatives with respect to μ allows the quantities β and γ_m to be determined by matching to (8.154). The equation for β in turn allows a determination of the dependence of g upon μ by solving the partial differential equation of (8.155).

For the case of a massless theory the renormalization group allows the scaling behavior of g, as a function of the Euclidean momentum, to be determined through the following argument. For the case of the ϕ^3 theory the renormalized self-energy for $m^2 = 0$ is given by

$$\Sigma = \frac{\hbar g^2}{32\pi^2}\left[\ln\left(\frac{-p^2}{\mu^2}\right) + \gamma_E - 2\right]. \tag{8.156}$$

From this expression it is apparent that the scaling of p^2, i.e., letting $p^2 \to \lambda p^2$, can be offset by scaling μ^2 according to $\mu^2 \to \lambda\mu^2$, so that the amplitudes become scale invariant with this prescription. Therefore, once the dependence of g upon μ is determined, the large p^2 behavior of physical quantities consistent with the renormalization group equation is determined by the large μ behavior of g. In effect, the renormalization group equation allows higher-order effects to be subsumed into the scaling behavior of the coupling constant.

The need to introduce a mass scale μ in order to renormalize a massless theory is brought about by the infrared divergences such theories possess. The bad infrared behavior will manifest itself as integrals that are logarithmically divergent in the vicinity of $p^2 = 0$. Of course, $p^2 = 0$ is precisely the mass-shell constraint for a massless theory. As a result, when such uncontrollable divergences are present, it is impossible to place diagrams on mass-shell. Such problems afflict massless nonabelian theories. Nevertheless, it is possible to renormalize such theories off mass-shell by maintaining $p^2 \neq 0$ and introducing the mass scale μ. Demanding that physical amplitudes are independent of the choice of mass scale μ allows the application of the renormalization group equation. Once the $\beta(g)$ function is identified, the definition (8.155) allows a solution for the behavior of the coupling constant as a function of the mass scale μ,

$$\int_{g_0}^{g} dg\, \beta^{-1}(g) = \ln\left(\frac{\mu}{\mu_0}\right), \qquad (8.157)$$

where g_0 is the value of the coupling constant at $\mu = \mu_0$. Because the simultaneous scaling of $\mu \to \infty$ and $p^2 \to \infty$ in the logarithms leaves the logarithm scale invariant, this gives the behavior of the effective coupling in the asymptotic region. Because large p is equivalent to small x in a Fourier transform, the scaling behavior of the effective coupling constant reveals the short-distance behavior of the theory. It is this approach which first demonstrated that the effective coupling in QCD tends to zero at high Euclidean momenta. This phenomenon is known as *asymptotic freedom* [26]. It means that, although QCD is a theory of strong interactions, at high Euclidean momenta the effective coupling constant between quarks and the gauge field tends to zero. For that reason, perturbative calculations of meaningful Euclidean processes, such as deep inelastic scattering amplitudes, are possible in QCD. Of course, since the effective coupling grows as $p^2 \to 0$, perturbation theory must break down, and it is therefore impossible to extrapolate the infrared behavior of the theory. While demonstrating asymptotic freedom for QCD is outside the scope of this book [2], the renormalization group will be applied to a simple model in Sec. 9.7 to demonstrate asymptotically free scaling behavior.

As a final note, useful for later applications, the renormalization conditions may be phrased in terms of the effective potential. For the case of the ϕ^3 theory, the mass may be renormalized by evaluating the inverse propagator at zero momentum, so that

$$\frac{\partial^2 V_{\text{eff}}(\phi_0)}{\partial \phi_0^2} = m^2. \qquad (8.158)$$

The coupling constant renormalization condition may be chosen to be the value of the three-point vertex at zero momentum, and in terms of the effective potential this would be expressed as

$$\frac{\partial^3 V_{\text{eff}}(\phi_0)}{\partial \phi_0^3} = g. \qquad (8.159)$$

The counterparts for the ϕ^4 theory will be used in the next chapter.

References

[1] The most cited example of perturbative failure is the single scalar field theory with an interaction described by $\lambda\phi^4$. Such a theory has been shown to be trivial in three and four space-time dimensions in the sense that it is equivalent to a free scalar field theory, and this is in direct contradiction of the low-order perturbative results. The original proofs of triviality are found in M. Aizenmann, Phys. Rev. Lett. **47**, 1 (1981); J. Fröhlich, Nucl. Phys. **B200** [FS4], 281 (1982).

[2] See, for example, C. Itzykson and J.-B. Zuber, *Quantum Field Theory*, McGraw-Hill, New York, 1980; G. Ross, *Grand Unified Theories*, Benjamin-Cummings, Menlo Park, California, 1985; P. Ramond, *Field Theory: A Modern Primer, Second Edition*, Addison-Wesley, Redwood City, California, 1989.

[3] J. Schwinger, Proc. Natl. Acad. Sci. (USA) **37**, 452 (1951); J. Schwinger, Phys. Rev. **82**, 664 (1951); J. Schwinger, *Particles and Sources*, Gordon and Breach, New York, 1969.

[4] K. Symanzik, Z. Natürforschung **9A**, 10 (1954).

[5] G. Jona-Lasinio, Nuovo Cimento **34**, 1790 (1964).

[6] An early review of these techniques is presented by B. Zumino in *Lectures on Elementary Particles and Quantum Field Theory, 1970 Brandeis University Summer Institute in Theoretical Physics*, ed. S. Deser, M. Grisaru, and H. Pendleton, MIT Press, Cambridge, Massachusetts, 1970.

[7] The extension of the effective action to the case of composite operators is found in J.M. Cornwall, R. Jackiw, and E. Tomboulis, Phys. Rev. **D10**, 2428 (1974).

[8] C. Becchi, A. Rouet, and R. Stora, Ann. Phys. **98**, 287 (1976); I.V. Tyupin, Lebedev preprint, FIAN No. 39 (1975), unpublished.

[9] J.C. Ward, Phys. Rev. **77**, 2931 (1950); Y. Takahashi, Nuovo Cimento **6**, 370 (1957).

[10] L. Baulieu, Phys. Rep. **129**, 1 (1985).

[11] N. Nakanishi and I. Ojima, *Covariant Operator Formalism of Gauge Theories and Quantum Gravity*, World Scientific, Singapore, 1990.

[12] See for example G. Barton, *Introduction to Advanced Field Theory*, Wiley-Interscience, New York, 1963; J.D. Bjorken and S.D. Drell, *Relativistic Quantum Fields*, New York, 1965.

[13] R.P. Feynman, Rev. Mod. Phys. **20**, 367 (1948).

[14] An example of an exactly solvable nonquadratic path integral appears in the treatment of infrared divergences in QED. See, for example, M.S. Swanson, Phys. Rev. **D28**, 798 (1983).

[15] A discussion of such use of different gauge conditions is found in G. Leibbrandt, Rev. Mod. Phys. **59**, 1067 (1987).

[16] F.J. Dyson, Phys. Rev. **75**, 1736 (1949); J. Schwinger, Proc. Nat. Acad. Sci. **37**, 452 (1951).

[17] The technique of renormalization stems from the early work of H. Bethe, Phys. Rev. **72**, 339 (1947); F.J. Dyson, Phys. Rev. **75**, 486 (1949); A. Salam, Phys. Rev. **82**, 217 (1951).

[18] J.C. Collins, *Renormalization*, Cambridge University Press, Cambridge, 1984. See also the texts cited in [2].

[19] Early work on the effective scaling of coupling constants was done by E.C.G. Stueckelberg and A. Peterman, Helv. Phys. Acta **26**, 499 (1953); M. Gell-Mann and F.E. Low, Phys. Rev. **95**, 1300 (1954).

[20] See, for example, J.D. Jackson, *Classical Electrodynamics, Second Edition*, Wiley, New York, 1975.

[21] A.A. Slavnov, Teoret. i Mat. Fiz. **10**, 99 (1972); J.C. Taylor, Nucl. Phys. **B33**, 436 (1971).

[22] The renormalizability of nonabelian gauge theories was first demonstrated by G. 't Hooft and M. Veltman, Nucl. Phys. **B44**, 189 (1972); *ibid.* **B50**, 318 (1972). The renormalization of gauge theories was discussed within the path integral formalism in B.W. Lee and J. Zinn-Justin, Phys. Rev. **D5**, 3121 (1972).

[23] Dimensional regularization stems from the works cited in [22], as well as that of C.G. Bollini and J.T. Giabidgi, Phys. Lett. **B40**, 566 (1972); J.F. Ashmore, Nuovo Cimento Lett. **4**, 289 (1972).

[24] An excellent review of dimensional regularization techniques is found in G. Leibbrandt, Rev. Mod. Phys. **47**, 849 (1975).

[25] C. Callan, Phys. Rev. **D2**, 1542 (1970); K. Symanzik, Comm. Math. Phys. **18**, 227 (1970).

[26] The concept of asymptotic freedom in QCD was first discussed in G. 't Hooft, Report at the Conference on Lagrangian Field Theories, Marseille, 1972; D.J. Gross and F. Wilczek, Phys. Rev. Lett. **30**, 1343 (1973); H.D. Politzer, Phys. Rev. Lett. **30**, 1346 (1973).

Chapter 9

Nonperturbative Results

In this chapter the path integral will be used to evaluate field theories for nonperturbative effects. Field theories are quite different than quantum mechanical systems in that they possess an infinite number of degrees of freedom. This was first encountered in Sec. 6.1, where the first continuous field theory model was the limit of an infinite assembly of coupled harmonic oscillators. One of the possible consequences of an infinite number of degrees of freedom is that there may be unitarily inequivalent representations of the quantized theory. This manifests itself in the form of energetically degenerate but orthogonal ground states available to the system, and the properties of the ground states may be determined by dynamical effects. This is the field theoretic extension of the concept of spontaneously broken symmetry discussed for a quantum mechanical system in Sec. 4.5, and it is characterized by the presence of a nonzero vacuum expectation value for one or more of the scalar or composite spinor fields.

In the quantum mechanical case discussed in Sec. 4.5 the two degenerate ground states were not truly orthogonal, and vacuum tunneling could take place; in the field theory case the degenerate vacuums are truly orthogonal. There are two important aspects of spontaneously broken symmetry. First, this mechanism can be adapted to give mass to the fields to which the scalar field is coupled. In particular, gauge fields may be rendered massive through this mechanism, and this is highly desirable since renormalizable, interacting, massive gauge field theories are difficult to formulate. The second aspect is problematic because the Goldstone theorem predicts the existence of zero mass excitations, Goldstone bosons, in the event that a continuous symmetry is broken by a nonzero vacuum expectation value. Since there are no physical zero mass bosons, this appears

358   Nonperturbative Results

to doom this approach to mass generation. However, the situation is rescued by the Higgs–Kibble mechanism in gauge theories, which allows these unwanted Goldstone bosons to be converted into ghosts while creating the longitudinal component of the massive vector boson.

In Sec. 9.1 the Goldstone theorem is proved using the effective action generating functional. In Sec. 9.2 the method of the effective potential is used to determine the conditions under which the scalar fields may develop a nonzero vacuum expectation value. In Sec. 9.3 spontaneously broken symmetry is examined for the case of a complex scalar field coupled to an abelian gauge field to demonstrate the Higgs–Kibble mechanism. In Sec. 9.4 its extension to the nonabelian case of the Glashow–Salam–Weinberg model is discussed. In Sec. 9.5 the chiral anomaly is demonstrated using path integral techniques, and is discussed as an example of an index theorem. In Sec. 9.6 the path integral is used to discuss the role of classical solutions to the equations of motion, and it is applied to solitons and to the solutions of the Euclidean equations of motion known as instantons. The chapter closes in Sec. 9.7 with applications of the effective potential to demonstrate dynamical breakdown of symmetry and to demonstrate restoration of symmetry at finite temperature.

9.1   The Goldstone Theorem

It was shown in Sec. 8.2 that the presence of a symmetry in the action gives rise to a set of identities for the Green's functions of the theory. These are conveniently expressed as a set of relations for the functional derivatives of the effective action. The Goldstone theorem is an outgrowth of these identities.

For the sake of simplicity, a theory composed solely of a set of scalar fields, ϕ^a, is considered, where a is an isospin index. The theory is assumed to be invariant under the action of some Lie group, G, and the representation of G is associated with a set of generators T^i_{ab} that act upon the representation space formed by the fields. The transformations of the scalar field are given by the usual form, $\phi' = U\phi$, so that the infinitesimal form of the transformation is given by

$$\delta\phi^a = i\Lambda^j T^j_{ab}\phi^b \,, \tag{9.1}$$

where the Λ^j are infinitesimal but arbitrary.

The effective action is also invariant under this transformation, so that

$$\Gamma[\phi^a_o] = \Gamma[\phi^a_o + \delta\phi^a_o] \,. \tag{9.2}$$

Sec. 9.1 The Goldstone Theorem

where ϕ_0^a is the vacuum expectation value of ϕ^a. Using a functional power series and the arbitrariness of the Λ^j, this invariance immediately yields a set of identities equal in number to the order of the group:

$$\int d^4x\, \phi_0^b(x)\, T_{ab}^i\, \frac{\delta \Gamma}{\delta \phi_0^a(x)} = 0. \tag{9.3}$$

Taking the functional derivative of this expression gives the identity

$$T_{ac}^i \frac{\delta \Gamma}{\delta \phi_0^a(y)} + \int d^4x\, \phi_0^b(x)\, T_{ab}^i\, \frac{\delta^2 \Gamma}{\delta \phi_0^a(x)\delta \phi_0^c(y)} = 0. \tag{9.4}$$

In Sec. 8.1 it was shown that the second derivative of Γ is the inverse of the *exact* propagator,

$$-\frac{\delta^2 \Gamma}{\delta \phi_0^a(x) \delta \phi_0^c(y)} = \Delta_{ac}^{-1}(x-y) = \int \frac{d^4p}{(2\pi)^4}\, G_{ac}^{-1}(p^2) e^{ip(x-y)}, \tag{9.5}$$

where

$$\Delta_{ac}(x-y) = \langle 0 | T\{\phi_a(x) \phi_c(y)\} | 0 \rangle. \tag{9.6}$$

In the true vacuum of the theory the first term in (9.4) vanishes due to (8.8). It is further assumed that any nonzero vacuum expectation value for the theory is a constant, thereby preserving translational invariance. For these conditions (9.5) becomes

$$G_{ab}^{-1}(p^2 = 0)\, T_{bc}^i \phi_0^c = 0. \tag{9.7}$$

From Sec. 8.4 it is clear that $-G_{ab}^{-1}(p^2 = 0)$ are the poles of the propagator, and therefore $-G_{ab}^{-1}(p^2 = 0)$ constitutes the elements of the *mass matrix* of the theory. The masses of the physically observable excitations of the theory are found by diagonalizing this matrix in isospin space. If there are nonzero values for some of the ϕ_0^a, then there will be a nonempty subset of generators for which

$$u_b^i = T_{bc}^i \phi_0^c \neq 0. \tag{9.8}$$

This is true because the representation of the group G is assumed to be irreducible. From (9.7) it is clear that the nonzero u_b^i form the elements of an eigenvector of the mass matrix with eigenvalue *zero*. Each generator that does not leave the multiplet formed by the ϕ_0^a invariant corresponds to a *broken symmetry*, and for every broken symmetry there is a corresponding zero mass excitation. This is the content of the Goldstone theorem [1].

9.2 The Effective Potential

The discussion of the Goldstone theorem gave no indication of a mechanism for causing the vacuum expectation of the scalar fields to be nonzero. The method of the effective potential [2], encountered first in the quantum mechanical case in Sec. 4.5 and defined for the field theory case in Sec. 8.1, is a convenient tool for probing the vacuum structure of a field theory. While it can be constructed by using its definition in terms of the zero momentum vertices, its evaluation is greatly simplified by using path integral techniques.

For the sake of simplicity a theory involving only a single scalar field will be considered initially. It is assumed that the Lagrangian has the form

$$\mathcal{L} = \tfrac{1}{2}\partial_\mu\phi\,\partial^\mu\phi - V(\phi)\,, \tag{9.9}$$

where the function $V(\phi)$ includes any mass terms present. If the theory possesses nontrivial vacuum expectation values, then the limits on the path integral must reflect this. The form for the transition element between coherent states, derived in Sec. 6.8, is given by

$$Z_{ab}[J] = \int_{\phi_a,\pi_a}^{\phi_b,\pi_b} \mathcal{D}\phi\,\mathcal{D}\pi\,\exp\left\{\frac{i}{\hbar}\int d^4x\left[\tfrac{1}{2}\dot\phi\pi - \tfrac{1}{2}\dot\pi\phi - \mathcal{H} + J\phi\right]\right\}, \tag{9.10}$$

The presence of nonzero limits requires some care, since the integration by parts on the $\dot\pi\phi$ term to put the action into its standard form creates an overall phase that depends on the limits. Recalling those properties of the coherent state that resulted in the limits of the path integral, it follows that the limits on the integral are precisely the vacuum expectations of the field and its conjugate momentum.

It is not difficult to show that the phase resulting from an integration by parts is zero for the cases that either the vacuum expectation value of the field is constant, so that $\pi_a = \pi_b = 0$, or that the values are the same for both the incoming and outgoing vacuums, so that $\pi_b\phi_b - \pi_a\phi_a = 0$. For the remainder of this section it will be assumed that the functional forms of the incoming and outgoing coherent states are the same, so that $\phi_a = \phi_b = \phi_0$, and that $\pi_a = \pi_b = 0$. For such a case the momentum may be integrated from the theory with the result that

$$Z[J] = \int_{\phi_0(x_-)}^{\phi_0(x_+)} \mathcal{D}\phi\,\exp\left\{\frac{i}{\hbar}\int d^4x(\mathcal{L}(\phi) + J\phi)\right\}, \tag{9.11}$$

where $\phi_0(x_\pm)$ denotes the form of the vacuum expectation value at the respective time limits. The limits on the path integral may now be absorbed into the action by translating the integration variables by $\phi(x) \to \phi(x) + \phi_c(x)$, where $\phi_c(x)$ is a solution to the equation of motion,

$$\Box \phi_c(x) + \frac{\partial V(\phi_c)}{\partial \phi_c(x)} = J(x) , \qquad (9.12)$$

that has the limits $\phi_c(x_\pm) = \phi_0(x_\pm)$. The resulting path integral is

$$Z[J] = \int_0^0 \mathcal{D}\phi \, \exp\left\{\frac{i}{\hbar} \int d^4x \, [\mathcal{L}(\phi + \phi_c) + J(\phi + \phi_c)]\right\} , \qquad (9.13)$$

and the method of steepest descent, a variant of the saddle-point approximation, is usually employed to evaluate (9.13). In this approximation the action is expanded around the value ϕ_c, and due to the limits on the path integral, integrations by parts are valid. The action appearing in the path integral becomes

$$\begin{aligned}\mathcal{L}(\phi + \phi_c) + J(\phi + \phi_c) &= \mathcal{L}(\phi_c) + J\phi_c - \phi\left(\Box\phi_c + \frac{\partial V(\phi_c)}{\partial \phi_c} - J\right) \\ &\quad + \tfrac{1}{2}\partial_\mu \phi \partial^\mu \phi - \tfrac{1}{2}\phi^2 \frac{\partial^2 V(\phi_c)}{\partial \phi_c^2} - \cdots ,\end{aligned} \qquad (9.14)$$

where the ellipsis refers to terms cubic and higher in ϕ as well as any counterterms necessary to renormalize the propagator. Recalling the equation of motion (9.12), it is apparent that the term linear in ϕ in (9.14) vanishes. Denoting

$$\frac{\partial^2 V(\phi_c)}{\partial \phi_c^2} = V''(\phi_c) , \qquad (9.15)$$

the path integral is approximately given by

$$\begin{aligned}Z[J] &\approx \int_0^0 \mathcal{D}\phi \, \exp\Big\{\frac{i}{\hbar}\int d^4x \, \Big[\mathcal{L}(\phi_c) + \tfrac{1}{2}\partial_\mu\phi\,\partial^\mu\phi \\ &\qquad -\tfrac{1}{2}V''(\phi_c)\phi^2 + J\phi_c\Big]\Big\} .\end{aligned} \qquad (9.16)$$

Form (9.16) is continued to Euclidean space and integrated to obtain

$$Z[J] \approx N \det\left[\Box_E + V''(\phi_c)\right]^{-\frac{1}{2}} \exp\left\{\frac{1}{\hbar}\int d^4x \, [\mathcal{L}_E(\phi_c) + J\phi_c]\right\} , \qquad (9.17)$$

where N is a constant independent of ϕ_c. The determinant appearing in (9.17) will be evaluated shortly. However, it must be once again noted that result (9.17) contains the same inherent flaw that was discussed in Sec. 4.5

for the quantum mechanical analog. The Gaussian approximation is valid only so long as the coefficient of the quadratic term remains positive. There may be values of ϕ_c for which this condition is not satisfied. As a result, the extension of the Gaussian approximation for the effective potential into those regions where the quadratic approximation breaks down must be viewed as an analytic continuation and not necessarily reliable.

A short digression demonstrates an important point. When the effective action is considered for constant sources $J(x) = j$, the generating functional $W[J]$ becomes the function $W[j]$ of the variable j. It is then a straightforward exercise to show that

$$\frac{\partial^2 W[j]}{\partial j^2} = \langle \left(\int d^4x\, \phi(x)\right)^2 \rangle - \left(\langle \int d^4x\, \phi(x)\rangle\right)^2 , \qquad (9.18)$$

where $\langle \cdots \rangle$ stands for the functional expectation value of the expression. Using the obvious inequality

$$\langle \left(\alpha - \int d^4x\, \phi(x)\right)^2 \rangle \geq 0 , \qquad (9.19)$$

where α is an arbitrary real number, it is easy to find a value for α such that (9.19) demonstrates that the right-hand side of (9.18) is a negative number, thereby giving the inequality

$$\frac{\partial^2 W[j]}{\partial j^2} \leq 0 . \qquad (9.20)$$

In Sec. 8.1 it was shown that the second derivative of the effective action is the inverse of the second derivative of W, as well as establishing the relation of the effective potential to the effective action. Using these results it can be shown that

$$\frac{\partial^2 V_{\text{eff}}(\phi_0)}{\partial \phi_0^2} \geq 0 . \qquad (9.21)$$

Result (9.21) shows that the *true* effective potential, calculated without approximation, must be a *convex* function of ϕ_0 [3].

The approximate form for the effective action is obtained by continuing result (9.17) back to Minkowski space and using the definitions of Sec. 8.1, so that

$$\Gamma[\phi_0] = -i\hbar \ln Z[J] - \int d^4x\, J\phi_0$$

$$\approx \int d^4x\, [\mathcal{L}(\phi_c) + J(\phi_c - \phi_0)] + \tfrac{1}{2} i\hbar \ln \det [\Box + V''(\phi_c)] . \qquad (9.22)$$

Sec. 9.2 The Effective Potential   363

The method of steepest descent has given the effective action to $O(\hbar)$.

The effective potential is obtained from (9.22) by setting $J = 0$ and treating ϕ_c and ϕ_0 as a constant. For the case that ϕ_c is a constant, it is clear that, since it interpolates between the incoming and outgoing form for ϕ_0, ϕ_c must coincide with ϕ_0. It is now necessary to evaluate the logarithmic term. This is accomplished easily in Euclidean space, where the effective potential is given by

$$V_{\text{eff}}(\phi_0) \int d^4x = V(\phi_0) \int d^4x + \tfrac{1}{2}\hbar \ln \det \left[\Box_E + V''(\phi_0)\right] , \qquad (9.23)$$

where the volume elements are Euclidean, related to the Minkowski volume by a factor of i. The logarithm of the determinant of this expression is obtained in the usual functional manner by using the identity for Hermitian differential operators derived in Sec. 1.4,

$$\det D = \exp\left\{\sum_j \ln \lambda_j\right\} \Rightarrow \ln(\det D)^{-\tfrac{1}{2}} = -\tfrac{1}{2} \sum_j \ln \lambda_j , \qquad (9.24)$$

where the λ_j are the eigenvalues of D. For the case under consideration,

$$\lambda_j = p_E^2 + V''(\phi_0) . \qquad (9.25)$$

The sum over j is converted to an integral over p in the usual manner. Denoting the volume of Euclidean space-time as L^4, the sum becomes

$$\sum_j = \lim_{L \to \infty} \frac{L^4}{(2\pi)^4} \sum_j \frac{(2\pi)^4}{L^4} \to \int d^4x \int \frac{d^4p}{(2\pi)^4} \qquad (9.26)$$

showing the emergence of an overall factor of the Euclidean volume. Dividing out this volume of space-time in accordance with the definition of the effective potential, the final form of the logarithmic contribution appearing in (9.22) to the effective potential is given by

$$V_1 = \tfrac{1}{2}\hbar \int \frac{d^4p}{(2\pi)^4} \ln\left[p_E^2 + V''(\phi_0)\right] . \qquad (9.27)$$

Nonperturbative Results

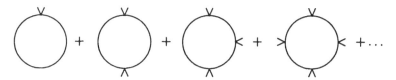

Fig. 9.1. The one-loop series for the effective potential.

It can be shown that this $O(\hbar)$ contribution is in fact the sum of all Euclidean one-loop diagrams of Fig. 9.1 with external momenta zero generated using the vertex

$$\mathcal{L}_I = -\tfrac{1}{2}\phi^2 V''(\phi_c) \,, \tag{9.28}$$

although obtaining the combinatoric factors that lead to the logarithm is subtle [4]. The restriction to zero external momenta for the vertex is attained by making the graph a closed loop. This gives the general technique to calculate the effective potential via graphical methods as a power expansion in $\hbar$, representing it as a sum of closed loops generated by the translated action. The number of loops, as discussed in Sec. 8.3, determines the power of $\hbar$ associated with the graph, and therefore the diagram's contribution to the effective potential. In a similar way the appearance of $V(\phi_0)$ in the effective potential can be understood as the sum of all zero-loop or *tree* graphs.

The integral appearing in (9.27) is divergent and requires renormalization. In evaluating the integral it is important to remember that only its derivatives with respect to ϕ_0 correspond to physically meaningful quantities, and therefore it is possible to ignore terms that are independent of ϕ_0. The integral is regularized by placing a cutoff Λ on the momentum limits. Denoting $M^2 = V''(\phi_0)$, the integration over $k_0 = \omega$ gives

$$\int \frac{d^3k}{(2\pi)^3} \frac{d\omega}{2\pi} \ln(\omega^2 + k^2 + M^2) =$$
$$\int \frac{d^3k}{(2\pi)^3} \tfrac{1}{2}\sqrt{k^2 + M^2} + \int \frac{d^3k}{(2\pi)^4} \left\{\omega \ln(\omega^2 + k^2 + M^2) - 2\omega\right\}\bigg|_{-\Lambda}^{\Lambda}. \tag{9.29}$$

Exercise 9.1: Prove that the second integral on the right-hand side of (9.29) is independent of M^2 in the limit that $\Lambda \to \infty$ and may therefore be discarded.

Using spherical coordinates and placing the cutoff Λ on the k appearing in the remaining integral gives

$$V_1 = \frac{\hbar}{32\pi^2}\left[2\Lambda^4\left(1+\frac{M^2}{\Lambda^2}\right)^{3/2} - M^2\Lambda^2\left(1+\frac{M^2}{\Lambda^2}\right)^{1/2}\right.$$
$$\left. - M^4\ln\left(\Lambda+\Lambda\sqrt{1+\frac{M^2}{\Lambda^2}}\right) + \tfrac{1}{2}M^4\ln M^2\right]. \quad (9.30)$$

Result (9.30) is expanded in a power series in M^2/Λ^2, and all terms that vanish in the limit $\Lambda \to \infty$ are discarded. The result is

$$V_1 = \frac{\hbar\Lambda^4}{16\pi^2} + \frac{\hbar\Lambda^2 M^2}{16\pi^2} + \frac{\hbar M^4}{64\pi^2}\ln\left(\frac{M^2}{4\Lambda^2}\right). \quad (9.31)$$

Discarding the first term as irrelevant since it is independent of M, the final form for the regularized effective potential is

$$V_{\text{eff}}(\phi_0) = V(\phi_0) + \frac{\hbar\Lambda^2}{16\pi^2}V''(\phi_0) + \frac{\hbar}{64\pi^2}[V''(\phi_0)]^2\ln\left(\frac{V''(\phi_0)}{4\Lambda^2}\right). \quad (9.32)$$

At this point the problem will be specialized to the case of the potential

$$V(\phi) = \tfrac{1}{2}m^2\phi^2 + \tfrac{1}{4}\lambda\phi^4. \quad (9.33)$$

The theory described by (9.33) has been shown to be trivial in three spatial dimensions in the sense that its associated S-matrix is unity [5]. In effect, the coupling constant is forced to zero by high-order renormalization effects. As a result, the manipulations of the potential (9.33) are purely formal if the theory is restricted to a single scalar field. It is widely believed that the theory is not trivial if the scalar field is coupled to other fields, such as gauge fields, so that in such a more general case the contribution to the effective potential derived here is relevant.

The effective potential is renormalized by imposing the renormalization conditions

$$\left.\frac{\partial^2 V(\phi_0)}{\partial \phi_0^2}\right|_{\phi_0=0} = m^2, \quad \left.\frac{\partial^4 V(\phi_0)}{\partial \phi_0^4}\right|_{\phi_0=0} = 6\lambda. \quad (9.34)$$

These conditions match the form of the original potential (9.33). It should be noted that the m and λ appearing in (9.33) are the bare mass and bare coupling constant. This is because the form $V(\phi_0)$ appearing in the effective potential arises from the bare Lagrangian. However, the form $V''(\phi_0)$ appearing in the one-loop correction to the effective potential contains only the renormalized coupling constants. This occurs because the

366 Nonperturbative Results

one-loop corrections are calculated by separating the mass, coupling constant, and wave-function renormalization counterterms and treating them as part of the higher-order interactions in (9.14) that were ignored in the Gaussian approximation. There is no need for wave-function renormalization at this level of approximation because the vacuum expectation value has been chosen to be a constant.

The bare mass and bare coupling constant are written $m_0^2 = m^2 + \delta m^2$ and $\lambda_0 = \lambda + \delta\lambda$, respectively. The effective potential (9.32) is evaluated using

$$V(\phi_0) = \tfrac{1}{2}(m^2 + \delta m^2)\phi_0^2 + \tfrac{1}{4}(\lambda + \delta\lambda)\phi_0^4 \,,$$
$$V''(\phi_0) = m^2 + 3\lambda\phi_0^2 \,. \tag{9.35}$$

It is tedious but straightforward to show that the renormalization conditions of (9.34) are satisfied if

$$\delta m^2 = -\frac{3\hbar}{32\pi^2}\left[4\lambda\Lambda^2 + 2\lambda^2 m^2 \ln\left(\frac{m^2}{4\Lambda^2}\right) + \lambda m^2\right],$$
$$\delta\lambda = -\frac{9\hbar\lambda^2}{32\pi^2}\left[2\ln\left(\frac{m^2}{4\Lambda^2}\right) + 3\right]. \tag{9.36}$$

Substituting the values (9.36) into the original form of the effective potential and taking the limit $\Lambda \to \infty$ gives the renormalized form

$$V_{\text{eff}}(\phi_0) = \tfrac{1}{2}m^2\left(1 - \frac{3\hbar\lambda}{32\pi^2}\right)\phi_0^2 + \tfrac{1}{4}\lambda\left(1 - \frac{27\hbar\lambda}{32\pi^2}\right)\phi_0^4$$
$$+ \frac{\hbar}{64\pi^2}(m^2 + 3\lambda\phi_0^2)^2 \ln\left(1 + \frac{3\lambda\phi_0^2}{m^2}\right). \tag{9.37}$$

If m^2 and λ are positive and λ is small, as it must be in order for the higher-order terms to be neglected, then it follows that the effective potential of (9.37) has its minimum at $\phi_0 = 0$. However, if $m^2 = -\mu^2$, where $\mu^2 > 0$, the true minima of the effective potential are no longer zero. This is trivial to see in the limit that $\hbar \to 0$, where only the tree approximation survives. Then there are two minima, given by

$$\phi_0 = \pm\sqrt{\frac{\mu^2}{\lambda}} \equiv \pm\sigma \,, \tag{9.38}$$

and the mass of the field, defined as the value of the second derivative in the true vacuum, is given by

$$\left.\frac{\partial^2 V(\phi_0)}{\partial \phi_0^2}\right|_{\phi_0} = 2\mu^2 \,. \tag{9.39}$$

Sec. 9.2 The Effective Potential

For this particular case there is no Goldstone boson since the original theory had no continuous symmetry, only the discrete symmetry $\phi \to -\phi$. However, this discrete symmetry is now broken since the theory oscillates around the vacuum expectation value. It follows that the effective interaction in the broken vacuum is given by

$$\mathcal{L}_I = \pm\sqrt{\mu^2\lambda}\,\phi^3 + \tfrac{1}{4}\lambda\phi^4 , \tag{9.40}$$

showing that the symmetry $\phi \to -\phi$ has disappeared.

However, it is trivial to see that the effective potential calculated in this approximation is not convex. At this point, it is important to show that the two coherent states, denoted $|\sigma\rangle$ and $|-\sigma\rangle$ and corresponding to the harmonic oscillator approximation to the ground states, are orthogonal.

Exercise 9.2: Using the results obtained in Sec. 6.8 for the inner product of two coherent states, show that $\langle\sigma|-\sigma\rangle = 0$.

It is also true that these two states must be energetically degenerate, so that $\langle\sigma|H|\sigma\rangle = \langle-\sigma|H|-\sigma\rangle = E_0$. Using the orthogonality and degeneracy of these two states, it is straightforward to show that the normalized state constructed from them,

$$|\phi_0\rangle = \alpha^{\frac{1}{2}}|\sigma\rangle + (1-\alpha)^{\frac{1}{2}}|-\sigma\rangle , \tag{9.41}$$

where α is an arbitrary number between zero and unity, is energetically degenerate to the two states and gives the expectation value

$$\langle\phi_0|\phi|\phi_0\rangle = (1-2\alpha)\sigma . \tag{9.42}$$

The fact that there exists a sequence of energetically degenerate states between the values $-\sigma$ and σ indicates that the convexity dilemma is associated with the breakdown of the Gaussian approximation that occurs for small magnitudes of ϕ_0. The standard solution is to employ the Maxwell construction for V_{eff}, and this simply sets the value of V_{eff} at all points between σ and $-\sigma$ equal to the value at σ [6, 7].

The presence of other types of fields in the theory may lead to additional contributions to the effective potential. For example, the spinor field action given by

$$\mathcal{L}_f = \bar{\psi}(i\slashed{\partial} - m)\psi - g\phi\bar{\psi}\psi \tag{9.43}$$

is Lorentz invariant and describes the nonlinear interaction of the spinor field ψ with the scalar field ϕ through what is known as a *Yukawa vertex*.

After shifting the scalar field by ϕ_0 and discarding all terms higher than quadratic in the action, the fermionic contribution to the effective action is found from the path integral,

$$V_f(\phi_0) \int d^4x = i\hbar \ln \int_0^0 \mathcal{D}\bar{\psi}\mathcal{D}\psi \exp\left\{\frac{i}{\hbar}\int d^4x\, \bar{\psi}(i\slashed{\partial} - m - g\phi_0)\psi\right\}. \quad (9.44)$$

This integral is evaluated by the familiar technique of continuing the integral to Euclidean time and using the rules for Grassmann integration. After discarding the irrelevant constant terms, the resulting contribution of the fermionic sector to the effective potential is

$$V_f(\phi_0)\int d^4x = -\hbar \ln \det(i\slashed{\partial}_E - m - g\phi_0), \quad (9.45)$$

where the fermionic determinant appears with the power unity, rather than the power of $-\frac{1}{2}$ as in the bosonic case. Using the identity (1.65) and denoting $M_f = m + g\phi_0$, the logarithmic term can be written using a functional Taylor series as

$$\begin{aligned}
\ln\det(i\slashed{\partial}_E - m - g\phi_0) &= \text{Tr}\ln(i\slashed{\partial}_E - M_f) \\
&= \text{Tr}\ln(i\slashed{\partial}_E) + \text{Tr}\ln\left(1 + \frac{iM_f}{\slashed{\partial}_E}\right) \\
&= \text{Tr}\ln(i\slashed{\partial}_E) - \sum_{n=1}^{\infty}\frac{(-iM_f)^n}{n}\text{Tr}\frac{1}{(\slashed{\partial}_E)^n}, \quad (9.46)
\end{aligned}$$

where the trace is over the spinor indices as well as the space-time arguments. The first trace may be dropped since it contains no reference to ϕ_0. The second trace may be performed by using the representation

$$\frac{1}{(\slashed{\partial}_E)} = \int \frac{d^4p}{(2\pi)^4}\frac{\slashed{p}_E}{p_E^2}e^{ip(x-y)}. \quad (9.47)$$

Using the Euclidean spinors derived in Sec. 6.8.4, it follows that

$$\text{Tr}\slashed{p}_E^{\,n} = \begin{cases} 4(p_E^2)^{n/2} & \text{if } n \text{ is even} \\ 0 & \text{if } n \text{ is odd.} \end{cases} \quad (9.48)$$

Combining this result with the rules for performing the trace over the space-time arguments gives

$$\text{Tr}\frac{1}{(\slashed{\partial}_E)^n} = \int d^4x \int \frac{d^4p}{(2\pi)^4}\frac{4}{(p_E^2)^{n/2}}, \quad (9.49)$$

if n is even, and zero if n is odd. The overall volume of Euclidean space has emerged and is cancelled by the definition of the effective potential. The final result is

$$V_f(\phi_0) = 4\hbar \sum_{n=2,4,...}^{\infty} \int \frac{d^4p}{(2\pi)^4} \frac{(-iM_f)^n}{n\,(p_E^2)^{n/2}}$$

$$= -2\hbar \int \frac{d^4p}{(2\pi)^4} \ln(p_E^2 + M_f^2). \quad (9.50)$$

The familiar integral, already analyzed earlier in this section, has reappeared, but this time with a negative coefficient. Its regularization is identical to that of the scalar contribution. The form prior to renormalization, but after dropping irrelevant terms, is given by

$$V_f(\phi_0) = -\frac{\hbar \Lambda^2 M_f^2}{4\pi^2} - \frac{\hbar M_f^4}{16\pi^2} \ln\left(\frac{M_f^2}{4\Lambda^2}\right). \quad (9.51)$$

As in the bosonic case, result (9.50) can also be shown to be the sum of all one-loop fermionic diagrams generated by the vertex

$$\mathcal{L}_f = -M_f \bar\psi \psi, \quad (9.52)$$

Comparing the effective potential for the two cases shows the interesting result that the fermionic loops enter into Feynman graphs with a sign opposite to that of the bosonic loops. This mechanism is used in supersymmetric field theories to soften divergences by forcing equal numbers of fermionic and bosonic degrees of freedom to be present. Loop diagrams, the source of divergent integrals, occur in supersymmetric theories in such a way that their infinite parts are equal and opposite due to the difference in fermionic and bosonic loops. This is also the same mechanism that allows Faddeev–Popov ghosts to eliminate the negative norm contribution of gauge field loop diagrams to the perturbation series, which threaten unitarity, by generating equal fermionic loop contributions with the opposite sign.

It is also possible for gauge fields to contribute to the effective potential, and the discussion of this problem sets the stage for the next section where the Higgs–Kibble mechanism is presented. A U(1) gauge invariant action for the complex scalar field, the result of Exercise 7.3, is given in the Feynman gauge by

$$\mathcal{L} = D_\mu^* \phi^* D^\mu \phi - V(\phi^*, \phi) - \tfrac{1}{4} F_{\mu\nu} F^{\mu\nu} - \tfrac{1}{2}(\partial_\mu A^\mu)^2, \quad (9.53)$$

where the gauge covariant derivative D_μ is given by

$$D_\mu = \partial_\mu + ieA_\mu, \quad (9.54)$$

370   Nonperturbative Results

and, for the moment, the other solely scalar terms in the action are denoted by the potential V. If the scalar field is shifted by the complex constant ϕ_0, then a mass-like term, $e^2|\phi_0|^2 A_\mu A^\mu$, appears for the vector field in the shifted Lagrangian.

Again, the method of steepest descent may be used to evaluate the gauge field contribution to the effective potential. It follows from the definition of the effective potential that this contribution is given by the Feynman gauge path integral

$$V_g(\phi_0) \int d^4x =$$
$$i\hbar \ln \int_0^\infty \mathcal{D}A\, \delta(\partial_\mu A^\mu) \exp\left\{\frac{i}{\hbar}\int d^4x\, \tfrac{1}{2}A_\mu(\Box + M_g^2)A^\mu\right\}, \qquad (9.55)$$

where $M_g^2 = 2e^2|\phi_0|^2$. The evaluation of this path integral is almost identical to that used to derive the Feynman gauge propagator in Sec. 8.3. It is left as an exercise to show that, after discarding terms independent of M_g^2, the Euclidean form of the path integral gives

$$V_g(\phi_0)\int d^4x = \tfrac{1}{2}\hbar \operatorname{Tr}\ln(\delta_{\mu\nu} + M_g^2 \Delta_{\mu\nu}^E), \qquad (9.56)$$

where the Euclidean form of the Feynman gauge propagator is given by

$$\Delta_{\mu\nu}^E = \int \frac{d^4k}{(2\pi)^4}\left(\frac{\delta_{\mu\nu}}{k_E^2} - \frac{k_\mu k_\nu}{k_E^4}\right)e^{ik(x-y)}. \qquad (9.57)$$

This form is the functional inverse of the differential operator $\delta_{\mu\nu}\Box_E - \partial_\mu \partial_\nu$ and is derived by recalling that the Euclidean limit of $g_{\mu\nu}$ is $-\delta_{\mu\nu}$, while $k^2 \to -k_E^2$. The trace appearing in (9.56) is over the space-time indices as well as the space-time arguments.

Exercise 9.3: Verify (9.56).

The trace is evaluated by expanding the logarithm in a functional power series and using the trace identity

$$\operatorname{Tr}(\Delta_{\mu\nu}^E)^n = 3\int d^4x \int \frac{d^4k}{(2\pi)^4}\left(\frac{1}{k_E^2}\right)^n. \qquad (9.58)$$

The functional power series can now be resummed to obtain

$$\operatorname{Tr}\ln(\delta_{\mu\nu} + M_g^2 \Delta_{\mu\nu}^E) = 3\int d^4x \int \frac{d^4k}{(2\pi)^4}\ln(k_E^2 + M_g^2), \qquad (9.59)$$

with the final result that

$$V_g(\phi_0) = \tfrac{3}{2}\hbar \int \frac{d^4k}{(2\pi)^4} \ln(k_E^2 + M_g^2) . \qquad (9.60)$$

Once again, the familiar integral has appeared, and its regularized form is given by

$$V_g(\phi_0) = \frac{3\hbar\Lambda^2 M_g^2}{16\pi^2} + \frac{3\hbar M_g^4}{64\pi^2} \ln\left(\frac{M_g^2}{4\Lambda^2}\right) . \qquad (9.61)$$

The final step is to generalize these results to the case that there are multiplets of scalar, spinor, and gauge fields. For the case of a real scalar multiplet the act of shifting the fields gives a quadratic action of the form

$$\mathcal{L} = \tfrac{1}{2}\partial_\mu \phi^a \partial^\mu \phi^a - \tfrac{1}{2}\frac{\partial^2 V(\phi_0)}{\partial \phi_0^a \, \partial \phi_0^b}\phi^a \phi^b , \qquad (9.62)$$

so that the mass matrix $\mathbf{M}$, whose elements are

$$M_{ab} = \frac{\partial^2 V(\phi_0)}{\partial \phi_0^a \, \partial \phi_0^b} , \qquad (9.63)$$

has appeared. The method of steepest descent may be still be employed by diagonalizing the mass matrix, assumed to be Hermitian, with a unitary transformation $\mathbf{U}$. The measure of the path integral is invariant under such a transformation, so that the path integral is reduced to a product of path integrals over the respective modes that diagonalize the mass matrix. If $\mathcal{M}_a^2$ represents the eigenvalues of $\mathbf{M}$, which are possibly complicated functions of the vacuum expectation values of the fields, then the regularized scalar contribution to the effective potential is given by

$$V_s(\phi_0) = V(\phi_0) + \sum_a \left[\frac{\hbar\Lambda^2 \mathcal{M}_a^2}{16\pi^2} + \frac{\hbar(\mathcal{M}_a^2)^2}{64\pi^2} \ln\left(\frac{\mathcal{M}_a^2}{4\Lambda^2}\right)\right] . \qquad (9.64)$$

By using the properties of the trace it is straightforward to show that (9.64) can be rewritten as

$$V_s = V(\phi_0) + \text{Tr}\left[\frac{\hbar\Lambda^2 \mathbf{M}}{16\pi^2} + \frac{\hbar \mathbf{M}^2}{64\pi^2} \ln\left(\frac{\mathbf{M}}{4\Lambda^2}\right)\right] . \qquad (9.65)$$

Identical results are found for the contributions from the other fields in the theory. For example, a generalized Yukawa vertex takes the form

$$\mathcal{L}_I = -G^a_{jk} \phi^a \bar{\psi}^j \psi^k , \qquad (9.66)$$

and, in the absence of initial masses for the spinors, gives the regularized spinor contribution to the effective potential,

$$V_s = -\text{Tr}\left[\frac{\hbar\Lambda^2 M_f M_f^\dagger}{4\pi^2} + \frac{\hbar(M_f M_f^\dagger)^2}{16\pi^2}\ln\left(\frac{M_f M_f^\dagger}{4\Lambda^2}\right)\right], \quad (9.67)$$

where M_f, assumed to be a Hermitian matrix, has the elements

$$(M_f)_{jk} = G_{jk}^a \phi_o^a . \quad (9.68)$$

The generalization of the gauge field contribution is obtained by forming the nonabelian extension of the action (9.53). A multiplet of complex scalar fields, ϕ^a, is introduced, and the gauge covariant derivative is written

$$D_\mu = \delta_{ab}\partial_\mu + ig T_{ab}^j A_\mu^j \equiv \partial_\mu + ig A_\mu , \quad (9.69)$$

where the T^j are the Hermitian generators of an appropriate Lie group. The gauge invariant scalar action is then given by

$$\mathcal{L}_s = D_\mu^\dagger \phi^\dagger D^\mu \phi . \quad (9.70)$$

Shifting the scalar fields by the constant ϕ_o^a gives the masslike term for the gauge fields:

$$g^2\left(\phi_o^{c*}T_{ca}^k T_{ab}^j \phi_o^b\right)A_\mu^j A^{k\mu} \equiv \tfrac{1}{2}M_{jk}^2 A_\mu^j A^{k\mu} . \quad (9.71)$$

Defining the gauge field mass matrix M_g whose elements are M_{jk}^2, the effective potential contribution from the gauge sector is given by

$$V_g(\phi_o) = \text{Tr}\left[\frac{3\hbar\Lambda^2 M_g}{16\pi^2} + \frac{3\hbar M_g^2}{64\pi^2}\ln\left(\frac{M_g}{4\Lambda^2}\right)\right] . \quad (9.72)$$

9.3 The Higgs–Kibble Mechanism

In the previous section it was seen that a masslike term could be derived for a gauge field if the scalar field to which it is coupled exhibits a spontaneously broken symmetry. At this point it is legitimate to question why the vector fields considered so far have all been massless, since there is no apparent impediment to putting in a mass term for the gauge fields by hand. Such an action for a vector field would take the form

$$\mathcal{L} = -\tfrac{1}{4}F_{\mu\nu}F^{\mu\nu} + \tfrac{1}{2}m^2 A_\mu A^\mu , \quad (9.73)$$

Sec. 9.3 The Higgs–Kibble Mechanism

and gives, in the absence of a gauge fixing term, the equation of motion

$$[g_{\mu\nu}(\Box + m^2) - \partial_\mu \partial_\nu] A^\nu = 0 . \tag{9.74}$$

A vector field satisfying this equation of motion is known as a *Proca boson*. It is apparent that the presence of the mass term in the action (9.73) spoils gauge invariance. As a result, the generalized Ward–Takahashi identities, which enable the systematic renormalization of the theory, are lost.

This equation of motion (9.74) can be used to derive the Proca propagator. The propagator in every theory encountered so far is given by the inverse of the differential operator that appears in the linear part of the equation of motion, up to subtleties involving possible constraints. The unregulated inverse of the differential operator appearing in (9.74) is given by

$$\Delta_{\mu\nu}(x - y) = -\int \frac{d^4k}{(2\pi)^4} \left(\frac{g_{\mu\nu}}{k^2 - m^2} - \frac{k_\mu k_\nu}{m^2(k^2 - m^2)} \right) e^{ik(x-y)} . \tag{9.75}$$

Proving that this is the inverse simply amounts to the observation that (9.75) satisfies

$$[(\Box + m^2)g_{\mu\nu} - \partial_\mu \partial_\nu] \Delta^{\nu\rho}(x - y) = \delta^\rho{}_\mu \delta^4(x - y) . \tag{9.76}$$

The difficulty with Proca bosons as interacting fields is now apparent in the form of the propagator. The Fourier transform of the propagator tends to a constant as $k \to \infty$, and this heralds great difficulty in renormalizing a theory where a Proca boson appears. In effect, the inverse mass appearing in the propagator plays the role of a coupling constant with positive dimension of length. As a result, self-energy diagrams for fields coupled to the Proca boson possess arbitrarily high degrees of divergence, rendering such theories, in the absence of miraculous cancellations, nonrenormalizable.

However, there are physical systems that are best modelled by coupling the currents present to a massive vector field. An example of such a system is given by the weak decays of hadrons, e.g., the decay of a neutron into a proton, an electron, and a neutrino. This decay violates parity since the neutrino that appears is always left-handed. In early attempts to model this by a phenomenological field theory, the weak interaction was described by the Fermi current–current coupling [8], written as

$$\mathcal{L}_I = -G_f J_w^\mu J_{w\mu} , \tag{9.77}$$

where J_w^μ are the weak currents. Some of the weak currents must be left-handed in order that the observed parity violations may be modelled. For example, the neutron decay was assumed to be driven by a current–current coupling of the general form

$$\mathcal{L}_I = -\frac{1}{\sqrt{2}} G_F (\bar{\psi}_p \gamma^\mu \psi_n)(\bar{\psi}_e \gamma_\mu (1-\gamma_5)\psi_\nu) \,, \tag{9.78}$$

where the subscripts n, p, e, and ν refer to the neutron, proton, electron, and neutrino, respectively. It is obvious by using free field expansions for the fields appearing in the interaction (9.78) that it can connect a state with a neutron to a state with a proton, an electron, and an antineutrino. The presence of the chiral projection operator $1-\gamma_5$, introduced in Sec. 6.3, ensures the violation of parity in the observed left-handed manner.

Because the Fermi coupling constant, G_F, appearing in (9.78) has the units (length)^2, the theory is nonrenormalizable by the power-counting arguments of Sec. 8.4. However, when viewed in momentum space, the Fermi current–current coupling bears a striking resemblance to the low-energy limit of a current–current coupling mediated by a Proca propagator. This is easily seen by comparing its form to the one that results from coupling a vector field to an external current, as was done to derive the Feynman rules in Sec. 8.3. The result, after integrating the Proca field out of the path integral, is an effective interaction of the form

$$\mathcal{L}_{\text{eff}} = \int d^4x\, d^4y\, J^\mu(x)\Delta_{\mu\nu}(x-y)J^\nu(y) \,. \tag{9.79}$$

When viewed in momentum space, (9.79) reduces to

$$\mathcal{L}_{\text{eff}} = \int d^4p\, \tilde{J}^\mu(p)\tilde{\Delta}_{\mu\nu}(p)\tilde{J}^\nu(-p) \,. \tag{9.80}$$

In the limit that this vertex is used to connect low-energy states, where $p \approx 0$, the Proca propagator reduces to $\approx m^{-2}$. Thus, if the Fermi interaction is a low-energy approximation to a coupling between weak currents and a Proca boson, the strength of the low-energy coupling is, up to an overall dimensionless factor, determined by the mass of the Proca boson. For such a case, the Fermi current–current coupling could be replaced by a vertex of the form $gJ_w^\mu A_\mu$, where A_μ is the field of a Proca boson. Unfortunately, from the arguments made at the beginning of this section, a theory where the Proca boson begins with a mass is also nonrenormalizable, and therefore constitutes no real calculational improvement over the original Fermi vertex.

It is at this point that the method of spontaneously broken symmetry enters. A gauge invariant theory with the vector field coupled to a

Sec. 9.3 The Higgs–Kibble Mechanism

scalar field and with no initial mass for the vector field is renormalizable. The vertices of the theory may be evaluated in the $\phi_0 = 0$ case and rendered finite through a well-defined renormalization procedure. Using the effective action, these same vertices can be used to find the vertices of the theory for the case that $\phi_0 \neq 0$ through a functional power series. There is therefore reason to expect that the $\phi_0 \neq 0$ sector of the theory is also renormalizable. However, such a sector for the theory describes a Proca boson, since the shifting of the scalar field generates a masslike term for the vector field. Therefore, a massless gauge field theory that undergoes spontaneously broken symmetry describes a renormalizable theory containing a massive Proca boson [9].

Of course, there is a stumbling block, and that is the Goldstone theorem of Sec. 9.1, which predicts the presence of a massless Goldstone boson in the event that the continuous gauge symmetry is broken. Since there are no observed massless scalar particles, this apparently dooms the whole mechanism of mass generation to failure. Fortunately, there is a loophole in the Goldstone theorem. While the appearance of a massless scalar field is inevitable, there remains the possibility in a manifestly covariant approach that these modes are unphysical, i.e., that they are ghosts and exist only in the zero norm sector of the theory. Such a result would make the Goldstone bosons unobservable by decoupling them from the S-matrix and thereby rescue the procedure.

However, there is an additional point to be considered. If the theory is renormalized in the $\phi_0 = 0$ sector, the gauge field is transverse, possessing no longitudinal spin polarization. If a Proca boson is truly present in the theory, then it must possess a longitudinal component of polarization by the arguments of Sec. 7.1. Therefore, the $\phi_0 \neq 0$ sector also faces the problem of providing the longitudinal degree of freedom to the Proca boson.

The Higgs–Kibble mechanism [10, 11] solves both problems simultaneously. It allows the would-be Goldstone boson in the broken sector of the theory to provide the longitudinal component of the Proca boson by forcing the Goldstone mode into the unphysical subspace. The remainder of this section is a demonstration of this mechanism in action for the simple abelian model introduced in the previous section. Its generalization to nonabelian models is straightforward, and is used in the standard electroweak model of Glashow, Salam, and Weinberg (GSW) to be discussed briefly in the next section. While it is possible to present the Higgs–Kibble mechanism in the Coulomb gauge [12], it will be demonstrated here in a manifestly covariant approach.

The starting point is the classical action for the gauge invariant coupling of a single charged scalar field to an abelian gauge field, first discussed in the previous section. This is given by

$$\mathcal{L} = D_\mu^* \phi^* D^\mu \phi - \tfrac{1}{4} F_{\mu\nu} F^{\mu\nu} - V(\phi^*, \phi), \quad (9.81)$$

where the gauge covariant derivative denotes $D_\mu = \partial_\mu + ieA_\mu$. The potential V is chosen to drive a breakdown of symmetry, with an explicit gauge invariant form given by

$$V(\phi^*, \phi) = -\mu^2 \phi^* \phi + \tfrac{1}{2} \lambda (\phi^* \phi)^2, \quad \mu^2, \lambda > 0. \quad (9.82)$$

At the classical level it is easy to discuss the Higgs–Kibble mechanism. This is accomplished by the simultaneous polar decomposition and translation of the scalar field,

$$\phi = e^{i\theta/\nu}(\rho + \nu), \quad (9.83)$$

where ρ and θ are two real-valued fields and ν is a constant to be determined. Simple substitution into the action gives the complicated set of terms

$$\begin{aligned}\mathcal{L} =\ & 2(\mu^2 \nu - \lambda \nu^3)\rho + \partial_\mu \rho \partial^\mu \rho + \partial_\mu \theta \partial^\mu \theta + 2e\nu A^\mu \partial_\mu \theta + e^2 \nu^2 A_\mu A^\mu \\ & - (3\lambda \nu^2 - \mu^2)\rho^2 + \frac{2}{\nu} \rho \partial_\mu \theta \partial^\mu \theta + 2e^2 \nu \rho A_\mu A^\mu + 4e\, \rho A^\mu \partial_\mu \theta \\ & + \frac{1}{\nu^2} \rho^2 \partial_\mu \theta \partial^\mu \theta + 2\frac{e}{\nu} \rho^2 A^\mu \partial_\mu \theta + e^2 \rho^2 A_\mu A^\mu \\ & - 2\lambda \nu \rho^3 - \tfrac{1}{2} \lambda \rho^4 - \tfrac{1}{4} F_{\alpha\beta} F^{\alpha\beta}. \end{aligned} \quad (9.84)$$

As discussed for quantum mechanical systems in Sec. 4.5, the minimal criteria for the stability of a theory are that, first, all terms linear in the fields must vanish and, second, all masses must be real. For the action of (9.84) this leads to the identification

$$\nu = \pm \sqrt{\frac{\mu^2}{\lambda}}. \quad (9.85)$$

For this case it is easy to see that ρ corresponds to an excitation with mass $m^2 = 2\mu^2$, while θ corresponds to the massless Goldstone mode. The vector field obtains the mass $M^2 = 2e^2 \nu^2$.

For the value of ν given by (9.85), it is simple algebra to show that the action (9.84) can be rewritten as

$$\begin{aligned}\mathcal{L} =\ & \partial_\mu \rho \partial^\mu \rho - \tfrac{1}{4} F_{\alpha\beta} F^{\alpha\beta} - 2\mu^2 \rho^2 - 2\lambda\nu \rho^3 - \tfrac{1}{2}\lambda \rho^4 \\ & + (e^2 \nu^2 + 2e^2 \nu \rho + e^2 \rho^2)\left(A_\mu + \frac{\partial_\mu \theta}{e\nu}\right)\left(A^\mu + \frac{\partial^\mu \theta}{e\nu}\right). \end{aligned} \quad (9.86)$$

Sec. 9.3 The Higgs–Kibble Mechanism

It is apparent that all vestiges of the Goldstone mode θ may be removed from the theory by making the redefinition

$$A_\mu \to A_\mu - \frac{\partial_\mu \theta}{e\nu}, \tag{9.87}$$

which leaves $F_{\alpha\beta} F^{\alpha\beta}$ invariant because of its formal resemblance to a gauge transformation. The final form for the action is

$$\begin{aligned}\mathcal{L} &= \partial_\mu \rho \, \partial^\mu \rho - 2\mu^2 \rho^2 - \tfrac{1}{4} F_{\alpha\beta} F^{\alpha\beta} + e^2 \nu^2 A_\mu A^\mu \\ &\quad + (2e^2 \nu \rho + e^2 \rho^2) A_\mu A^\mu - 2\lambda\nu\rho^3 - \tfrac{1}{2}\lambda \rho^4 \,.\end{aligned} \tag{9.88}$$

The theory now describes a single massive scalar particle interacting with a massive Proca boson.

While all this is true at the classical level, demonstrating the quantum counterpart is slightly more subtle. The polar decomposition of the fields is not useful in the quantized theory since it generates a nontrivial and intractable Jacobian in the path integral. However, the basic idea of performing a gauge transformation to rid the theory of the unwanted Goldstone mode can be carried over to the quantum field theory. The close relationship between gauge transformations and ghost structure in covariant formulations of electrodynamics gives a strong indication that the Goldstone mode will disappear into the unphysical subspace.

The quantized Higgs model, as it has come to be known, can be analyzed by path integral techniques. For simplicity, in the remainder of this section only the tree approximation to the effective potential will be considered, so that $V_{\text{eff}}(\phi_0) = V(\phi_0)$, where ϕ_0 is a constant. For convenience, the definition of the constants appearing in the initial potential will be altered to read

$$V(\phi) = -\mu^2 \phi^* \phi + \lambda(\phi^* \phi)^2 \,. \tag{9.89}$$

The minima of the effective potential must satisfy

$$\frac{\partial V_{\text{eff}}(\phi_0)}{\partial \phi_0} = -\mu^2 \phi_0^* + 2\lambda(\phi_0^* \phi_0)\phi_0^* = 0 \,, \tag{9.90}$$

and it is clear that this possesses the solution

$$\phi_0 = e^{i\alpha} \sqrt{\frac{\mu^2}{2\lambda}} \equiv \frac{\nu}{\sqrt{2}} e^{i\alpha}\,, \tag{9.91}$$

where α is an arbitrary constant. This arbitrary phase occurs because the potential contains a infinite degeneracy in its minima.

Since α is an arbitrary constant, it can play no role in the dynamics of the theory. This will be demonstrated momentarily. The theory oscillates around the vacuum expectation value of the scalar field, and this is reflected by transforming the scalar field according to

$$\phi \to \phi e^{i\alpha} + \frac{\nu}{\sqrt{2}} e^{i\alpha} . \tag{9.92}$$

The original phase invariance of the theory causes α to disappear from the resulting action, as promised. Therefore, the breakdown of symmetry is entirely equivalent to translating the scalar field by the *real* number ν, according to

$$\phi \to \phi + \frac{\nu}{\sqrt{2}} . \tag{9.93}$$

It is useful to represent the complex field in terms of two real fields,

$$\phi = \frac{1}{\sqrt{2}}(\phi_1 + i\phi_2) , \tag{9.94}$$

and this change of variables generates a trivial Jacobian in the path integral. Since the vacuum expectation value of the field (9.93) is real, it follows that the field ϕ_1 is the component that develops the nonzero vacuum expectation value ν. The path integral representation of the vacuum transition amplitude is therefore given in the Feynman gauge by

$$_+\langle \nu | \nu \rangle_- = \int_\nu^\nu \mathcal{D}\phi_1 \mathcal{D}\phi_2 \mathcal{D}A\, \delta(\partial_\mu A^\mu) \exp\left\{ i \int d^4x\, \mathcal{L}(\phi_1, \phi_2, A_\mu) \right\}, \tag{9.95}$$

where the gauge volume has been dropped as irrelevant. The action is given explicitly by

$$\begin{aligned}
\mathcal{L} &= \tfrac{1}{2}\partial_\mu \phi_1 \partial^\mu \phi_1 + \tfrac{1}{2}\partial_\mu \phi_2 \partial^\mu \phi_2 - \tfrac{1}{4}F_{\alpha\beta}F^{\alpha\beta} - \tfrac{1}{2}(\partial_\mu A^\mu)^2 \\
&\quad + e\phi_2 A^\mu \partial_\mu \phi_1 - e\phi_1 A^\mu \partial_\mu \phi_2 + \tfrac{1}{2}e^2 A_\mu A^\mu (\phi_1{}^2 + \phi_2{}^2) \\
&\quad + \tfrac{1}{2}\mu^2(\phi_1{}^2 + \phi_2{}^2) - \tfrac{1}{4}\lambda(\phi_1{}^2 + \phi_2{}^2)^2 .
\end{aligned} \tag{9.96}$$

This path integral form is readily derived by the usual analysis of the Hamiltonian for the gauge theory. It can be verified that the whole procedure of Sec. 7.2 goes through, allowing the Gauss's law constraint to be absorbed into the action and the momenta to be integrated.

The limits on the path integral may be absorbed into the action by translating the ϕ_1 field by ν. Using the tree level value for ν generates the quadratic terms

$$\begin{aligned}
\mathcal{L}_0 &= \tfrac{1}{2}\partial_\mu \phi_1 \partial^\mu \phi_1 - \mu^2 \phi_1{}^2 + \tfrac{1}{2}\partial_\mu \phi_2 \partial^\mu \phi_2 - e\nu A^\mu \partial_\mu \phi_2 \\
&\quad - \tfrac{1}{4}F_{\alpha\beta}F^{\alpha\beta} - \tfrac{1}{2}(\partial_\mu A^\mu)^2 + \tfrac{1}{2}e^2 \nu^2 A_\mu A^\mu .
\end{aligned} \tag{9.97}$$

Sec. 9.3 The Higgs–Kibble Mechanism

There are many higher-order terms involving all three fields. However, it is apparent from the action (9.97) that ϕ_2 has emerged as the Goldstone boson since it possesses no mass term. The Higgs mechanism allows ϕ_2 to be removed from the quadratic part of the action by making the gaugelike transformation

$$A_\mu \to A_\mu + \frac{1}{M}\partial_\mu \phi_2 , \qquad (9.98)$$

where $M = e\nu$. It is easy to see that the quadratic part of the action becomes

$$\begin{aligned}\mathcal{L}_0 &= \tfrac{1}{2}\partial_\mu \phi_1 \partial^\mu \phi_1 - \mu^2 \phi_1{}^2 + -\tfrac{1}{4}F_{\alpha\beta}F^{\alpha\beta} + \tfrac{1}{2}M^2 A_\mu A^\mu \\ &\quad - \tfrac{1}{2}(\partial_\mu A^\mu + \frac{1}{M}\Box \phi_2)^2 .\end{aligned} \qquad (9.99)$$

The last term will vanish when the gauge condition is taken into account, since the transformation on A_μ has caused the Dirac delta in the measure of the path integral to become

$$\delta(\partial_\mu A^\mu) \to \delta\left(\partial_\mu A^\mu + \frac{1}{M}\Box \phi_2\right) . \qquad (9.100)$$

Of course, there are still many higher-order terms with ϕ_2 appearing in them. However, the Dirac delta in the measure now allows ϕ_2 to be easily integrated from the theory, since it enforces the identity

$$\Box \phi_2 = -M\partial_\mu A^\mu . \qquad (9.101)$$

As a result, all occurrences of ϕ_2 in the higher-order terms are replaced with the solution to (9.101), given by the formal expression

$$\phi_2 = -M \left(\frac{1}{\Box}\right)\partial_\mu A^\mu . \qquad (9.102)$$

The final form of the path integral for the transformed theory is

$$_+\langle \nu | \nu \rangle_- = \int_0^0 \mathcal{D}\phi \mathcal{D}A \, \exp\left\{i\int d^4x (\mathcal{L}_0 + \mathcal{L}_I)\right\} , \qquad (9.103)$$

where

$$\mathcal{L}_0 = \tfrac{1}{2}\partial_\mu \phi \, \partial^\mu \phi - \mu^2 \phi^2 - \tfrac{1}{4}F_{\alpha\beta}F^{\alpha\beta} + \tfrac{1}{2}M^2 A_\mu A^\mu . \qquad (9.104)$$

The Goldstone boson has disappeared into the unphysical subspace via the gauge condition, leaving behind an action describing a single massive scalar boson interacting with an unconstrained Proca boson.

380   Nonperturbative Results

There are two important points to be made about the higher-order terms. Due to the transformation (9.98) and the identity (9.102) there are many nonlocal terms possessing coefficients with positive powers of length. A naive power counting argument would indicate that the theory is non-renormalizable. However, these apparently divergent terms are needed precisely to enable the theory to control the bad ultraviolet behavior of the Proca propagator. This is a case where "miraculous cancellations" among the higher-order terms do take place, and the theory remains renormalizable. While such an outcome might be expected on the basis that the underlying symmetric theory was renormalizable, the verification of this property is beyond the scope of this book, and the reader is recommended to the references [9, 13, 14].

The second point is that the S-matrix elements involving $\partial_\mu A^\mu$ are no longer zero. This is to be expected since the Proca boson must have a longitudinal component. The original physical spectrum of the gauge field, i.e., the states allowed in the asymptotic spectrum, consisted of the two transverse modes and the putative ghosts, constructed from $\partial_\mu A^\mu$. While the ghosts have a vanishing S-matrix in the massless case, the massive case allows the excitation associated with $\partial_\mu A^\mu$ to have a nontrivial S-matrix. A demonstration of this is based on the properties of the Proca propagator (9.75), which can be derived from the quadratic part of the path integral.

Exercise 9.4: Verify that (9.75) is indeed the propagator for the Proca boson appearing in the shifted theory.

Since the propagator is the vacuum expectation value of the time-ordered product of fields, it follows that

$$\langle 0|T\{A^\mu(x)A^\nu(y)\}|0\rangle = i\Delta^{\mu\nu}(x-y) . \tag{9.105}$$

Using the form (9.75) for $\Delta_{\mu\nu}$ immediately yields

$$\langle 0|T\{\partial_\mu A^\mu(x)A^\nu(y)\}|0\rangle = i\partial_\mu \Delta^{\mu\nu}(x-y) = \frac{i}{M^2}\partial^\nu \delta^4(x-y) . \tag{9.106}$$

The excitations built from $\partial_\mu A^\mu$ have transmuted into physical modes with positive norm. The equation of motion for the Proca boson,

$$(\Box + M^2)A^\mu - \partial^\mu \partial_\nu A^\nu = J^\mu , \tag{9.107}$$

immediately yields the result that the current to which A^μ is coupled is no longer conserved:

$$\partial_\mu J^\mu = M^2 \partial_\mu A^\mu . \tag{9.108}$$

Sec. 9.3 The Higgs–Kibble Mechanism

The Higgs mechanism is readily extended to the nonabelian case [15], for which the action is given in the complex basis for the scalars by (9.70) of the previous section. The details of this are relatively easy to see in a real basis for the scalar fields. It should be noted that changing to such a basis requires doubling the size of the representation of the group. This is done in the following way. If Φ represents the multiplet of complex scalar fields, a new multiplet twice as large is defined by

$$R = \begin{pmatrix} \Phi \\ \Phi^* \end{pmatrix}, \qquad (9.109)$$

and the new gauge covariant derivative $\mathcal{D}_\mu$ is written in terms of the old gauge covariant derivative D_μ as

$$\mathcal{D}_\mu = \begin{pmatrix} D_\mu & 0 \\ 0 & D_\mu^\dagger \end{pmatrix}, \qquad (9.110)$$

so that the size of the matrix $\mathcal{D}_\mu$ matches dimension of R. The original action for the gauge field–scalar coupling is then written

$$\tfrac{1}{2} \mathcal{D}_\mu^\dagger R^\dagger \mathcal{D}^\mu R, \qquad (9.111)$$

and this may be unitarily transformed to a real basis for the scalar fields. As an example, for the simplest possible case of one complex scalar field, as in the Higgs model, the matrix transformation would take the form

$$\begin{pmatrix} \phi_1 \\ \phi_2 \end{pmatrix} = \frac{1}{\sqrt{2}} \begin{pmatrix} 1 & 1 \\ -i & i \end{pmatrix} \begin{pmatrix} \phi \\ \phi^* \end{pmatrix} \equiv \mathbf{U}\Phi. \qquad (9.112)$$

The same unitary transformation $\mathbf{U}$ applied to $\mathcal{D}_\mu$ yields the gauge covariant derivative for the real representation,

$$\mathcal{D}_\mu^R = \mathbf{U} \mathcal{D}_\mu \mathbf{U}^{-1}. \qquad (9.113)$$

> **Exercise 9.5**: Construct the real representation of SU(2).

The first major step in analyzing the nonabelian case is to verify that the number of Goldstone bosons is identical to the number of massive vector fields. The mass matrix for the gauge fields in the complex representation was derived in (9.71) and is given here for the real representation by

$$\tfrac{1}{2} g^2 \left(\phi_0^c T_{ca}^k T_{ab}^j \phi_0^b \right) A_\mu^j A^{k\mu} \equiv \tfrac{1}{2} M_{jk}^2 A_\mu^j A^{k\mu}, \qquad (9.114)$$

where the generators T^j act on the real representation space of the scalars. It is assumed that this matrix is symmetric and real and is therefore Hermitian. From inspection of (9.114) it is evident that the gauge boson A_μ^j will actually appear in the mass matrix, and therefore obtain a mass, *if and only if* the vector in isospin space, $T_{ab}^j \phi_o^b$, is nonzero. Using the terminology of Sec. 9.1, this means that every broken generator is accompanied by the appearance of the associated vector field in the quadratic mass terms of the shifted Lagrangian. However, the Goldstone theorem of Sec. 9.1 proved that every broken generator is also accompanied by a Goldstone boson. Therefore, in the general nonabelian case, no matter how the symmetry is broken, there will be an identical number of massive vector fields and Goldstone bosons. This is an absolute necessity, for otherwise the Higgs mechanism could not provide a longitudinal piece for all the massive vector bosons.

The second step is to gauge away the Goldstone bosons. This starts by noting that, after the fields are translated, there will be quadratic cross terms of the form

$$g A_\mu^j \partial^\mu \phi^a T_{ab}^j \phi_o^b \; . \tag{9.115}$$

It is recalled from the Goldstone theorem that if $g T_{ab}^j \phi_o^b$ is nonzero, then it forms the components of an isospin eigenvector of the scalar mass matrix with the eigenvalue zero. Because the scalar mass matrix is Hermitian, it follows that these nonzero eigenvectors must be orthogonal, so that the nonzero eigenvectors must satisfy

$$g^2 \phi_o^c T_{ca}^k T_{ab}^j \phi_o^b = M_{(j)}^2 \delta_{jk} \; , \tag{9.116}$$

where use has been made of the Hermiticity of the generators. Comparison of (9.116) with (9.114) immediately establishes that in the real representation the gauge field mass matrix is automatically diagonal, and that the normalization factor in (9.116) is precisely the gauge field mass eigenvalue. Therefore, the cross terms can be written

$$g A_\mu^j \partial^\mu \phi^a T_{ab}^j \phi_o^b \equiv M_{(j)} A_\mu^j \partial^\mu \phi_G^j \; , \tag{9.117}$$

where the ϕ_G^j are the Goldstone modes of the theory:

$$\phi_G^j = \frac{g}{M_{(j)}} \phi_o^a T_{ab}^j \phi^b \; . \tag{9.118}$$

The Higgs mechanism consists of removing the Goldstone modes from the quadratic part of the action by the redefinition

$$A_\mu^j \to A_\mu^j + \frac{1}{M_{(j)}} \partial_\mu \phi_G^j \; . \tag{9.119}$$

This translation causes the gauge constraints of the massive vector fields to become

$$\delta(\partial_\mu A^{j\mu}) \to \delta\left(\partial_\mu A^{j\mu} + \frac{1}{M_{(j)}}\Box \phi_G^j\right), \qquad (9.120)$$

and this allows all the Goldstone modes to be integrated from the theory.

9.4 The $SU(2)_L \times U(1)$ Electroweak Model

The current–current Fermi model of weak interactions discussed at the beginning of the previous section may be united with the electromagnetic properties of the participating particles by modelling the system with a nonabelian gauge theory, known as the Glashow–Salam–Weinberg (GSW) model [16]. The spontaneous breakdown of symmetry in this renormalizable field theory results in massive Proca bosons that act as the intermediaries of the weak interactions [17], as well as leaving one massless vector field that is understood to be the photon. The resulting theory is renormalizable and predicts the existence of two charged Proca bosons, $W_\mu^\pm$, and one neutral Proca boson, Z_μ. The masses of these particles were originally inferred from low-energy data, and their actual observation capped the triumph of the model.

The model begins by using quarks to build the *hadrons*, defined as any particle that participates in the strong force. As discussed in Sec. 7.5, the strong interactions are modelled by QCD, and all known hadrons can be viewed as bound states of two or three quarks. The two-quark bound states form the lighter hadrons, such as pi mesons, while the three-quark bound states are the heavier hadrons, such as the neutron and proton, referred to as *baryons*. The mechanics of this is complicated and not completely understood, but certain simple aspects of the quark model lie within the purview of this book. In presenting this material many subtleties and nuances will be omitted, and the reader is urged to consult more extensive treatments [18].

As of this writing, the quarks are believed to be organized into three generations of two quarks each, with each generation more massive than the previous. Each generation results in the existence of more massive hadrons as their composites. However, the hadrons constructed from each generation participate in *both* the weak and strong interactions in an identical manner, in effect realizing the universality of the weak Fermi coupling G_F. As mentioned in Sec. 7.5, in order that the hadrons can be constructed with their known electric charges, the quarks must carry a fractional electric charge. For example, suppressing their color index, the two light quarks

u and d form the first generation, so that these two quarks are written as a doublet,

$$Q = \begin{pmatrix} u \\ d \end{pmatrix} . \qquad (9.121)$$

The u quark possesses electric charge $\frac{2}{3}e$, while the d quark has charge $-\frac{1}{3}e$. The composite state $u\bar{d}$ therefore has charge e, and constitutes a charged pi meson. The composite state uud has charge e, and forms the proton, while the state udd has charge zero, and is the neutron. The baryon number of each composite is derived by associating baryon number $\frac{1}{3}$ with each quark, and therefore baryon number $-\frac{1}{3}$ with each antiquark. Thus, $u\bar{d}$ has baryon number zero, while uud has baryon number one. In effect, baryon number is one-third of the difference between the number of quarks and the number of antiquarks participating in a process. Experimental evidence shows that violation of baryon number conservation is extremely small. However, this does not mean that the conservation of the baryon number is exact, and further remarks on this can be found in the next section where chiral anomalies are discussed.

There is a second general type of particle called a *lepton*. Leptons participate only in the electromagnetic and weak interactions. They also appear in generations that can be written as doublets. The first generation is composed of the electron and its neutrino, so that the doublet is written

$$L = \begin{pmatrix} \nu_e \\ e \end{pmatrix} . \qquad (9.122)$$

Placing these two particles together might seem questionable, since the neutrino is only left-handed and apparently massless, while the electron is observed with both chiralities and carries mass. Furthermore, the neutrino is electrically neutral while the electron carries electric charge e. Nevertheless, this arrangement will be seen to create great simplifications.

In what follows only the first generation of leptons and quarks will be treated. Despite many subtleties involving mixing of bare quarks from different generations and the associated problems of quark mass generation and CP violation, treating a single generation captures many of the weak interaction properties of the standard model. In terms of the quarks and leptons the effective Fermi current–current model of weak interactions can be written

$$\mathcal{L}_w = \frac{1}{\sqrt{2}} G_F J_\mu^\dagger J^\mu , \qquad (9.123)$$

where the current is entirely left-handed,

$$J_\mu = \cos\theta_c \, \bar{u}\gamma_\mu(1-\gamma_5)d + \bar{\nu}_e \gamma_\mu(1-\gamma_5)e . \qquad (9.124)$$

Sec. 9.4 The $SU(2)_L \times U(1)$ Electroweak Model

where θ_c is the *Cabibbo angle*. In what follows θ_c will be set to zero for simplicity. It is the appearance of ν_e and e in this current that suggests placing them together in the doublet, and it is the phenomenological success and structure of this effective vertex that motivates the construction of the GSW model. However, from the standpoint of the weak interactions it is only the left-handed chiralities of the quarks and leptons that are important, while both chiralities of the quarks and the electron participate in the electromagnetic interaction. Therefore, if the current–current interaction is to be understood as the low-energy limit of a gauge theory, the gauge fields responsible for the weak interactions must couple only to the left-handed parts of each generation.

The gauge theory approach to uniting the weak interactions with the electromagnetic interactions therefore requires that the left-handed pieces of the fields be treated differently than the right-handed pieces. Because of the doublet nature of the generations, the simplest possible nonabelian gauge group is $SU(2)$ in its fundamental representation. Because the weak interactions involve only the left-handed pieces, this gauge group will act only on the left-handed quarks and leptons, and hence the distinction $SU(2)_L$.

Using $SU(2)$ introduces three gauge bosons. The presence of intermediate vector bosons in the weak interactions requires that at least two of them carry electric charge. It would be tempting to associate the third neutral gauge boson with the photon. However, the experimental observation of *neutral* current events in weak processes requires the use of four gauge fields so that two electrically charged and one electrically neutral Proca bosons can occur in addition to the photon. The solution is to introduce a $U(1)$ gauge field that interacts with both the left and right sectors in order to obtain the desired electromagnetic coupling. The strength of the $U(1)$ coupling to the various left and right sectors can then be varied to make contact with the observed electromagnetic charges of each field.

With these restrictions and ideas in mind the basic model is constructed by first forming the left-handed pieces of the quark and lepton doublets,

$$L_L = \tfrac{1}{2}(1-\gamma_5)L, \quad Q_L = \tfrac{1}{2}(1-\gamma_5)Q, \qquad (9.125)$$

and these will transform under $SU(2)_L$. The left-handed sector is therefore described by the action

$$\begin{aligned}\mathcal{L}_L &= i\bar{Q}_L \gamma^\mu (\partial_\mu + \tfrac{1}{2}ig W_\mu^j \sigma_j + \tfrac{1}{2}ig' Y_Q B_\mu) Q_L \\ &\quad + i\bar{L}_L \gamma^\mu (\partial_\mu + \tfrac{1}{2}ig W_\mu^j \sigma_j + \tfrac{1}{2}ig' Y_L B_\mu) L_L .\end{aligned} \qquad (9.126)$$

The factor of $\tfrac{1}{2}$ is for later convenience. The right-handed sector is formed from the three SU(2) singlets,

$$e_R = \tfrac{1}{2}(1+\gamma_5)e, \quad u_R = \tfrac{1}{2}(1+\gamma_5)u, \quad d_R = \tfrac{1}{2}(1+\gamma_5)d, \quad (9.127)$$

and the action consistent with their electric charges is given by

$$\begin{aligned}\mathcal{L}_R &= i\bar{e}_R\gamma^\mu(\partial_\mu + ig'B_\mu)e_R + i\bar{u}_R\gamma^\mu(\partial_\mu - \tfrac{2}{3}ig'B_\mu)u_R \\ &\quad + i\bar{d}_R\gamma^\mu(\partial_\mu + \tfrac{1}{3}ig'B_\mu)d_R .\end{aligned} \quad (9.128)$$

In order to determine the strength Y of the abelian gauge field coupling to the left-handed sector it is noted that each of the states in the left-handed sector is labelled by the eigenvalues of $\tfrac{1}{2}(\sigma_3 + Y)$. Therefore, the sign of the electric charge of the leptons, where the electromagnetic interactions are assumed to correspond to an unbroken abelian gauge symmetry, must be given by $Y_L = -1$, since this assignment will result in zero charge for the neutrino and -1 for the electron. Likewise, the assignment of $Y_Q = \tfrac{1}{3}$ results in the charges $\tfrac{2}{3}$ and $-\tfrac{1}{3}$ for the u and d quarks, respectively.

The gauge fields, in addition to possessing the standard free actions, are coupled to a doublet of complex scalar fields, written as

$$\phi = \begin{pmatrix} \phi_+ \\ \phi_0 \end{pmatrix} . \quad (9.129)$$

Since the Goldstone modes must be combined with the gauge fields to form the Proca bosons, two of the degrees of freedom for ϕ must carry electric charge, while the other two must be neutral. This is accomplished by choosing the action for the gauge–scalar coupling to be

$$\mathcal{L}_s = (\partial_\mu - \tfrac{1}{2}igW_\mu^j\sigma_j - \tfrac{1}{2}ig'B_\mu)\phi^\dagger(\partial_\mu + \tfrac{1}{2}igW_\mu^j\sigma_j + \tfrac{1}{2}ig'B_\mu)\phi . \quad (9.130)$$

It is possible to include a gauge invariant Yukawa coupling for the leptons to the scalar fields,

$$\mathcal{L}_I = -G_L\left[\bar{e}(\phi^\dagger L_L) + (\bar{L}_L\phi)e\right] . \quad (9.131)$$

A set of similar terms for the quark fields can be written down, but requires far more development and discussion than the space available to this book permits. The reader is recommended to the many excellent treatments of the standard model in texts [18].

All that remains is to break the symmetry by assuming that the effective potential, at tree level, is given by

$$V = -\mu^2\phi^\dagger\phi + \lambda(\phi^\dagger\phi)^2 . \quad (9.132)$$

Sec. 9.4 The SU(2)$_L$ × U(1) Electroweak Model

The vacuum expectation of ϕ is assumed to be real in order that the vacuum carry no electric charge, so that the vacuum expectation value must take the form

$$\langle \phi \rangle = \frac{1}{\sqrt{2}} \begin{pmatrix} 0 \\ v \end{pmatrix}, \tag{9.133}$$

where

$$v = \sqrt{\frac{\mu^2}{\lambda}}. \tag{9.134}$$

Translating the scalar field by the vacuum expectation value generates mass terms for the vector fields. It is straightforward to see that

$$\left[(\tfrac{1}{2}gW_\mu^j \sigma^j + \tfrac{1}{2}g' B_\mu)\langle\phi\rangle\right]^2 = \tfrac{1}{4}g^2 v^2 W_\mu^+ W^{-\mu} + \tfrac{1}{8}(g^2 + g'^2)v^2 Z_\mu Z^\mu, \tag{9.135}$$

where

$$W_\mu^\pm = \frac{1}{\sqrt{2}}(W_\mu^{(1)} \pm i W_\mu^{(2)}),$$

$$Z_\mu = \frac{1}{\sqrt{g^2 + g'^2}}(g' W_\mu^{(3)} - g B_\mu). \tag{9.136}$$

The form for Z_μ defines a normalized vector in isospin space

$$\frac{1}{\sqrt{g^2 + g'^2}} \begin{pmatrix} g' \\ -g \end{pmatrix}. \tag{9.137}$$

It follows that the fourth gauge boson must be the combination of $W_\mu^{(3)}$ and B_μ that is orthonormal to Z_μ in isospin space, and this is given by

$$A_\mu = \frac{1}{\sqrt{g^2 + g'^2}}(g W_\mu^{(3)} + g' B_\mu). \tag{9.138}$$

The field A_μ is massless and therefore corresponds to the photon field.

It is relatively easy to connect the parameters appearing in the theory to measured quantities. Clearly, the masses of the gauge fields are given by

$$M_W = \tfrac{1}{2}g v, \quad M_Z = \tfrac{1}{2}v\sqrt{g^2 + g'^2}. \tag{9.139}$$

The electromagnetic coupling is found by examining the coefficient of the term $A_\mu \bar{e} \gamma^\mu e$ in the broken theory, and this coefficient must be the electric charge q, so that

$$q = \frac{g g'}{\sqrt{g^2 + g'^2}}. \tag{9.140}$$

The coupling of the $W^\pm$ bosons to the weak current (9.124) is given by

$$\mathcal{L}_I = \frac{g}{2\sqrt{2}}(W_\mu^+ J^\mu + W_\mu^- J^{\dagger\mu}) \,. \qquad (9.141)$$

Using the arguments of the previous section, this leads to an effective current–current interaction of the form

$$\mathcal{L}_{\text{eff}} = \tfrac{1}{8}g^2 \int d^4x\, d^4y\, J_\mu^\dagger(x)\Delta^{\mu\nu}(x-y)J_\nu(y) \,, \qquad (9.142)$$

where $\Delta^{\mu\nu}$ is the Proca propagator for the $W^\pm$ bosons. In the low-energy limit this expression is dominated by the mass of the $W^\pm$ particles, so that

$$\mathcal{L}_{\text{eff}} \approx \frac{g^2}{8M_W^2} J_\mu^\dagger J^\mu \,. \qquad (9.143)$$

Comparing this to (9.123) and using (9.139) gives

$$G_F = \frac{g^2\sqrt{2}}{8M_W^2} = \frac{\sqrt{2}}{2v^2} \,. \qquad (9.144)$$

The Fermi constant G_F is well known from low energy data, so that the vacuum expectation value is fixed to be $\simeq 246$ GeV. The final piece necessary to fix the masses uniquely comes from the prediction of the existence of the electrically neutral Z_μ boson, which must be coupled to an electrically neutral current. The effective current–current interaction generated by the Z makes predictions for scattering, describing events where the individual electric charges of the participants are conserved, such as $\nu + \pi \to \nu + \pi$. These so-called neutral current events are observed, and their strength uniquely fixes the masses of the W and Z bosons to be $\simeq 80$ GeV and $\simeq 90$ GeV respectively.

After the Higgs mechanism has disposed of three of the four degrees of scalar freedom, one scalar particle remains as a physically excitable mode of the theory, and its mass is given by $\sqrt{2}\mu$. The so-called Higgs boson is the only fundamental scalar particle that appears in the standard model, since all other scalars are quark composites. Unfortunately, its mass is undetermined, and as of this writing, no direct observation of it has been made. It has been argued that mass generation via the Higgs mechanism, involving the elusive Higgs boson, may represent some low-energy approximation of a more complicated mechanism of *dynamical* symmetry breaking. A dynamical mechanism might involve the formation of a condensate of spinor particle pairs, as in the case of superconductivity, that behaves as a scalar field at low energies. Simple models where this occurs will be discussed briefly in Sec. 9.5 and Sec. 9.7.

9.5 Chiral Anomalies

In the discussion of the Ward–Takahashi identities associated with a classical invariance of the action, it was pointed out that the measure of the path integral must also remain invariant if the current associated with the invariance is to be conserved in the quantized theory. In all of the examples treated so far this has been true. However, there is an important exception to this result, and its existence is tied to deep topological properties of gauge fields.

In Sec. 6.4.3 it was pointed out that the massless Dirac field action possesses invariance under chiral transformations,

$$\psi \to \exp(i\alpha\gamma_s)\psi . \qquad (9.145)$$

The requirement for masslessness arises from the transformation property of $\bar\psi$,

$$\bar\psi \to \bar\psi \exp(i\alpha\gamma_s) , \qquad (9.146)$$

so that a mass term $\bar\psi\psi$ is not invariant.

However, there is a deeper problem with chiral invariance in quantized theories, and this can be seen from the path integral measure of a Dirac field discussed for the Euclidean case in Sec. 6.8.3. In the configuration space form the measure is written

$$\mathcal{D}\psi\,\mathcal{D}\bar\psi = \prod_x d\psi(x)\,d\bar\psi(x) , \qquad (9.147)$$

so that a chiral transformation will generate a nontrivial Jacobian. In the terminology of Sec. 4.3 this means that the current associated with the classical invariance under chiral transformations will possess an anomaly, referred to as the *chiral anomaly* [19]. This Jacobian will now be analyzed in the context of the path integral [20] for a general gauge theory coupled to massless spinors.

The starting point is the observation that the path integral for a general gauge theory factorizes into the product of two path integrals:

$$\begin{aligned}
{}_+\langle 0|0\rangle_- &= \int_0^0 \mathcal{D}A\,\Delta_\chi\,\delta(\chi)\exp\left\{i\int d^4x\,\mathcal{L}_g(A)\right\} \\
&\quad \times \int_0^0 \mathcal{D}\psi\,\mathcal{D}\bar\psi\,\exp\left\{i\int d^4x\,\mathcal{L}_f(\psi,A)\right\} .
\end{aligned} \qquad (9.148)$$

The gauge field action $\mathcal{L}_g$ contains no reference to the spinor field. The integration over the gauge fields does, however, appear in the second path integral in the form of the gauge configuration entering the spinor action, assumed to be of the form

$$\mathcal{L}_f = i\bar{\psi}\gamma^\mu(\partial_\mu + igA_\mu)\psi \equiv i\bar{\psi}\slashed{D}\psi. \tag{9.149}$$

The notation in (9.149) is general enough to cover both the abelian and nonabelian cases. Because the action $\mathcal{L}_f$ is quadratic in the spinor fields, the second path integral may be formally evaluated.

The theory is continued to Euclidean time by the usual prescription of $t \to -it$ and $A_0 \to iA_0$. In Sec. 6.8.3 it was shown that the Euclidean measure for the spinor fields is given by integrating over the Grassmann coefficients of the expansions

$$\psi(x) = \int \frac{d^4p}{(2\pi)^2} \alpha_\lambda(p) w_\lambda(p) e^{ipx},$$

$$\bar{\psi}(x) = \int \frac{d^4p}{(2\pi)^2} \beta_\lambda(p) w_\lambda^\dagger(p) e^{-ipx}, \tag{9.150}$$

The $w_\lambda(p)$ represent the four orthonormal solutions of

$$i\slashed{p}_E w_\lambda(p) = \lambda_p w_\lambda(p), \tag{9.151}$$

where the eigenvalues are $\lambda_p = \pm p_E$. The explicit form for these solutions is given in Sec. 6.8.3. The path integral measure is then written, up to an overall normalization factor, as

$$\mathcal{D}\bar{\psi}\mathcal{D}\psi = \prod_{p,\lambda} d\beta_\lambda(p)\, d\alpha_\lambda(p). \tag{9.152}$$

In the Euclidean region the resulting differential operator $i\slashed{D}_E$ is self-adjoint. Therefore, given a gauge field configuration, the Euclidean spinor eigenfunction equation, written

$$i\slashed{D}_E W(x) = \lambda_E W(x), \tag{9.153}$$

can be assumed to possess a complete set of orthonormal eigenspinors. For simplicity of notation these eigenspinors will be given a discrete index, $W_n(x)$, and they satisfy

$$i\slashed{D}_E W_n(x) = \lambda_n W_n(x), \tag{9.154}$$

$$\int d^4x\, W_n^\dagger(x) W_m(x) = \delta_{nm}, \tag{9.155}$$

$$\sum_n W_n^{\dagger a}(x) W_n^b(y) = \delta_{ab}\delta^4(x-y), \tag{9.156}$$

where the spinor indices have been displayed in (9.156). The generalization of these formulas to the nonabelian case is accomplished by the addition of isospin indices to the fields and the differential operator. Of course, the subscript n may actually be a combination of many indices, some of which may be continuous as in the free eigenmode equation (9.151). The eigenvalues λ_n are implicit functionals of the gauge field configuration used to solve the eigenvalue problem.

As in (9.150) the spinor field, or fields as the case may be, are given the general expansions

$$\psi(x) = \sum_n a_n W_n(x), \qquad (9.157)$$

$$\bar\psi(x) = \sum_n b_n W_n^\dagger(x), \qquad (9.158)$$

and the spinor measure is written

$$\mathcal{D}\bar\psi \mathcal{D}\psi = \prod_n db_n \, da_n . \qquad (9.159)$$

For the moment it will be assumed that the two measures, (9.152) and (9.159), are equivalent. It is then possible to expand, at least formally, the eigenspinors W_n in terms of the complete set of free Euclidean eigenspinors,

$$W_n^a(x) = \int \frac{d^4p}{(2\pi)^4} \tilde{W}_n(p,\lambda) w_\lambda^a(p) e^{ipx} \equiv \int \frac{d^4p}{(2\pi)^4} U_n^a(p) e^{ipx}, \qquad (9.160)$$

where the coefficient $\tilde{W}_n$ is given by

$$\tilde{W}_n(p,\lambda) = \int d^4x \, w_\lambda^\dagger(p) W_n(x) e^{-ipx} . \qquad (9.161)$$

so that, on the face of it, the two sets of eigenspinors span the same function space and therefore are equivalent.

Exercise 9.6: Show that the coefficients $U_n^a(p)$ satisfy a completeness relation of the form

$$\sum_n U_n^{\dagger a}(p) U_n^b(k) = (2\pi)^4 \delta_{ab} \delta^4(p-k) . \qquad (9.162)$$

The spinor part of the path integral is evaluated by using the expansions of (9.157) in the action, so that

$$\int d^4x\, \bar{\psi} i \slashed{D}_E \psi = \sum_n \lambda_n b_n a_n \,. \tag{9.163}$$

Performing the Grassmann integrations over a_n and b_n immediately gives the spinor part of the path integral,

$$\prod_n \lambda_n = \det(i \slashed{D}_E) \,. \tag{9.164}$$

The ramifications of the failure of the measure to be invariant under chiral transformations is exposed by performing a spatially dependent chiral transformation on the path integral. As was the case in Sec. 8.2 for gauge invariance, this generates the divergence of the current associated with the chiral transformation. The transformation is written

$$\psi(x) \to \exp[i\alpha(x)\gamma_5]\psi(x) \,, \tag{9.165}$$

and it is easy to see that the spinor action transforms according to

$$\int d^4x\, \mathcal{L}_f \to \int d^4x\, (\mathcal{L}_f + \alpha \partial_\mu J^{5\mu}) \,. \tag{9.166}$$

The *axial current* J_μ^5 is given by

$$J_\mu^5 = \bar{\psi}\gamma_\mu\gamma_5\psi \,. \tag{9.167}$$

In the classical theory J_μ^5 is conserved by virtue of the invariance of the Lagrangian under the global chiral transformation.

The measure is not invariant, and using (9.156), it follows that the a_n and b_n transform according to

$$\begin{aligned}
a_n &= \int d^4x\, W_n^\dagger(x)\psi(x) \to \\
a_n' &= \int d^4x\, W_n^\dagger(x) \exp[i\alpha(x)\gamma_5]\psi(x) \\
&= \sum_m \int d^4x\, d^4y\, W_n^\dagger(x) \exp[i\alpha(x)\gamma_5] W_m(x) W_m^\dagger(y)\psi(y) \\
&= \sum_m C_{nm} a_m \,,
\end{aligned} \tag{9.168}$$

where

$$C_{nm} = \int d^4x\, W_n^\dagger(x) \exp[i\alpha(x)\gamma_5] W_m(x) \,. \tag{9.169}$$

It is straightforward to show that the identical result is obtained for the transformation of b_n. Recalling that a_n and b_n are Grassmann variables, the Jacobian generated by this transformation is given by

$$\prod_n da_n\, db_n \;\to\; (\det C)^{-2} \prod_n da'_n\, db'_n \;. \tag{9.170}$$

This determinant may be absorbed into the action by rewriting it with the familiar identity

$$(\det C)^{-2} = \exp(-2\,\text{Tr}\,\ln C) \;. \tag{9.171}$$

This is especially easy to evaluate for the case that α is chosen to be infinitesimal. For such a case it follows that

$$C_{nm} \approx \delta_{nm} + i\int d^4x\, \alpha(x) W_n^\dagger(x) \gamma_s W_m(x) \;. \tag{9.172}$$

Defining $A(x)$ by

$$A(x) = \sum_n W_n^\dagger(x) \gamma_s W_n(x) \;, \tag{9.173}$$

and dropping terms $O(\alpha^2)$ and higher gives

$$\text{Tr}\,\ln C \approx i\int d^4x\, \alpha(x) \sum_n W_n^\dagger(x)\gamma_s W_n(x) = i\int d^4x\, \alpha(x) A(x) \;. \tag{9.174}$$

Combining this with (9.166) gives the effective action that appears in the path integral as a result of the chiral transformation,

$$\int d^4x\, \mathcal{L}_f \;\to\; \int d^4x\, \left[\mathcal{L}_f + \alpha(\partial_\mu J^{5\mu} - 2A)\right] \;. \tag{9.175}$$

Clearly, as in Sec. 8.2, the path integral must be independent of α, and this immediately gives the expectation value

$$\langle \partial_\mu J^{5\mu}(x) \rangle = \langle 2A(x) \rangle \;. \tag{9.176}$$

If the right-hand side of (9.176) does not vanish, then the axial current is not conserved in the quantized case, and the theory is said to possess a *chiral anomaly*. The actual presence and effects of an anomaly depend upon the details of the theory under examination. However, there is one general case where anomalies must be avoided. If the theory begins as a massless nonabelian gauge theory with the gauge fields coupled to only one of the spinor chiralities, such as the GSW model of the previous section, then the Ward–Takahashi identities of the theory depend critically upon the conservation of the axial current. If the theory possesses an anomaly, the Slavnov–Taylor identities no longer hold, and it is impossible to prove the general renormalizability of the theory. On the other hand, if the theory is renormalizable without depending upon the conservation of the axial current, then an anomaly can create physically measurable phenomena. Such is the case in the decay of the neutral pi meson into two photons, although discussing this lies outside the scope of this book.

Because there is no mass term present in the differential operator $i\not{D}_E$, it is clear that it anticommutes with γ_5. The solutions are broken into left and right pieces,

$$W_n(x) = W_n^{(R)}(x) + W_n^{(L)}(x) , \qquad (9.177)$$

and it follows that

$$i\not{D}_E W_n^{(R)} = \tfrac{1}{2} i\not{D}_E (1+\gamma_5) W_n = \tfrac{1}{2}(1-\gamma_5) i\not{D}_E W_n = \lambda_n W_n^{(L)} . \qquad (9.178)$$

An identical argument gives

$$i\not{D}_E W_n^{(L)} = \lambda_n W_n^{(R)} , \qquad (9.179)$$

and this relation, along with the Hermiticity of $i\not{D}_E$, gives

$$W_n^{(L)\dagger} i\not{D}_E = \lambda_n W_n^{(R)\dagger} . \qquad (9.180)$$

Putting this together gives

$$\begin{aligned} \lambda_n \int d^4x\, W_n^{(R)\dagger} W_n^{(R)} &= \int d^4x\, W_n^{(L)\dagger} (i\not{D}_E) W_n^{(R)} \\ &= \lambda_n \int d^4x\, W_n^{(L)\dagger} W_n^{(L)} . \end{aligned} \qquad (9.181)$$

From this it follows that, *if λ_n is nonzero,*

$$\int d^4x\, W_n^{(L)\dagger} W_n^{(L)} = \int d^4x\, W_n^{(R)\dagger} W_n^{(R)} \qquad (9.182)$$

The orthogonal nature of the two chiral projections results in

$$W_n^{(L)\dagger} W_m^{(R)} = 0 , \qquad (9.183)$$

so that

$$\int d^4x\, W_n^\dagger(x) W_m(x) = \int d^4x\, (W_n^{(R)\dagger} W_m^{(R)} + W_n^{(L)\dagger} W_m^{(L)}) = \delta_{nm} , \qquad (9.184)$$

Sec. 9.5 Chiral Anomalies

and this yields the final result that, *if* $\lambda_n \neq 0$,

$$\int d^4x \, W_n^{(L)\dagger} W_n^{(L)} = \int d^4x \, W_n^{(R)\dagger} W_n^{(R)} = \tfrac{1}{2} \,. \tag{9.185}$$

However, if $\lambda_n = 0$, i.e., W_n is a *zero mode*, this proof breaks down and the norms of the left and right pieces of the eigenspinor do *not* need to coincide.

> **Exercise 9.7**: Prove that a basis for the zero modes may be chosen such that each one is either left- or right-handed.

When the chiral decomposition of the fields is substituted into the expression for the anomaly and (9.185) is used the result is

$$\int d^4x \sum_n W_n^\dagger(x) \gamma_s W_n(x) = n_+ - n_- \,, \tag{9.186}$$

where $n_\pm$ is the total number of right/left zero modes. This number will now be related to the form of the gauge field used to calculate the eigenvalues. This will be done by calculating the total number of right and left eigenmodes and finding the difference. Both numbers are infinite, but because the numbers of nonzero right and left modes match, the difference between the two numbers can be finite.

However, because both sums are divergent, subtracting them to get a sensible number requires that each of the two sums must be regulated, i.e., made finite, in a consistent way. A simple method [20] for doing this is to exponentially damp the contribution of nonzero modes by replacing the sum with

$$A_M(x) = \sum_n W_n^{aj\dagger}(x) \gamma_{ab}^s W_n^{bj}(x) \exp\left(-\frac{\lambda_n^{j\,2}}{M^2}\right) , \tag{9.187}$$

where M is a large mass, and the spinor and isospin indices have been explicitly displayed. Since $i\not{D}_E$ is Hermitian, all the eigenvalues are real. The anomaly is regained at the end of the calculation by the limit $M \to \infty$. The damping factor may be written in terms of $\not{D}_E$, so that the anomaly becomes

$$A_M(x) = \sum_n W_n^{aj\dagger}(x) \gamma_{ab}^s \exp\left(\frac{\not{D}_E^{\;2}}{M^2}\right)^{(jk)}_{(bc)} W_n^{ck}(x) \,. \tag{9.188}$$

The representation (9.160) may be used to write the unknown functions, W_n, in terms of plane waves. The result is

$$A_M(x) = \sum_n \int \frac{d^4p}{(2\pi)^4} \frac{d^4k}{(2\pi)^4} e^{-ipx} U_n^{ja\dagger}(p) \gamma_{ab}^5 \exp\left(\frac{\not{D}_E^2}{M^2}\right)_{(bc)}^{(jk)} U_n^{ck}(k) e^{ikx} \ . \tag{9.189}$$

Using the nonabelian version of identity (9.162) reduces this expression to

$$\begin{aligned}A_M(x) &= \int \frac{d^4k}{(2\pi)^4} e^{-ikx} \gamma_{ab}^5 \exp\left(\frac{\not{D}_E^2}{M^2}\right)_{(ba)}^{(jj)} e^{ikx} \\ &= \text{Tr} \int \frac{d^4k}{(2\pi)^4} e^{-ikx} \gamma^5 \exp\left(\frac{\not{D}_E^2}{M^2}\right) e^{ikx} \ , \end{aligned} \tag{9.190}$$

where the trace is over both isospin and spinor indices.

The operator $\not{D}_E^2$ has the form

$$\begin{aligned}\not{D}_E^2 &= \gamma^\mu \gamma^\nu D_\mu D_\nu = \tfrac{1}{4}\{\gamma^\mu, \gamma^\nu\}\{D_\mu, D_\nu\} + \tfrac{1}{4}[\gamma^\mu, \gamma^\nu][D_\mu, D_\nu] \\ &= -\Box_E - g^2 A^\mu A_\mu + \tfrac{1}{2} g \sigma^{\mu\nu} F_{\mu\nu} \ , \end{aligned} \tag{9.191}$$

where definitions (6.274) and (7.173) have been used, and the Lorentz gauge condition has been employed. It can be shown that the results are independent of the gauge condition chosen. The momentum k appearing in the integral (9.190) is now written $k = Mz$, so that z is a dimensionless variable. The resulting form of the anomaly is

$$A_M(x) = \text{Tr} \int \frac{d^4z}{(2\pi)^4} M^4 e^{-iMzx} \gamma^5 \exp\left(\frac{\not{D}_E^2}{M^2}\right) e^{iMzx} \ , \tag{9.192}$$

The exponential is now expanded and the trace taken. It is obvious that the nth term in the expansion must yield no worse than M^{-4} in order to avoid being discarded in the $M \to \infty$ limit. It is left as an exercise to use (9.191) to show that the only terms that meet this criterion are given by

$$e^{-izMx} \frac{M^4}{n!} \left(\frac{\not{D}_E^2}{M^2}\right)^n e^{izMx} =$$

$$\frac{M^4}{n!}(-z_E^2)^n + \frac{gM^2}{(n-1)!}(-z_E^2)^{n-1}(\tfrac{1}{2}\sigma^{\mu\nu} F_{\mu\nu} + g^2 A_\mu A^\mu)$$

$$+ \frac{g^2}{2(n-2)!}(-z_E^2)^{n-2}(\tfrac{1}{2}\sigma^{\mu\nu} F_{\mu\nu} + g^2 A_\mu A^\mu)^2 \ . \tag{9.193}$$

Sec. 9.5   Chiral Anomalies   397

> **Exercise 9.8**: Verify (9.193).

Using the properties of γ^5 shows that only the last term survives, since $\text{Tr}\,\gamma^5 = 0$ and $\text{Tr}\,\gamma^5 \sigma^{\mu\nu} = 0$. Using the normalization of the Lie generators, $\text{Tr}\,(T^a T^b) = \frac{1}{2}\delta_{ab}$, and the Euclidean identity

$$\text{Tr}\,(\gamma^5 \sigma^{\mu\nu} \sigma^{\alpha\beta}) = 4\varepsilon^{\mu\nu\alpha\beta}\,, \tag{9.194}$$

the anomaly becomes

$$\begin{aligned}
\lim_{M\to\infty} A_M(x) &= \tfrac{1}{4}\sum_{n=2}^{\infty}\int \frac{d^4 z}{(2\pi)^4}\,\frac{g^2}{(n-2)!}(-z_E^2)^{n-2}\varepsilon^{\mu\nu\alpha\beta} F^a_{\mu\nu} F^a_{\alpha\beta}\\
&= \tfrac{1}{4} g^2 \varepsilon^{\mu\nu\alpha\beta} F^a_{\mu\nu} F^a_{\alpha\beta} \int \frac{d^4 z}{(2\pi)^4}\, e^{-z_E^2}\\
&= \frac{g^2}{16\pi^2} \text{Tr}\,(F_{\mu\nu}{}^*F^{\mu\nu})\,. \tag{9.195}
\end{aligned}$$

Recalling the relation of the divergence of the axial current to the anomaly, it follows that

$$\partial_\mu J^{5\mu} = \frac{g^2}{16\pi^2} F^a_{\mu\nu}{}^*F^{a\mu\nu}\,. \tag{9.196}$$

The Pontryagin density, first defined in (7.253), has emerged, and this shows that the anomaly relates the difference in the number of left and right zero modes of the Dirac operator to the topological index of the gauge configuration. It was shown in Sec. 7.6 that there are pure gauge configurations characterized by a winding number n for the integral of the Pontryagin density. It follows for the case of pure gauge configurations that

$$n_+ - n_- = \frac{1}{16\pi^2}\int d^4x\,\text{Tr}\,(F_{\mu\nu}{}^*F^{\mu\nu}) = n\,. \tag{9.197}$$

Result (9.197) is a specific case of a general class of *index theorems*. These theorems relate the dimension of the kernels of the left and right chirality Dirac operators in the presence of topologically nontrivial gauge connections. The kernel of a differential operator is the space spanned by the eigenfunctions associated with eigenvalue zero. The kernel corresponds to the zero modes of the two chiral Dirac operators in this case, and the anomaly, by virtue of counting the number of orthogonal zero modes, is simply the difference in the dimension of the left and right chirality kernels. This result also shows that the simple pure gauge configurations spoil the chiral invariance of the theory.

In the GSW model and many of its extensions, the different fermion couplings occur for the left- and right-handed multiplets. In the GSW model the left-handed sector is coupled to the SU(2) gauge fields, while the right-handed sector is not. This alters the structure of the anomaly equation in these theories. For example, in the case where the left- and right-handed sectors for the fermions have different coupling, the action is written

$$\mathcal{L} = \mathcal{L}_R + \mathcal{L}_L = i\bar{\psi}_R \slashed{D}_R \psi_R + i\bar{\psi}_L \slashed{D}_L \psi_L , \qquad (9.198)$$

where $\slashed{D}_L$ and $\slashed{D}_R$ may involve different gauge fields. The eigenvalue equations for the respective sectors become

$$\slashed{D}_R W_n = \lambda_n^W W_n , \quad \slashed{D}_L V_n = \lambda_n^V V_n . \qquad (9.199)$$

where each of the differential operators, $\slashed{D}_{R,L}$, defines a complete set of *both* left- and right-handed solutions. The two sets of solutions, W and V, are determined by the gauge field configuration used in the respective differential operator and are in no way related. Of course, only the respective chirality solution is to be employed in the left- and right-handed sectors.

The classical action (9.198) is invariant under a chiral transformation involving either or both the left- and right-handed sectors. For instance, the transformation, given by

$$\psi \rightarrow \exp\left[\tfrac{1}{2}i\alpha(1-\gamma_5)\right]\psi , \qquad (9.200)$$

acts only on the left-handed sector. A repetition of the earlier steps used to derive the chiral anomaly shows that the left-handed current associated with the classical invariance (9.200),

$$J_L^\mu = \tfrac{1}{2}\bar{\psi}\gamma^\mu(1-\gamma_5)\psi , \qquad (9.201)$$

satisfies an anomalous conservation law,

$$\partial_\mu J_L^\mu = -\sum_n V_n^\dagger \gamma^5 V_n , \qquad (9.202)$$

where *both* chiralities of the V solutions appear in the anomaly.

Exercise 9.9: Verify result (9.202).

Likewise, the right-handed current

$$J_R^\mu = \tfrac{1}{2}\bar{\psi}\gamma^\mu(1+\gamma_5)\psi , \qquad (9.203)$$

also satisfies an anomalous conservation law,

$$\partial_\mu J_R^\mu = \sum_n W_n^\dagger \gamma^5 W_n ,\qquad (9.204)$$

where both chiralities of the W solutions also appear in the anomaly.

The vector current is the sum of the left- and right-handed currents,

$$J^\mu = \bar\psi \gamma^\mu \psi = J_L^\mu + J_R^\mu ,\qquad (9.205)$$

and by virtue of the different couplings, possesses the anomaly

$$\partial_\mu J^\mu = \sum_n \left[W_n^\dagger \gamma^5 W_n - V_n^\dagger \gamma^5 V_n \right] .\qquad (9.206)$$

Result (9.206) reveals a means to avoid the anomaly if the theory requires conservation of the vector current. The two contributions appearing on the right-hand side of (9.206) will cancel if the spinors of the right-handed sector are coupled to gauge fields in a manner identical to the spinors of the left-handed sector, even if the spinors of the left- and right-handed sectors are different. This is because the value of the anomaly depends only on the gauge field configuration.

In the event the couplings are different, the vector charge operator Q is no longer conserved. Instead, it satisfies the equation

$$\begin{aligned} Q(t_+) - Q(t_-) &= \int_{t_-}^{t_+} dt\, \dot Q = \int d^4 x\, \partial_\mu J^\mu \\ &= \int d^4 x\, \mathrm{Tr} \left[\frac{g_R^2}{16\pi^2} F_{\mu\nu}^R {}^*F^{R\mu\nu} - \frac{g_L^2}{16\pi^2} F_{\mu\nu}^L {}^*F^{L\mu\nu} \right] . \end{aligned} \qquad (9.207)$$

It was seen in Sec. 7.1 that the Q operator has as eigenvalues the difference between the total number of particles and the total number of antiparticles. If the coupling is different between the left- and right-handed sectors, it follows that processes exist that violate conservation of particle-antiparticle number.

A specific case of this is the GSW model. There, the left-handed quarks and leptons are coupled to the SU(2) gauge fields, and these fields possess topologically nontrivial configurations. The right-handed quarks and leptons are coupled only to the U(1) gauge field, and it possesses no topological properties. It was pointed out in Sec. 9.4 that the baryon number B is associated with the conservation of the quark minus antiquark number. This is precisely given by $\frac{1}{3}Q$. Thus, by virtue of (9.207), the conservation of baryon number is violated in the GSW model. An evaluation of the rate

400   Nonperturbative Results

using instantons, to be discussed in the next section, shows that baryon number violation is extremely suppressed at currently available energies [21].

Exercise 9.10: Show that baryon number minus lepton number, $B-L$, is exactly conserved in the GSW model.

An extremely interesting model where the chiral anomaly plays a central role is by Schwinger [22]. The Schwinger model describes a massless spinor field minimally coupled to an abelian gauge field and is exactly solvable if attention is restricted to 1+1 space-time dimensions. The exact operator and state solution [23] displays many remarkable properties, including dynamical mass generation for the vector field, confinement of the spinor particles and a θ-vacuum structure. The presentation here will be limited to discussing the role of the chiral transformation in the path integral approach to solving the model.

The action for the abelian Schwinger model in the Feynman gauge is given by

$$\mathcal{L} = i\bar{\psi}\gamma^\mu(\partial_\mu + igA_\mu)\psi - \tfrac{1}{4}F_{\mu\nu}F^{\mu\nu} - \tfrac{1}{2}(\partial_\mu A^\mu)^2 \ . \tag{9.208}$$

Because the Schwinger model owes its solvability to its space-time dimensionality, it is important to note the peculiarities of 1+1 dimensional spinors. The gamma algebra for 1+1 spinor fields is easily represented by the following variants of the Pauli spin matrices,

$$\gamma^0 = \begin{pmatrix} 0 & 1 \\ 1 & 0 \end{pmatrix}, \ \gamma^1 = \begin{pmatrix} 0 & 1 \\ -1 & 0 \end{pmatrix}, \tag{9.209}$$

and it follows that

$$\gamma^{\mu\dagger} = \gamma^0\gamma^\mu\gamma^0 \ . \tag{9.210}$$

The 1+1 dimensional form for γ_s is defined as

$$\gamma_s = \gamma^0\gamma^1 = \begin{pmatrix} -1 & 0 \\ 0 & 1 \end{pmatrix} \ . \tag{9.211}$$

The 1+1 Lorentz group is characterized by a single parameter, and the single generator for the Lorentz group is given by

$$T = \tfrac{1}{2}i[\gamma^0,\gamma^1] = \begin{pmatrix} -i & 0 \\ 0 & i \end{pmatrix} \ . \tag{9.212}$$

Therefore, a finite Lorentz transformation is given by

$$S(\Lambda) = \exp\{i\alpha T\} = \begin{pmatrix} e^\alpha & 0 \\ 0 & e^{-\alpha} \end{pmatrix}. \tag{9.213}$$

The two-component 1+1 spinor field ψ is introduced. The Lorentz transformation shows that the bilinear form $\bar\psi\psi$, where $\bar\psi = \psi^\dagger\gamma^0$, is a Lorentz invariant.

The absence of a mass term in the spinor action makes the classical theory invariant under chiral transformations. The quantized theory however possesses an anomaly. It is straightforward to repeat the steps used earlier in this section to evaluate the 3+1 chiral anomaly to determine the effect of a space-time dependent chiral transformation, given by

$$\psi(x) \to \exp[ig\phi(x)\gamma_5]\psi(x), \tag{9.214}$$

where $\phi(x)$ is an *arbitrarily large* space-time dependent function.

Exercise 9.11: Show that the tranformation (9.214) induces the 1+1 dimension action in the path integral to change to

$$\mathcal{L} \to \mathcal{L} - \partial_\mu\phi J_s^\mu - \frac{g^2}{4\pi}\phi\varepsilon_{\mu\nu}F^{\mu\nu}, \tag{9.215}$$

where $\varepsilon_{\mu\nu}$ is the Levi-Civita symbol appropriate to two dimensions and $J_s^\mu = g\bar\psi\gamma^\mu\gamma_5\psi$.

The term proportional to $\partial_\mu\phi$ generated by the chiral transformation may be used to cancel the coupling between the spinor and gauge field. This is accomplished by first noting that the gauge condition $\partial_\mu A^\mu = 0$ reduces the number of gauge degrees of freedom from two to one, and so the gauge field can be simultaneously tranformed according to

$$A_\mu = -\varepsilon_{\mu\nu}\partial^\nu\phi. \tag{9.216}$$

The antisymmetry of $\varepsilon_{\mu\nu}$ assures that the gauge condition is satisfied. The magic of the Pauli spin matrices gives the equality

$$gA_\mu\bar\psi\gamma^\mu\psi = -g\varepsilon_{\mu\nu}\bar\psi\gamma^\nu\psi\,\partial^\mu\phi = g\bar\psi\gamma^\mu\gamma_5\psi\,\partial_\mu\phi = \partial_\mu\phi\,J_s^\mu, \tag{9.217}$$

so that the spinor–gauge coupling is exactly canceled, while the remainder of the gauge field action becomes

$$-\tfrac{1}{4}F_{\mu\nu}F^{\mu\nu} = \tfrac{1}{2}\partial_\mu\partial_\nu\phi\,\partial^\mu\partial^\nu\phi. \tag{9.218}$$

The chiral anomaly term reduces to

$$-\frac{g^2}{4\pi}\epsilon_{\mu\nu}F^{\mu\nu} = \frac{g^2}{2\pi}\partial_\mu\phi\,\partial^\mu\phi\,, \qquad (9.219)$$

after an integration by parts.

> **Exercise 9.12**: Verify (9.217), (9.218), and (9.219).

The path integral measure reduces to an integration over the scalar mode ϕ, while the action appearing in the path integral becomes

$$\mathcal{L} = \tfrac{1}{2}\partial_\mu\partial_\nu\phi\,\partial^\mu\partial^\nu\phi - \frac{g^2}{2\pi}\partial_\mu\phi\,\partial^\mu\phi\,, \qquad (9.220)$$

so that the saddle-point evaluation corresponds to using the solutions of the equation of motion,

$$\Box\left(\Box + \frac{g^2}{\pi}\right)\phi = 0\,. \qquad (9.221)$$

The solution takes the form

$$\left(\Box + \frac{g^2}{\pi}\right)\phi = \Lambda\,, \qquad (9.222)$$

where Λ satisfies $\Box\Lambda = 0$ and serves as a source term for ϕ. A mass term, g^2/π, has appeared in this fourth-order equation of motion, and it is this term that corresponds to a dynamical mechanism for mass generation.

The resulting path integral has an action that is quadratic in ϕ and ψ and is therefore exactly integrable. The time-ordered products of the spinor fields in the original coupled theory can be evaluated by the following technique. The chiral transformation on the spinor fields used to decouple them from the gauge field leaves a phase on the spinor fields. Written in terms of the two components of the spinor field, this phase is given by

$$\langle\psi_a(x)\cdots\rangle \to \langle\exp[\pm ig\phi(x)]\psi_a(x)\cdots\rangle\,, \qquad (9.223)$$

where the minus sign occurs for $a = 1$ and the plus sign for $a = 2$. This phase can be absorbed into the action of the path integral by writing it as

$$\exp[\pm ig\phi(x)] = \exp\left\{i\int d^4z\, g\delta^4(z-x)\phi(z)\right\}\,. \qquad (9.224)$$

Therefore, each occurrence of a spinor field in a time-ordered product generates a sourcelike term in the action, where the source term is a Dirac delta at the space-time position of the spinor field. Thus, the spinor fields serve as sources of the remnant of the gauge field, which has become massive. Since the phase term is linear in the field ϕ the path integral remains exactly solvable. This result can be used to show that the ground state of the Schwinger model is characterized by the formation of bound states of the spinor particles, i.e., a fermionic condensate. Demonstration of these properties of the Schwinger model lies outside the scope of this book.

9.6 Classical Solutions

In the treatment of quantum mechanical path integrals in Chapter 4 the role of classical solutions to the equations of motion was discussed. It was pointed out that they can describe possible ground state configurations for the system, and they can also be used to interpolate between two different ground states at different times to give an approximate ground state tunnelling or oscillation amplitude. The same uses occur for field theoretic systems, and this is again referred to as the saddle-point approximation. Because classical solutions to the equations of motion capture nonlinear aspects of the dynamics, it is widely believed that many of their properties carry over into the quantized system. In this respect it is possible to view the tree approximation to the vacuum expectation value in scalar theories as the use of a constant, translationally invariant classical solution to model the nonlinear behavior of the ground state. In the case of nonabelian gauge theories it has already been seen that the pure gauge configurations, which are classical solutions to the gauge field equations of motion, break the measure of the path integral into disjoint homotopic sectors and demonstrate the existence of the θ vacuum. In this section both aspects of classical solutions will be discussed.

9.6.1 The Kink Solution

In the first case, there exist numerous simple $1+1$ dimensional scalar field theories that possess static classical solutions with nontrivial properties. An example of such a theory is given by the familiar action

$$\mathcal{L} = \tfrac{1}{2}\partial_\mu\phi\,\partial^\mu\phi - V(\phi) = \tfrac{1}{2}\partial_\mu\phi\,\partial^\mu\phi + \tfrac{1}{2}\mu^2\phi^2 - \tfrac{1}{4}\lambda\phi^4 , \qquad (9.225)$$

which gives the equation of motion

$$\Box\phi + \frac{\partial V}{\partial \phi} = \left(\frac{\partial^2}{\partial t^2} - \frac{\partial^2}{\partial x^2}\right)\phi - \mu^2\phi + \lambda\phi^3 = 0 . \qquad (9.226)$$

This equation possesses the familiar constant solutions,

$$\phi_0 = \pm\sqrt{\frac{\mu^2}{\lambda}}. \qquad (9.227)$$

However, in addition, it admits the static solution

$$\phi_c(x) = \sqrt{\frac{\mu^2}{\lambda}} \tanh\left(\frac{\mu}{\sqrt{2}}(x - x_0)\right), \qquad (9.228)$$

where x_0 is arbitrary. It also admits an infinite family of static spatially periodic solutions, the Jacobi elliptic functions, but these will be considered later in this section in the context of Euclidean quantum mechanics.

The nature of the hyperbolic tangent shows that the solution (9.228) interpolates spatially between the two constant solutions of (9.227). For this reason (9.228) is usually referred to as the *kink solution*. Other field theories give rise to classical solutions that describe objects colliding and emerging from the collision with no deviation from their initial shape. These are known as *solitary waves* or *solitons* [24]. The sine-Gordon theory, described by $V(\phi) = \beta \cos \alpha\phi$, possesses solutions with this property. The kink solution does not have this property, and so it is not a soliton.

It is also possible to find a time-dependent solution to (9.226) by boosting the solution (9.228) into a moving frame, so that

$$\phi_c(x,t) = \sqrt{\frac{\mu^2}{\lambda}} \tanh\left(\frac{\mu}{\sqrt{2}}\gamma(x - x_0 - vt)\right), \qquad (9.229)$$

where $\gamma = (1-v^2)^{-1/2}$. This simply describes a kink travelling at the constant velocity v. Using this solution it is possible to choose the parameters x_0 and v so that the center of the kink is at $x = a$ when $t = t_-$ and is at $x = b$ when $t = t_+$. For the case that the $t_\pm$ are chosen so that $t_\pm = \pm\frac{1}{2}T$, these parameters are given by

$$v = \frac{b-a}{T}, \quad x_0 = \tfrac{1}{2}(a+b). \qquad (9.230)$$

Of course, in the limit that $T \to \infty$ this reduces to the $v = 0$ case, which corresponds classically to a stationary kink.

It is natural to consider a quantum mechanical state with a kink present as a possible alternative ground state to the translationally invariant ground states associated with (9.227) [25]. Such a ground state will be written $|\phi_c, a, t\rangle$, where a denotes the center of the kink and has the property that

$$\langle \phi_c, a, t | \phi(x,t) | \phi_c, a, t \rangle = \phi_c(x,t). \qquad (9.231)$$

The nature of such a state will be discussed momentarily, but some ramifications can be immediately seen. It is straightforward, using the canonical form for the Hamiltonian and the solution (9.229), to show that the difference in energy between the kink ground state, modelled with the solution (9.229), and the translationally invariant ground state is

$$H_{\text{kink}} - H_{\text{constant}} = \frac{M}{\sqrt{1-v^2}}, \qquad (9.232)$$

where $M = 2\sqrt{2}\,\mu^3/3\lambda$. The unusual form for the mass is induced by the fact that λ has the units of (length)$^{-2}$ in the 1+1 dimensional theory.

> **Exercise 9.13**: Verify (9.232).

Result (9.232) has the same form as a particle of mass M moving at a constant velocity. If λ is small, the mass of this particle is much greater than $\sqrt{2}\mu$, the mass of the particles associated with the constant ground states. In this simple model the kink is therefore understood as a heavy, fundamental excitation available to the system, distinct from the lighter particle excitations. Of course, the spectrum of the theory in the presence of the kink must be determined.

Such a state is difficult to construct in a canonical quantization procedure. Once the spectrum of the free field consistent with the presence of the kink is determined, the kink state may be realized as a coherent state using the appropriate extension of the formulas introduced in Sec. 6.8. The coherent state can then be used to represent the asymptotic kink ground state, allowing the development of a self-consistent form for S-matrix elements. However, this procedure forces the kink to behave according to the time dependence of the solution chosen to represent it. In effect, this determines in advance whether the kink is allowed to interact with the particles of the system. For example, the form (9.229) describes a kink moving at a constant velocity, and this, at least intuitively, forces the kink to be a *free* object since it prevents the kink from undergoing a change of momentum.

The solution to this dilemma is to introduce a quantum mechanical coordinate Q, whose eigenvalues describe the center of the kink [26]. The form of the classical solution is amended to $\phi_c(x - Q)$. The in and out states of the theory are written as a direct product:

$$|P\rangle_{\text{in}} = |p\rangle \otimes |a\rangle. \qquad (9.233)$$

The state $|p\rangle$ is a Fock-like particle state, while $|a\rangle$ is an eigenstate of the operator Q, which is referred to as the *collective coordinate*. The eigenvalue of Q then becomes a variable of integration in the path integral measure, and the integration over this eigenvalue restores translational invariance and, therefore, conservation of momentum to the system.

To see how this works in the simple case of the theory (9.225), the path integral representation for the transition amplitude between the states $|\phi_c, t_-, a\rangle$ and $|\phi_c, t_+, b\rangle$ is written

$$\langle \phi_c, t_+, b | \phi_c, t_-, a \rangle = \int_{\phi_c(a)}^{\phi_c(b)} \mathcal{D}\phi \, \exp\left\{ i \int d^2x \left[\mathcal{L}(\phi) + J\phi\right] \right\} . \quad (9.234)$$

This form is obtained by integrating the momentum from the path integral (6.237). It is easy to see that the phase associated with the presence of the soliton vanishes, since

$$\int dx \left[\phi_c(x-b)\dot{\phi}_c(x-b) - \phi_c(x-a)\dot{\phi}_c(x-a)\right] = 0 . \quad (9.235)$$

The limits on the path integral of (9.234) may be absorbed by translating ϕ by the solution (9.229) subject to the choices for the parameters given by (9.230). For reasons that will become evident shortly, terms of that are dependent on the parameters a and b must be retained. It is easy to see that the expanded action has the form

$$\int d^2x \, \mathcal{L}(\phi + \phi_c) = \int d^2x \left[\mathcal{L}(\phi_c) - \tfrac{1}{2}\phi\left(\Box + \frac{\partial^2 V}{\partial \phi_c^2}\right)\phi + \ldots\right] , \quad (9.236)$$

where, for the action under consideration,

$$\frac{\partial^2 V}{\partial \phi_c^2} = 2\mu^2 - 3\mu^2 \text{sech}^2(x - \tfrac{1}{2}(a+b) - vt) . \quad (9.237)$$

The presence of the kink has induced the potential (9.237), and the spectrum of the fluctuation field ϕ is determined by it. However, it is clear from the limits on the spatial part of the action integral that the first term on the right-hand side of (9.236) has no reference to a or b, and can therefore be dropped.

The standard procedure for evaluating the path integral is to solve the Euclidean eigenvalue problem

$$\left(\Box_E + \frac{\partial^2 V}{\partial \phi_c^2}\right)\phi_n = \lambda_n \phi_n , \quad (9.238)$$

expand the fluctuation fields as a Fourier series in the orthonormal eigenmodes, and integrate over the coefficients. A difficulty immediately arises for the case that the kink has time dependence, since the Euclidean form of the potential (9.237) will be complex. A complex potential spoils the Hermiticity of the eigenvalue equation, and the eigenmodes will no longer be complete. For the case under consideration the limit $v \to 0$, induced by the limit $T \to \infty$, salvages the Hermiticity. The modes are found by assuming the form

$$\phi_n = u_n(x) e^{i\omega\tau}, \qquad (9.239)$$

and this reduces the eigenvalue problem to the Hermitian form

$$\left(-\frac{d^2}{dx^2} + \frac{\partial^2 V}{\partial \phi_c^2}\right) u_n(x) = (\lambda_n - \omega^2) u_n(x) \equiv \omega_n^2 u_n(x). \qquad (9.240)$$

By redefining the variable appearing in (9.240) according to $\mu x/\sqrt{2} = z$, the mode equation in the presence of the kink becomes

$$\left[-\tfrac{1}{2}\frac{d^2}{dz^2} + (3\tanh^2 z - 1)\right] u(z) = \frac{\omega^2}{\mu^2} u(z). \qquad (9.241)$$

This eigenvalue problem is exactly solvable. In terms of the dimensionless coordinate, there are two discrete forms,

$$\begin{aligned} \omega_0^2 &= 0, & u_0 &= \operatorname{sech}^2(z), \\ \omega_1^2 &= \tfrac{3}{2}\mu^2, & u_1 &= \tanh(z)\operatorname{sech}(z), \end{aligned} \qquad (9.242)$$

and a set of continuous eigenfunctions,

$$\omega_p^2 = p^2 + 2\mu^2,$$
$$u_p = \exp\left(\frac{i\sqrt{2}pz}{\mu}\right)\left[3\tanh^2(z) - 1 - \frac{2p^2}{\mu^2} - i\frac{3\sqrt{2}p}{\mu}\tanh(z)\right], \qquad (9.243)$$

where the solutions have not been normalized. The existence of the zero mode signals trouble for the Gaussian approximation to the path integral, since the integration over that mode will diverge.

The presence of the zero mode is not a feature specific to the kink, but is present when any classical solution is used. Regardless of the form for V there is always one eigenfunction that possesses eigenvalue zero and this zero mode is proportional to $\partial_x \phi_c$. This is readily seen from the static equation of motion (9.226), which shows that

$$\frac{\partial}{\partial x}\left(-\frac{\partial^2 \phi_c}{\partial x^2} + \frac{\partial V}{\partial \phi_c}\right) = \left(-\frac{\partial^2}{\partial x^2} + \frac{\partial^2 V}{\partial \phi_c^2}\right)\frac{\partial \phi_c}{\partial x} = 0. \qquad (9.244)$$

Nonperturbative Results

Denoting M^2 as the value of the integral

$$M^2 = \int dx \left(\frac{\partial \phi_c}{\partial x}\right)^2 , \qquad (9.245)$$

the normalized eigenfunction $u_o = \partial_x \phi_c / M$ has the eigenvalue zero, and is orthogonal to all other solutions to (9.240). Its existence is a result of the fact that the value x_o appearing in (9.228) is arbitrary. This is an outgrowth of the translational invariance of the original theory, and for that reason this eigenfunction is referred to as the *translation mode*.

For the moment, the presence of this zero mode will be ignored in order to expose the problems that arise from applying the Gaussian approximation when it is not well defined. Following the standard procedure the fluctuation field is expanded in the Euclidean Fourier series,

$$\begin{aligned}\phi &= \sum_{n\neq 0} \int \frac{d\omega}{2\pi} [a_n(\omega) u_n(x) e^{i\omega t} + a_n^*(\omega) u_n^*(x) e^{-i\omega t}] \\ &+ \int \frac{d\omega}{2\pi} q(\omega) u_o(x) \cos \omega t ,\end{aligned} \qquad (9.246)$$

and the measure of the path integral is defined as

$$\mathcal{D}\phi = \prod_\omega dq(\omega) \prod_n da_n^*(\omega) \, da_n(\omega) . \qquad (9.247)$$

Appending a source term to the action and evaluating the Euclidean path integral by the method of steepest descent yields the result

$$\langle \phi_c, t_+, b | \phi_c, t_-, a \rangle_J =$$
$$\left[\det\left(\Box_E + \frac{\partial^2 V}{\partial \phi_c^2}\right)\right]^{-\frac{1}{2}} \exp\left\{\tfrac{1}{2} \int d^2x \, d^2y \, J(x) \Delta_E(x-y) J(y)\right\} . \quad (9.248)$$

The Euclidean propagator appearing in (9.248) is given by

$$\begin{aligned}\Delta_E(x-x') &= \sum_{n\neq 0} \int \frac{d\omega}{2\pi} \frac{e^{i\omega(t-t')}}{\omega^2 + \omega_n^2} u_n^*(x) u_n(x') \\ &+ \int \frac{d\omega}{2\pi} \frac{u_o(x) u_o(x')}{\omega^2} \cos\omega t \, \cos\omega t' .\end{aligned} \qquad (9.249)$$

Exercise 9.14: Verify result (9.249).

By inspection it is obvious that the second integral in (9.249), the contribution from the translation mode, is infrared divergent. At the same time, the zero eigenvalue present in the determinant causes the result to be undefined. Both problems are resolved by introducing the collective coordinate [26]. In the original expansion of the Lagrangian, the field was translated to $\phi \to \phi + \phi_c$. If the classical solution is written

$$\phi_c(x) = \phi_c\left(x - \frac{q}{M}\right), \qquad (9.250)$$

then a Taylor series expansion yields

$$\phi_c\left(x - \frac{q}{M}\right) = \phi_c(x) - \frac{q}{M}\partial_x\phi_c(x) + \ldots = \phi_c(x) - q u_o(x) + \ldots, \quad (9.251)$$

where the ellipsis refers to terms of order q^2 and higher in the expansion. These terms will become part of the interaction terms, and will not appear in the integrations that determine the propagator. Using the Fourier decomposition of q,

$$q(t) = \int \frac{d\omega}{2\pi} q(\omega) \cos\omega t, \qquad (9.252)$$

it follows that the first term in the Taylor series cancels the translation mode in the Fourier decomposition (9.246) of ϕ. The field translation can therefore be replaced with

$$\phi + \phi_c = \tilde{\phi} + \tilde{\phi}_c, \qquad (9.253)$$

where

$$\begin{aligned}
\tilde{\phi} &= \sum_{n\neq 0} \int \frac{d\omega}{2\pi} \left[a_n(\omega) u_n(x) e^{i\omega t} + a_n^*(\omega) u_n^*(x) e^{-i\omega t}\right], \\
\tilde{\phi}_c &= \phi_c\left(x - \frac{q}{M}\right) + \frac{q}{M}\partial_x\phi_c(x).
\end{aligned} \qquad (9.254)$$

The infrared divergence in the propagator is removed, and the zero eigenvalue no longer appears in the determinant, so that the path integral is once again well defined.

Of course, such a form for the classical solution should have been used from the start of the analysis. The field translation can be written

$$\phi + \phi_c\left(x - \frac{q}{M}\right) = \phi - \frac{q}{M}\partial_x\phi_c(x) + \phi_c(x) + \phi_r(x,q), \qquad (9.255)$$

where ϕ_r are the terms quadratic and higher in q that result from a Taylor series expansion. It is not difficult to show that the action may be expanded about the solution $\phi_c(x)$, which has no reference to q, and a consistent set of Feynman rules for ϕ and q derived. The collective coordinate q acts as the position operator for the kink, and the vacuum transition element has as its limits the in and out position of the kink. The presence of the collective coordinate serves to reinstate translational invariance.

> **Exercise 9.15**: Show that $\langle \phi_c, a, t_+ | \phi_c, b, t_- \rangle$ is a function solely of $a - b$ in the absence of a source term.

The asymptotic ground states may be projected onto momentum eigenstates by the usual procedure:

$$\langle \phi_c, p, t_+ | \phi_c, p', t_- \rangle = \int \frac{da}{2\pi} \frac{db}{2\pi} \langle \phi_c, a, t_+ | \phi_c, b, t_- \rangle e^{ip'b} e^{-ipa} . \quad (9.256)$$

Using the fact that the transition element is solely a function of $a - b$ in the absence of a source term, it follows that

$$\langle \phi_c, p, t_+ | \phi_c, p', t_- \rangle = \\ \frac{1}{2} \int \frac{dx}{2\pi} \frac{dy}{2\pi} \langle \phi_c, x, t_+ | \phi_c, 0, t_- \rangle e^{-i\frac{1}{2}(p-p')y} e^{-i\frac{1}{2}(p+p')x} . \quad (9.257)$$

The integral over y gives $\delta(p-p')$, so that, in the absence of scattering with the particle modes of the theory, the momentum of the kink is conserved. The presence of additional particles in the asymptotic states changes this result, but the development of these ideas is outside the scope of this book.

9.6.2 Vacuum Tunnelling

Another use for classical solutions is to derive transition elements between ground states. For simplicity, this idea will be demonstrated in the context of quantum mechanical systems, although it finds its most important applications in quantum field theory and current models of the early universe [6]. The idea is simple, and was discussed cursorily in Sec. 4.5. As a system is cooled a new ground state may become available, one whose energy is lower than the current state of the system. As a result, the system will transit to this lower energy ground state. However, many times there exists a potential barrier to the transition, and the width and height of this barrier will affect the transition rate.

In standard wave mechanics barrier penetration occurs because the wave function is nonzero on both sides of the barrier. In order to determine the path integral counterpart of this phenomenon it is necessary to understand the behavior of an unstable system. The easiest example is that of the quantum mechanical system where $V(q) = -\frac{1}{2}m\omega^2 q^2$, corresponding to a repulsive potential. Since the potential is not bounded from below there can be no quantum mechanical ground state. However, the position $q = 0$ satisfies $\partial V/\partial q = 0$, and so constitutes an unstable ground state.

The path integral formalism may be applied to determine the unstable vacuum persistence functional $\langle q = 0, T | q = 0, 0 \rangle$. This matrix element is available from the standard attractive harmonic oscillator matrix element by analytic continuation in the frequency ω. Clearly, if ω is continued into imaginary values the original theory of an attractive potential becomes the repulsive potential of interest here. Using the results of Sec. 3.3, the matrix element is given by

$$\langle q = 0, T | q = 0, 0 \rangle = \left(\frac{im\omega}{2\pi \sinh \omega T} \right)^{\frac{1}{2}}, \qquad (9.258)$$

and this is clearly proportional to $\exp -\frac{1}{2}\omega T$ for large times. Thus, the state $|q = 0\rangle$ is unstable quantum mechanically, and its decay rate is ω. It is important to note that the property of instability of the state was derived from the prefactor of the path integral, and this in turn is the determinant of the differential operator appearing in the action.

It is here that the Euclidean formalism enters the discussion. The Wick rotated path integral has the action

$$i \int dt \, [\tfrac{1}{2}m\dot{q}^2 - V(q)] \to -\int d\tau \, [\tfrac{1}{2}m\dot{q}^2 + V(q)]. \qquad (9.259)$$

The configurations that minimize the Euclidean action satisfy the equation

$$m \frac{d^2 q}{d\tau^2} - \frac{\partial V}{\partial q} = 0. \qquad (9.260)$$

Equation (9.260) is the Euler–Lagrange equation for the system where $V(q)$ has been replaced by $-V(q)$. The boundary conditions on the solution must match the limits on the path integral. Expanding the Euclidean action around the Euclidean solution, referred to as an *instanton*, using the method of steepest descent, and analytically continuing the result back to real time gives an approximation to the determinant that determines the stability of the system. This approach has the advantage that properties of the solutions to the Euclidean equations of motion can be deduced by classical mechanical energy arguments using the potential $-V(q)$. The term *instanton* has come to mean any finite action Euclidean solution to the equations of motion. The original instanton, a Euclidean solution to the Yang–Mills theory, will be discussed later in this section.

This method will be demonstrated in the case of the unstable potential $-\frac{1}{2}m\omega^2 q^2$. The equation of motion is given by

$$m\frac{d^2q}{d\tau^2} - \frac{\partial V}{\partial q} = m\frac{d^2q}{d\tau^2} + m\omega^2 q = 0 , \qquad (9.261)$$

subject to the boundary conditions $q = 0$ at $\tau = 0$ and $\tau = T$, and this has the solution $q_c = 0$. Expanding the solution around this gives the differential operator

$$D = -m\frac{d^2}{d\tau^2} - m\omega^2 . \qquad (9.262)$$

The eigenvalues of this operator are determined subject to the boundary condition that the eigenfunctions vanish at $\tau = 0$ and $\tau = T$ since the dependence upon the end points of the integration has vanished. The eigenvalues are easily determined to be

$$\lambda_n = m\frac{n^2\pi^2}{T^2} - m\omega^2 . \qquad (9.263)$$

Regardless of the value of n and ω, there exist values of T for which the eigenvalue λ_n is negative. As discussed in Sec. 4.5, zero or negative eigenvalues in the Euclidean region indicate a breakdown of the Gaussian approximation for the path integral, since the integration for that mode is no longer defined. The path integral mechanism for representing the instability of a quantum state is therefore the existence of negative Euclidean eigenvalues of D for some range of T in the evaluation of its propagator.

This problem is related to the determination of those classical trajectories that correspond to minima of the action, as opposed to the larger set of trajectories that correspond to extrema. For the purposes of simplicity discussion will be limited to the case that the Lagrangian density has the form $\mathcal{L} = \frac{1}{2}m\dot{x}^2 - V(x)$. To expose this relationship, the non-Euclidean action is given a functional power series expansion around a classical solution, $x = x_c + q$, where x_c is a classical trajectory chosen to satisfy the boundary conditions and q represents the fluctuations around the classical trajectory. At the classical level q can be thought of as a slight deviation from the classical trajectory, and it is a means to test the stability of the extremum. If x_c represents at least a local minimum of the action, then the small deviation q should increase the value of the action. The action has the expansion

$$S = S|_{x_c} + \delta S|_{x_c} + \frac{1}{2}\delta^2 S|_{x_c} + \ldots . \qquad (9.264)$$

By virtue of using a classical solution, the second term in the expansion vanishes. After an integration by parts the third term is given by

$$\tfrac{1}{2}\delta^2 S\big|_{x_c} = -\int_{t_a}^{t_b} dt\, q \left(m\ddot{q} - \frac{\partial^2 V(q)}{\partial q^2}\bigg|_{x_c} q \right). \tag{9.265}$$

Result (9.265), along with the boundary conditions $q(t_a) = q(t_b) = 0$, defines the eigenvalue equation

$$m\ddot{q}_n - \frac{\partial^2 V(q)}{\partial q^2}\bigg|_{x_c} q_n = -\lambda_n q_n, \tag{9.266}$$

which is assumed to possess a set of orthornormal solutions. The general fluctuation q is built as a linear superposition of the solutions of (9.266):

$$q = \sum_n a_n q_n. \tag{9.267}$$

Inserting (9.267) into (9.266) and using orthonormality gives

$$\delta^2 S\big|_{x_c} = \sum_n \tfrac{1}{2}\lambda_n a_n^2. \tag{9.268}$$

As long as all the λ_n are positive-definite the classical trajectory corresponds to at least a local minimum, since all fluctuations about the classical trajectory increase the value of the action. However, if any of the eigenvalues are negative, there exist trajectories nearby to x_c that lower the value of the action, indicating that the trajectory does not minimize the action.

The problem of minimizing the classical trajectory bears a striking resemblance to the quantum mechanical problem of determining the stability of the path integral. Therefore, any results bearing on the nature of the classical problem can be adapted to the quantum mechanical case. Several results from Morse theory [27] are especially relevant. In particular, the following simple criterion exposes the presence of negative eigenvalues for a given problem. Every classical trajectory between two points can be characterized by the initial momentum p required to begin and end at the respective points in the time interval $T = t_b - t_a$, so that the classical trajectory can be written $x_c(p,t)$. A negative eigenvalue occurs in (9.266) for each value of t in the interval $t_b - t_a$ such that there is a *node*,

$$\frac{\partial x_c(p,t)}{\partial p} = 0. \tag{9.269}$$

414 Nonperturbative Results

For a general proof the reader should consult the references [27]. However, the potential $V(q) = -\frac{1}{2}kq^2$ is a simple case that demonstrates this result. The classical Euclidean trajectory from $q = 0$ to $q = 0$ in the interval $T = t_b - t_a$ is given by

$$x_c = A_n \sin\left(\frac{n\pi t}{T}\right), \tag{9.270}$$

where T must be such that $n\pi = \omega T$. Therefore, n is an integer, given by $n = \omega T/\pi$. It is not difficult to see that x_c has n turning points or nodes in the time interval T, and this predicts n negative eigenvalues for (9.266). Examination of (9.263) verifies this result.

This technique can be applied to an arbitrary potential $V(q)$, although the interpretation of the results requires some care. The Euclidean form of the transition element will be dominated at large times by the ground state of the system. Ignoring complications due to ground state degeneracy it is straightforward to see that, as $\tau \to \infty$, the Euclidean transition element becomes

$$\langle q, \tau | q, 0 \rangle = \sum_n \langle q | n \rangle \langle n | q \rangle e^{-E_n \tau} \to |\psi_g(q)|^2 e^{-E_g \tau}, \tag{9.271}$$

where E_g is the ground state energy of the system and $\psi_g(q)$ is the ground state wave function. In the quantum mechanical case the instanton method allows an approximate determination of the ground state properties.

As a specific example, the potential

$$V(q) = \tfrac{1}{2}\lambda(q^2 - a^2)^2, \tag{9.272}$$

will be examined for its behavior at $q = 0$. Despite the fact that $\partial V/\partial q = 0$ at $q = 0$, it is clear that any wave packet centered around $q = 0$ will eventually disperse into the two wells centered at $\pm a$.

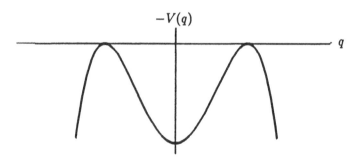

Fig. 9.2. The potential $-V(q) = -\tfrac{1}{2}\lambda(q^2 - a^2)^2$.

Sec. 9.6 Classical Solutions

The Euclidean version of this potential, $-V(q)$, is depicted in Fig. 9.2. This shows that the relative maximum at $q = 0$ has become a well, while the wells at $q = \pm a$ have become peaks in the Euclidean region. The transition element $\langle q = 0, T | q = 0, 0 \rangle$, which will be used to determine the ground state wave function at $q = 0$, is dominated by the Euclidean solution corresponding to the particle rolling up the hill from the bottom of the well at $q = 0$ toward one of the peaks, rebounding before or upon reaching the top of the peak, and returning back to the original well at $q = 0$ after time T. This is the so-called bounce instanton solution, and it must be a solution to the Euclidean equation of motion

$$m\frac{d^2q}{d\tau^2} + 2\lambda a^2 q - 2\lambda q^3 = 0 , \tag{9.273}$$

with the period T. Because the solution has a bounce point, there will be a negative eigenvalue associated with such a trajectory.

The solution to (9.273) is given by one of the Jacobi elliptic functions, the modular sine, denoted $\operatorname{sn}(t | T)$ [28]. These solutions may be given an integral representation by using energy techniques. The equation of motion shows that the Euclidean energy,

$$\tfrac{1}{2}m\left(\frac{dq}{d\tau}\right)^2 - V(q) = E , \tag{9.274}$$

must be constant along the trajectory. Result (9.274) gives a first-order differential equation that can be formally integrated, as in classical mechanics, to express the solution as a function of E. Of course, the value of E appearing in (9.274) is determined from the period T.

> **Exercise 9.16**: Use the properties of the modular sine to show that there is no nonzero solution to (9.273) for periods $T < (m/2\lambda a^2)^{1/2}$.

The fact that there is no solution for small values of T shows that the Gaussian approximation to the path integral for small values of T is derived from the Euclidean action

$$\mathcal{L}_E = \tfrac{1}{2}m\dot{q}^2 - \lambda a^2 q^2 . \tag{9.275}$$

As a result, for small values of T, the Gaussian approximation gives

$$\langle q = 0, T | q = 0, 0 \rangle \propto \exp\left(-\sqrt{\frac{\lambda a^2}{2m}}\, T\right) . \tag{9.276}$$

Determining the negative eigenvalue associated with the Jacobi elliptic function will not be presented here [29]. However, the classical solution (9.273) has particularly simple properties when $T \to \infty$, since this limit corresponds to $E = 0$. For this case there is a Euclidean trajectory from $q = a$ to $q = -a$ given by

$$q_c(\tau) = a \tanh\left(\sqrt{\frac{\lambda a^2}{m}}\,\tau\right). \qquad (9.277)$$

and this trajectory possesses no nodes. For this special case, the Euclidean quantum mechanical theory has begun to mimic the simple scalar field theory in the presence of a kink solution, discussed earlier in this section. The differential equation for the eigenmodes u of the theory becomes

$$\left(-m\frac{d^2}{d\tau^2} + 2\lambda a^2 - 6\lambda a^2 \tanh^2(\tau)\right) u(\tau) = \omega u(\tau). \qquad (9.278)$$

The solutions (9.242) and (9.243) can be adapted to this case to calculate the instanton approximation to the long-term ground state transition element $\langle a, T | -a, -T \rangle$.

> **Exercise 9.17**: Solve the associated zero mode problem and use the previous results to calculate the behavior of the Euclidean transition element $\langle a, T | -a, -T \rangle$ in the limit $T \to \infty$.

9.6.3 Yang–Mills Instantons

In Sec. 7.6 the pure gauge solutions to nonabelian gauge theories was discussed and shown to possess nontrivial properties. In the previous section it was pointed out that the violation of baryon number in the standard model can be triggered by topologically nontrivial Euclidean field configurations. The path integral has been constructed to be manifestly gauge invariant by evaluating the θ vacuum transition amplitude, and so pure gauge solutions can make no contribution to physical processes since they correspond to gauge transformations. It is then natural to ask if there are other solutions to the Euclidean equations of motion, not simply pure gauge solutions, that possess nontrivial topological properties. Because the gauge field θ vacuum persistence element is evaluated after a Wick rotation, such field configurations could contribute to the transition amplitude and result in baryon nonconservation.

Sec. 9.6 Classical Solutions 417

The Euclidean path integral is damped by the value of the action evaluated for the gauge field configuration,

$$_+\langle 0,\theta|0,\theta\rangle_- = \sum e^{-S_E} . \qquad (9.279)$$

Therefore, it is generally believed that classical solutions to the Euclidean gauge field equations of motion provide the dominant contribution to the transition element, since these configurations minimize the Euclidean action. In what follows an SU(2) gauge field theory will be considered, and the gauge fixing term will be set to zero. The value of the Euclidean action is positive definite, and this can be seen from the fact that it takes the form

$$S_E = \int d^4x \left(\tfrac{1}{2} F^a_{0j} F^a_{0j} + \tfrac{1}{4} F^a_{ij} F^a_{ij} \right) \equiv \int d^4x \, \tfrac{1}{4} (F_{\mu\nu})^2 \geq 0 . \qquad (9.280)$$

The Euclidean equation of motion will be written in matrix form as

$$D_\mu F_{\mu\nu} = \partial_\mu F_{\mu\nu} + ig[A_\mu, F_{\mu\nu}] , \qquad (9.281)$$

where the Euclidean result $\partial_\mu F^\mu{}_\nu = g^{\mu\rho} \partial_\mu F_{\rho\nu} = -\delta_{\mu\rho} \partial_\mu F_{\rho\nu} = -\partial_\mu F_{\mu\nu}$ has been used, with a similar result for the commutator in the covariant derivative.

An important inequality for the gauge field action comes from the positive-definiteness of the Euclidean integral given by

$$\int d^4x \, \mathrm{Tr} \left(F_{\mu\nu} \pm {}^*F_{\mu\nu} \right)^2 \geq 0 . \qquad (9.282)$$

Using the result that ${}^*F_{\mu\nu} {}^*F_{\mu\nu} = F_{\mu\nu} F_{\mu\nu}$, this inequality gives the result that

$$S_E = \int d^4x \, \tfrac{1}{2} \mathrm{Tr} \, (F_{\mu\nu})^2 \geq \pm \tfrac{1}{2} \int d^4x \, \mathrm{Tr} \, (F_{\mu\nu} {}^*F_{\mu\nu}) = \pm \frac{8\pi^2}{g^2} n , \qquad (9.283)$$

where the gauge field configuration has been assumed to belong to the nth homotopic sector. Because the inequality must hold for either choice of sign, the Euclidean gauge field action has the lower bound

$$S_E \geq \frac{8\pi^2}{g^2} |n| . \qquad (9.284)$$

The equality certainly holds for the case that the argument of the integral in (9.282) is zero, and for that case it must be that

$$F_{\mu\nu} = \pm {}^*F_{\mu\nu} . \qquad (9.285)$$

418 Nonperturbative Results

A field configuration such that (9.285) holds is said to be *self-dual* for the positive sign, or *anti-self-dual* for the negative sign. Therefore, self-dual or anti-self-dual gauge field configurations yield an absolute minimum of the Euclidean action [30].

The relation of self-dual field configurations to the solutions of the equations of motion is established in the following way. For a Euclidean field configuration A_μ, the associated dual tensor, defined by

$$^*F_{\mu\nu} = \varepsilon_{\mu\nu\rho\sigma} F_{\rho\sigma} = \varepsilon_{\mu\nu\rho\sigma} \left(\partial_\rho A_\sigma - \partial_\sigma A_\rho + ig[A_\rho, A_\sigma] \right) , \qquad (9.286)$$

is automatically a solution of the equations of motion:

$$D_\mu {}^*F_{\mu\nu} = 0 . \qquad (9.287)$$

Exercise 9.18: Prove statement (9.287).

If the field configuration is self-dual or anti-self-dual, then (9.287) immediately implies that

$$D_\mu F_{\mu\nu} = 0 . \qquad (9.288)$$

Therefore, (anti)self-dual field configurations are solutions of the equations of motion that minimize the Euclidean action in the topological sector to which the field configuration belongs. These solutions are referred to as Yang–Mills instantons.

The nature of these solutions has been intensively studied. For the case of SU(2) a general form for instantons in the nth topological sector has been found. Only the $n = 1$ case will be discussed here; the general method for constructing and evaluating the other cases is presented in the book by Rajaraman [24]. First, the antisymmetric SU(2) element $\Sigma_{\mu\nu}$ is defined in terms of the Pauli spin matrices σ_j,

$$\begin{aligned}
\Sigma_{ij} &= -\tfrac{1}{2}\varepsilon_{ijk}\sigma_k & (i,j) = (1,2,3) , \\
\Sigma_{0j} &= -\Sigma_{j0} = -\tfrac{1}{2}\sigma_j & j = (1,2,3) , \\
\Sigma_{00} &= 0 .
\end{aligned} \qquad (9.289)$$

By construction, the element $\Sigma_{\mu\nu}$ satisfies $\Sigma_{\mu\nu} = \tfrac{1}{2}\varepsilon_{\mu\nu\rho\sigma}\Sigma_{\rho\sigma}$ and is therefore self-dual. Using $\Sigma_{\mu\nu}$, the $n = 1$ instanton gauge field configuration can be written

$$A_\mu = \frac{1}{g}\Sigma_{\mu\nu} \frac{2(x_\nu - a_\nu)}{(x-a)^2 + \lambda^2} , \qquad (9.290)$$

Sec. 9.6 Classical Solutions 419

where a_μ is an arbitrary constant Euclidean four-vector, and λ is an arbitrary constant with units of length.

Using (9.290) in the definition of $F_{\mu\nu}$ gives

$$F_{\mu\nu} = \frac{1}{g}\Sigma_{\mu\nu}\frac{4\lambda^2}{[(x-a)^2+\lambda^2]^2} . \qquad (9.291)$$

Because $\Sigma_{\mu\nu}$ is manifestly self-dual, the form for $F_{\mu\nu}$ is self-dual, verifying that (9.290) is indeed a solution of the Euclidean equations of motion. In the asymptotic region, $x \to \infty$, the solution (9.290) reduces to the winding number one Euclidean pure gauge solution displayed in Sec. 7.6:

$$\lim_{x \to \infty} = \frac{1}{ig}\partial_\mu \mathbf{U}_1 \mathbf{U}_1^{-1} , \qquad (9.292)$$

where the matrix $\mathbf{U}_1$ is given by

$$\mathbf{U}_1 = \frac{1}{r}(x^0 + ix^j\sigma^j) . \qquad (9.293)$$

This is corroborated by showing that the solution satisfies the equality (9.284). Evaluating the Euclidean action using (9.291) gives

$$S_E = \int d^4x \tfrac{1}{2}\mathrm{Tr}\,(F_{\mu\nu}F_{\mu\nu}) = \frac{48\lambda^4}{g^2}\int d^4x\,\frac{1}{(x^2+\lambda^2)^4} = \frac{8\pi^2}{g^2} , \qquad (9.294)$$

where the translational invariance of Euclidean space has been used to remove the appearance of a. The fact that (9.290) is localized in time as well as space is the origin of the name *instanton*.

Because the solution (9.290) is pure gauge only asymptotically allows it to contribute to the vacuum transition element. Because it is topologically nontrivial it can cause baryon number conservation to be violated. In the Euclidean region, the contribution of the instanton to the path integral is damped by the factor [21]

$$_+\langle 0,\theta|0,\theta\rangle_- \propto e^{-S_E} . \qquad (9.295)$$

where $S_E = 8\pi^2/g^2$. For the case of the SU(2) gauge fields of the GSW model, the coupling constant is found to have the value $g^{-2} \approx 2.3$ from measurements of electroweak processes. This gives the value

$$e^{-S_E} \approx 10^{-80} , \qquad (9.296)$$

an extremely low rate for baryon violation. There is current research that indicates high temperature effects significantly increase the rate [31], but this lies beyond the scope of this book.

9.6.4 The Abelian Magnetic Monopole

In Sec. 7.6 it was shown that Maxwell's equation $\nabla \cdot \mathbf{B} = 0$ is a consequence of the geometric nature of gauge fields. If the gauge field is nonsingular, then this equation will always hold. However, relaxing the conditions on $\mathbf{A}$ to include singular functions allows $\mathbf{B}$ to develop a nonzero divergence, corresponding to the existence of a magnetic monopole.

A static magnetic monopole would necessarily give rise to a static radial magnetic field,

$$\mathbf{B} = \frac{g}{r^2} \hat{r}, \qquad (9.297)$$

where r is the spatial radius, and this corresponds to a solution of the equation

$$\nabla \cdot \mathbf{B} = 4\pi g \, \delta^3(\mathbf{x}). \qquad (9.298)$$

The magnetic field can still be derived from the gauge field by the usual relation $\mathbf{B} = \nabla \times \mathbf{A}$, but $\mathbf{A}$ must be given, in spherical coordinates, by the singular form [32]

$$A_r = A_\theta = 0, \quad A_\phi = \frac{g}{r} \frac{(1 - \cos\theta)}{\sin\theta}. \qquad (9.299)$$

Exercise 9.19: Verify that the gauge potential of (9.299) gives a magnetic field of the form (9.297).

Deriving magnetic monopoles from the vector potential allows the incorporation of magnetic monopoles into the standard electromagnetic couplings in quantum mechanics and quantum field theory without disrupting the structure of gauge invariance.

The price paid for maintaining gauge invariance in the presence of magnetic monopoles is the singular nature of the gauge field. The gauge field (9.299) is singular for the value $\theta = \pi$, so that there is a "string" of singularities along the $-z$ axis. On the face of it, this singularity appears to make volume and closed surface integrations impossible, since the divergences are not isolated, but extend throughout half of all space. Of course, the so-called Dirac string can be rotated to lie along the $+z$ axis, and this corresponds to the form

$$A'_r = A'_\theta = 0, \quad A'_\phi = -\frac{g}{r} \frac{(1 + \cos\theta)}{\sin\theta}. \qquad (9.300)$$

Sec. 9.6 Classical Solutions

The key step in rendering the Dirac string physically irrelevant comes by noting that the difference between the two forms is given by

$$\mathbf{A} - \mathbf{A}' = \frac{2g}{r \sin \theta} \hat{\mathbf{e}}_\phi . \tag{9.301}$$

Using the form of the gradient in spherical coordinates shows that this difference can be written as

$$\mathbf{A} - \mathbf{A}' = \nabla \chi , \tag{9.302}$$

where χ is a function solely of the azimuthal angle ϕ,

$$\chi = 2g\phi . \tag{9.303}$$

Therefore, the two forms, $\mathbf{A}$ and $\mathbf{A}'$, may be related by a gauge transformation,

$$\mathbf{A}' = \mathbf{A} - \nabla \chi . \tag{9.304}$$

It is easy to see from the form of the gradient and the function χ that the other components are unaffected by the gauge transformation.

The Dirac string may be eliminated from calculations in the following way [33]. When calculating in the upper hemisphere, the form $\mathbf{A}$, given by (9.299), is used. When calculating in the lower hemisphere, a gauge transformation is made so that the form $\mathbf{A}'$, given by (9.300), can be used. As a result, the form of the gauge field used throughout all of space is free of the Dirac string. However, the gauge transformation is not without ramifications. If the gauge field is coupled to a matter field, the matter field must also undergo the same gauge transformation in order to leave the equations of motion invariant, so that the gauge transformed matter field ψ' is given by

$$\psi' = e^{ie\chi}\psi . \tag{9.305}$$

This new function must be single valued in order to satisfy the original boundary conditions placed upon it. Therefore, it must satisfy

$$e^{ie\chi(\phi)}\psi(r,\theta,\phi) = e^{ie\chi(\phi+2\pi)}\psi(r,\theta,\phi+2\pi) . \tag{9.306}$$

This means that the phase must be identical after a rotation of 2π, and using the form for χ shows that this can be true only if

$$i4\pi eg = ie\chi(\phi + 2\pi) - ie\chi(\phi) = i2\pi n , \tag{9.307}$$

where n is an arbitrary integer. The Dirac quantization condition $eg = \frac{1}{2}n$ has again emerged, but this time the explicit form of the magnetic field has dictated its necessity. The act of gauge transforming ψ when passing between the two hemispheres is at the basis of a general mathematical structure known as a *fiber bundle* [34]. Very loosely speaking, a fiber bundle consists of a base space, a fiber, and a connection. In this case the sphere on which the wave function is defined is the base space, the wave function itself is the fiber, and the gauge field is the connection that gives the change in the fiber as it is dragged along paths. A fiber bundle is topologically nontrivial when a gauge transformation is required as the fiber is dragged over the open sets that cover the base space. A realistic development of these ideas is outside the scope of this book [34].

Magnetic monopoles, even in the abelian case discussed here, have many remarkable properties, and the verified experimental observation of such an object would be of tremendous importance. For example, the bound state of an electric monopole and a magnetic monopole, known as a *dyon*, could behave as a fermion, even if both of the original particles were bosons. To see this, it is only necessary to calculate the angular momentum present in the electromagnetic field surrounding the pair. This is calculated from the usual expression,

$$\mathbf{L}_{em} = \int d^3x \left[\mathbf{x} \times (\mathbf{E} \times \mathbf{B}) \right] . \tag{9.308}$$

It is left as an exercise to use the forms of $\mathbf{E}$ and $\mathbf{B}$ associated with a static electric monopole and magnetic monopole separated by the distance $\mathbf{a}$ to show that

$$\mathbf{L}_{em} = -eg\,\hat{\mathbf{a}} , \tag{9.309}$$

where $\hat{\mathbf{a}}$ is a unit vector pointing from the electric charge to the magnetic charge.

Exercise 9.20: Verify result (9.309).

For the case that the value of n in the Dirac quantization condition is odd, the resulting angular momentum is a half integer multiple of $\hbar$, corresponding to a fermionic system [35].

These results have their nonabelian counterparts. In nonabelian theories solutions corresponding to dyons can be obtained and these solutions possess nontrivial topological properties. As an example of the effects of such an object, the presence of monopoles in many nonabelian extensions of the GSW model can serve to enhance the decay rate of baryons induced through the anomaly. The subject of nonabelian monopoles will not be presented here [36].

9.7 Applications of the Effective Potential

In this section several applications of the effective potential will be made. The first step is to introduce rules for calculating at finite temperature in the Euclidean region. These rules are combined with the effective potential to demonstrate restoration of symmetry at a critical temperature in a simple model. The second application is to evaluate the Higgs model in the limit that the symmetric theory is renormalized to zero mass for the scalar field. This demonstrates the dynamical breakdown of symmetry known as the Coleman–Weinberg mechanism. The third application is to the Gross–Neveu model. This simple 1+1 dimensional model of fermions undergoes a dynamical breakdown of symmetry for a composite field structure.

9.7.1 Finite Temperature and Symmetry Restoration

The quantum mechanical partition function was introduced for bosonic and fermionic degrees of freedom in Chapters 4 and 5. The extension of finite temperature effects to field theories is accomplished by introducing the grand canonical ensemble,

$$Z_\beta = \sum_n \langle n | e^{-\beta H} | n \rangle \equiv \text{Tr}\, e^{-\beta H}, \qquad (9.310)$$

where the sum is over the entire physical subspace of allowed particle states.

As was the case for the partition function, the grand canonical ensemble may be related to the path integral at Euclidean time. Initially the case of a single scalar field will be considered. Using the completeness of the coherent states $|\phi\rangle$, it follows that (9.310) can be rewritten as

$$\begin{aligned} Z_\beta &= \sum_n \int \mathcal{D}\phi \, \langle n | \phi \rangle \langle \phi | e^{-\beta H} | n \rangle \\ &= \int \mathcal{D}\phi \sum_n \langle \phi | e^{-\beta H} | n \rangle \langle n | \phi \rangle = \int \mathcal{D}\phi \, \langle \phi | e^{-\beta H} | \phi \rangle. \end{aligned} \qquad (9.311)$$

Recalling the form for the transition element between coherent states,

$$Z = \langle \phi, t | \phi, 0 \rangle = \langle \phi | e^{-iHt} | \phi \rangle, \qquad (9.312)$$

it is apparent that this transition element becomes the grand canonical ensemble by continuing it to the Euclidean time $t \to -i\beta$ and integrating over ϕ.

Of course, the transition element (9.312) has a path integral representation,

$$Z = \int_\phi^\phi \mathcal{D}\phi \exp\left\{ i \int dx \int_0^t d\tau\, \mathcal{L}(\phi) \right\} , \qquad (9.313)$$

where the momentum has been integrated since the in and out states coincide. Wick rotating to the time $-i\beta$ and summing over all field configurations that are periodic therefore gives the grand canonical ensemble.

This gives the finite temperature rules for the bosonic system. The requirement of periodicity in τ for the field configurations means that the Euclidean fields must have a Fourier series representation [37],

$$\phi(\mathbf{x}, \tau) = \frac{1}{\beta} \sum_{n=-\infty}^{\infty} \exp\left(i2\pi n \frac{\tau}{\beta} \right) \phi_n(\mathbf{x}) , \qquad (9.314)$$

rather than the usual Fourier transform. Therefore, for the bosonic scalar field the finite temperature effects are given by the following replacements in the Euclidean region:

$$\omega \to \frac{2\pi n}{\beta}, \quad \int \frac{d\omega}{2\pi} \to \frac{1}{\beta} \sum_n . \qquad (9.315)$$

The fermionic field rules are derived similarly. The grand canonical ensemble is projected onto the Grassmann-valued coherent states $|\eta\rangle$, and these have the usual Grassmann property that

$$\langle n|\eta\rangle\langle\eta|n\rangle = \langle -\eta|n\rangle\langle n|\eta\rangle , \qquad (9.316)$$

due to the anticommutativity of the Grassmann variables. The grand canonical ensemble then becomes

$$\begin{aligned} Z_\beta &= \sum_n \int \mathcal{D}\eta \, \langle n|\eta\rangle\langle\eta|e^{-\beta H}|n\rangle \\ &= \int \mathcal{D}\eta \sum_n \langle -\eta|e^{-\beta H}|n\rangle\langle n|\eta\rangle = \int \mathcal{D}\eta \, \langle -\eta|e^{-\beta H}|\eta\rangle . \quad (9.317) \end{aligned}$$

Again, the transition element between the Grassmann-valued coherent states is given by

$$Z = \langle -\eta, t|\eta, 0\rangle = \langle -\eta|e^{-iHt}|\eta\rangle , \qquad (9.318)$$

so that the replacement $t \to -i\beta$ along with an integration over η gives the grand canonical ensemble. Therefore, the path integral representation

Sec. 9.7 Applications of the Effective Potential

of this transition element,

$$Z = \int_{\eta}^{-\eta} \mathcal{D}\eta \, \exp\left\{i \int dx \int_0^t d\tau \, \mathcal{L}(\eta)\right\}, \qquad (9.319)$$

gives the grand canonical ensemble when continued to Euclidean times. This result shows that the Euclidean path integral is to be summed over all fermionic configurations that are antiperiodic in the Euclidean time interval β. The Euclidean field must therefore be given the Fourier series expansion

$$\eta(\mathbf{x}, \tau) = \frac{1}{\beta} \sum_{n=-\infty}^{\infty} \exp\left(i 2\pi(n + \tfrac{1}{2})\frac{\tau}{\beta}\right) \eta_n(\mathbf{x}), \qquad (9.320)$$

in order to ensure the antiperiodicity property. This leads to the fermionic finite temperature replacements in the Euclidean region:

$$\omega \to \frac{2\pi(n + \tfrac{1}{2})}{\beta}, \quad \int \frac{d\omega}{2\pi} \to \frac{1}{\beta} \sum_n. \qquad (9.321)$$

There is a subtlety in the case of gauge fields. There, the sum over the physically allowed states includes zero norm ghosts that must be prevented from contributing to the grand canonical ensemble. However, it is important to note that all gauge field modes are initially bosonic. Therefore, the projection of the grand canonical ensemble onto the gauge field coherent states does not have the Grassmann property (9.316). The grand canonical ensemble is therefore given by the Euclidean form of the path integral

$$Z_\beta = \int_{A_\mu}^{A_\mu} \mathcal{D}A \, \delta(\chi) \, \Delta_\chi \, \exp\left\{i \int dx \int_0^t d\tau \, \mathcal{L}_\chi(A)\right\}. \qquad (9.322)$$

The Faddeev–Popov determinant Δ_χ is absorbed into the action by using scalar Grassmann variables $\bar{c}$ and c. However, these Grassmann variables must be periodic in the interval β since the original determinant was periodic. Therefore, both the Faddeev–Popov ghosts and the gauge fields undergo the Euclidean replacements [38]

$$\omega \to \frac{2\pi n}{\beta}, \quad \int \frac{d\omega}{2\pi} \to \frac{1}{\beta} \sum_n. \qquad (9.323)$$

These rules can be combined with the effective potential in the Euclidean region to demonstrate the existence of a phase transition in the simple scalar model [2, 39]. The Euclidean result (9.27) for the effective potential is reevaluated using the finite temperature rules (9.315) appropriate for scalar fields. The one-loop contributions become, for $\hbar = 1$,

$$V_1 = \frac{1}{2\beta} \sum_{n=-\infty}^{\infty} \int \frac{d^3k}{(2\pi)^3} \ln\left(\frac{4\pi^2 n^2}{\beta^2} + \mathbf{k}^2 + M^2\right) . \tag{9.324}$$

The sum over n is evaluated by noting that

$$\frac{\partial}{\partial R} \sum_{n=-\infty}^{\infty} \ln(n^2\pi^2 + R^2) = \sum_{n=-\infty}^{\infty} \frac{2R}{n^2\pi^2 + R^2} = 2\coth R . \tag{9.325}$$

Integrating (9.325) and dropping the infinite constant, irrelevant to the effective potential since it is independent of R, gives

$$\sum_{n=-\infty}^{\infty} \ln(n^2\pi^2 + R^2) = 2\ln\sinh R . \tag{9.326}$$

Applying result (9.326) to (9.324) and dropping all terms independent of M gives

$$\begin{aligned} V_1 &= \frac{1}{\beta} \int \frac{d^3k}{(2\pi)^3} \ln\sinh\left(\frac{\beta}{2}\sqrt{\mathbf{k}^2 + M^2}\right) \\ &= \tfrac{1}{2} \int \frac{d^3k}{(2\pi)^3} \varepsilon_k + \frac{1}{\beta} \int \frac{d^3k}{(2\pi)^3} \ln\left(1 - e^{-\beta\varepsilon_k}\right) , \end{aligned} \tag{9.327}$$

where

$$\varepsilon_k = \sqrt{\mathbf{k}^2 + M^2} . \tag{9.328}$$

The first integral in (9.327) is independent of β and is the previous zero temperature result (9.29). The remaining integral is not divergent, and since it vanishes in the $\beta \to \infty$ limit, it represents the finite temperature contribution to the effective potential. That all divergences are present in the zero temperature form of the effective potential is a general result, so that finite temperature effects cannot and do not alter the renormalization of the theory.

This result will now be specialized to the case that

$$V(\phi) = -\tfrac{1}{2}\mu^2\phi^2 + \tfrac{1}{4}\lambda\phi^4 . \tag{9.329}$$

The mechanism for spontaneous breakdown of symmetry in this model is the presence of a negative coefficient for the renormalized mass term. If this coefficient were to become positive then only the symmetric vacuum,

Sec. 9.7 Applications of the Effective Potential

characterized by $\phi_0 = 0$, would be available. The finite temperature contribution to the $\phi_0{}^2$ term will now be calculated and shown to force the coefficient to become positive for sufficiently high temperature.

For the model under consideration the value of M^2 is given by

$$M^2 = \frac{\partial^2 V(\phi_0)}{\partial \phi_0{}^2} = -\mu^2 + 3\lambda \phi_0{}^2 \,. \tag{9.330}$$

It is apparent that as $\phi_0 \to 0$ the value of M^2 becomes negative, and the Gaussian approximation of the path integral used to derive the effective potential must break down. This difficulty was discussed for the related quantum mechanical problem in Sec. 4.5 and for the convexity of the effective potential in Sec. 9.2. There, as here, it is argued that although the Gaussian approximation breaks down, the theory must be stabilized by higher-order terms in ϕ, and that the Gaussian approximation is still valid near the minima of the effective potential. It is clear from the integrals appearing in (9.327) that, at the point where $M^2 \approx 0$, the theory must begin to transit toward the symmetric vacuum, and a much more careful analysis of the effective potential is needed to determine the details of the phase transition taking place.

Using this simplistic assumption it is necessary to calculate only the lowest-order finite temperature correction to M^2 in the effective potential, and this is done by using a Taylor series expansion of the second integral appearing in (9.327), denoted V_β. It is straightforward to show that

$$\begin{aligned}
\left.\frac{\partial V_\beta}{\partial M^2}\right|_{M^2=0} &= \tfrac{1}{2} \int \frac{d^3 p}{(2\pi)^3} \frac{1}{|p|} \frac{e^{-\beta|p|}}{1 - e^{-\beta|p|}} \\
&= \frac{1}{4\pi^2} \int_0^\infty dp \, \frac{p e^{-\beta p}}{1 - e^{-\beta p}} \\
&= \frac{1}{4\pi^2 \beta^2} \sum_{n=1}^\infty \frac{1}{n^2} = \frac{1}{24\beta^2} \,.
\end{aligned} \tag{9.331}$$

Putting this together with the form for M^2 gives the coefficient of the quadratic part of the effective potential in this approximation:

$$V(\phi_0) = -\tfrac{1}{2}\left(\mu^2 - \frac{\lambda}{4\beta^2}\right)\phi_0{}^2 + \cdots \,. \tag{9.332}$$

This shows that the coefficient of the quadratic term becomes positive for temperatures greater than the critical temperature $\beta_c^{-1} = 2\mu/\sqrt{\lambda}$.

Of course, other fields in the theory may contribute to the effective potential. For the case that the scalar field is coupled to a spinor field of mass m through the Yukawa vertex,

$$\mathcal{L}_I = -g\phi\bar{\psi}\psi , \qquad (9.333)$$

it is possible to find the fermionic contribution,

$$V_f(\phi_0) = -2\int \frac{d^3p}{(2\pi)^3}\varepsilon_p - \frac{4}{\beta}\int \frac{d^3p}{(2\pi)^3}\ln\left(1+e^{-\beta\varepsilon_p}\right) , \qquad (9.334)$$

where

$$\varepsilon_p = \sqrt{\mathbf{p}^2 + (m+g\phi_0)^2} . \qquad (9.335)$$

Exercise 9.21: Verify (9.334) and calculate the lowest-order contribution of a massless fermion to the term quadratic in ϕ_0 appearing in the expansion of the effective potential.

9.7.2 The Coleman–Weinberg Mechanism

Many attempts to generalize the GSW model of Sec. 9.4 suffer from the so-called fine-tuning problem [40]. This arises from the fact that any attempt to unify all forces, QCD with electroweak, must contend with the very large difference in the scales of QCD and electroweak. Renormalization group analysis of the separate theories indicates that the effective coupling constants of the two theories achieve the same value at or around energies of 10^{15} GeV. If there is a larger unified theory of all forces, it must explain why the electroweak sector has a mass scale of order 10^2 GeV, a magnitude of 10^{13} lower than the unification scale. This is complicated by radiative corrections to masses in such grand unified models, which typically cause the energy scales of the two sectors to approach one another, rather than become further apart. This requires adjusting parameters in these theories to be preposterously close.

The Coleman–Weinberg mechanism [41] generates such widely separated mass scales in a much more natural way. The basic idea is to renormalize the effective potential so that the mass of the scalar field is exactly zero in the symmetric vacuum. In order to demonstrate that such a renormalization condition can still lead to spontaneous breakdown of symmetry it is necessary to couple the scalar field to a gauge field. This is done in the standard gauge invariant way,

$$\mathcal{L}_g = D_\mu^* \phi^* D^\mu \phi - \tfrac{1}{4} F_{\mu\nu}F^{\mu\nu} , \qquad (9.336)$$

Sec. 9.7 Applications of the Effective Potential

where the abelian gauge covariant derivative is given by

$$D_\mu = \partial_\mu - ieA_\mu . \tag{9.337}$$

The scalar field is given the standard bare gauge invariant potential,

$$V(\phi) = m_0^2 \phi^*\phi + \lambda_0 (\phi^*\phi)^2 . \tag{9.338}$$

The theory is best analyzed using a real basis for the scalar fields,

$$\phi = \frac{1}{\sqrt{2}}(\phi_1 + i\phi_2) , \tag{9.339}$$

and due to the phase invariance of the potential, it is always possible to choose the vacuum expectation value of ϕ to be real, so that

$$\langle 0|\phi|0\rangle = \frac{\phi_0}{\sqrt{2}} . \tag{9.340}$$

Assuming that the renormalized mass of the scalar field is zero in the symmetric vacuum, it is easy to see that the mass matrices for the scalar and gauge field sector are given by

$$M_s^2 = \frac{\partial^2 V(\phi)}{\partial \phi_0^2} = 3\lambda \phi_0^2 ,$$
$$M_g^2 = e^2 \phi_0^2 . \tag{9.341}$$

Using the results of Sec. 9.2, the regularized effective potential, containing contributions from both the scalar and gauge sectors, is given by

$$V_{\text{eff}}(\phi_0) = \tfrac{1}{2}\delta m^2 \phi_0^2 + \tfrac{1}{4}(\lambda + \delta\lambda)\phi_0^4 + \frac{3\hbar\Lambda^2}{16\pi^2}(\lambda + e^2)\phi_0^2$$
$$+ \frac{3\hbar}{64\pi^2}\phi_0^4 \left[3\lambda^2 \ln\left(\frac{3\lambda\phi_0^2}{4\Lambda^2}\right) + e^4 \ln\left(\frac{e^2\phi_0^2}{4\Lambda^2}\right) \right] . \tag{9.342}$$

The effective potential is renormalized so that the mass of the scalar field is exactly zero in the symmetric vacuum, and this is represented by the condition

$$\left. \frac{\partial^2 V_{\text{eff}}}{\partial \phi_0^2} \right|_{\phi_0 = 0} = 0 . \tag{9.343}$$

This condition is easily satisfied by the identification

$$\delta m^2 = -\frac{3\hbar\Lambda^2}{8\pi^2}(\lambda + e^2) . \tag{9.344}$$

430   Nonperturbative Results

However, renormalizing the coupling constant is a different matter. The absence of a mass scale has introduced infrared divergences in the form of the logarithms in (9.342), and it is precisely these infrared divergences that drive the theory away from the symmetric vacuum. It is straightforward to show that

$$\frac{\partial^4 V_{\text{eff}}}{\partial \phi_0^4} = 6(\lambda + \delta\lambda) + \frac{75\hbar}{16\pi^2}(3\lambda^2 + e^4)$$
$$+ \frac{24\hbar}{64\pi^2}\left[\lambda^2 \ln\left(\frac{3\lambda\phi_0^2}{4\Lambda^2}\right) + e^4 \ln\left(\frac{e^2\phi_0^2}{4\Lambda^2}\right)\right]. \quad (9.345)$$

It is clear that this expression cannot be evaluated at $\phi_0 = 0$. Instead, it must be assumed that the coupling constant is defined only for the physical vacuum, which is characterized by $\phi_0 = \nu \neq 0$. The renormalization condition then becomes

$$\left.\frac{\partial^4 V_{\text{eff}}}{\partial \phi_0^4}\right|_{\phi_0=\nu} = 6\lambda, \quad (9.346)$$

which is satisfied if

$$\delta\lambda = -\frac{\hbar}{16\pi^2}\left[\lambda^2 \ln\left(\frac{3\lambda\nu^2}{4\Lambda^2}\right) + e^4 \ln\left(\frac{e^2\nu^2}{4\Lambda^2}\right)\right]$$
$$- \frac{25\hbar}{32\pi^2}(3\lambda^2 + e^4). \quad (9.347)$$

Inserting (9.344) and (9.347) into the original form (9.342) gives the renormalized effective potential,

$$V_{\text{eff}}(\phi_0) = \tfrac{1}{4}\lambda\phi_0^4 - \frac{25\hbar}{128\pi^2}(3\lambda^2 + e^4)\phi_0^4$$
$$+ \frac{3\hbar}{64\pi^2}(3\lambda^2 + e^4)\phi_0^4 \ln\left(\frac{\phi_0^2}{\nu^2}\right). \quad (9.348)$$

The condition for the physical vacuum is now given by

$$\left.\frac{\partial V_{\text{eff}}}{\partial \phi_0}\right|_{\phi_0=\nu} = \left[\lambda - \frac{11\hbar}{16\pi^2}(3\lambda^2 + e^4)\right]\nu^3 = 0. \quad (9.349)$$

This is satisfied if the renormalized coupling constants are chosen, ignoring the term $O(\lambda^2)$, so that

$$\lambda = \frac{11\hbar}{16\pi^2}e^4. \quad (9.350)$$

Sec. 9.7 Applications of the Effective Potential

When the coupling constants satisfy this relationship, the contributions of the gauge field and the scalar field to the effective potential balance, allowing the cancellation of the infrared divergences. Of course, because the theory possesses no intrinsic length scale, the vacuum expectation value is indeterminate.

Nevertheless, the masses of the scalar and gauge fields are related. It follows that the mass of the scalar field Higgs particle is given by

$$M_H{}^2 = \left.\frac{\partial^2 V_{\text{eff}}}{\partial \phi_0{}^2}\right|_{\phi_0 = \nu} = \left(3\lambda - \frac{27\hbar e^4}{16\pi^2}\right)\nu^2 = \frac{3\hbar e^4}{8\pi^2}\nu^2 , \qquad (9.351)$$

where (9.350) has been used. The mass of the Proca boson is given by $M_g{}^2 = e^2\nu^2$, and so the ratio of their masses is fixed to be the small value

$$\frac{M_H{}^2}{M_g{}^2} = \frac{3e^2\hbar}{8\pi^2} . \qquad (9.352)$$

It can be shown that this entirely quantum effect persists when higher-order terms are included in the effective potential. The Higgs mechanism is free to work in such a system, leading to a consistent theory with a natural solution to the fine-tuning problem.

> **Exercise 9.22**: Apply the Coleman–Weinberg mechanism to an SU(2) invariant nonabelian gauge theory in the fundamental representation to determine the mass ratio.

9.7.3 The Gross–Neveu Model

The Gross–Neveu model [42] is a simple theory involving a 1+1 dimensional spinor field with an attractive four-point interaction. The restriction to 1+1 dimensions ensures renormalizability. The gamma matrix algebra for 1+1 dimensional spinors has already been developed in Sec. 9.5, where the Schwinger model was discussed.

Using the results from Sec. 9.5 shows that a possible Lorentz invariant action for a massive 1+1 spinor field is given by

$$\mathcal{L} = i\bar{\psi}(\gamma^\mu \partial_\mu + im)\psi + g(\bar{\psi}\psi)^2 . \qquad (9.353)$$

Because the sign of g is assumed to be positive, the four-point potential is attractive since larger values of $\bar{\psi}\psi$ correspond to lower values of the energy. From the dimensionality of the system it is clear that ψ has the units of (length)$^{-1/2}$, so that g is a dimensionless coupling constant. From power counting arguments the theory is therefore renormalizable. It should also be apparent that any attempt to extend the four-point interaction to higher dimensions will render the theory nonrenormalizable, as was the Fermi interaction of Sec. 9.3.

At the classical level an equivalent action may be obtained by introducing the scalar field σ and rewriting the action as

$$\mathcal{L} = i\bar{\psi}(\gamma^\mu \partial_\mu + im)\psi - \tfrac{1}{2}\alpha\sigma^2 + \sqrt{2\alpha g}\,\sigma\bar{\psi}\psi \; . \tag{9.354}$$

The equation of motion for σ is given by

$$\sigma = \sqrt{\frac{2g}{\alpha}}\,\bar{\psi}\psi \; , \tag{9.355}$$

and substituting (9.355) into (9.354) reproduces (9.353) for an arbitrary real and positive value of the dimensionless constant α. In effect, σ represents the composite field $\bar{\psi}\psi$.

That this is also true at the quantum level is readily seen from the path integral. The vacuum transition element for the theory of (9.354) is given by

$$Z = \int_{\sigma_0}^{\sigma_0} \mathcal{D}\sigma\,\mathcal{D}\psi\,\exp\left\{i\int d^2x\,\mathcal{L}(\psi,\sigma)\right\} \; , \tag{9.356}$$

where the possibility that σ has a nonvanishing vacuum expectation value is being considered. The action involving σ may be written as a quadratic form,

$$-\tfrac{1}{2}\alpha\sigma^2 + \sqrt{2\alpha g}\,\sigma\bar{\psi}\psi = -\tfrac{1}{2}\alpha\left(\sigma - \sqrt{\frac{2g}{\alpha}}\,\bar{\psi}\psi\right)^2 + g(\bar{\psi}\psi)^2 \; , \tag{9.357}$$

and in the Wick rotated path integral becomes a pure Gaussian term. Integrating σ from the theory reduces the action and measure appearing in the path integral back to that of the original theory (9.353).

The action with σ present can now be analyzed by the standard techniques of the effective potential. The field σ is translated by σ_0, and the quadratic part of the action is given by

$$\mathcal{L}_q = i\bar{\psi}(\gamma^\mu \partial_\mu + im - i\sqrt{2\alpha g}\,\sigma_0)\psi - \tfrac{1}{2}\alpha\sigma^2 \; , \tag{9.358}$$

and this can be integrated by the methods of Sec. 9.2. There are minor differences from the results of Sec. 9.2 induced by the difference in dimensionality. It is not difficult to show that the effective potential is given in

Sec. 9.7 Applications of the Effective Potential

the Euclidean region by

$$V_{\text{eff}}(\sigma_0) = \tfrac{1}{2}\alpha\sigma_0^2 - \tfrac{1}{2}\hbar \int \frac{d^2k}{(2\pi)^2} \ln(k_E^2 + M^2) , \qquad (9.359)$$

where the mass M is given by

$$M^2 = (m - \sqrt{2\alpha g}\,\sigma_0)^2 . \qquad (9.360)$$

Exercise 9.23: Verify (9.359).

At this point the specific case of a massless fermion will be analyzed, so that $m = 0$. The regularized form for (9.359) is then given by

$$V_{\text{eff}}(\sigma_0) = \tfrac{1}{2}\left(\alpha + \delta\alpha - \frac{\hbar\alpha g}{2\pi}\right)\sigma_0^2 + \frac{\hbar\alpha g}{4\pi}\sigma_0^2 \ln\left(\frac{\alpha g \sigma_0^2}{\Lambda^2}\right) , \qquad (9.361)$$

where Λ is the cutoff on the momentum space integrations. Again, as in the case of the Coleman–Weinberg mechanism, the absence of a mass scale has introduced infrared divergences that prevent renormalizing the theory in the symmetric vacuum. However, unlike the Coleman–Weinberg mechanism, it is impossible to renormalize the theory to have zero mass. The solution is to introduce the arbitrary renormalization point μ. The massless theory is renormalized by choosing the renormalized coupling α to be unity, so that

$$\left.\frac{\partial^2 V_{\text{eff}}}{\partial \sigma_0^2}\right|_{\sigma_0 = \mu} = 1 . \qquad (9.362)$$

The demand that the results be independent of μ will lead to a renormalization group equation for the theory, as discussed in Sec. 8.4. This gives the renormalization factor

$$\delta\alpha = -\frac{\hbar g}{\pi} - \frac{\hbar g}{2\pi}\ln\left(\frac{\alpha g \mu^2}{\Lambda^2}\right) . \qquad (9.363)$$

Inserting (9.363) into (9.361) gives the renormalized effective potential:

$$V_{\text{eff}}(\sigma_0) = \tfrac{1}{2}\left(1 - \frac{3\hbar g}{2\pi}\right)\sigma_0^2 + \frac{\hbar g}{4\pi}\sigma_0^2 \ln\left(\frac{\sigma_0^2}{\mu^2}\right) . \qquad (9.364)$$

An examination of the effective potential shows that it always possesses an absolute minimum for nonzero σ_0, regardless of the choice for μ. The criterion for the physical vacuum then yields

$$\frac{\partial V_{\text{eff}}}{\partial \sigma_0} = 1 - \frac{\hbar g}{\pi} + \frac{\hbar g}{\pi}\ln\left(\frac{\sigma_0}{\mu}\right) = 0 , \qquad (9.365)$$

434   Nonperturbative Results

and this has the solution

$$\sigma_0 = \mu \exp\left(1 - \frac{\pi}{\hbar g}\right). \tag{9.366}$$

Because inverse powers of g occur in this solution, it is a manifestly nonperturbative result.

Once the value for μ is chosen, it fixes the value of σ_0, which in turn can be related to physically observable aspects of the model. Returning to the original Lagrangian, it is apparent that the mass of the fermion is given in the shifted vacuum by

$$M_f = \sqrt{2g}\,\sigma_0 = \sqrt{2g}\,\mu \exp\left(1 - \frac{\pi}{\hbar g}\right). \tag{9.367}$$

The value of g is not fixed; however, the demand that all these results be independent of the choice of μ fixes g to be a function of μ. This is determined by solving the renormalization group equation

$$\mu \frac{dV_{\text{eff}}}{d\mu} = 0 \tag{9.368}$$

and assuming that all the parameters appearing in V_{eff} are implicit functions of μ. The derivative of the effective potential will be evaluated using the solution (9.366), so that

$$\frac{d\sigma_0}{d\mu} = \exp\left(1 - \frac{\pi}{\hbar g}\right) + \frac{\mu g}{\hbar g^2}\frac{dg}{d\mu}\exp\left(1 - \frac{\pi}{\hbar g}\right) \tag{9.369}$$

and

$$\ln\left(\frac{\sigma_0}{\mu}\right) = 1 - \frac{\pi}{\hbar g}. \tag{9.370}$$

Using these relations, the differential equation generated by implementing the demand (9.368) is one involving only derivatives of g. A straightforward evaluation of (9.368) using (9.369), (9.370), and the fact that (9.366) satisfies (9.365), gives

$$\frac{dg}{d\mu} = -\frac{2\hbar g^2}{(\hbar g + 2\pi)\mu}. \tag{9.371}$$

This equation can be solved by integration, giving

$$2\left(1 - \frac{\pi}{\hbar g}\right) + \ln\left(\frac{\hbar g}{\pi}\right) = \ln\left(\frac{\mu_0^{\,2}}{\mu^2}\right), \tag{9.372}$$

Sec. 9.7 Applications of the Effective Potential

where μ_0 is the value of μ for which $g = \pi/\hbar$. The value of μ_0 can be related to the physical mass M_f in the following way. Setting $g = \pi/\hbar$ in (9.367) gives

$$\mu = \sqrt{\frac{\hbar}{2\pi}} M_f \ . \tag{9.373}$$

However, if $g = \pi/\hbar$, then $\mu = \mu_0$, so that

$$\mu_0 = \sqrt{\frac{\hbar}{2\pi}} M_f \ . \tag{9.374}$$

Once M_f is inserted, the behavior of the coupling constant is determined for all values of μ. Equation (9.372) shows that the $\mu \to \infty$ limit gives $g \to 0$. The theory is therefore asymptotically free.

The pole of the composite field propagator,

$$\langle 0 | T\{\sigma(x)\sigma(y)\} | 0 \rangle = 2g \langle 0 | T\{\bar{\psi}(x)\psi(x)\bar{\psi}(y)\psi(y)\} | 0 \rangle \ , \tag{9.375}$$

has been set to unity in this renormalization scheme. The value of σ_0 is the value of the composite field in the ground state,

$$\sigma_0 = \sqrt{2g} \langle 0 | \bar{\psi}\psi | 0 \rangle \ . \tag{9.376}$$

A nonzero value for σ_0 shows that the ground state of the model is such that fermion pairs have condensed to create a state with lower energy than the normal vacuum. This follows from the fact that $\bar{\psi}\psi$ would vanish in a state with no fermions.

Such a mechanism is observed in normal superconductors. There the charges of the electrons are screened by the background lattice through which they are moving while they simultaneously undergo an attractive force mediated by phonon exchange corresponding to lattice distortion induced by electrostatic forces. Due to the charge screening, this relatively weak attractive force can overcome the repulsive Coulomb potential, and the resulting net attractive potential can be approximated by a four-point vertex of the form present in (9.353). The ground state of the superconductor is characterized by the formation of a condensate of paired electrons, known as Cooper pairs, and this ground state is lower in energy than the normal ground state. This condensate of Cooper pairs may be modeled by a charged scalar field, the complex version of (9.354),

$$\sigma = \psi^\dagger \psi^\dagger \ . \tag{9.377}$$

This field also interacts with the electromagnetic field in much the same manner as the abelian Higgs model. Therefore, in the interior of the normal superconductor the photon field is effectively massive, leading to the many remarkable effects observed. The Higgs model, and the Gross–Neveu model of this section, are relativistic versions of these phenomena. It is not surprising that treatments of superconductors using path integrals are very similar to the relativistic models of this book [43].

An important extension of the Gross–Neveu model is to give a simple model for the breakdown of chiral symmetry in the QCD sector of the standard model. There, the gluon mediated forces between quarks produce attractive potentials between color singlets, and an effective field theory describing this phenomenon would naturally use an attractive four-point vertex between the quark fields. Doing so gives rise to a condensate of quark pairs in the ground state that breaks chiral invariance and induces effective masses for the quarks and their bound states. Numerical analysis of QCD verifies this phenomenon.

References

[1] J. Goldstone, Nuovo Cimento **19**, 154 (1961); Y. Nambu and G. Jona-Lasinio, Phys. Rev. **122**, 345 (1961); J. Goldstone, A. Salam, and S. Weinberg, Phys. Rev. **127**, 965 (1962).

[2] L. Dolan and R. Jackiw, Phys. Rev. **D9**, 3320 (1974); S. Coleman in *Laws of Hadronic Matter*, ed. A. Zichichi, Academic Press, New York, 1975; J. Iliopoulos, C. Itzykson, and A. Martin, Rev. Mod. Phys. **47**, 165 (1975).

[3] K. Symanzik, Commun. Math. Phys. **16**, 48 (1970).

[4] R. Jackiw, Phys. Rev. **D9**, 661 (1974).

[5] M. Aizenmann, Phys. Rev. Lett. **47**, 1 (1981); J. Fröhlich, Nucl. Phys. **B200** [FS4], 281 (1982).

[6] A discussion of this, along with many other aspects of the effective potential and its applications to problems in modern cosmology, is found in R.H. Brandenberger, Rev. Mod. Phys. **57**, 1 (1985).

[7] R.J. Rivers, *Path Integral Methods in Quantum Field Theory*, Cambridge University Press, Cambridge, 1987.

[8] E. Fermi, Z. Physik **88**, 161 (1934).

[9] For a thorough discussion of this point see S. Coleman in *Laws of Hadronic Matter*, ed. A. Zichichi, Academic Press, New York, 1975.

[10] P.W. Higgs, Phys. Lett. **12**, 132 (1964); Phys. Rev. Lett. **13**, 508 (1964); Phys. Rev. **145**, 1156 (1966).

[11] R. Brout and F. Englert, Phys. Rev. Lett. **13**, 321 (1964); G.S. Guralnik, C.R. Hagen, and T.W.B. Kibble, Phys. Rev. Lett. **13**, 585 (1964); T.W.B. Kibble, Phys. Rev. **155**, 1554 (1967).

[12] The Higgs mechanism in the Coulomb gauge is discussed extensively in J. Bernstein, Rev. Mod. Phys. **46**, 7 (1974).

[13] J.C. Collins, *Renormalization*, Cambridge University Press, Cambridge, 1984.

[14] L.H. Ryder, *Quantum Field Theory*, Cambridge University Press, Cambridge, 1985.

[15] The group theory of spontaneously broken symmetry was first discussed in S. Bludman and A. Klein, Phys. Rev. **131**, 2363 (1962).

[16] S.L. Glashow, Nucl. Phys. **22**, 579 (1961); A. Salam, in *Proceedings of the Eighth Nobel Symposium*, ed. N. Svartholm, Almqvist and Wiksell, Stockholm, 1968; S. Weinberg, Phys. Rev. Lett. **19**, 1264 (1967).

[17] The intermediate vector boson hypothesis for weak interactions was first proposed by R.P. Feynman and M. Gell-Mann, Phys. Rev. **109**, 193 (1958).

[18] See, for example, L.B. Okun, *Leptons and Quarks*, North-Holland, Amsterdam, 1982; F. Halzen and A.D. Martin, *Quarks and Leptons*, Wiley, New York, 1984; O. Nachtmann, *Elementary Particle Physics: Concepts and Phenomena*, Springer-Verlag, New York, 1985.

[19] J. Schwinger, Phys. Rev. **82**, 664 (1951); S. Adler, Phys. Rev. **177**, 2426 (1969); J.S. Bell and R. Jackiw, Nuovo Cimento **60A**, 47 (1969); W.A. Bardeen, Phys. Rev. **184**, 1848 (1969).

[20] K. Fujikawa, Phys. Rev. **D21**, 2848 (1980).

[21] Baryon number violation was originally calculated by G. t' Hooft, Phys. Rev. Lett. **37**, 8 (1976); Phys. Rev. **D14**, 3432 (1976).

[22] J. Schwinger, Phys. Rev. **128**, 2425 (1962).

[23] J. Lowenstein and J. Swieca, Comm. Math. Phys. **24**, 1 (1971); A.Z. Capri and R. Ferrari, Nuovo Cimento **62A**, 273 (1981).

[24] An outstanding monograph on solitons is R. Rajaraman, *Solitons and Instantons: An Introduction to Solitons and Instantons in Quantum Field Theory*, Elsevier, New York, 1987.

[25] The technique for quantizing the soliton sector presented here was originally developed by N.H. Christ and T.D. Lee, Phys. Rev. **D12**, 1606 (1975).

[26] C.G. Callan and D.J. Gross, Nucl. Phys. **B93**, 29 (1975).

[27] J. Milnor, *Morse Theory*, Princeton University Press, Princeton, N.J., 1963. A very readable discussion of the Morse theory is found in L.S. Schulman, *Techniques and Applications of Path Integration*, Wiley, New York, 1981.

[28] See for example M. Abramowitz and I. Stegun, *Handbook of Mathematical Functions*, Dover, New York, 1965.

[29] For a full discussion of this point see S. Coleman in *The Whys of Subnuclear Physics*, ed. A. Zichichi, Plenum Press, New York, 1979.

[30] A.A. Belavin, A.M. Polyakov, A.S. Schwartz, and Y.S. Tyupkin, Phys. Lett. **59B**, 85 (1975); R. Jackiw and C. Rebbi, Phys. Rev. Lett. **37**, 172 (1976). For general reviews see [29], [24], and A.M. Polyakov, *Gauge Fields and Strings*, Harwood, Chur, 1987.

[31] Enhancement of baryon violation by instanton-like effects is discussed in H. Aoyama and H. Goldberg, Phys. Lett. **B188**, 506 (1987); T. Banks, G. Farrar, M. Dine, and B. Sakita, Nucl. Phys. **B347**, 581 (1990); H. Aoyama and H. Kikuchi, Phys. Rev. **D43**, 1999 (1991).

[32] P.A.M. Dirac, Proc. Roy. Soc. (London) **A133**, 60 (1931).

[33] T.T. Wu and C.N. Yang, Phys. Rev. **D12**, 3845 (1975).

[34] See for example T. Eguchi, P.B. Gilkey, and A.J. Hanson, Phys. Rep. **66**, 213 (1980); C. Nash and S. Sen, *Topology and Geometry for Physicists*, Academic Press, New York, 1983; M. Nakahara, *Geometry, Topology, and Physics*, Adam Hilger, Bristol, 1990.

[35] An excellent discussion of monopoles and dyons is given by S. Coleman in *The Unity of the Fundamental Interactions*, ed. by A. Zichichi, Plenum Press, New York, 1983.

[36] See the original papers by G. 't Hooft, Nucl. Phys. **B79**, 276 (1974); A.M. Polyakov, JETP Lett. **20**, 194 (1974); or the reviews [24] and [35].

[37] T. Matsubara, Prog. Theor. Phys. **14**, 351 (1955).

[38] C. Bernard, Phys. Rev. **D9**, 3312 (1974).

[39] A comprehensive study of quantum field theory and phase transitions is presented in J. Zinn-Justin, *Quantum Field Theory and Critical Phenomena*, Oxford University Press, Oxford, 1989.

[40] For discussions of the many attempts to embed the GSW model into a larger and more comprehensive theory see P. Langacker, Phys. Rep. **72C**, 185 (1981); C. Quigg, *Gauge Theories of the Strong, Weak, and Electromagnetic Interactions*, Benjamin-Cummings, New York, 1983; G. Ross, *Grand Unified Theories*, Benjamin-Cummings, Menlo Park,

California, 1985.

[41] S. Coleman and E. Weinberg, Phys. Rev. **D7**, 1888 (1973).

[42] D. Gross and A. Neveu, Phys. Rev. **D10**, 3235 (1974).

[43] An excellent presentation of path integral techniques applied to condensed matter problems is found in V.N. Popov, *Functional Integrals and Collective Excitations*, Cambridge University Press, Cambridge, 1987.

Index

abelian gauge group, 271
action
    classical fields, 170
    classical mechanics, 12
Aharanov–Bohm effect, 304
analytic continuation, 24, 66, 411
annihilation operators, 93, 140, 192, 194, 243
anomalies, 87
    chiral, 389
antiparticles, 196
asymptotic
    fields, 198
    freedom, 354, 435
    states, 199, 203

Baker–Campbell–Hausdorff
    and coherent states, 93, 208
    and Grassmann variables, 120
    theorem, 19
baryon number, 384
    violation, 399, 419
Bianchi identities, 291
Bose–Einstein statistics, 140, 196
BRST symmetry, 248, 323
Brownian motion, 48

Callan–Symanzik equation, 353
canonical quantization of
    scalar field, 191
    spinor field, 194
    vector field, 239
canonical transformations, 86
    and gauge conditions, 114, 251
Casimir operator, 269

charge
    conjugation, 181
    electric, 235
    generalized, 85, 185
    generators, 91, 147, 190
Chern–Simons form, 292
chirality operator, 180
chiral symmetry, 189, 389
classical solutions
    and path integral, 62, 403
    limit, 334
    trajectory, 12
coherent states
    field theoretic, 208, 215
    Grassmann oscillator, 142
    harmonic oscillator, 92
cohomology, 288
Coleman–Weinberg mechanism, 428
collective coordinate, 404
color, 285
commutator, 15, 33
completeness, 3
completing the square, 26
complex scalar field, 172, 235
confinement, 203, 285
connected Green's function, 59, 314
connection, 303
conservation laws
    see charge, generalized
constraints, 105, 138, 237
    first-class, second-class, 107, 109
coordinate ordering, 38
Coulomb
    gauge, 241, 245, 260
    Green's function, 236

counterterms, 341, 344
covariant
    component, 159
    derivative, 235, 273
    equation, 170
CPT theorem, 197
creation operators, 93, 192, 243
critical temperature, 102, 427
currents, 185

determinant of
    differential operator, 22
    function, 21
    matrix, 16
Dirac
    delta, 2, 124
    brackets, 107, 245
    condition, 291, 305, 421
    equation, 177
    spinor field, 169
    string, 420
distribution, 2
dimensional regularization, 350
dual tensor, 291
dynamical symmetry breaking, 388, 402, 436
dyons, 422
Dyson–Schwinger equation, 341
Dyson–Wick expansion, 207

effective action
    quantum field, 315, 358
    quantum mechanical, 82
effective coupling, 341
effective potential
    and free energy, 99
    convexity of, 362, 367
    field theoretic, 317, 360
    one-loop approximation, 364
    quantum mechanical, 98
Ehrenfest's theorem, 87
eigenvalues
    and stability, 412
    differential operator, 8
    matrix, 16

energy-momentum vector, 162, 187, 193
equal-time anticommutation
    relation, 136, 195, 258
equal-time commutation
    relation, 155, 192
Euler–Lagrange equation for
    classical fields, 171
    classical mechanics, 13
exterior derivative, 287

Faddeev–Popov
    determinant, 260, 283
    ghosts, 257, 260, 284, 369
Fermi current–current coupling, 373
Fermi–Dirac statistics, 140, 196, 425
Feynman
    gauge, 239
    rules, 206, 331
fiber bundle, 422
forms, 286
Fourier
    series, 4
    transform, 6
free energy, 78
functional derivative, 9, 130
functionals, 9

gauge transformations
    first kind, 189
    large, 298
    nonabelian, 272
    second kind, 90, 233
gauge
    constraints, 112, 237, 277
    fields, 233, 272
    fixing, 238, 277
    invariance, 233
Gaussian integrals
    bosonic, 22, 225
    Grassmann, 125, 226
Gauss's law, 232, 240, 251, 277, 297
generating functional, 58, 313
ghosts, 134, 243, 338, 377, 425
gluons, 285

Index

Goldstone
  theorem, 358
  bosons, 359, 376
Grassmann
  integrals, 121
  classical mechanics, 128
  quantum mechanics, 133
Green's function
  connected, 206, 328
  quantum mechanical, 54
Gribov ambiguity, 278, 301
Gross–Neveu model, 431
groups, 162
  nonabelian, 164, 263
GSW model, 375, 383
Gupta–Bleuler condition, 244

Haar measure, 281, 295
Hamiltonian, 13, 186, 192, 194
Hamilton's equations
  bosonic, 14, 106
  fermionic, 130
harmonic oscillator, 65, 131
  approximation, 103
Heisenberg picture, 35
helicity, 175, 195, 247
Hermitian
  matrix, 17
  operator, 7
Higgs
  mechanism, 372
  boson, 388
Hilbert space, 32
holonomy, 306
homology, 288
homotopy, 70, 295

indefinite metric, 238, 243
index theorems, 397
infrared divergences, 203, 343, 353
inner product, 4, 32
interaction picture, 204
instanton, 294, 411
  baryon number violation, 419

  bounce solution, 415
  Yang–Mills, 416
isospin, 270

Jacobi identity, 265
Jacobian
  and chiral transformation, 393
  bosonic, 25, 64
  fermionic, 122

kink solution, 403
  quantization, 405
Klein–Gordon equation, 172
KMS condition, 79, 424

Lagrangian density
  see action
leptons, 384
Levi–Civita symbol, 16
Lie algebra, 263
  Cartan subalgebra, 267
  Casimir operator, 269
  compact, 263, 281
  Dynkin diagrams, 269
  generators, 264
  representations, 265
  structure constants, 265
Lorentz transformation, 160, 400
  relation to $SL(2,C)$, 165
LSZ reduction, 201, 338

magnetic monopole, 289, 305, 420
Majorana spinor, 181
mass matrix, 359, 382
Maxwell field, 232
measure of path integral
  quantum field, 158, 209, 218
  quantum mechanical, 40, 46
metric tensor, 159
mode expansion, 140, 191, 194
momentum density, 154
Morse theory, 413

neutrinos, 179, 373, 384
Noether's theorem, 185

parity transformation, 160
partition function, 77, 145, 423
perturbation theory, 61, 325
photon, 247
physical subspace, 108, 238
Planck's constant, 75, 330
Poincaré transformation, 160
Poisson
    bracket, 14, 131, 227, 279
    resummation, 68
polarization vectors, 242
Pontryagin index, 295, 397
power counting, 347
Proca boson, 373
propagator, 37, 314, 329, 334

quantum
    chromodynamics, 285, 354
    effects, 38
    electrodynamics, 249
quarks, 285, 354, 383

Rarita–Schwinger field, 169
renormalization group, 347, 353, 433
renormalized
    coupling, 347
    effective potential, 366
    mass, 345
    wave function, 199, 347
rotations, 88, 161, 188

saddle-point expansion, 28, 104, 361, 402, 405
Schrödinger picture, 34
Schwinger model, 400
self-consistency conditions, 103, 376
self-dual gauge configurations, 418
self-energy, 332, 340
Slavnov–Taylor identities, 349
S-matrix, 197, 249
solitons, 404
source function, 55, 81, 312, 319
spin, 188
    and statistics, 196
spontaneously broken symmetry
    quantum field, 359, 367, 378
    quantum mechanical, 96, 103
    restoration of, 102, 427

steepest descent, 100, 361
step function, 34
Stokes's theorem, 288
summation convention, 11, 159
superconductivity, 435
supersymmetry, 146, 369
symmetry, 86, 319

temperature, finite, 78, 424
temporal gauge, 237, 284
theta vacuum, 299
time ordering, 34, 58, 200, 312, 329, 336
time reversal, 182
topology, 67, 286, 397, 422
Trotter product formula, 39
tunnelling, 410

ultraviolet divergences, 343
unitary
    matrix, 17
    theory, 200, 249
    transformation, 17

vacuum, 92, 97, 140, 192
    expectation value, 313, 373
vertices, 314
    bare, 326
    one-particle-irreducible, 316

Ward–Takahashi identities, 323, 394
W-boson, 387
weak interactions, 373
wedge product, 286
Weyl spinor, 169, 173
Wick rotation, 47, 252
Wilson loop, 306
winding number, 70

Yang–Mills fields
    classical, 270
    quantized, 276
Yukawa vertex, 367, 371

Z-boson, 387
zero modes, 301, 395, 407

A CATALOG OF SELECTED
DOVER BOOKS
IN SCIENCE AND MATHEMATICS

CATALOG OF DOVER BOOKS

Engineering

FUNDAMENTALS OF ASTRODYNAMICS, Roger R. Bate, Donald D. Mueller, and Jerry E. White. Teaching text developed by U.S. Air Force Academy develops the basic two-body and n-body equations of motion; orbit determination; classical orbital elements, coordinate transformations; differential correction; more. 1971 edition. 455pp. 5 3/8 x 8 1/2. 0-486-60061-0

INTRODUCTION TO CONTINUUM MECHANICS FOR ENGINEERS: Revised Edition, Ray M. Bowen. This self-contained text introduces classical continuum models within a modern framework. Its numerous exercises illustrate the governing principles, linearizations, and other approximations that constitute classical continuum models. 2007 edition. 320pp. 6 1/8 x 9 1/4. 0-486-47460-7

ENGINEERING MECHANICS FOR STRUCTURES, Louis L. Bucciarelli. This text explores the mechanics of solids and statics as well as the strength of materials and elasticity theory. Its many design exercises encourage creative initiative and systems thinking. 2009 edition. 320pp. 6 1/8 x 9 1/4. 0-486-46855-0

FEEDBACK CONTROL THEORY, John C. Doyle, Bruce A. Francis and Allen R. Tannenbaum. This excellent introduction to feedback control system design offers a theoretical approach that captures the essential issues and can be applied to a wide range of practical problems. 1992 edition. 224pp. 6 1/2 x 9 1/4. 0-486-46933-6

THE FORCES OF MATTER, Michael Faraday. These lectures by a famous inventor offer an easy-to-understand introduction to the interactions of the universe's physical forces. Six essays explore gravitation, cohesion, chemical affinity, heat, magnetism, and electricity. 1993 edition. 96pp. 5 3/8 x 8 1/2. 0-486-47482-8

DYNAMICS, Lawrence E. Goodman and William H. Warner. Beginning engineering text introduces calculus of vectors, particle motion, dynamics of particle systems and plane rigid bodies, technical applications in plane motions, and more. Exercises and answers in every chapter. 619pp. 5 3/8 x 8 1/2. 0-486-42006-X

ADAPTIVE FILTERING PREDICTION AND CONTROL, Graham C. Goodwin and Kwai Sang Sin. This unified survey focuses on linear discrete-time systems and explores natural extensions to nonlinear systems. It emphasizes discrete-time systems, summarizing theoretical and practical aspects of a large class of adaptive algorithms. 1984 edition. 560pp. 6 1/2 x 9 1/4. 0-486-46932-8

INDUCTANCE CALCULATIONS, Frederick W. Grover. This authoritative reference enables the design of virtually every type of inductor. It features a single simple formula for each type of inductor, together with tables containing essential numerical factors. 1946 edition. 304pp. 5 3/8 x 8 1/2. 0-486-47440-2

THERMODYNAMICS: Foundations and Applications, Elias P. Gyftopoulos and Gian Paolo Beretta. Designed by two MIT professors, this authoritative text discusses basic concepts and applications in detail, emphasizing generality, definitions, and logical consistency. More than 300 solved problems cover realistic energy systems and processes. 800pp. 6 1/8 x 9 1/4. 0-486-43932-1

THE FINITE ELEMENT METHOD: Linear Static and Dynamic Finite Element Analysis, Thomas J. R. Hughes. Text for students without in-depth mathematical training, this text includes a comprehensive presentation and analysis of algorithms of time-dependent phenomena plus beam, plate, and shell theories. Solution guide available upon request. 672pp. 6 1/2 x 9 1/4. 0-486-41181-8

Browse over 9,000 books at www.doverpublications.com

Physics

THEORETICAL NUCLEAR PHYSICS, John M. Blatt and Victor F. Weisskopf. An uncommonly clear and cogent investigation and correlation of key aspects of theoretical nuclear physics by leading experts: the nucleus, nuclear forces, nuclear spectroscopy, two-, three- and four-body problems, nuclear reactions, beta-decay and nuclear shell structure. 896pp. 5 3/8 x 8 1/2. 0-486-66827-4

QUANTUM THEORY, David Bohm. This advanced undergraduate-level text presents the quantum theory in terms of qualitative and imaginative concepts, followed by specific applications worked out in mathematical detail. 655pp. 5 3/8 x 8 1/2.
0-486-65969-0

ATOMIC PHYSICS AND HUMAN KNOWLEDGE, Niels Bohr. Articles and speeches by the Nobel Prize–winning physicist, dating from 1934 to 1958, offer philosophical explorations of the relevance of atomic physics to many areas of human endeavor. 1961 edition. 112pp. 5 3/8 x 8 1/2. 0-486-47928-5

COSMOLOGY, Hermann Bondi. A co-developer of the steady-state theory explores his conception of the expanding universe. This historic book was among the first to present cosmology as a separate branch of physics. 1961 edition. 192pp. 5 3/8 x 8 1/2.
0-486-47483-6

LECTURES ON QUANTUM MECHANICS, Paul A. M. Dirac. Four concise, brilliant lectures on mathematical methods in quantum mechanics from Nobel Prize–winning quantum pioneer build on idea of visualizing quantum theory through the use of classical mechanics. 96pp. 5 3/8 x 8 1/2. 0-486-41713-1

THE PRINCIPLE OF RELATIVITY, Albert Einstein and Frances A. Davis. Eleven papers that forged the general and special theories of relativity include seven papers by Einstein, two by Lorentz, and one each by Minkowski and Weyl. 1923 edition. 240pp. 5 3/8 x 8 1/2. 0-486-60081-5

PHYSICS OF WAVES, William C. Elmore and Mark A. Heald. Ideal as a classroom text or for individual study, this unique one-volume overview of classical wave theory covers wave phenomena of acoustics, optics, electromagnetic radiations, and more. 477pp. 5 3/8 x 8 1/2. 0-486-64926-1

THERMODYNAMICS, Enrico Fermi. In this classic of modern science, the Nobel Laureate presents a clear treatment of systems, the First and Second Laws of Thermodynamics, entropy, thermodynamic potentials, and much more. Calculus required. 160pp. 5 3/8 x 8 1/2. 0-486-60361-X

QUANTUM THEORY OF MANY-PARTICLE SYSTEMS, Alexander L. Fetter and John Dirk Walecka. Self-contained treatment of nonrelativistic many-particle systems discusses both formalism and applications in terms of ground-state (zero-temperature) formalism, finite-temperature formalism, canonical transformations, and applications to physical systems. 1971 edition. 640pp. 5 3/8 x 8 1/2. 0-486-42827-3

QUANTUM MECHANICS AND PATH INTEGRALS: Emended Edition, Richard P. Feynman and Albert R. Hibbs. Emended by Daniel F. Styer. The Nobel Prize–winning physicist presents unique insights into his theory and its applications. Feynman starts with fundamentals and advances to the perturbation method, quantum electrodynamics, and statistical mechanics. 1965 edition, emended in 2005. 384pp. 6 1/8 x 9 1/4. 0-486-47722-3

Browse over 9,000 books at www.doverpublications.com

CATALOG OF DOVER BOOKS

Physics

INTRODUCTION TO MODERN OPTICS, Grant R. Fowles. A complete basic undergraduate course in modern optics for students in physics, technology, and engineering. The first half deals with classical physical optics; the second, quantum nature of light. Solutions. 336pp. 5 3/8 x 8 1/2. 0-486-65957-7

THE QUANTUM THEORY OF RADIATION: Third Edition, W. Heitler. The first comprehensive treatment of quantum physics in any language, this classic introduction to basic theory remains highly recommended and widely used, both as a text and as a reference. 1954 edition. 464pp. 5 3/8 x 8 1/2. 0-486-64558-4

QUANTUM FIELD THEORY, Claude Itzykson and Jean-Bernard Zuber. This comprehensive text begins with the standard quantization of electrodynamics and perturbative renormalization, advancing to functional methods, relativistic bound states, broken symmetries, nonabelian gauge fields, and asymptotic behavior. 1980 edition. 752pp. 6 1/2 x 9 1/4. 0-486-44568-2

FOUNDATIONS OF POTENTIAL THERY, Oliver D. Kellogg. Introduction to fundamentals of potential functions covers the force of gravity, fields of force, potentials, harmonic functions, electric images and Green's function, sequences of harmonic functions, fundamental existence theorems, and much more. 400pp. 5 3/8 x 8 1/2.
0-486-60144-7

FUNDAMENTALS OF MATHEMATICAL PHYSICS, Edgar A. Kraut. Indispensable for students of modern physics, this text provides the necessary background in mathematics to study the concepts of electromagnetic theory and quantum mechanics. 1967 edition. 480pp. 6 1/2 x 9 1/4. 0-486-45809-1

GEOMETRY AND LIGHT: The Science of Invisibility, Ulf Leonhardt and Thomas Philbin. Suitable for advanced undergraduate and graduate students of engineering, physics, and mathematics and scientific researchers of all types, this is the first authoritative text on invisibility and the science behind it. More than 100 full-color illustrations, plus exercises with solutions. 2010 edition. 288pp. 7 x 9 1/4. 0-486-47693-6

QUANTUM MECHANICS: New Approaches to Selected Topics, Harry J. Lipkin. Acclaimed as "excellent" (*Nature*) and "very original and refreshing" (*Physics Today*), these studies examine the Mössbauer effect, many-body quantum mechanics, scattering theory, Feynman diagrams, and relativistic quantum mechanics. 1973 edition. 480pp. 5 3/8 x 8 1/2. 0-486-45893-8

THEORY OF HEAT, James Clerk Maxwell. This classic sets forth the fundamentals of thermodynamics and kinetic theory simply enough to be understood by beginners, yet with enough subtlety to appeal to more advanced readers, too. 352pp. 5 3/8 x 8 1/2. 0-486-41735-2

QUANTUM MECHANICS, Albert Messiah. Subjects include formalism and its interpretation, analysis of simple systems, symmetries and invariance, methods of approximation, elements of relativistic quantum mechanics, much more. "Strongly recommended." – *American Journal of Physics*. 1152pp. 5 3/8 x 8 1/2. 0-486-40924-4

RELATIVISTIC QUANTUM FIELDS, Charles Nash. This graduate-level text contains techniques for performing calculations in quantum field theory. It focuses chiefly on the dimensional method and the renormalization group methods. Additional topics include functional integration and differentiation. 1978 edition. 240pp. 5 3/8 x 8 1/2.
0-486-47752-5

Browse over 9,000 books at www.doverpublications.com

CATALOG OF DOVER BOOKS

Physics

MATHEMATICAL TOOLS FOR PHYSICS, James Nearing. Encouraging students' development of intuition, this original work begins with a review of basic mathematics and advances to infinite series, complex algebra, differential equations, Fourier series, and more. 2010 edition. 496pp. 6 1/8 x 9 1/4. 0-486-48212-X

TREATISE ON THERMODYNAMICS, Max Planck. Great classic, still one of the best introductions to thermodynamics. Fundamentals, first and second principles of thermodynamics, applications to special states of equilibrium, more. Numerous worked examples. 1917 edition. 297pp. 5 3/8 x 8. 0-486-66371-X

AN INTRODUCTION TO RELATIVISTIC QUANTUM FIELD THEORY, Silvan S. Schweber. Complete, systematic, and self-contained, this text introduces modern quantum field theory. "Combines thorough knowledge with a high degree of didactic ability and a delightful style." – *Mathematical Reviews.* 1961 edition. 928pp. 5 3/8 x 8 1/2. 0-486-44228-4

THE ELECTROMAGNETIC FIELD, Albert Shadowitz. Comprehensive undergraduate text covers basics of electric and magnetic fields, building up to electromagnetic theory. Related topics include relativity theory. Over 900 problems, some with solutions. 1975 edition. 768pp. 5 5/8 x 8 1/4. 0-486-65660-8

THE PRINCIPLES OF STATISTICAL MECHANICS, Richard C. Tolman. Definitive treatise offers a concise exposition of classical statistical mechanics and a thorough elucidation of quantum statistical mechanics, plus applications of statistical mechanics to thermodynamic behavior. 1930 edition. 704pp. 5 5/8 x 8 1/4.
0-486-63896-0

INTRODUCTION TO THE PHYSICS OF FLUIDS AND SOLIDS, James S. Trefil. This interesting, informative survey by a well-known science author ranges from classical physics and geophysical topics, from the rings of Saturn and the rotation of the galaxy to underground nuclear tests. 1975 edition. 320pp. 5 3/8 x 8 1/2.
0-486-47437-2

STATISTICAL PHYSICS, Gregory H. Wannier. Classic text combines thermodynamics, statistical mechanics, and kinetic theory in one unified presentation. Topics include equilibrium statistics of special systems, kinetic theory, transport coefficients, and fluctuations. Problems with solutions. 1966 edition. 532pp. 5 3/8 x 8 1/2.
0-486-65401-X

SPACE, TIME, MATTER, Hermann Weyl. Excellent introduction probes deeply into Euclidean space, Riemann's space, Einstein's general relativity, gravitational waves and energy, and laws of conservation. "A classic of physics." – *British Journal for Philosophy and Science.* 330pp. 5 3/8 x 8 1/2. 0-486-60267-2

RANDOM VIBRATIONS: Theory and Practice, Paul H. Wirsching, Thomas L. Paez and Keith Ortiz. Comprehensive text and reference covers topics in probability, statistics, and random processes, plus methods for analyzing and controlling random vibrations. Suitable for graduate students and mechanical, structural, and aerospace engineers. 1995 edition. 464pp. 5 3/8 x 8 1/2. 0-486-45015-5

PHYSICS OF SHOCK WAVES AND HIGH-TEMPERATURE HYDRO DYNAMIC PHENOMENA, Ya B. Zel'dovich and Yu P. Raizer. Physical, chemical processes in gases at high temperatures are focus of outstanding text, which combines material from gas dynamics, shock-wave theory, thermodynamics and statistical physics, other fields. 284 illustrations. 1966–1967 edition. 944pp. 6 1/8 x 9 1/4.
0-486-42002-7

Browse over 9,000 books at www.doverpublications.com